Lecture Notes in Civil Engineering 609

Lecture Notes in Civil Engineering (LNCE) publishes the latest developments in Civil Engineering—quickly, informally and in top quality. Though original research reported in proceedings and post-proceedings represents the core of LNCE, edited volumes of exceptionally high quality and interest may also be considered for publication. Volumes published in LNCE embrace all aspects and subfields of, as well as new challenges in, Civil Engineering. Topics in the series include:

- Construction and Structural Mechanics
- Building Materials
- Concrete, Steel and Timber Structures
- Geotechnical Engineering
- Earthquake Engineering
- Coastal Engineering
- Ocean and Offshore Engineering; Ships and Floating Structures
- Hydraulics, Hydrology and Water Resources Engineering
- Environmental Engineering and Sustainability
- Structural Health and Monitoring
- Surveying and Geographical Information Systems
- Indoor Environments
- Transportation and Traffic
- Risk Analysis
- Safety and Security

To submit a proposal or request further information, please contact the appropriate Springer Editor:

- Pierpaolo Riva at pierpaolo.riva@springer.com (Europe and Americas);
- Swati Meherishi at swati.meherishi@springer.com (Asia—except China, Australia, and New Zealand);
- Wayne Hu at wayne.hu@springer.com (China).

All books in the series now indexed by Scopus and EI Compendex database!

Edmund Lorencowicz · Bruno Huyghebaert ·
Jacek Uziak
Editors

Farm Machinery and Processes Management in Sustainable Agriculture

XII International Scientific Symposium 2024

Springer

Editors
Edmund Lorencowicz
Faculty of Production Engineering
University of Life Sciences
Lublin, Poland

Bruno Huyghebaert
Walloon Agricultural Research Centre
Gembloux, Belgium

Jacek Uziak
College of Management and Enterprise
Wałbrzych, Poland

ISSN 2366-2557 ISSN 2366-2565 (electronic)
Lecture Notes in Civil Engineering
ISBN 978-3-031-70954-8 ISBN 978-3-031-70955-5 (eBook)
https://doi.org/10.1007/978-3-031-70955-5

This Springer imprint is published by the registered company Springer Nature Switzerland AG
The registered company address is: Gewerbestrasse 11, 6330 Cham, Switzerland

Preface

The International Scientific Symposium "Farm Machinery and Processes Management in Sustainable Agriculture"—FMPMSA 2024 was held for the twelfth time. Thirty years have passed since the first meeting of researchers and experts attracted to development in technical processes in agriculture. The idea was born in 1994, when the International Seminar "Rational Mechanization of Family Farms" was organized at the Agricultural University of Lublin. It was attended by 30 local scientists and only one from abroad-Switzerland.

The current thematic formula was developed gradually. In 2006, the main organizer—the Department of Machine Operation and Production Processes Management from Faculty of Production Engineering, University of Life Sciences in Lublin, started cooperation with Walloon Agricultural Research Center CRA-W from Gembloux in Belgium. This enhanced the international reach of the symposium and also identified new areas of interest. The title of the symposium was changed to include the issues of sustainable agriculture. The following years have allowed to establish new and expand already existing international contacts. Symposia were held not only in Lublin, but also in other centers—in 2008 and 2015 at CRA-W in Gembloux in Belgium and in 2022 at the University of Bari Aldo Moro in Italy.

Over the last thirty years, the understanding of the concept of "sustainable development" has changed significantly. That change, in the context of agriculture and its processes, has been reflected in papers presented at subsequent symposia.

Main activities related to sustainable agriculture have been traditionally reduced to the negative impact on the environment. However, social and economic aspects are also very important. The development of sustainable agriculture should have a positive impact on the development of local societies and the level of income. That involves not only a direct impact on farmers and their families, but also on the environment and rural areas infrastructure.

Additional topics included in the themes of the symposia have been Agriculture 4.0 (or Farming 4.0) with Precision Agriculture (PA). Integrating digital technology into agriculture has become the most influential trend in the sector. With the possibility of farmers applying water, fertilizers, and pesticides only in the required quantities and in very specific target areas, plus technologies such as robots, temperature and moisture sensors, aerial images, and GPS technology should allow the farms to be more profitable, efficient, safe, and environmentally friendly.

This year's twelfth symposium was held in Lublin, at the University of Life Sciences. Over 80 people from 16 countries registered for the symposium. The topics of the presented papers included issues related to the sustainability of processes and technology in the agri-food economy. Traditionally, an away session was also organized-this time in one of the most modern beverage-can factories in Europe-Ball Packaging Europe Lublin Sp. z o. o.

This publication is a compilation of selected articles presented at the Symposium and then subjected to a verification and review process. After approval and correction, they were qualified for publication. Published articles are arranged alphabetically by the first author.

Detailed information about the presented papers and publications from the period of 1994–2024 can be found at www.up.lublin.pl/fmpmsa.

Edmund Lorencowicz
Jacek Uziak
Bruno Huyghebaert

Organization

Scientific Committee

Chairmen

Edmund Lorencowicz	University of Life Sciences in Lublin, Poland
Bruno Huyghebaert	CRA-W Gembloux, Belgium

Members

Alex Folami Adisa	Federal University of Agriculture, Abeokuta, Nigeria
Arlindo Almeida	Polytechnic Institute of Bragança, Portugal
Atanas Zdravkov Atanasov	"Angel Kanchev" University of Ruse, Bulgaria
Fatima Baptista	University of Évora & MED– Mediterranean Institute for Agriculture, Environment and Development, Portugal
Volodymyr Bulgakov	University of Life and Environmental Sciences of Ukraine, Kyiv, Ukraine
Philippe Burny	Walloon Agricultural Research Centre & Gembloux Agro-Bio Tech, University of Liege, Gembloux, Belgium
Karl-Heinz Dammer	ATB Leibnitz-Institut für Agrartechnik und Bioökonomie, Potsdam, Germany
Ester Foppa Pedretti	Marche Polytechnic University, Ancona, Italy
Fran Gjoka	Agricultural University of Tirana, Albania
Sławomir Kocira	University of Life Sciences in Lublin/Poland & University of South Bohemia in České Budějovice, Czechia
Milan Koszel	University of Life Sciences in Lublin, Poland
Artur Kraszkiewicz	University of Life Sciences in Lublin, Poland
José Rafael Marques da Silva	University of Évora, Portugal
Radko Mihaylov	Technical University of Varna, Bulgaria
Paula A. Misiewicz	Harper Adams University, UK
Gerhard Moitzi	BOKU-University of Natural Resources and Life Sciences, Vienna, Austria
Janusz Nowak	University of Life Sciences in Lublin, Poland

Jüri Olt	Estonian University of Life Sciences, Tartu, Estonia
Taskin Oztas	Atatürk University, Erzurum, Turkey
Athanassios Papageorgiou	Technological Educational Institute of Peloponnese, Kalamata, Greece
Stanisław Parafiniuk	University of Life Sciences in Lublin, Poland
Simone Pascuzzi	University of Bari Aldo Moro, Italy
Fabienne Rabier	CRA-W Gembloux, Belgium
Francesco Santoro	University of Bari Aldo Moro, Italy
Giacomo Scarascia-Mugnozza	Polytechnic of Bari, Italy
Yves Schenkel	CRA-W Gembloux, Belgium
Enkeleda Shkurta	National Environmental Agency & Polytechnic University of Tirana, Albania
Alaa Subr	College of Agricultural Engineering Sciences, University of Baghdad, Iraq
Hop Tho Hi Min	Gembloux Agro-Bio Tech, University of Liege, Belgium
Jacek Uziak	College of Management and Enterprise, Wałbrzych, Poland
Viktor Vojnich	University of Szeged, Hungary
Jens Karl Wegener	JKI Institute for Application Techniques in Plant Protection, Braunschweig, Germany

Contents

PATAT'UP: Towards a Low-Input Potato

Feriel Ben Abdallah[1(✉)], Florine Decruyenaere[1], Charlotte Bataille[1], Vincent César[1], Fabienne Rabier[1], and Vincent Berthet[2]

[1] Walloon Agricultural Research Centre (CRA-W), Gembloux, Belgium
f.benabdallah@cra.wallonie.be
[2] Walloon Potato Organization (FIWAP), Gembloux, Belgium

Abstract. Potato cultivation in Wallonia holds economic and cultural significance, but its heavy reliance on inputs poses risks at environmental, health, and economic levels. The Patat'Up project aims to address these challenges by identifying and integrating levers to create low-input production systems.

The use of late blight-resistant varieties is one of the chosen levers to reduce the high fungicide use. The objective of the 2023 trial was to determine the possibilities of delaying fungicide interventions, according to the variety's resistance, without affecting yield. The trial investigated four different varieties with four modalities: untreated modality; control modality with full treatments based on a warning monitoring system of late blight; and two delayed modalities where the first treatment was postponed by 2 or 4 applications. Results showed that delaying treatments is not possible for sensitive varieties (Fontane and Agria) in the 2023 context with high late blight pressure. Delaying the first intervention for these varieties induces foliage destruction from 25 to 100% and yield reduction from 7 to 25 t/ha compared to the control. With the use of mid-resistant variety (Louisa), delaying only 2 applications is possible. Extending the delay to 4 applications induces significant foliage destruction and a yield loss of 10 t/ha compared to the control. For the resistant variety (Acoustic), both delayed treatments resulted in similar yields to the control with no or few symptoms of foliage destruction. The use of a resistant variety enabled savings of 66% to 84% of active ingredients.

Keywords: potato · late blight · resistant varieties · fungicides treatments

1 Introduction

Potato cultivation in Wallonia holds economic and cultural significance, but its heavy reliance on inputs poses risks at environmental, health, and economic levels. Among the plant protection products, fungicides are the most commonly used in potatoes in order to control late blight caused by the naturally occurring agent *Phytophthora infestans*. Late blight is a devastating disease in our regions with the ability to colonize all parts of the plant [1]. It can destroy foliage within a few days. In the case of an early attack, yield losses can be significant. Attacks on tubers lead to storage problems. Since its appearance around 1845 in Europe, its impact on crop productivity and quality has been steadily increasing. Because of the emergence of new strains, late blight has a greater aggressiveness and manifests earlier in the season [2].

E. Lorencowicz et al. (Eds.): FMPMSA 2024, LNCE 609, pp. 1–10, 2024.
https://doi.org/10.1007/978-3-031-70955-5_1

Late blight control relies on the use of regular applications of fungicides particularly as a preventative measure in advance of any disease symptoms. Late blight still requires between 10 and 20 applications of fungicides over the growing season depending on the potential infection risks and inoculum pressure [3, 4]. In addition to the economic costs, regular use of fungicides can result in adverse environmental impacts whereas their mechanical application leads to soil compaction and additional CO_2 emission, thereby detrimentally affecting the ecological footprint of potato farming [5]. Thus, a more sustainable control strategy for late blight is urgently needed. The future control strategy needs to integrate the use of resistant varieties and chemical control linked to Decision Support Systems (DSS). DSS are available in order to help farmers to schedule fungicide spraying based on the disease risk. In Belgium, VigiMAP is an interactive web-based platform developed by CARAH (agronomical research center of Hainaut Province, Belgium). It can predict if the crop needs to be sprayed preventively according to weather data, potato variety and level of fungicide protection of the crop. The use of cultivars with different level of resistance can reduce both fungicide quantities and application frequencies.

The Patat'Up project (funded by the Wallonia Recovery Plan, Belgium) aims to identify and combine levers to create low-input production systems. In the case of the late blight trial, the goal is to reduce the number of fungicide treatments by using potato varieties that are more resistant to late blight. The idea is that a resistant or a medium resistant variety can tolerate foliar late blight symptoms without compromising tuber yield (loss in quantity and quality).

2 Material and Methods

2.1 Experimental Design

The trial was conducted in complete randomized blocks with 4 repetitions (Fig. 1). The trial cropping characteristics are given in Table 1. Each experimental plot was 6 m (corresponding to 17 tubers) long and 4 rows large, with an inter-row of 0.75 m. The trial was surrounded by two rows of Bintje, sensitive variety, to serve as inoculation strips. The harvest yield was estimated on the base of the two central rows of each plot.

The trial investigated four varieties with increasing levels of resistance to late blight and four modalities of fungicides treatments (Table 1).

The four investigated modalities are: untreated modality (modality 1); control modality with full treatments based on the VigiMAP DSS (modality 2); and two delayed modalities where the first treatment was postponed by 2 or 4 applications compared to the VIGIMAP recommendations (modality 3 and 4 respectively). The dates and applied fungicides products are summarized in Table 2.

2.2 Inoculation

Inoculations were performed with a detached infected leaf (with 2 strains of *P. infestans* 13_A2 and 36_A2) showing abundant sporulation. The leaves were placed at regular intervals along the trial, in the inoculation strips (Bintje row in the center of the varieties). Inoculations were performed on three dates (12/07, 24/07, and 27/07).

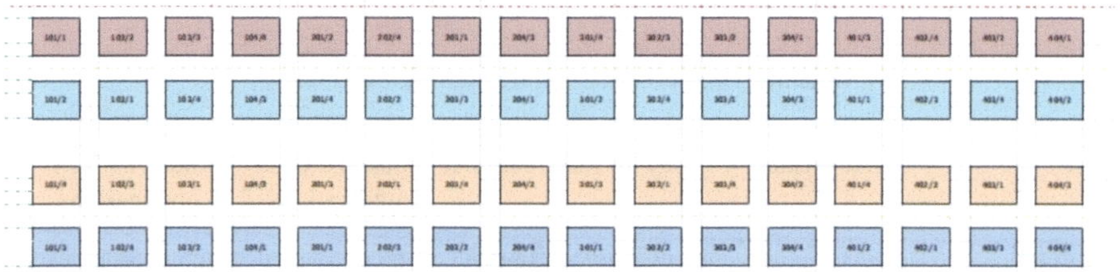

Fig. 1. Plan of the Patat'Up late blight 2023 trial. Plots of Fontane are in red, those of Agria in light blue, those of Acoustic in orange and those of Louisa in blue. The two yellow dotted lines, one between the Fontane and Agria plots and the other between the Acoustic and Louisa plots, are the inoculum strips of Bintje.

Table 1. Cropping characteristics of the potato trials realized in 2023 in loam soil of the Gembloux area in Belgium

Cropping characteristics	
Planting date	20-05-2023
Emergence	15-06-2023
Haulm killing date	15-09-2023
Harvest date	04-10-2023
Variety	Fontane* (2)
	Agria* (4)
	Louisa* (7)
	Acoustic* (8.8)
Fungicides modalities	Modality 1: untreated modality
	Modality 2: control with full treatments
	Modality 3: delayed by 2 applications
	Modality 4: delayed by 4 applications

*Late blight resistance rating (from 1 sensitive variety to 9 resistant variety) rating on the basis of MILVAR trials (annually trials on two sites in Belgium).

Irrigation was applied 7 times (10 L/irrigation) during the season (between 14/06 and 25/07) to promote the development of late blight.

2.3 Disease Estimate

The phytosanitary monitoring of the plot was carried out from mid-June to early September on a rating scale (Table 3) from 0 to 10 (10 = no symptoms/disease; 0 = vegetation completely destroyed) on the two central rows of the 6 m plots.

Table 2. Dates and products applied by modality

	Modalities and products applied during treatment			
Dates	Modality 1: untreated	Modality 2: control with full treatments	Modality 3: delayed by 2 applications	Modality 4: delayed by 4 applications
07-07-23	–	Infinito 1,6 L/ha	–	–
26-07-23	–	Ranman Top 0,5 L/ha	–	–
		Sacron 0,22 kg/ha		
04-08-23	–	Infinito 1,6 L/ha	Infinito 1,6 L/ha	–
		Sacron 0,22 kg/ha	Sacron 0,22 kg/ha	
09-08-23	–	Infinito 1,6 L/ha	Infinito 1,6 L/ha	–
		Sacron 0,22 kg/ha	Sacron 0,22 kg/ha	
18-08-23	–	Ranman Top 0,5 L/ha	Ranman Top 0,5 L/ha	Ranman Top 0,5 L/ha
		Sacron 0,22 kg/ha	Sacron 0,22 kg/ha	Sacron 0,22 kg/ha
23-08-23	–	Infinito 1,6 L/ha	Infinito 1,6 L/ha	Infinito 1,6 L/ha
		Narita 0,5 L/ha	Narita 0,5 L/ha	Narita 0,5 L/ha
31-08-23	–	Ranman Top 0,5 L/ha	Ranman Top 0,5 L/ha	Ranman Top 0,5 L/ha
06-09-23	–	Infinito 1,6 L/ha	Infinito 1,6 L/ha	Infinito 1,6 L/ha
Total of active ingredient (g/ha)	–	6136	2107	1004

To compare the severity of infection by variety, the RAUDPC (Relative Area Under the Disease Progress Curve) was calculated for each variety using the formula:

$$\text{RAUDPC} = \Sigma(\text{xi} * \text{ti})/(\text{n} * 10 * \max(\text{ti})) \quad (1)$$

where:

xi = rating obtained at ti = observation days, n = total number of observations.

The values of AUDPC were normalized by dividing the AUDPC value by the total area (= the number of days between the first appearance of late blight symptoms and the end of observation * 10). The unitless value ranges from 0 to 1. The closer the variety's value is to 0, the more severe the infection, and the closer it is to 1, the more resistant the variety is.

Table 3. Rating scale for foliage destruction by late blight (CARAH scale based on the Wageningen scale used by EuroBlight)

% destroyed	Description	Points
0	No symptoms	10
0.01	1 to 2 leaflets per plot (1 leaflet/10,000 leaflets)	9.90
0.05	3 to 5 leaflets per plot (or one terminal shoot per plot)	9.75
0.1	Around 10 leaflets per plot. First sporulations	9.5
0.5	On average 1 leaflet per affected plant (around 25 to 40 leaflets per plot)	9
1	An average of 4 leaflets per plant affected (around 100 leaflets per plot). Light general infection	8
2.5	On average, 8 to 10 leaflets per plant affected (around 200 to 250 leaflets per plot)	7
10	On average 1 leaflet out of 5 per plant affected	6
25	25% of foliage is affected. Vegetation still appears green. All plants are infected	5
50	Around 50% of the foliage is affected, and the overall color of the crop is green-brown	4
75	75% of foliage is affected, leaving 25% of green parts	3
95	Still some green parts, especially some stems	2
99	A few green leaflets remain, while the stems turn brown or die	1
100	Vegetation completely destroyed	0

2.4 Weather Data Acquisition

Daily precipitation (L/m^2) and air temperature (°C) were recorded by the reference meteorological platform (Agromet) located at Gembloux - Ernage nearby the experimental field.

2.5 Statistical Analysis

The statistical analysis was performed with the SAS software package (SAS 9.2). The effects of fungicides modalities studied per variety were subjected to analysis of variance (ANOVA) using PROC MIXED (SNK).

3 Results

3.1 Local Weather Conditions

The month of May experienced cold and relatively well-watered weather. June was very hot and very dry. July experienced much more variable and wetter weather (over 90 L of precipitation) mainly at the end of the month. Monthly temperatures were around 17 °C. August was cool and very humid in the first period. The weather changed starting from the 10th with an increase in temperatures and a decrease in precipitation.

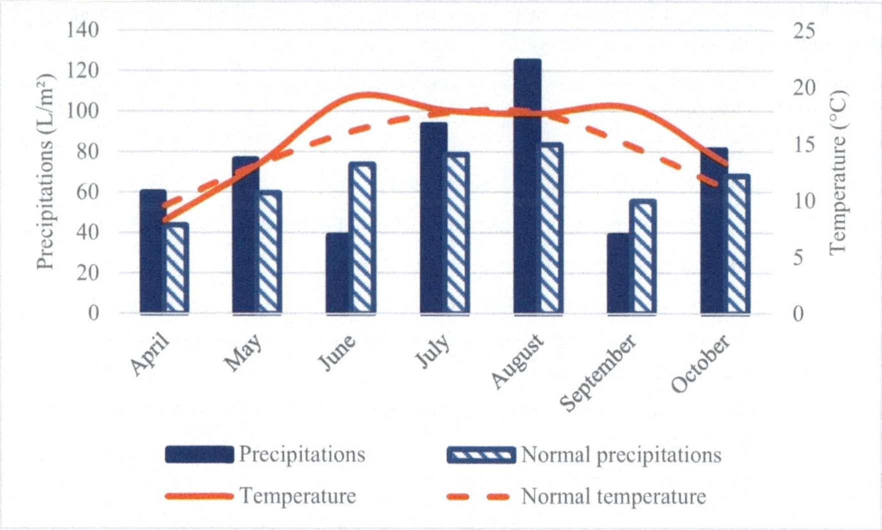

Fig. 2. Monthly rainfalls (L/m^2) and mean monthly temperature (°C) data (collected meteorological platform Agromet) in Gembloux-Ernage (nearby the trials location)

3.2 Late Blight Situation in Wallonia

The late blight pressure was low during the spring of 2023, with the first cases observed in the environment during the first half of May. The hot and dry weather in June then "cleared out" the environment of any new infections. The return of precipitation in July favored the development of new infections, and it was towards the end of the month that new cases were observed. During August, a strong late blight pressure was observed in the environment.

3.3 Late Blight Symptoms in the Experimental Trial

Figure 3 shows the evolution of late blight symptoms for the untreated fungicide modality. The first late blight symptoms were observed at the end of July for Fontane (rating of 9.41), Agria (rating of 9.69) and Louisa (rating of 9.98) with only 1 out of the 4 untreated plots showing 2 leaflets with late blight. These symptoms then progressed to complete foliage destruction for Fontane or nearly complete for Agria by the end of August (rating of 0.75). At the same time, Louisa had a rating of 4.25 with approximately 50% of the foliage affected. Acoustic showed some initial late blight symptoms at the beginning of August, for only one out of the 4 untreated plots, with no progression until the end of August. Then, Acoustic reached a rating of 6 (corresponding to less than 10% of foliage destroyed) at the beginning of September. At this last rating date, Louisa was rated at 2, and Agria had a value very close to 0.

The Area Under Disease Progress Curve (AUDPC) shows that Fontane and Agria present almost similar infection severity (with values of 0.31 and 0.38, respectively) (Fig. 3). Louisa shows a value above the average of 0.5, and Acoustic presents a value of 0.92 very close to 1, indicating good resistance to late blight foliage.

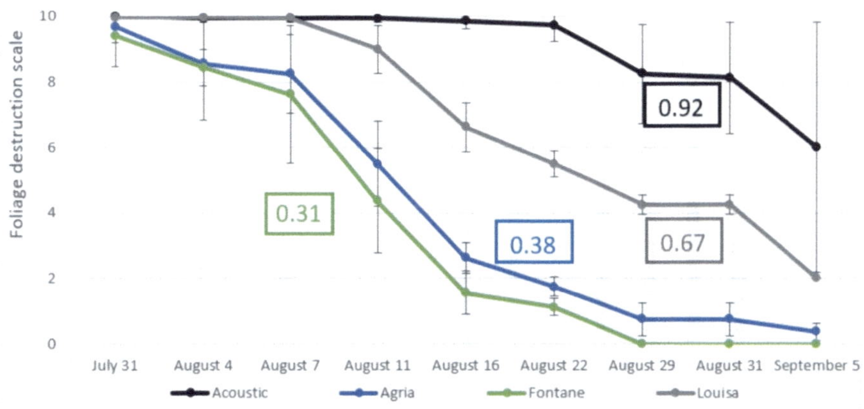

Fig. 3. Evolution of foliage destruction by late blight for each variety from late July to early September for modality 1 without fungicide protection. Foliage destruction ranged from 0 to 10 as explained in Table 3.

Figure 4 shows the last observation rating in early September per variety for each treatment modality. Despite a completely fungicide treatment, the modality 2 shows late blight symptoms (as observed for Agria, Fontane, and Louisa) with different levels of severity depending on the variety's sensitivity. This is due to several factors: the high late blight pressure observed from late July, intensified by the presence of inoculum bands; the positioning of plant protection products, which was challenging due to short optimal spraying windows (abundant precipitation in August) and the timing of fungicides application as, on certain dates, rain occurred just a few hours after the treatment was applied.

Acoustic shows at the beginning of September, a same rating of 10 for modality 2 (completely treated) and the modality with a delay of 2 treatments. Modality 4, with 4 treatments delayed, shows some initial spots with a rating of 9.9 but with no significant difference from modality 2 and 3. The untreated modality (modality 1) is not significantly different from the treated modalities for Acoustic. The large standard deviation observed for this modality 1 is due to the rating of 2 and 3.50 for two plots, showing significant foliage destruction symptoms, and 9 and 9.5 for the other two plots (Fig. 3).

The obtained rating scale for Louisa, showed no significant difference between modality 2 and 3 (Fig. 3). However, a delay of 4 treatments (modality 4) induces a rather severe infection in Louisa with a rating below 5.

For Agria and Fontane, the modality 4 showed a similar rating scale of foliage destruction than modality 1 (Fig. 3). The modality 2 is significantly different from modality 3, indicating that a delay of 2 treatments induces varying degrees of foliage destruction depending on the variety's sensitivity (Agria with a rating of 5.6 corresponding to less than 25% foliage destruction and Fontane with a rating of 3.9 corresponding to a destruction between 50 and 75%).

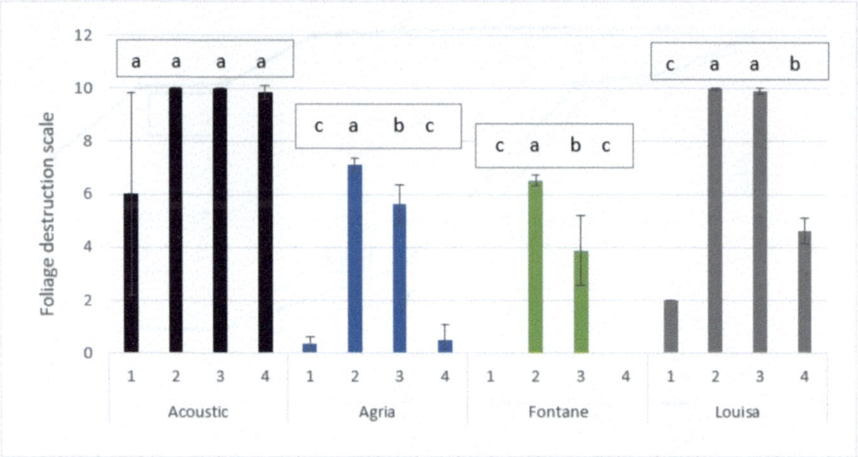

Fig. 4. Late blight foliage destruction scale by variety and fungicides treatments modalities obtained for the last observation date (04/09/23). 1 corresponds to untreated modality, 2 corresponds to control modality with full treatments based on VigiMAP, 3 corresponds to delayed modality where the first treatment was postponed by 2 applications and 4 corresponds to delayed modality where the first treatment was postponed by 4 applications. For each variety, modalities with identical letters have statistically equivalent estimated marginal means at the 5% probability threshold, according to the SNK test.

3.4 Tuber Yield

Considering the yields obtained per variety and per treatment modality, Fig. 5 shows that the variety Acoustic demonstrates interesting tubers yields across all tested modalities, with an average of 99.47 t/ha. Acoustic and Louisa show no significant differences in yield among the 3 modalities receiving fungicides treatments. Thus, the treatment delay does not induce a decrease in yield for both Acoustic and Louisa. Despite, relatively severe foliage destruction symptoms observed around mid-August (Fig. 3), Louisa was able to resist to *Phytophthora infestans* until the end of the season without any detrimental effect on yield. It is important to note that the disease occurred at the end of the season.

Despite the yield loss observed due to the absence of fungicides applications, Acoustic and Louisa (loss of 8 and 18 t/ha respectively) show no significant difference between modality 1 and 2 (Fig. 5).

The yield loss observed for Agria and Fontane in modality 3 compared to modality 2 reflects the foliage destruction observed in Fig. 2 and confirms the sensitivity of Agria and particularly of Fontane, with a loss of 8 and 17 t/ha, respectively (Fig. 5). Compared to the modality 2, the absence of fungicides applications (modality 1) results in a significant yield loss for Fontane and Agria (loss of 27 t/ha and 29 t/ha respectively). The delay of treatment by 4 applications for both these varieties resulted in similar yields than those obtained for modality 1.

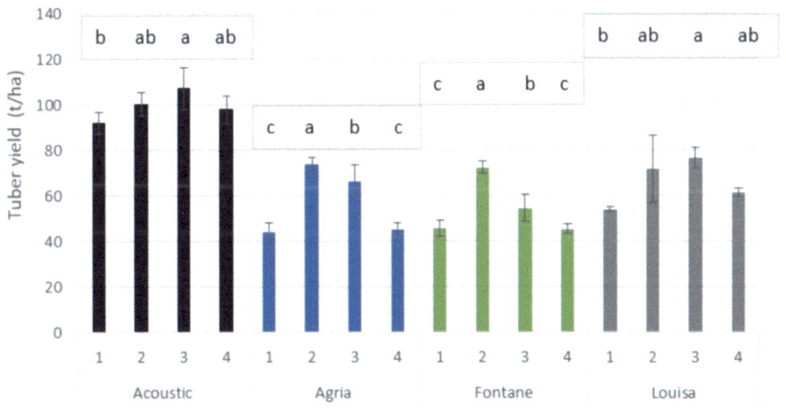

Fig. 5. Tubers' yield (calibers over 35 mm) obtained for the 4 varieties and according to the fungicides treatments modalities. 1 corresponds to untreated modality, 2 corresponds to control modality with full treatments based on VigiMAP, 3 corresponds to delayed modality where the first treatment was postponed by 2 applications and 4 corresponds to delayed modality where the first treatment was postponed by 4 applications. For each variety, modalities with identical letters have statistically equivalent estimated marginal means at the 5% probability threshold, according to the SNK test.

4 Conclusion

The high late blight pressure observed during the 2023 season highlighted differences between fungicide treatment modalities.

For sensitive varieties such as Fontane and Agria, reducing fungicides is not possible. Indeed, delaying fungicides treatments by two applications results in an average foliage destruction of 25% for Agria and about 75% for Fontane with more severe symptoms. Delaying fungicides treatments by two applications leads to complete foliage destruction for these two sensitive varieties. In terms of yield, delaying fungicides applications by two treatments results in a yield loss of 7 t/ha for Agria (compared to the fully protected modality) and a more significant loss of 17 t/ha for Fontane. Delaying fungicides applications by four treatments results in a higher loss of 27 to 28 t/ha respectively for Agria and Fontane.

Unlike the two sensitive varieties, Louisa can tolerate a delay of two fungicides treatments (RAUDPC value of 0.99 very close to that observed for modality 2). However, delaying fungicides applications by four dates is not possible since the foliage destruction observed in mid-August for modality 4 (which reached over 25% of foliage destruction by late August) resulted in a yield loss of 10 t/ha compared to the reference modality.

Acoustic shows similar yields obtained for the delayed modalities 3 and 4 (delay of 2 or 4 applications respectively) compared to the reference modality. In modality 3, no late blight symptoms were observed on the foliage. For modality 4, only a few initial symptoms (around 2 or 3 leaflets showing symptoms) were observed at the end of August without any impact on yield. The use of resistant varieties thus allows for a reduction in active ingredient usage (g/ha) ranging from 66% for modality 3 to 84% for modality 4. These results need to be confirmed under high late blight pressure observed at the beginning of the season, in contrast to the 2023 trial where the first symptoms appeared at the end of July.

Acknowledgments. The Walloon Recovery Plan (Belgium) funded this study. We would like to thank the Walloon Agricultural Research Centre and particularly the "Crop Production", "Crops and forest health" and "Biodiversity and plant and forest improvement" Units for their assistance with field operations, sampling and analysis.

References

1. Hooker, W.J.: Compendium of Potato Diseases. International Potato Center, Lima (1981)
2. Fry, W.E., Birch, P.R.J., Judelson, H.S., Grünwald, N.J., Danies, G., Everts, K.L., et al.: Five reasons to consider Phytophthora infestans a reemerging pathogen. Phytopathology **105**(7), 966–981 (2015)
3. Goffart, J.P., Haverkort, A., Storey, M., Haase, N., Martin, M., Lebrun, P., et al.: Potato production in northwestern Europe (Germany, France, the Netherlands, United Kingdom, Belgium): characteristics, issues, challenges and opportunities. Potato Res. **65**(3), 503–547 (2022)
4. Arora, R.K., Sharma, S., Singh, B.P.: Late blight disease of potato and its management. Potato J. **41**(1), 16 (2014)
5. Haverkort, A.J., Hillier, J.G.: Cool farm tool–potato: model description and performance of four production systems. Potato Res. **54**, 355–369 (2011)

An Image Processing Algorithm to Address the Problem of Stains Merge on Water Sensitive Papers and Its Impact on the Evaluation of Spray Quality Indicators

Ameer H. Al-Ahmadi[1], Alaa Subr[1(✉)], Stanisław Parafiniuk[2], and Marek Milanowski[3]

[1] Department of Agricultural Machines and Equipment, College of Agricultural Engineering Sciences, University of Baghdad, Baghdad 10071, Iraq
alaa.kamel@coagri.uobaghdad.edu.iq

[2] Department of Machinery Exploitation and Management of Production Processes, University of Life Sciences in Lublin, Głęboka 31, 20-612 Lublin, Poland

[3] Department of Thermal Technology and Food Process Engineering, University of Life Sciences in Lublin, Głęboka 31, 20-612 Lublin, Poland

Abstract. There are many techniques that can be used to estimate the spray quality traits such as the spray coverage, droplet density, droplet count, and droplet diameter. One of the most common techniques is to use water sensitive papers (WSP) as a spray collector on field conditions and analyzing them using several software. However, possible merger of some droplets could occur after they deposit on WSP, and this could affect the accuracy of the results. In this research, image processing technique was used for better estimation of the spray traits, and to overcome the problem of droplet merger. The droplets were classified as non-merged and merged droplets based on their roundness, then the merged droplets were separated based on the average non-merged droplets areas. The results were compared to those which were analyzed using different software. In addition, an analysis was done to samples that were created to simulate the water sensitive paper with pre-known results as a validation. The results showed that separating the droplet based on the developed software increased the coverage, droplet count, and the mean droplet diameter by an average of 1.9, 4.5, and 1.25 times, respectively. This technique showed better estimation of the spray traits using WSP.

Keywords: Droplet merger · Image processing · Water sensitive paper · Spray quality

1 Introduction

For pesticide applicators, it is essential to ensure that the spray solution reaches the target with high effectiveness and minimum losses. To evaluate this, there are different methods in laboratories. However, under field conditions using water sensitive paper (WSP) is

E. Lorencowicz et al. (Eds.): FMPMSA 2024, LNCE 609, pp. 11–22, 2024.
https://doi.org/10.1007/978-3-031-70955-5_2

still a popular method. This method is simple, quick, and low cost. After analyzing WSP, it reports data (with the aid of image processing software) about the stain's coverage percentage, stains number, and stains diameter Krawczuk et al. (2021). However, this method has a few limitations, one of them is the overlapping (merging) of adjacent stains especially on dense spray (high coverage percentage). This merging of stains usually affects the analyzing process of data especially when calculating the stains diameter and stains number. Sachithra and Subhashini, (2023) reported different problems with the use of WSP, among them is the difficulty to detects droplets with a diameter of less than 50 μm; the effect of liquid physical properties on the spread of droplets on WSP; the effect of high humidity; finally, the effect of stains merger (overlapping).

Privitera et al. (2023) stated that there is a high probability of occurring an error during analyzing WSP due to the overlapping of stains, especially if the sample is too large (dense spray). Actually, the problem of stains overlapping becomes more visible when the percentage of stains coverage on WSP is more than 20% (Brandoli et al., 2021; Özlüoymak and Bolat, 2020; Fox et al., 2003; Zhu et al., 2011). While He et al. (2024) stated that WSP could be used as a tool to assess the deposition of spray if the spray coverage percentage is below 30% with minimum stains overlap. The study of Cerruto et al., (2016) to simulate the WSP has revealed that stains overlapping was independent of spray features. For example, when the coverage percentage was in range from 4.7% to 61.6% the stains overlapping was in range from 0.3% up to 33.4%.

Moreover, stains overlapping (especially when coverage percentage is between 36% and 43%) could affect the results of stains density on the WSP (Cerruto et al., 2016). Cunha et al. (2013) stated in his study of testing different computer programs for analyzing WSP that one of the programs (CIR) gave a tendency to report a high number of stains density. They interpreted this to the higher sensitivity of this program to perform the separation of overlapping stains. In some cases, even when the percentage coverage of stains increased (due to the increase of spray pressure), the stains density decreased because of the stains overlapping (Cerruto et al., 2019).

In agriculture, using artificial intelligence (AI) and image processing techniques has become common and essential to improve the performance of agricultural operations, aiming in the end to achieve the sustainability of agriculture (Sachithra and Subhashini, 2023).

Different attempts were made to overcome the problem of stains overlapping on WSP. One of them is the study of Marcal and Cunha, (2008), they proposed a method that depends on image processing to evaluate the data of WSP. The method depends on counting the overlapped stains of WSP images (after converting them to binary images) based on mathematical morphology. They concluded that their method was effective in situations with considerable stains overlapping (high coverage percentage). However, they also reported that their method could not be able to solve the problem of full overlap between two stains. Sayinci et al. (2012) used a method for elimination of overlapped and non-circular stains on WSP depending on the roundness values of stains. Özlüoymak and Bolat (2020) used an advanced morphological segmentation technique (in image processing software) to evaluate the spray parameters on WSP (like droplet count). They reported that this technique was better in providing accuracy and more data of detecting overlapped stains compared with other image processing software. Xun and Gil, (2024) proposed a novel methodology by using the segmentation of merged stains based on concave point detection and ellipse fitting. They concluded that their method could be used to have more accurate indicators of spray quality from the WSP. Brandoli et al., (2021) developed a technology based on the custom image analysis software, aiming to use it as a real-time tool for analyzing the WSP. Besides this, they developed a smartphone application that uses this technology, this application could identify the overlapping stains better than former methods. They indicated that their method resulted in finer precision, with better accuracy, and more information. Yan et al. (2024) proposed a method to analyze the deposition of stains on WSP, it included combination of different algorithms and criteria functions. For example, the maximum between-cluster variance algorithm; criteria of area threshold; roundness factor, mathematical morphology operations; and the optimized watershed algorithm. They concluded a high consistency between their method and the manual method, with an advantage of convenient and short time to assess the spray quality on WSP for their method.

The main objective of this study was: firstly, to develop software to distinguish merged and non-merged stains on WSP depending on image processing software and contour detecting. Secondly, to count the number of stains that created the overlapping stains. Thirdly, to correct the results for the parameters of stains coverage percentage, stains density, and stains diameter after the segmentation of the overlapping of stains. Moreover, a validation for the effectiveness of this method was planned to be done.

2 Material and Methods

The results of Image-Pro Premier 3D software (IPP) were compared to the newly developed software before and after correction. The newly developed software was based on image processing and contour detecting of the droplets in addition to some statistical estimation as described below.

2.1 Proposed Approach for Analyzing Spray Traits

Image Processing Procedure

Images of water sensitive papers were processed using several procedures. Image processing was performed using OpenCV library via Python programming language. First, the RGB images were converted to grayscale using cv2.cvtColor(img, cv2.COLOR_BGR2GRAY). The grayscale image was blurred to remove undesired noise using cv2.blur() with a window of 3 by 3. Then the blurred image was converted into a binary image using threshold method with THRESH_BINARY_INV and cv2.THRESH_OTSU thresholding type. The Otsu thresholding estimates a threshold value that minimizes the weighted within-class variance.

Contour Detection

The method of cv2.findContours() was used to detect the boundaries of each droplet including the merged droplets. cv2.RETR_LIST was used as contour retrieval algorithm to record all the contours without any hierarchical relations. Since all the boundaries' points are needed, the cv2.CHAIN_APPROX_NONE approximation mode was passed to the findContour method.

Roundness Estimation

Roundness of each detected droplet weas estimated using the following equation:

$$Roundness = \frac{(4\pi\ Area)}{Perimeter^2}$$

The area and perimeter of each droplet was calculated using the cv2.contourArea() and cv2.arcLength(), respectively. The estimated roundness of each droplet was checked if it is more than a predefined threshold the droplet will be considered as a single droplet, otherwise it will be considered as a cluster of merged droplets. The areas of each group were recorded in a list to facilitate summing and averaging them.

Mean area of the non-merged droplet was calculated by dividing the total area of the non-merged droplet by the number of them.

$$A_{mean} = \frac{\sum_{i=0}^{n} A_i}{n}$$

where A_{mean} is the mean area of the non-merged droplet, A_i the area of the droplet i, and n is the total count of the non-merged droplet.

Estimate the Correct Merged Droplet Count

The droplet count was estimated by summing the number of non-merged droplets and the number of merged droplets. Before separating the merged droplets, the software will consider all of the droplets as non-merged, but after separating them, the software will calculate the droplet count based on the following equation:

$$count_{merged} = \frac{\sum_{i=0}^{n} A_{merged\,i}}{A_{mean}}$$

where A_{mean} is the average area of the non-merged droplets, A_{merged_i} the area of the merged droplet i.

The correct total droplet count will be the summation of the non-merged droplets count and the corrected droplet count.

Estimate the Spray Coverage

The spray coverage was estimated by dividing the summation of the total area of the non-merged droplets and the total area of the merged droplet by the image area all measured in squared pixel.

$$Coverage = \frac{\sum_{i=0}^{n} A_i + \sum_{i=0}^{n} A_{merged\,i}}{A_{image}} \times 100\%$$

where A_{image} is the image area (image height $\times$ image width) in unit of pixel.

Estimate the Droplet Density

Droplet density was estimated by dividing the summation of the total droplet count including the corrected non-merged droplet count and the merged droplet count by the WSP area in squared centimeter.

$$droplet_{density} = \frac{total_{droplet}}{Area_{WSP}}$$

Estimate droplet size (DV0.5)

Mean droplet diameter or $DV_{0.5}$ was calculated based on the following equation:

$$Droplet_{diameter} = \sqrt{\frac{4 \times A_{mean}}{\pi}} \times C$$

where C is a converter factor that converts pixels into micrometer.

a factor to convert the droplet diameter to micron the following equation was used (Fig. 1):

$$factor = \frac{image\ width(cm)}{image\ width\ (pixel)} \times 10000$$

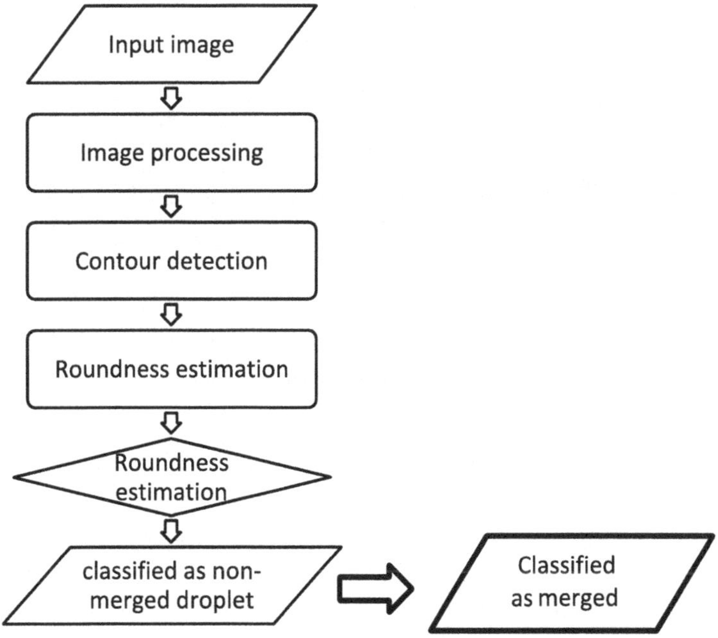

Fig. 1. The algorithm followed for the newly developed software

3 Results and Discussion

The data of the new developed software before (NDSB) and after (NDSA) correction in addition to the data obtained from IPP software were analyzed using the Tukey mean comparison test to investigate the difference among the means of the three groups.

3.1 Validation Test

To validate the results of the newly developed software three water sensitive papers were simulated using the Microsoft PowerPoint software. The Actual droplet count was calculated manually for each simulated paper. The actual droplet diameter was fixed for each simulated water sensitive paper. The actual coverage was calculated based on the following equation (Fig. 2 and Table 1):

$$Coverage = \frac{\frac{\pi (Droplet\ diameter\ (\mathrm{cm})^2)}{4} \times Droplet\ count}{Image\ length\ (\mathrm{cm}) \times Image\ width\ (\mathrm{cm})} \times 100\%$$

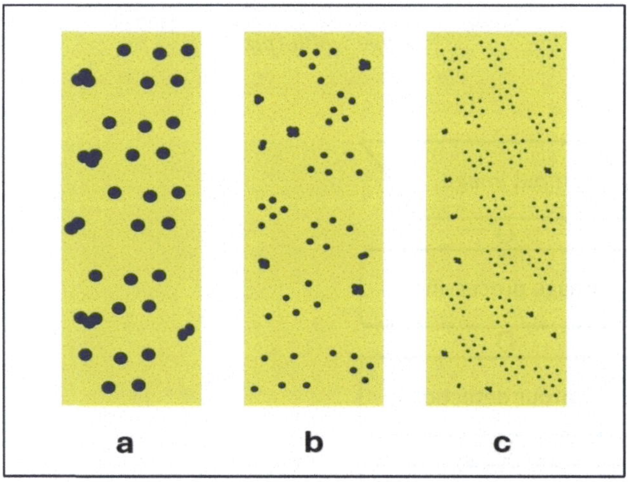

Fig. 2. Simulated water sensitive paper

3.2 Spray Coverage

The analysis showed that the spray coverage did not significantly differ among the three groups. However, there were numeric differences; the average coverage obtained by the newly developed software after correction, the newly developed software before correction, and the IPP software were 19.3%, 19.3%, and 12.5%, respectively. The equality of the coverage of the newly developed software before and after correction was expected since the coverage was calculated as the total area of the droplets divided by the total area of the WSP which was not affected by the correction (Fig. 3).

Table 1. Actual and predicted spray traits values for the simulated water sensitive papers

Trait	Simulated paper	Actual values	Predicted values
Coverage (%)	a	9.4	9.69
	b	4.25	3.85
	c	2.96	2.59
Droplet count	a	38	38
	b	63	61
	c	209	212
Droplet diameter (μm)	a	2500	2482
	b	1300	1243
	c	600	547

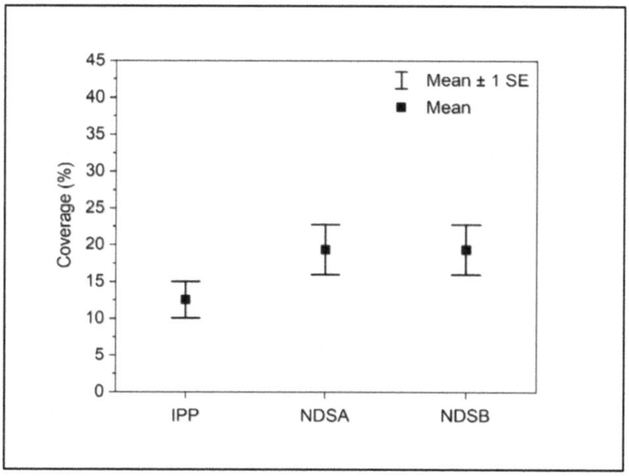

Fig. 3. Estimated spray coverage using the three methods

3.3 Droplet Count

The analysis showed that there were significant differences among the three methods ($F_{2,24} = 8.9, p = 0.001$). The Tukey test revealed a significant difference among the droplet counts of the three groups. The newly developed software after correction significantly differed from the developed software before correction and the IPP software which in turn did not significantly differ from each other. The droplet count obtained by the newly developed software after correction, the newly developed software before correction, and the IPP software were 1123, 568, and 389, respectively. The results showed that the correction succeeded in separating the merged droplets. These results express the extent to which the droplet count is affected by the droplet merger (Fig. 4).

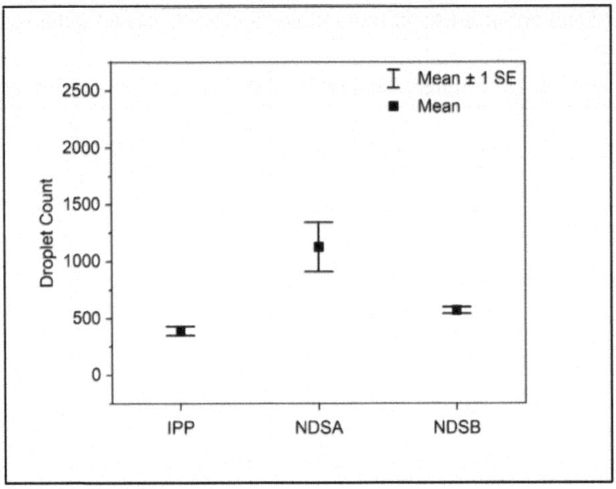

Fig. 4. Estimated droplet count using the three methods

3.4 Droplet Diameter

The analysis revealed that there was a significant difference among the three methods ($F_{2,24} = 31, p = 2.22 \times 10^{-7}$). The Tukey test expressed that there were significant differences among all the studied method. The droplet diameter obtained by the newly developed software after correction, the newly developed software before correction, and the IPP software were 665, 894, and 493 μm, respectively. This result showed that the mean droplet diameter decreased using the newly developed software after correction compared to the newly developed software before correction, but it is larger than IPP software (Fig. 5).

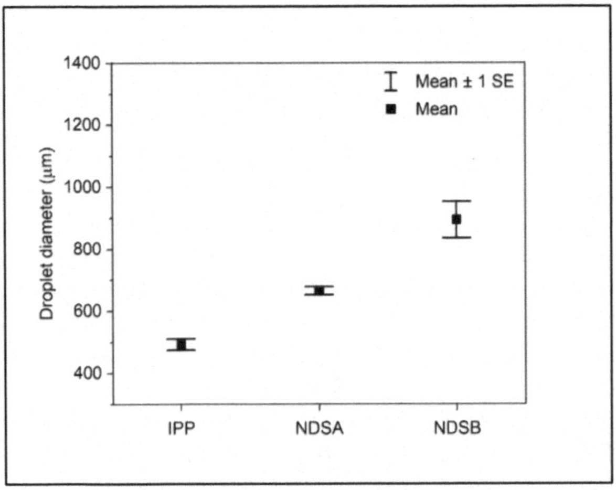

Fig. 5. Estimated droplet diameter using the three methods

3.5 Regression Analysis

Regression analysis was performed to investigate the relationship between the results of the studied traits using the IPP software and our developed software after correction, and the capability of predicting the correct values from the IPP software values.

For the coverage trait, a significant positive relationship between the IPP software data and our developed software after correction ($F_{1,7} = 198.8,\ p = 2.14 \times 10^{-6}$). The intercept of the model was not significant, whereas the slope was significant ($t_7 = 14.1,\ p = 2.14 \times 10^{-6}$). The R-Square was 0.97 indicating that this model was well performed to predict the corrected coverage from the non-corrected coverage (Fig. 6).

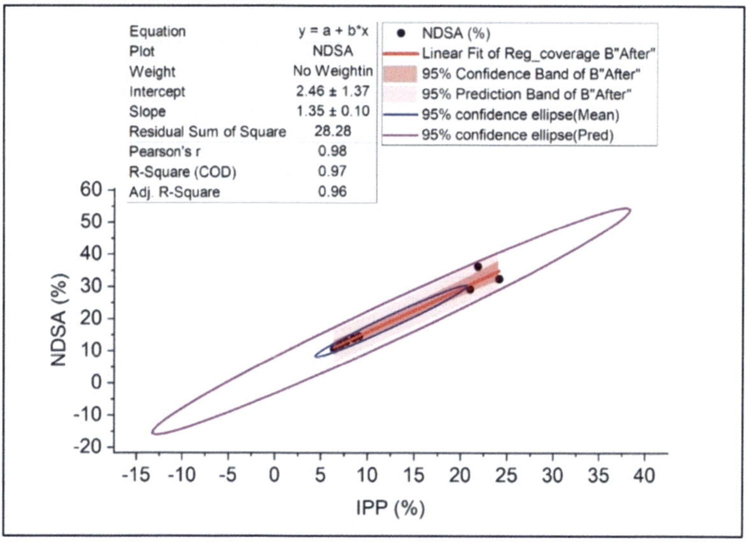

Fig. 6. Linear regression for coverage before and after correction

In relation to the droplet count trait, the analysis showed that there was a significant relationship between the droplet count data before and after correction ($F_{1,7} = 41.13,\ p = 3.63 \times 10^{-4}$). The intercept was significant ($t_7 = -2.45,\ p = 4.38 \times 10^{-2}$), and the slope of the model was highly significant ($t_7 = 6.55,\ p = 3.17 \times 10^{-4}$). The R-Square was 0.85 indicating that the model performance was slightly lower than that of the coverage trait.

For the droplet diameter, the regression analysis showed that there was not a significant relationship between the droplet diameter IPP software and the newly developed software after the correction (Fig. 8). However, a significant relationship between the IPP software and the newly developed software before the correction was found ($F_{1,7} = 14.7,\ p = 6.33 \times 10^{-3}$) with an R-Square of 0.68 (Figs. 7 and 9).

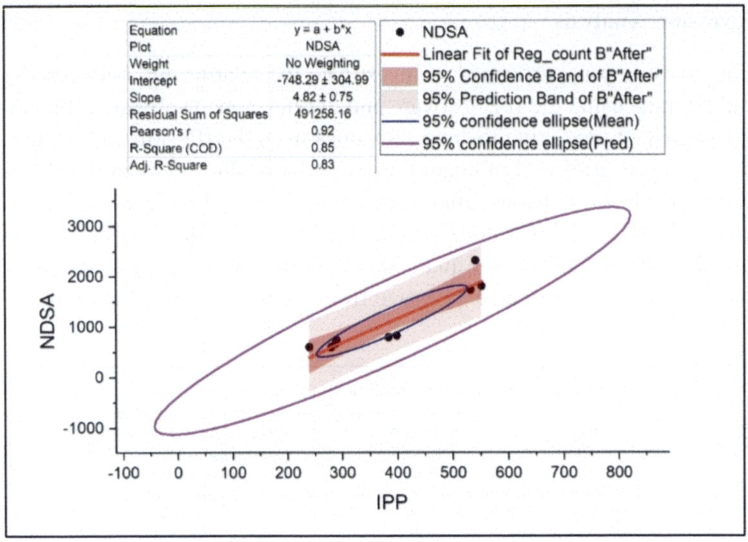

Fig. 7. Linear regression for droplet count before and after correction

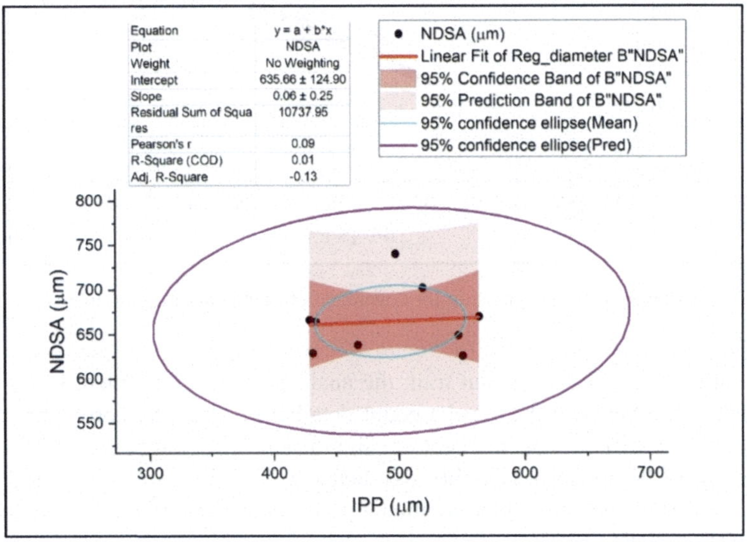

Fig. 8. Linear regression for droplet diameter IPP software and after correction

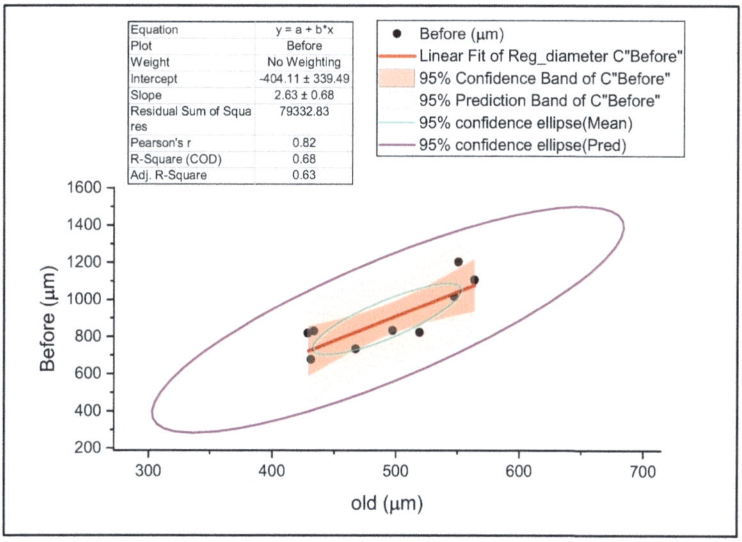

Fig. 9. Linear regression for droplet diameter obtained by IPP software and the newly developed software before correction

4 Conclusion

In this study, a personally developed software based on image processing and contour detecting was investigated for its accuracy to estimate some spray traits including coverage, droplet count, mean droplet diameter and its results were compared to that of Image-Pro Premier 3D software. The results showed that the coverage estimated by the newly developed software before and after correction did not significantly differ from that of the IPP software. The droplet counts increased when using the newly developed software after correction that may refer to better droplet separation. The droplet diameter using the newly developed software after correction was higher than that of the IPP software but less than that of the newly developed software before correction. Moreover, the regression analysis showed that the coverage of the newly developed software after correction was predictable by the IPP software results.

References

Brandoli, B., et al.: DropLeaf: a precision farming smartphone tool for real-time quantification of pesticide application coverage. Comput. Electron. Agric. **180**, 105906 (2021). https://doi.org/10.1016/j.compag.2020.105906

Cerruto, E., Failla, S., Longo, D., Manetto, G.: Simulation of water sensitive papers for spray analysis. Agric. Eng. Int. CIGR J. **18**(4), 22–29 (2016)

Cerruto, E., Manetto, G., Longo, D., Failla, S., Papa, R.: A model to estimate the spray deposit by simulated water sensitive papers. Crop Prot. **124**, 104861 (2019)

Cunha, J.P.A.R., Farnese, A.C., Olivet, J.J.: Computer programs for analysis of droplets sprayed on water sensitive papers. Planta Daninha **31**, 715–720 (2013)

Fox, R.D., Derksen, R.C., Cooper, J.A., Krause, C.R., Ozkan, H.E.: Visual and image system measurement of spray deposits using water–sensitive paper. Appl. Eng. Agric. **19**(5), 549 (2003)

He, M., Qi, P., Han, L., He, X.: Study on spray evaluation: the key role of droplet collectors. Agronomy **14**(2), 305 (2024)

Krawczuk, A., Parafiniuk, S., Przywara, A., Huyghebaert, B., Rabier, F., Limbourg, Q., Mostade, O., Kocira, S.: Technical parameters of biostimulant spraying a determinant of biometric traits and yield of soybean seeds. Agric. Eng. **25**, 171–179 (2021)

Marçal, A.R.S., Cunha, M.: Image processing of artificial targets for automatic evaluation of spray quality. Trans. ASABE Am. Soc. Agric. Biol. Eng. **51**, 811–821 (2008)

Özlüoymak, Ö.B., Bolat, A.: Development and assessment of a novel imaging software for optimizing the spray parameters on water-sensitive papers. Comput. Electron. Agric. **168**, 105104 (2020)

Privitera, S., Manetto, G., Pascuzzi, S., Pessina, D., Cerruto, E.: Drop size measurement techniques for agricultural sprays: a state-of-the-art review. Agronomy **13**(3), 678 (2023)

Sachithra, V., Subhashini, L.D.C.S.: How artificial intelligence uses to achieve the agriculture sustainability: systematic review. Artif. Intell. Agric. **8**, 46 (2023)

Sayinci, B., Bastaban, S. and Sánchez-Hermosilla, J.: Determination of optimal spot roundness variation interval for droplet size analysis on water sensitive paper (2012)

Xun, L., Gil, E.: A novel methodology for water-sensitive papers analysis focusing on the segmentation of overlapping droplets to better characterize deposition pattern. Crop Prot. **176**, 106492 (2024)

Yan, F., Zhang, Y., Zhu, Y., Wang, Y., Niu, Z., Abdukamolovich, J.A.: An image segmentation of adhesive droplets based approach to assess the quality of pesticide spray. Smart Agric. Technol. **8**, 100460 (2024)

Zhu, H., Salyani, M., Fox, R.D.: A portable scanning system for evaluation of spray deposit distribution. Comput. Electron. Agric.. Electron. Agric. **76**(1), 38–43 (2011)

Monitoring and Evaluation of the Moisture Retention of Leached Chernozem Under Different Types of Tillage

Atanas Z. Atanasov[1(✉)], Plamena D. Nikolova[1], and Boris I. Evstatiev[2]

[1] Department of Agricultural Machinery, Agrarian and Industrial Faculty, University of Ruse "Angel Kanchev", 7017 Ruse, Bulgaria
aatanasov@uni-ruse.bg

[2] Department of Electronics, Faculty of Electrical Engineering, Electronics and Automation, University of Ruse "Angel Kanchev", 7017 Ruse, Bulgaria

Abstract. The study discusses the influence of some soil treatments on its moisture storage, soil salinity and nutrient content. The application of remote sensing approach of the soil can provide accurate information for the preparation of maps of moisture storage and of the potential soil fertility in specific agricultural areas. The study is aimed monitoring and evaluation of the moisture retention of leached Chernozem under different types of tillage using device operating on the principle of electromagnetic induction (EMI). The monitoring was conducted on the 12th of October 2023 in agricultural cooperative "Zlatno zarno" near Kalipetrovo village. The geographical location of the village is 44°04′50.96″ N, 27°14′11.89″ E and at an altitudinal range of 82 m. Ploughing was found to increase soil moisture compared to disking, with uniform moisture distribution observed in plowed fields. Soil salinity was higher in areas with pH $<$6.5, leading to reduced moisture content and increased levels of certain nutrients. Plowing resulted in lower residual nutrient levels and less soil compaction compared to disking. Additionally, there was a significant correlation between soil water content and NH_3-N+NO_3-N content, with the Kohen's Kappa coefficient reaching 0.74. These results indicate that the soil water content could be used as a rough indicator for the soil NH_3-N+NO_3-N content and vice versa. These findings provide insights into optimizing sustainable moisture management in agricultural systems, particularly in Chernozem soils, to enhance water use efficiency and soil health.

Keywords: Investigation · Assessment · Electromagnetic induction (EMI) · Sustainability · Soil characteristics · Smart Farming

1 Introduction

Water is one of the important biological factors affecting plant growth. Insufficient soil moisture during the critical periods of plant vegetation is the cause of instability in the yields obtained from them. The provision of the necessary amount of water for the plants through rainfall is not always sufficient and evenly distributed over time. The possible

E. Lorencowicz et al. (Eds.): FMPMSA 2024, LNCE 609, pp. 23–36, 2024.
https://doi.org/10.1007/978-3-031-70955-5_3

ways to obtain the additional necessary amount of water is through different types of irrigation. According to [1] soil moisture is indicated as the most important factor in changing the maximum water holding capacity caused by soil compaction. The decrease in the maximum water holding capacity depends on the number of impacts on soil cover. The comparison of the influence of different types of tillage and no tillage on available soil water capacity was studied by [2]. Other authors [3] study changes in the fertility of a leached Chernozem under different primary tillage technologies. Monitoring and assessment of soil condition is very important in identifying appropriate agrotechnical measures to increase improving soil water retention capacity in non-irrigated areas. There are various methods for monitoring soil moisture content. The well-known methods of collecting soil samples from individual points in the field for gravimetric determination of soil moisture [4] are accurate but too laborious and time-consuming to process. A quick method to determine soil moisture can be through remote monitoring of water relations in the soil-plant-atmosphere system [5]. This method has been successfully used by [6, 7]. The use of sensors to monitor soil moisture at different depths [8] enables remote acquisition of accurate real-time information but cannot be effectively used for spatially unrelated information. The information obtained from the measurement of the spatial distribution of soil moisture is of great importance in the management of soil water reserves. One of the methods for non-contact measurement of the spatial distribution of soil moisture was proposed by Singenta Company, where gamma rays are used in the measurement. Other similar methods are proposed by [9–11] who use a device operating on the principle of electromagnetic induction (EMI).

The aim of our research is to determine the spatial distribution of soil moisture of leached Chernozem under different types of tillage.

2 Materials and Methods

The research employs a systematic approach, utilizing experimental plots to compare the moisture retention of leached Chernozem under different fertilization and tillage practices.

2.1 Study Area Description

The study area is located in the Northeast part of Bulgaria in Silistra region. The characteristic of this part of Bulgaria is that the soils are very fertile, but the rainfall is unevenly distributed throughout the year (Fig. 1) and is often the cause of unsustainable crop yields.

The study was carried out on the 12th of October 2023 in agricultural cooperative "Zlatno zarno" near Kalipetrovo village. The geographical location of the village is 44°04′50.96″ N, 27°14′11.89″ E and at an altitudinal range of 82 m above sea level. The experimental area included six fields with an approximate area of each field of 0.76 h., and the total area of all fields was 2.4 h. The predominant soil in all six fields was leached Chernozem. The slope of each of the experimental fields was below 0.3%. The fields are located within a radius of 3 km to 5 km around the village. In our study, we numbered the fields as Field-1 Field-2, Field-3, Field-4, Field-5 and Field-6. The crops grown on the

Fig. 1. Amount of precipitation in Silistra region for 2022 and 2023

experimental fields are wheat and corn. Crop rotation is done every year. In the previous year 2022, Field-1, Field-2, Field-3 and Field-4 were disked to a depth of 25 cm, after which wheat was sown. Field-5 and Field-6 were plowed to a depth of 35 cm and corn was sown in the spring. In the current year 2023, the cultivation of wheat, corn and sunflower is planned on the six experimental fields.

2.2 Soil Surveying Method

When conducting the experiment, a non-contact scanner Scanfield 5S of the company Scanfield 5S Smart System (e-mail: office@scanfield5s.com) was used, aggregated to a highly passable Toyota Tundra car (Fig. 2). The operation of the scanner is not affected by the climatic conditions and the surface condition of the scanned area. The working scanning speed is up to 50 km h^{-1}, the scanning depth is 1.1 m with a working scanning range width of 10 to 36 m. The sensor used in this type of device works on the principle of Electromagnetic Induction (EMI) without making direct contact with the soil [11].

Fig. 2. Scanfield 5S device for soil scanning

The method of operation of the device is based on spatial distribution of soil electrical conductivity (ECa). According to [12] ECa is a measurement that correlates with soil properties that affect crop productivity, including soil texture, cation exchange capacity (CEC), drainage conditions, organic matter level, salinity, soil characteristics, the individual layers of the soil and the presence of nutrients.

The soil signal penetration depth was in the range of 0–30 cm, which was sufficient for the analysis of the soil condition in the root-inhabited soil layer of the crops intended for sowing next year.

The examined fields were cleaned of any metallic items to prevent the introduction of artificial magnetic interference.

Scanning with a non-contact scanner Scanfield 5S was performed in the following sequence: 1) Traversing the scanned field and collecting data, 2) Processing the data by artificial intelligence Scanfield 5S, and determining the number of soil samples and GPS coordinates of the sampling locations, 3) Taking soil samples, 4) Processing of the soil samples in a laboratory for soil analysis of the company Scanfield 5S Smart System, 5) Entering the processed data into artificial intelligence Scanfield 5S, 6) Building a spectral map of the scanned field.

2.3 Data Analysis

In order to analyses the data and obtain some agreement between the different parameters, some of them are being compared using the following methodology.

1. Two parameters are chosen, which are to be compared with each other.
2. Their classes are presented in graphical form using a greyscale, in order to unify their appearance.
3. If necessary, the classes of one of the parameters is reorder in reverse order, in case the two variables have a negative correlation.
4. 20000 pixels are randomly chosen, which are distributed so that the number of pixels in each class is proportional to its relative area.
5. A confusion matrix is created, which presents how the pixels are classified for each class.
6. All pixels, which are classified outside the evaluated area are filtered out.
7. The precision and recall are calculated for each class, according to:

$$Acc = \frac{Correct\ number\ of\ pixels}{Total\ number\ of\ pixels} \tag{1}$$

8. The Kohen's Kappa is calculated, according to [13]:

$$k = \frac{N \cdot \sum_{i=1}^{n} m_{i,i} - \sum_{i=1}^{n} (G_i \cdot C_i)}{N^2 - \sum_{i=1}^{n} (G_i \cdot C_i)} \tag{2}$$

where i is the class number, N is the total number of pixels; m_i, i is the number of pixels, which belong to class i and were classified correctly; C_i is the total number of predicted pixels in the corresponding class; G_i is the total number of actual pixels in the corresponding class.

Kohen's Kappa could be interpreted as follows [13]:

- $k \leq 0$ – no agreement;
- $0.01 \leq k \leq 0.20$ – none to slight agreement;
- $0.21 \leq k \leq 0.40$ – fair agreement;
- $0.41 \leq k \leq 0.60$ – moderate agreement;
- $0.61 \leq k \leq 0.80$ – substantial agreement;
- $0.81 \leq k \leq 1.00$ – almost perfect agreement.

3 Results and Discussions

Preliminary findings suggest notable differences in moisture retention among the various tillage methods. The obtained results when disking tillage Field-1 is shown in Figs. 3, 4, 5, 6, 7 and 8. During the different phases of their development, plants need different amounts of water. From Fig. 3 it can be seen that the soil moisture in experimental Field 1 is between <65 L m^3 and >90 L m^3. The amount of moisture in the soil significantly influences the uptake of nutrients. Plants can only assimilate elements in water-soluble forms. When there is a shortage, nutrients can be supplied through fertilizers, but their absorption efficiency is closely linked to the presence of water in the root zone [12]. It is crucial to assess the soil moisture for individual fields and use this information to establish guidelines for nutrient application into the soil. The uneven distribution of soil moisture in the experimental field is a prerequisite for unevenness in the obtained yields from the cultivated crops.

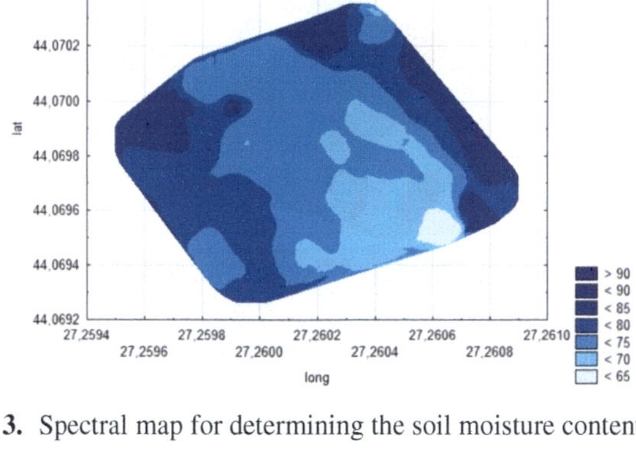

Fig. 3. Spectral map for determining the soil moisture content (W)

In Fig. 4 shows the salinity of the soil in the individual sections. From the analysis of the obtained results, it was found that in the areas with high soil moisture >90 L m^3, <90 L m^3 <85 L m^3 and <80 L m^3 of field moisture content (PPV) the prevailing pH >7.4; pH <7.4; pH <7.2. For the areas with soil moisture <70 L m^3 and <65 L m^3 of field moisture content (PPV), an acidic soil reaction with pH <6.8 and pH <6.6 was observed.

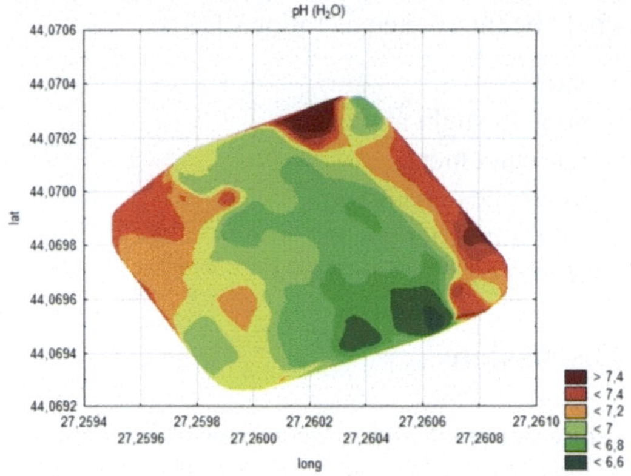

Fig. 4. Spectral map for determining the content of pH in the soil

The high acidity of the soil in the areas with low humidity indicates that we have unabsorbed mineral fertilizers from the plants, which is why soil salinity is observed. From Fig. 5 it can be seen that in the section with the presence of NH_3-N+NO_3-N >30 mg kg^{-1} in Fig. 6 the section with the presence of P_2O_5 >100 mg kg^{-1} and Fig. 7 the section with the presence of K_2O >115 mg kg^{-1} soil moisture is the lowest <70 L m^3 and <65 L m^3 which confirms the claims that the absorption efficiency is closely related to the presence of water in the root zone.

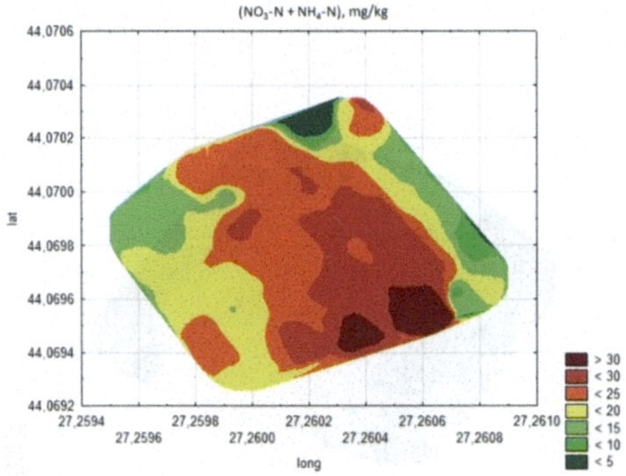

Fig. 5. Spectral map for determining the content of NH_3-N+NO_3-N mg kg^{-1} in the soil

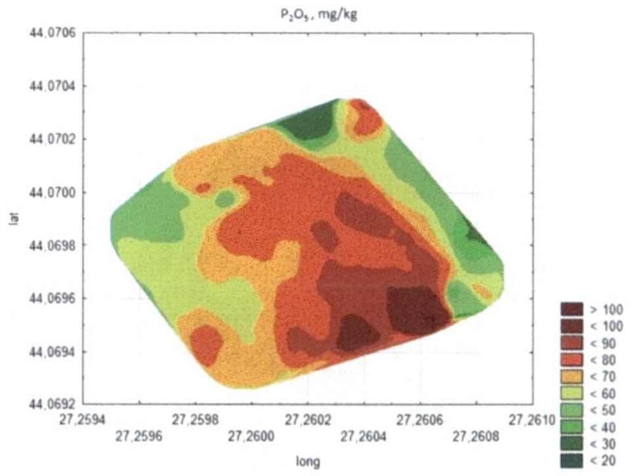

Fig. 6. Spectral map for determining the content of P_2O_5 mg kg^{-1} in the soil

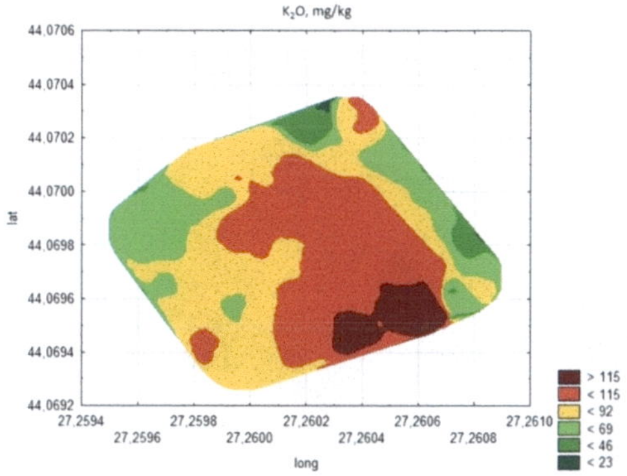

Fig. 7. Spectral map for determining the content of K_2O mg kg^{-1} in the soil

The compaction of the soil as a result of the technological operations carried out with the cultivated crops has an adverse effect on its porosity and moisture storage. From Fig. 8, it was found that in the sections of the field with soil compaction D2I >40 cm, we have soil moisture <70 L m^3 and <65 L m^3, which affects the uptake of nutrients in these sections.

Next, the data was analyzed according to the described methodology using the software ArcGIS Pro v. 2.8.6. Three classification maps were selected for comparison: the moisture content, the soil compaction D2I and the content of NH_3-N+NO_3-N mg kg^{-1} in the soil. Considering the D2I map has 10 classes instead of 7, the following classes were merged:

- <34 and <32;
- <30 and <28;
- <26 and <24.

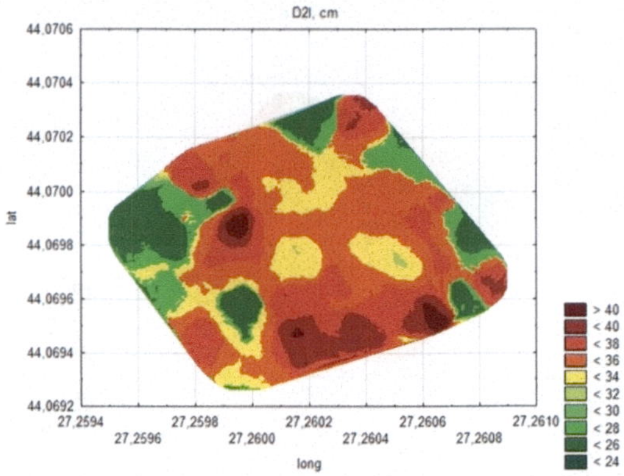

Fig. 8. Spectral map for determining soil compaction D2I cm

The three maps were converted to greyscale for easier comparison (Fig. 9).

a) b) c)

Fig. 9. Greyscale classification of the selected maps for analysis: a) water content; b) soil compaction D2I; c) NH_3-N+NO_3-N content.

Next, the dependency between the following maps were compared:

- Water content and D2I;
- Water content and NH_3-N+NO_3-N content.

For the comparison 20000 random points were chosen and compared for compliance in each situation. Thereafter, the points which fall outside the area of the field are filtered out and as a result the remaining analysis points are 13099 and 12888, respectively for the first and second comparison scenario.

In Table 1 is presented the confusion matrix for the first scenario, where the soil compaction D2I is compared with the soil water content. We can notice that even though there is some dependency, as the samples follow the diagonal, the correlation between the classes is poor. The precision does not surpass 58.2% and the recall – 35%. Furthermore, the resulting Cohen's Kappa is 0.077, which indicates a slight agreement between the two variables.

Table 1. Confusion matrix between the soil water content and the soil compaction D2I.

		Soil water content								
	Classes	>90	<90	<85	<80	<75	<70	<65	Total	Precision
Soil compaction D2I	<26	15	340	723	453	0	0	0	1531	0.010
	<30	12	48	574	700	0	0	0	1334	0.036
	<34	0	0	78	904	1060	101	0	2143	0.036
	<36	1	0	112	1302	2049	869	0	4333	0.300
	<38	9	0	72	286	1462	685	0	2514	0.582
	<40	5	0	0	70	388	487	111	1061	0.459
	>40	1	0	0	0	75	20	87	183	0.475
	Total	43	388	1559	3715	5034	2162	198	13099	
	Recall	0.349	0.124	0.050	0.350	0.290	0.225	0.439		Kappa: 0.077

Similarly, in Table 2 is presented the confusion matrix between the soil water content and the soil NH_3-N+NO_3-N content. We can notice that in this case the agreement is significantly better. The highest precision value reaches 95.9% and the highest recall value 0 98.4%. Furthermore, the Cohen's Kappa coefficient is 0.74, which indicates a substantial agreement between the two variables. In other words, the soil water content could be well used as an approximate indicator for the Soil NH_3-N+NO_3-N content and vice versa.

Table 2. Confusion matrix between the soil water content and soil NH_3-N+NO_3-N content.

		Soil water content								
	Classes	>90	<90	<85	<80	<75	<70	<65	Total	Precision
Soil NH_3-N+NO_3-N content	<5	12	203	3	0	0	0	0	218	0.055
	<10	26	174	137	1	0	0	0	338	0.515
	<15	4	5	1369	425	0	0	0	1803	0.759
	<20	1	5	22	3053	116	0	0	3197	0.955
	<25	3	1	0	145	3873	15	0	4037	0.959
	<30	0	0	0	0	978	1684	3	2665	0.632
	>30	0	0	0	0	0	441	189	630	0.300
	Total	46	388	1531	3624	4967	2140	192	12888	
	Recall	0.261	0.448	0.894	0.842	0.780	0.787	0.984		Kappa: 0.740

The obtained results when plowing Field-5 is shown in Figs. 10, 11, 12, 13, 14 and 15. From Fig. 10 it was found that when plowing the soil moisture in experimental Field-5 is between <66 L m^3 and >92 L m^3, and the difference compared to disking is 2 L m^3. There are no sharp areas with humidity <66 L m^3. A uniform distribution of humidity around <74 L m^3 and <82 L m^3 is observed throughout the field.

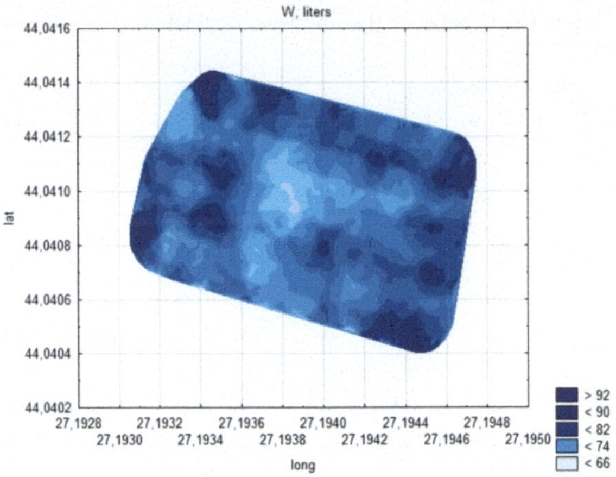

Fig. 10. Spectral map for determining the soil moisture content (W)

Unlike Fig. 2 at Fig. 11 it was found that in a large part of the cultivated area the pH of the soil is close to neutral from <6.82 and >6.96. There are no areas with an alkaline soil reaction. Some salinity was observed in the middle of the field with pH <6.7.

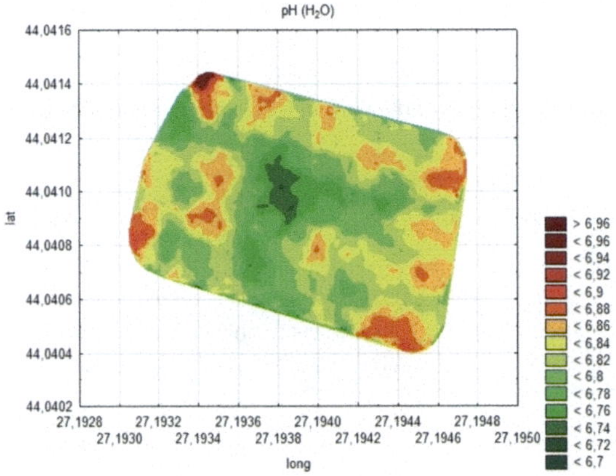

Fig. 11. Spectral map for determining the content of pH in the soil

The content of NH_3-N+NO_3 determined from the spectral map of Fig. 12 is between <20 mg kg^{-1} and >26 mg kg^{-1} which shows that the difference in the sections with the highest values NH_3-N + NO_3 compared with the spectral map of Fig. 5 was 4 mg kg^{-1} less in plowed soil than in discing. This can be explained by the washing of the fertilizer in the lower soil layers or better absorption by the root system of the plants.

With the obtained spectral map for P_2O_5 from Fig. 13, it was found that the amount of mineral fertilizer P_2O_5 available in the soil is between <64 mg kg^{-1} and >82 mg kg^{-1}, with the highest values in the middle of the field. Compared to the maximum values from the spectral map in Fig. 7, we have 18 mg kg^{-1} less P_2O_5 mg kg^{-1} at plow.

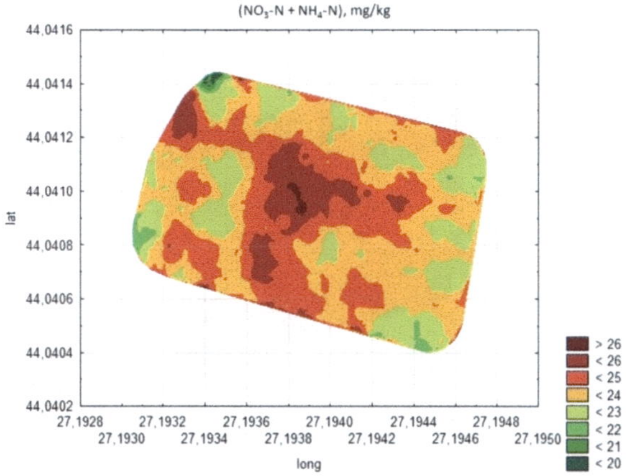

Fig. 12. Spectral map for determining the content of NH_3-N+NO_3-N mg kg^{-1} in the soil

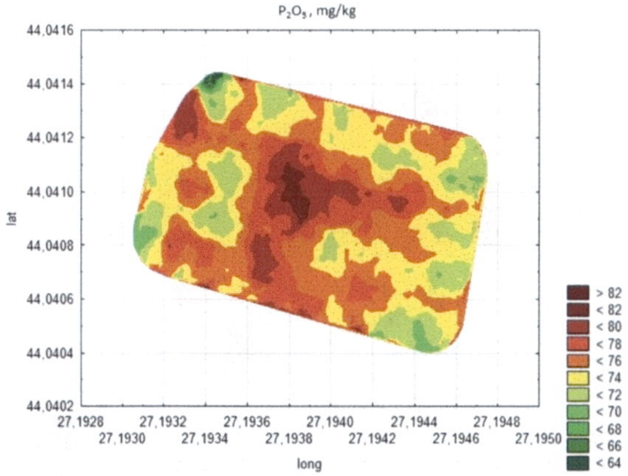

Fig. 13. Spectral map for determining the content of P_2O_5 mg kg-1 in the soil

The obtained values for K_2O are between <84 mg kg^{-1} and >108 mg kg^{-1} at plow (Fig. 14). The difference in maximum values between plowing and discing was 7 mg kg^{-1} more K_2O in discing.

From Fig. 15, it was found that soil compaction (D2I) is in insignificant limits between <14, 4 cm and >15 cm. Soil compaction during plowing is 62,5% less than when discing (Fig. 8). It was found that during plowing soil compaction D2I at a depth of 35 cm is not high and does not affect the moisture content of the soil. This can be explained by the fact that when plowing to a depth of 35 cm, the soil is loosened.

The obtained spectral maps for the remaining four fields are similar and are not shown in the present work.

No additional analysis was made for this fields according to the data analysis methodology, because most of its parameters vary in a very close range, which would not allow to obtain meaningful results.

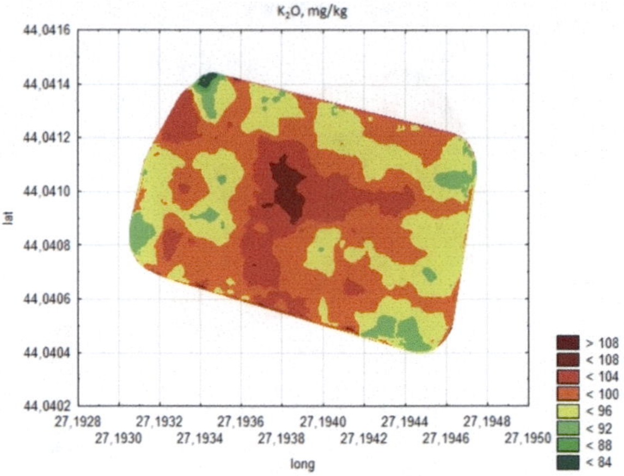

Fig.14. Spectral map for determining the content of K_2O mg kg^{-1} in the soil

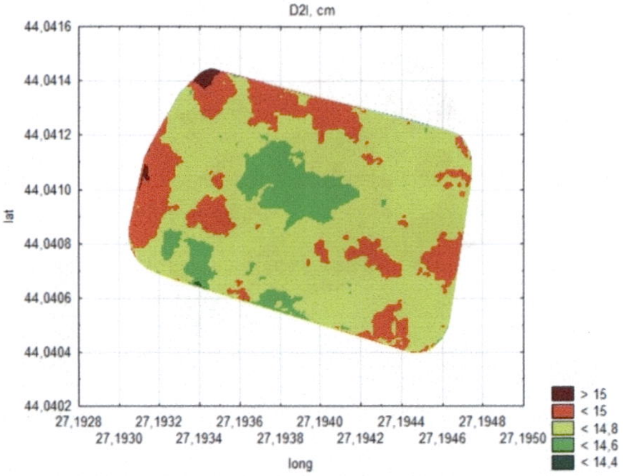

Fig. 15. Spectral map for determining D2I cm

4 Conclusions

Soil tillage affect its moisture storage, soil salinity and nutrient content. It was found that with ploughing, the maximum soil moisture was 2 L m^3 more than with disking. When plowing, a uniform moisture distribution between <74 L m^3 and <82 L m^3 was observed in a large part of the field. In both types of soil tillage, in the areas with increased soil salinity, pH <6.5, the moisture content is <65 L m^3. In the plots with pH <6.5, higher values of NH_3-N+NO_3-N, P_2O_5 and K_2O were observed than the other plots, which is due to the reduced absorption capacity of the plant root system. In the plowing treatment, the values of residual NH_3-N+NO_3-N, P_2O_5 and K_2O in the soil are lower than in the disking treatment. Soil compaction during plowing is 62,5 L m^3 less than when discing.

The obtained results also indicate that there is no significant agreement between the soil water content and the soil compaction D2I. However, for the first field the results showed that there was a strong agreement between the soil water content and the soil NH_3-N+NO_3-N content, with the Kohen's Kappa coefficient reaching 0.74. These results indicate that the soil water content could be used as a rough indicator for the soil NH_3-N+NO_3-N content and vice versa.

This study contributes valuable insights into optimizing moisture management in agricultural systems by evaluating the impact of various tillage practices on the moisture retention of leached Chernozem. The findings have implications for sustainable soil management practices and may guide farmers in making informed decisions to enhance water use efficiency and preserve the health of Chernozem soils.

Acknowledgments. The research reflects the results of the work on project No. 24-FAI-01, financed by the "Scientific Research" fund of the University of Ruse."

References

1. Bekhovykh, Y., Bekhovykh, L., Lyoevin, A., Sizov, E.: Changes in maximum water holding capacity of Chernozem soil caused by soil compaction. E3S Web Conf. **262**, 03003 (2021). https://doi.org/10.1051/e3sconf/202126203003
2. Bescansa, P., Imaz, M.J., Virto, I., Enrique, A., Hoogmoed, W.B.: Soil water retention as affected by tillage and residue management in semiarid Spain. Soil Tillage Res. **87**(1), 19–27 (2006). https://doi.org/10.1016/j.still.2005.02.028
3. Korolev, V.A., Gromovik, A.I., Borontov, O.K.: Changes in the fertility of a leached Chernozem under different primary tillage technologies. Eurasian Soil Sc. **49**, 95–101 (2016). https://doi.org/10.1134/S1064229316010099
4. Reynolds, S.G.: The gravimetric method of soil moisture determination. J. Hydrol. **11**, 258–273 (1970)
5. El-Hendawy, S.E., et al.: Potential of the existing and novel spectral reflectance indices for estimating the leaf water status and grain yield of spring wheat exposed to different irrigation rates. Agric. Water Manag. **217**, 356–373 (2019)
6. Crusiol, L.G.T., et al.: In-season monitoring of maize leaf water content using ground-based and UAV-based hyperspectral data. Sustainability **14**, 9039 (2022). https://doi.org/10.3390/su14159039
7. Zhou, J.-J., et al.: Evaluating the performance of hyperspectral leaf reflectance to detect water stress and estimation of photosynthetic capacities. Remote Sens. **13**, 2160 (2021). https://doi.org/10.3390/rs13112160
8. Shen, X., et al.: Optimizing the positioning of soil moisture monitoring sensors in winter wheat fields. Water **10**, 1707 (2018). https://doi.org/10.3390/w10121707
9. Shauka, H.T., Flower, K.C., Leopold, M.: Quasi-3D mapping of soil moisture in agricultural fields using electrical conductivity sensing. Agric. Water Manag. **259**, 107246 (2022). https://doi.org/10.1016/j.agwat.2021.107246
10. Brogi, C., Huisman, J.A., Pätzold, S., von Hebel, C., Weihermüller, L., Kaufmann, M.S., van der Kruk, J., Vereecken, H.: Large-scale soil mapping using multi-configuration EMI and supervised image classification. Geoderma **335**, 133–148 (2019). https://doi.org/10.1016/j.geoderma.2018.08.001
11. Hossain, M.B., Lamb, D.W., Lockwood, P.V., Frazier, P.: EM38 for volumetric soil water content estimation in the root-zone of deep vertosol soils. Comput. Electron. Agric. **74**(1), 100–109 (2010). https://doi.org/10.1016/j.compag.2010.07.003

12. Bratoev, K., Beloev, H., Mitkov, A., Mitev, G.: On the possibility of conducting fast and reliable soil tests. Agric. Mach. **1**, 75–80 (2020)
13. McHugh, M.L.: Interrater reliability: the kappa statistics. Biochem. Med. (Zagreb.) **22**(3), 276–282 (2012)

Measurement of Soil Density

Asparuh Atanasov(✉), Svilen Stoyanov, and Stefan Tenev

Technical University of Varna, Varna, Bulgaria
asparuh.atanasov@tu-varna.bg

Abstract. The main parameter of the soil is its density, as it directly affects the development of the root system of agricultural crops. The article examines a method for testing soil density by equipping a cultivator with strain gauges to measure linear deformations in the working organs. By considering the elastic deformation, it is possible to determine the shear resistance of the soil. The obtained results were compared with the results of soil density measurement with a standard penetrometer. Based on the obtained results, a soil compaction map was created, showing similar areas of increased and decreased density. The method allows the retrofitting of any standard cultivator by measuring the load on each individual working organ and allows the creation of a continuous soil density map. To obtain accurate results, it is necessary to calibrate the installed sensors before each measurement.

Keywords: Plow · Soil · Cultivator · Penetrometer · Bulk density

1 Introduction

Soil density is a very important parameter that directly affects the development of agricultural crops and determines the correct planning of agrotechnical measures. The term "soil density" encompasses a number of indicators characteristic of a given soil - volume and specific density, relative weights and porosity, between which there are certain dependencies. Mechanical measurement of bulk density is labor-intensive and expensive but provides the most accurate information. The most common measurement method is a standard penetration test. Soil density is measured using a standard vertical cone penetration test - SPT, (Standard Penetration Test), [1, 2]. The accuracy of the measurement is directly dependent on the type of soil, and the highest accuracy of the result is for fine-grained soils. The test is more accurate with a low clay content, and with a high composition of gravel fractions it is unsatisfactory. Soils in areas of drought may exhibit natural cementation – ie. Strong hardening. This condition often changes the force required for penetration. Penetration resistance depends on the humidity at which field measurements are made.

To calculate the resistance to penetration (penetration) with dimension (kg/cm^2) the formula is used:

$$P_{pen} = n\frac{mgh_1}{Sh_2} \tag{1}$$

E. Lorencowicz et al. (Eds.): FMPMSA 2024, LNCE 609, pp. 37–45, 2024.
https://doi.org/10.1007/978-3-031-70955-5_4

where $\mathbf{P_{pen}}$ is penetration resistance;
n is the number of presses;
m is the mass of the impactor;
h1 is the height of the impact body;
S is the area of the rod embedded in the soil;
h2 is rod penetration depth.

A disadvantage of the method is the obtaining of discrete values and not a complete map of the tested field.

A sensor developed by [3] is based on a cultivator working body on which an octagonal transducer measuring the force is mounted. When cutting the soil aggregate by the working body, part of the arm of the sensor (cultivator) interacts with the ground, the resistive cutting force tends to deform the octagonal transducer. A strain gauge mounted on the arm measures the force required to cut the soil along a vertical and horizontal axis. Through this, the transducer can also sense a bending moment.

Another system is a group of heel tools attached to the backside of a vertical smooth blade was developed and tested by [4]. The system is capable of evaluating the mechanical resistance of the soil at three different depths. In this system, the relatively low signal-to-noise ratio makes it difficult to estimate the mechanical resistance of the soil near the surface. Another prototype of the vertical blade equipped with a set of cutting tools was used to evaluate a spatial model of soil resistance to identify the trend of soil resistance change with depth, assuming a linear variation of the resistance force with depth [5]. The system is prone to errors due to the closeness of the useful signal to interference. An advantage is the estimation of soil density at three different depths.

The reviewed mechanical sensors provide an opportunity to evaluate compactions in the soil layer by means of direct measurement. The soil contact of the mechanical density sensors gives them high measurement accuracy. A disadvantage is the need for an additional agrarian technical measure in which the density of the tested field is taken into account [6–8].

The study is based on a standard cultivator organ equipped with a strain gauge that measures the resistive cutting effort. The possibility of multiplying the measuring body with the reading of readings from several cultivators at the same time was tested.

2 Instrumentation and Methods

When tillage is carried out with a harrow or a plough, it is possible to calculate the density of the soil as a function of the tractive force.

The drag force is calculated from:

$$P_t = k.V_p \tag{2}$$

where: k is the soil specific resistance (kN/m); VP – the working speed of the machine (m/s) [9].

$$k = \frac{P_t}{V_P} \tag{3}$$

A disadvantage of this method is the impossibility of determining the exact boundaries of the cemented area. With a cultivator with a width of 9 m, it is not possible to report compacted areas with a size smaller than the width of the working body. Increasing the accuracy of measurement can be achieved by taking into account the pulling force on each of the working organs. In this way, it is possible to measure every 10 cm in width of the cultivated area. Their measurement in the direction of the processing performed depends on the frequency of taking the readings during movement.

Arrangement of the measuring elements of a cultivator for combined cultivation "classic", KSPN 92 shed [10]. Working width 9.2 m, Number of working bodies – 37, Fig. 1.

Fig. 1. KSPN 92 [10] 4 indicates the location of the strain gauge for each working organ

Thirty-seven measuring elements are required, and the working width is 9.2 m. It is necessary to install 37 strain gauges between the supports and the blades [11].

For the experiment, an experimental set-up was created by means of simultaneous measurement of two working bodies. Its scheme is shown in Fig. 2.

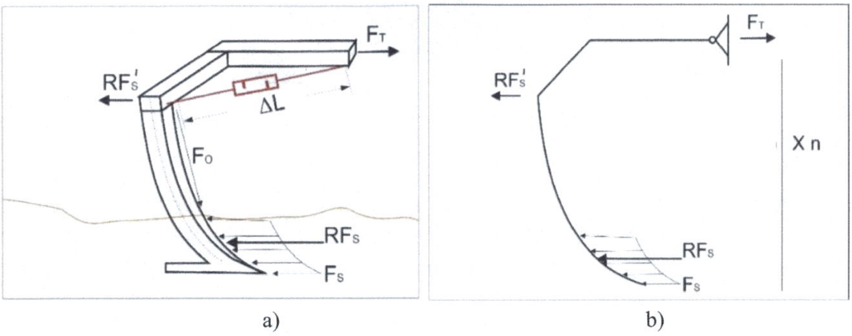

Fig. 2. Scheme of the experimental setup. Structural model a), kinematic diagram b)

Traction force $\mathbf{F_T}$ obtained during the agrotechnical event. When the cultivator-type working body is pulled below the surface of the ground, the counteracting force $\mathbf{F_S}$ is created over the entire working area that is in contact with the soil.

The resistance to shearing of the soil aggregate due to the frictional force is caused by the particles contacting the working body. This force is applied along the entire cutting edge and it is usually uneven due to the structure of the soil, in Fig. 2 it is conventionally denoted as distributed load $\mathbf{F_S}$. Its equivalent is a horizontal vector $\mathbf{RF_S}$, the point of application of which varies with soil composition and compacted zones. With the cultivator, the working depth is up to 20 cm and the exact determination is not necessary.

This produces a stretching force in the $\mathbf{F_O}$ blade that tends to straighten the working body. The $\mathbf{RF_S}$ force projection is $\mathbf{RF_S}$'. The deformation of the supporting frame and the working body during normal operation should be in the elastic range. The measurement of this deformation $\mathbf{\Delta L}$ between the frame and the cultivator allows the resistance force to be taken into account. As a result of the calculation, the density of the soil through which the plow passes is obtained. The scheme created in this way can be repeated for each working body, in which case we will get a continuous map of the compacted areas when conducting a standard agrotechnical event. With red color in Fig. 3 shows the mounting location of the strain gauge. This position is preferable to avoid accidental contact with the ground and defects in the gauges, as well as to prevent the strain gauges from measuring small forces.

In this case, 2 S-shaped strain gauges model DEE300 are used, with a maximum load of 300 kg, resistance of the strain gauges 350 Ω, accuracy 1%. The strain gauges are manufactured by Keli Senziting Technology Co LTD [12] and are certified according to standard GB/T7551-2008, ISO 9001:2008.

Fig. 3. Strain gauges DEE300kg

A NAD24-12-LX 24-bit two-channel analog-to-digital converter was used. The device is a two-channel 24-bit analog-to-digital converter, with a fixed measurement speed and the ability to control the gain of the built-in, ultra-low-noise DC amplifier. It is powered by the USB port.

The device is designed to measure slowly changing quantities with high accuracy. An ADC converts the amplified difference between the positive and negative inputs of the respective channel of a differential amplifier into a digital code. The output values of the different channels are read sequentially, and the output result is obtained in a 24-bit digital code.

The device powered the strain gauges. Analog device ADR292 stabilizer is used in the device, with an accuracy of 0.07% and inherent noises not greater than 4 μV RMS, in addition, the same is used as a reference source of the ADC, which results in reducing the influence of internal noises on measurement time.

For ambient temperature changes, especially when using an uncompensated bridge, it is imperative to account for the change in bridge component values from temperature.

3 Results and Discussion

An experiment was conducted at the experimental field of the Dobrudja Technological College. The cultivator's working body is attached to a motor tiller according to the diagram in Fig. 2. Two strain gauges are attached to the cultivator, connected to the two channels of the ADC (Fig. 4).

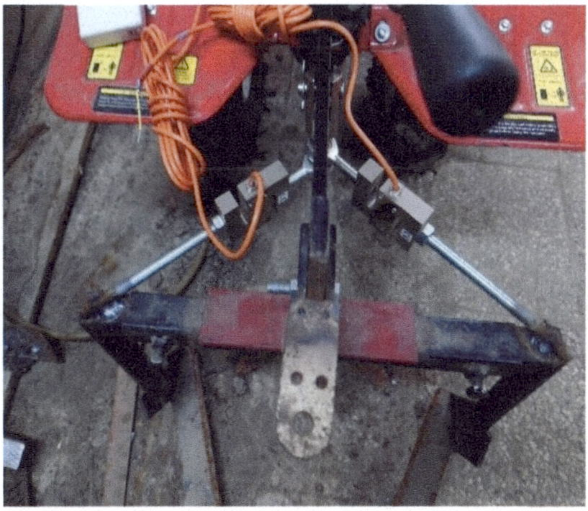

Fig. 4. Location of strain gauges

Software compiled on LabView was used to extract the information. Data transmission rate 57600 bit, time between sensor readings 1 s is selected. At the movement speed of the motor cultivator between 1–2 km/h. The actual resulting interval is a 0.5 cm offset between measurements. The readings of the sensors are reset after installation and the reading of the software is calibrated to 0. To check the accuracy of the result, tare is performed, shown in Table 1. The front resistance area of the working body is 2200 mm^2.

Table 1. Tare the system

kg	1	3	4	6	8	9	10
кPa	0,0857	0,087	0,091	0,94	0,15	0,2	0,23

To compare the result, soil density testing was performed for the experimental area with a standard cone penetrometer AG-BS-70 with a working range of 0–400 PSI (0–3500 kPa or kN/m^2). The test result is shown in Table 2. The measurement was made in 3 m in length and width.

Table 2. Soil density test, values are in kN/m^2

	0 m	3 m	6 m
15 m	1312,5	1750	1750
12 m	1050	962,5	787,5
9 m	1312,5	612,5	1312,5
6 m	1312,5	1312,5	1312,5
3 m	525	918,75	612,5
0 m	875	612,5	1312,5

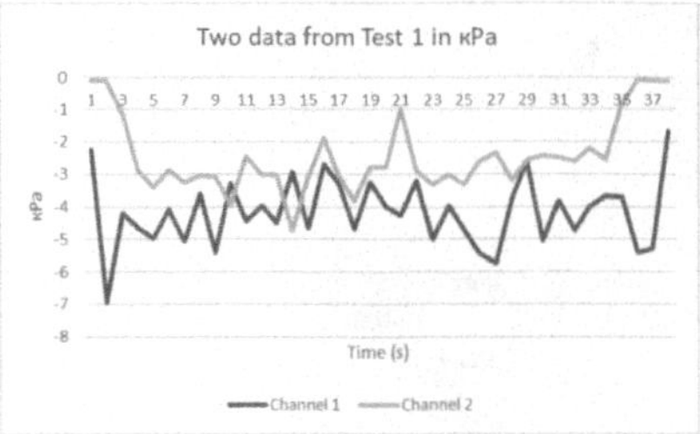

Fig. 5. The result of test 1

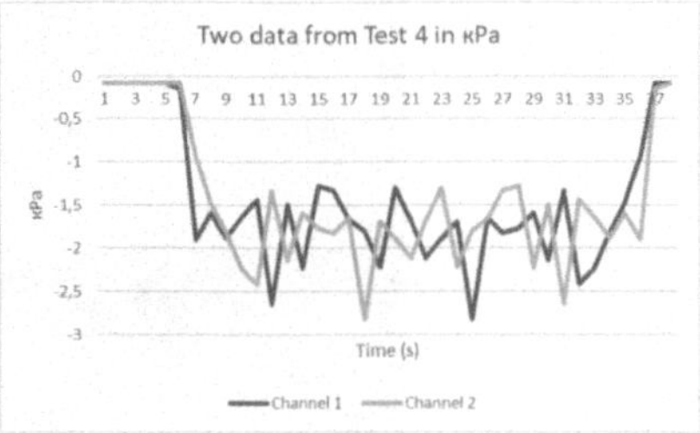

Fig. 6. The result of test 4

The results of tests 1 (Fig. 5) and 2 (Fig. 6) are the cultivation of one of the lengths of the experimental field. The zero sections are before the start of the movement of the working body. The change in values is in a negative direction, which indicates that there is tension on the strain gauge - **ΔL** increases. Resistance is a positive number and must be taken in modulus.

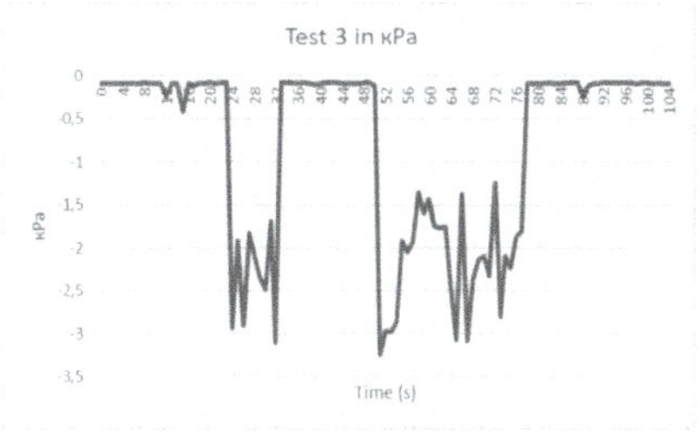

Fig. 7. The result of test 3

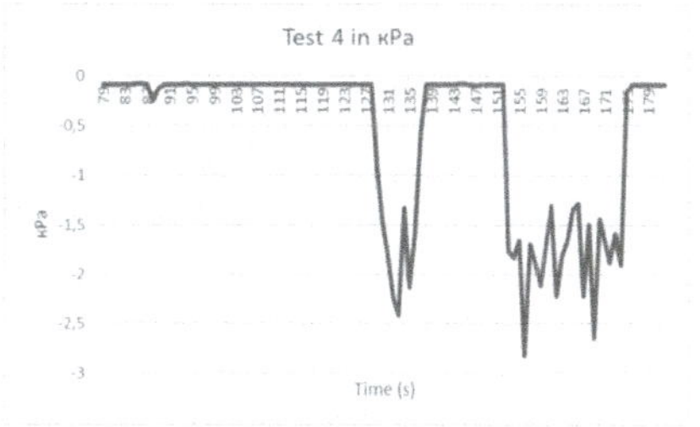

Fig. 8. The result of test 4

Tests 3 (Fig. 7) and 4 (Fig. 8) have a zone with zero values in the graph due to stopping the movement of the tiller. It is necessary to cut these areas, but they do not affect the measured values. It is necessary to consider the dwell time to determine the distance to the compacted areas.

After correcting for the variation trends for the tested field shown in Fig. 9. Figure 10 shows a map of the field with the seal zones. When using a fusion cultivator, we actually get a continuous map of the field.

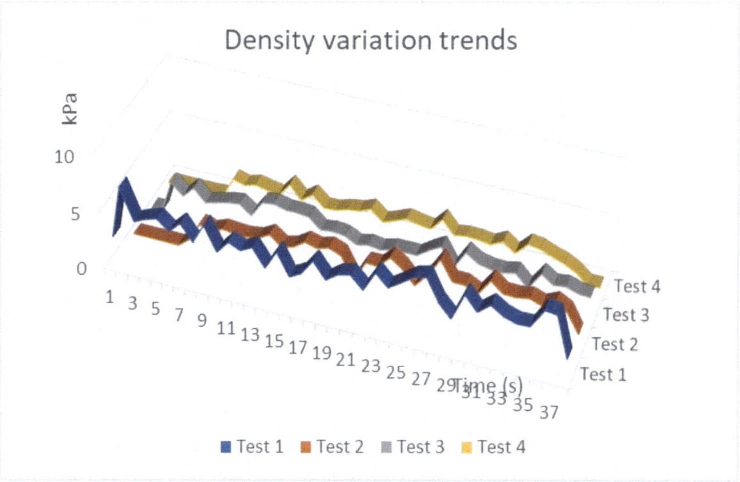

Fig. 9. Density variation trends

When calculating the coordinates of the tested area, a measurement is used by reading time and converting it to distance. This is a drawback, and if a stop is necessary, the dwell time must be measured. Also, the variable speed of movement would make it difficult to accurately determine the distance. The obtained density values are in the range.

The results obtained for the different densification zones using the penetrometer and the ΔL measurement in the cultivator test show a similar pattern of greater compaction in test zone 4 and least in test zone 1.

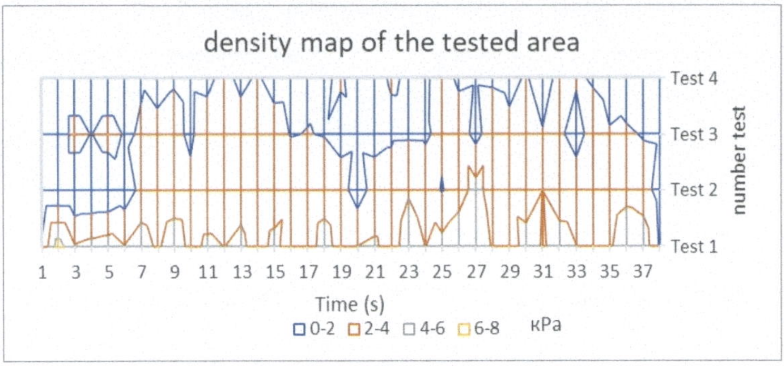

Fig. 10. Density map of the tested area

4 Conclusions

The measurement approach provides accurate information, and it is possible to create a continuous map of the cultivated area. The advantage is that the measurement is carried out during the planned agrotechnical measures, and it is possible to read the width values every 10 cm. And no further processing of the obtained results is necessary.

The system can be mounted to any standard tillage machine. After calibration, it is possible to adjust the readings according to the working body on which it is mounted. Allows simultaneous reading of the readings of each working body.

A disadvantage is the measurement of the distance over time, which can give a deviation in the obtained dimensions with a variable speed of the machine. A GPS mount is required to record the georeferenced coordinates at the measurement points.

References

1. ASAE Standards: S313.2. Soil Cone Penetrometer, 49th edn. ASAE, St. Joseph (2002)
2. Baumgardner, M.F., Silva, L.F., Beihl, L.L., Stoner, E.R.: Reflectance properties of soils. Adv. Agron. **38**, 1–44 (1985)
3. Suárez, M., Coronel, C., Coba, D., Nava, F., Vega, J., Cueto, O., Bravo, E.: Sensor para la medición continúa de la compactación del suelo. Revista Mexicana de Ciencias Agrícolas Pub. Esp. Núm. 4 1 de noviembre - 31 de diciembre, pp. 800–806 (2012)
4. Adamchuk, V.I., Morgan, M.T., Sumali, H.: Application of a strain gauge array to estimate soil mechanical impedance on-the-go. Trans. ASAE **44**(6), 1377–1383 (2001a)
5. Adamchuk, V.I., Morgan, M.T., Sumali, H.: Mapping of spatial and vertical variation of soil mechanical resistance using a linear pressure model. Paper No. 01-1019, ASAE, St. Joseph, Michigan (2001b)
6. Atanasov, A., Stoyanov, S., Mihaylova, D.: Cultivator-based soil density measurement method. In: 2023 International Conference Automatics and Informatics (ICAI), Varna, Bulgaria, pp. 455–458 (2023). https://doi.org/10.1109/ICAI58806.2023.10339088
7. Andrade, P., Aguera, J., Upadhyaya, S.K., Jenkins, B.M., Rosa, U.A., Josiah, M.: Evaluation of a dielectric based moisture and salinity sensor for in-situ applications. Paper No. 01-1010, ASAE, St. Joseph, Michigan (2001a)

8. Andrade, P., Rosa, U.A., Upadhyaya, S.K., Jenkins, B.M., Aguera, J., Josiah, M.: Soil profile force measurements using an instrumented tine. Paper No. 01-1060, ASAE, St. Joseph, Michigan (2001b)
9. Irinchev, D.: Farm tractor power in traction operations. Agric. Sci. **IV**(12), 109–111 (2013)
10. Madaraagro (2023). https://www.madaraagro.com/bg/
11. Bankova, A.: Investigation of the qualitative dependence between the character of wear and the mutual location of wearing supports. In: 2022 International Conference on Communications, Information, Electronic and Energy Systems (CIEES), Veliko Tarnovo, Bulgaria, pp. 1–4 (2022). https://doi.org/10.1109/CIEES55704.2022.9990870
12. http://en.kelichina.com

Information Processing Systems to Support Integrated Crop Protection

Marcin Baran(✉), Kamila Roik, and Anna Tratwal

Institute of Plant Protection – National Research Institute, Władysława Węgorka 20, 60-318 Poznań, Poland
M.Baran@iorpib.poznan.pl

Abstract. In the near future, Polish agriculture will enter the era of digital agriculture, integrating, with the help of the latest technologies, the characteristics of a particular farm, field, crop making it possible to prepare and carry out in a minimum number of protection treatments (chemical, non-chemical) with the right plant protection products. This can have a significant impact on reducing the number of chemical treatments, which is clearly stated in the provisions of integrated pest management. Decision support systems are dynamic software tools whose recommendations, indications, vary depending on feeding them with measurement data, which can propose an optimal decision path in choosing when to carry out a chemical treatment. Decision support systems are not only tools based on various types of computer systems and algorithms, they are also many other sources of information (meteorological data) that can be used for decision-making processes. One of the key solutions in this area is monitoring and signaling of agrophages and controlling them only at specific developmental stages for diseases and pests sensitive to a specific active substance, after exceeding thresholds of economic harmfulness. By analyzing the field data collected by the field informers of the two largest advisory platforms through the applications of the Edwin advisory platform and the Pest Warning System, we have the opportunity to assess the correctness of the field recommendations and the decision support system generated regarding the application of chemical protection once the economic damage thresholds are exceeded.

Keywords: pest monitoring · integrated control · On-line Pest Warning System · field control · digital agriculture

1 Introduction

Agriculture as a branch of the economy is subject to relationships and interactions with other markets. The agricultural model must therefore evaluate and adapt to the needs of different markets, consumer needs, climate change. It must be a form of agriculture that is not really modern, but wise, so that it does not lose in this quest the farmer who builds it with his daily being. In the world of the enormity of information delivered to the common man every day, one can easily get lost. Every day we are bombarded by the mass media, with millions of pieces of information. Therefore, the information that

E. Lorencowicz et al. (Eds.): FMPMSA 2024, LNCE 609, pp. 46–51, 2024.
https://doi.org/10.1007/978-3-031-70955-5_5

is delivered must be properly targeted to the recipient, profiled and contain up-to-date information that can be helpful in the moment. After that, it is historical information. That's why it's so important to provide short messages that keep you informed about what's happening right now in agriculture, or even in a specific field.

It is impossible to directly tell a farmer what to do with his field. However, it is possible to help him in his endeavor by supporting his decision-making. Decision support systems are dedicated to almost all mass-produced crops on an industrial scale. Any specialist in his field can thus predict the threat to his crop, from a disease or pest. There are many systems, as well as guides, cultivation methodologies, advice, recommendations and also the preferences of the specific user. There are also a number of portals that offer their resources in paid, subscription form or directly online. All portals by virtue of their industry offer similar information, which is presented in a fairly straightforward manner. A standout in this regard is the portal Pest Warning System, which, in terms of its database, offers a fairly sizable library of all available materials and updated publications of the most well-known agricultural institutions in the country (IOR-PIB, COBORU, IHAR, IO-Skierniewice, IUNG-Pulawy).

The materials are available to any user without the need to log in or provide data. In addition, nationwide monitoring and signaling of basic agricultural crops (winter wheat, corn, rapeseed, potato, sugar beet, bean crops) is carried out. This is a unique map of signaling points in which specialists, after inspecting specific plantations of a given crop, provide up-to-date information via a website about observed threats from basic diseases and plant pests. The points are located on the country's grid, divided into provinces located in smaller or larger clusters covering fairly evenly the entire country, depending on the location of crops and the area of operation of the signaling device. After opening the point, the user receives specific information about the number of vetting, the type of crop, the monitored threat and the preference to carry out a chemical treatment, after exceeding the threshold of economic harmfulness for a given agrophage. This is up-to-date, important information for him about the expected or already existing threat in a particular vantage point. This information is not related only to this particular point (field).

The user (no login required) gets information that in the vicinity of the field of interest, depending on the occurrence of a disease or insect gradation, the threat may also appear on his surrounding crops. During mass appearances, agrophages do not attack individual plants on a micro-scale. Such phenomena always occur on a larger scale mastering quite large area. For potatoes, the appearances of potato beetle and potato blight are monitored. The user gets information about the intensity of the occurrence of the agrophage with a message about the proposed implementation of chemical protection measures in a given area, presented in a simple graphic way on a map of Poland. The results of observations are an important element of the decision support system, but the final point, deciding on the need to carry out treatment, regardless of whether these are indications of the computer system or visual inspections of plantations, is the fact of exceeding the threshold of economic harmfulness. The tools created, visualization and facilitation of access to the database will greatly facilitate the decision to carry out chemical treatment.

2 Tools in Monitoring and Signaling

The signaling points are distributed based on a given crop in an assigned locality and the type of crop (winter wheat, corn, rapeseed, potato, sugar beet, bean crops) and the range of a given beacon. After vetting a given field, the whistleblower enters its observations via the website, which automatically appears on the Pest Warning System page with the monitored threat from an agrophage on a given crop. The points are located on the country's grid, divided into provinces located in smaller or larger clusters covering the entire country to a fairly uniform extent. The course of field data entry by beacons is analyzed on an ongoing basis. Field data are entered in two ways through the applications of the Edwin advisory platform, the Eodr platform and the Pest Warning System.

In 2023, field observations were conducted for fungal diseases on wheat leaves (*Zymoseptoria tritici* - septoria, *Parastagonospora nodorum* - septoria - symptoms on leaves, *Pyrenophora tritici-repensis* - brown leaf spot) and European corn borer (*Ostrinia nubilalis*) on corn in designated agroclimatic zones.

The decision support system for winter wheat takes into account meteorological data from measuring stations, among other things, based on air temperature and meteorological precipitation occurrences during the growing season. The decision support system for winter wheat, as for corn, is based on 5 climate zones distributed and visualized (Fig. 1).

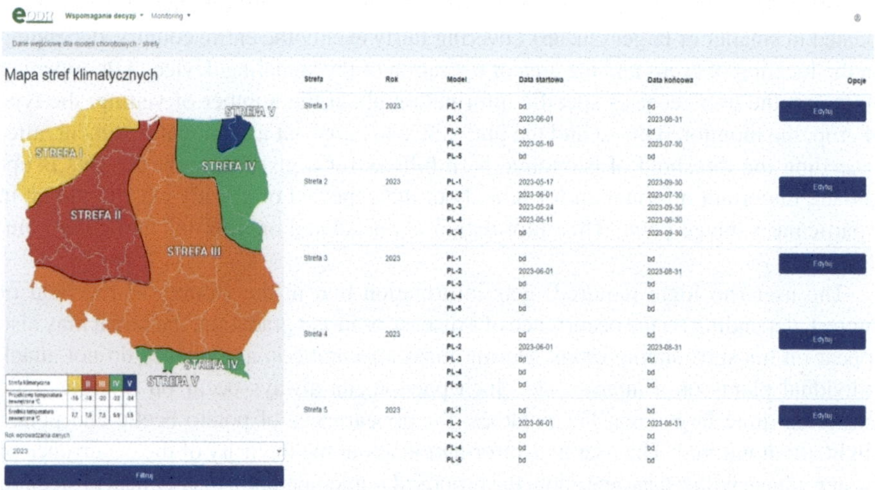

Fig. 1. Map of climate zones on which decision support systems are based.

Winter wheat crops were analyzed in the same way for the incidence of foliar diseases (septoria, brown leaf spot) across the country. Field data were downloaded from the Edwin advisory platform and the Pest Warning System beacon database. Depending on the degree of infestation, fungal diseases of leaves could cause yields to drop from 10% to as much as 60%. Due to the high threat and the economic losses increasing year after year, it is very important to monitor the occurrence of septoria, brown leaf spot, and to set an appropriate date for their control. It is important to properly select fungicide preparations, active substances and use the recommended dose, so that fungal pathogens do not become resistant to the active substance. Fungicide treatments must be adapted to the growth stage of plants and meteorological conditions, so the basis for proper control of septoria and brown leaf spot is their early and good recognition. Septoria takes a long time to develop hidden in the internal tissues of the plant. Symptoms are similar to cereal brown leaf spot, leaf fusarium or striped leaf septoria. The last symptom of the disease is brown spots on the husks, so often these diseases can be confused with each other.

In the case of corn crops, the formats of available observation point data were standardized for DSS model analysis (Fig. 2). There was also a focus on the methodology (unified) for monitoring the flight dynamics of the European corn borer in corn crops in order to provide the right data for decision-support systems. By defining uniform guidelines regarding crop monitoring, field data were homogeneous. Methods were consulted on methods for capturing European corn borer flight activity/dynamics peaks in corn crops for proper data feed to the decision support system. A standardized calculator of economic harmfulness thresholds was created for the proper application of DSS in agricultural crops. Harm thresholds are an essential element in the course of dealing with chemical plant protection products. Due to the different course of meteorological conditions in different regions of Poland, and the different intensity of pest occurrence in the country, climatic zones related to the earliness of sowing and further development of host plants were determined, on which the analyzed DSS were based.

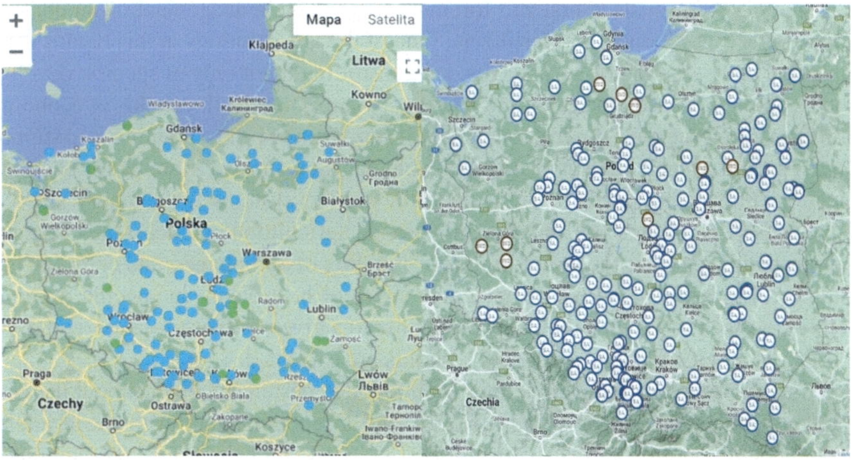

Fig. 2. Distribution of corn crop monitoring points from the Edwin advisory platform and the Pest Warning System.

3 Discussion

After confronting and testing the results of field observations and the indications of advisory systems, it was found that the guidelines and recommendations of the wheat disease decision support system did not take into account changes in canopy moisture and the susceptibility of specific wheat varieties to fungal pathogen infestation. This could have a significant impact on the advisory system's recommendations. Therefore, the results are inconclusive and require continuation and validation of the decision support system. In addition, often the symptoms on wheat leaves were the result of infestation by a complex of pathogens, thus the results of field observations in some cases could be subject to error.

In the case of the European corn borer on corn, the indications of the decision support system in many cases, depending on the climate zone, coincided with the actual need for chemical treatment. In the case of the occurrence of the European corn borer on corn, the indications of the decision support system in many cases, depending on the climate zone, coincided with the actual need for chemical treatment. The system, after appropriate data entry and development based on meteorological data, taking into account field observations, can be used to determine the timing of chemical treatment for the European corn borer in corn crops. It should be noted, however, that the analyzed period of observations and indications of the decision support system was quite short, and the results may be subject to error, hence the need to continue research in subsequent growing seasons.

The requirements of integrated conservation, future changes related to the guidelines and recommendations of the European Green Deal clearly indicate the need for increased work and activity in this area. The main thrust of the planned changes and, above all, of the "Field to Table" and "For Biodiversity" strategies is to combat climate change, protect natural resources, and increase biodiversity.

Achieving these goals is first and foremost responsible management of inputs such as crop protection products and fertilizers. One of the key solutions in this regard is the monitoring and signaling of agrophages, i.e. controlling them only at specific developmental stages of diseases and pests and their host plants after the thresholds of economic harmfulness are exceeded. An important part of supporting proper signaling is to support farmers, producers and advisors using available decision support systems for chemical protection. Expand the availability of decision support systems for advisory services and the possibility of their use and dissemination in agricultural practice. One of the key solutions in this regard is the monitoring and signaling of agrophages, and consequently their control only at specific developmental stages of host plants, at specific developmental stages of pests, when they are susceptible to a given active substance, and, most importantly, when economic damage thresholds are exceeded.

There is a provision in the guidelines for integrated pest management about the need for monitoring and signaling of agrophages, and a recommendation to use available DSS in chemical protection. Decision support systems can help reduce the amount of chemicals applied by up to 50% compared to a standard treatment, provided that the treatment is carried out in a timely manner. However, this requires validation of decision support systems based on measurement data obtained in subsequent growing seasons, as well as feeding them with data from weather stations.

4 Summary

Decision support systems can help improve the overall sustainable use of crop protection products. After confronting the results of field observations and the indications of advisory systems, it is possible to determine how and when to make a chemical treatment. Improving field observations and working with decision support systems will help refine the precise timing of treatment at the specific developmental stage of the plant and the developmental stage of the insect that is sensitive to the active substance. Chemical protection should be initiated when the economic threshold of a particular pest is exceeded. When planning an insecticide treatment, one must take into account thermal conditions and the presence of beneficial insects. Preventive non-chemical measures are a very important element to secure the yield. The mathematical models used for the aforementioned decision support system were developed at the Plant Protection Institute - PIB and are the unit's own achievement. The use of the Polish model developed on the basis of experiments carried out under our climatic conditions, so unlike models developed in Denmark or Germany, for example, the danger of inaccurate indications resulting from different input data (e.g., calculation of the rate of development of pests under specific meteorological conditions) will be eliminated.

References

1. Baran, M., Roik, K., Bocianowski, J.: Analiza monitoringu uszkodzeń kukurydzy powodowanych przez omacnicę prosowiankę (*Ostrinia nubilalis* HBN.) w Polsce w latach 2006–2016. Prog. Plant Prot. **62**, 4 (2022)
2. Jakubowska, M., Tratwal, A., Magdalena, K.M.: The weather as an indicator for decision-making support systems regarding the control of cutworms in beets and cereal leaf beetles in cereals and their adoption in farming practice. Agron. Sect. Precis. Digit. Agric. **13**(3), 786 (2023)
3. Tratwal, A., Baran, M.: The role of guidelines on pest monitoring and warning systems in integrated pest management. J. Plant Prot. Res. **58**(3), 211–214 (2018)
4. Tratwal, A., Baran, M.: Platforma Sygnalizacji Agrofagów – nauka praktyce rolniczej. Prog. Plant Prot. **59**(2), 119–125 (2019). https://doi.org/10.14199/ppp-2019-016
5. Tratwal, A., Baran, M., Jakubowska, M.: Online pest warning system as a tool supporting agricultural practice in the integrated control requirements. In: Lorencowicz, E., Uziak, J., Huyghebeaert, B. (eds.) Farm Machinery and Processes Management in Sustainable Agriculture X International Scientific Symposium FMPMSA, Lublin, Poland, pp. 179–182 (2019). ISBN 978-83-66017-74-0. https://doi.org/10.24326/fmpmsa.2019.1

PRIOR'eau, a Tool for Spatializing the Detection of Plant Protection Products in Walloon Water Resources for Prevention Purposes

Guillaume Bergiers(✉), Bastien Durenne, Bernard Weickmans, and Bruno Huyghebaert

Unité Sols, Eaux et Productions Intégrées, Département Durabilité, Systèmes et Prospectives, Centre Wallon de Recherches Agronomiques (CRA-W), 5030 Gembloux, Belgique
g.bergiers@cra.wallonie.be

Abstract. The PRIOR'eau tool exploits the Walloon Region's databases of plant protection products (PPP) detection in groundwater (GW) and surface water (SW). It can be used to define priority areas, where awareness-raising actions by field advisors need to be targeted. This tool is developed in collaboration with the Public Service of Wallonia (PSW) to support the actions of the regional water resource protection operator PROTECT'eau. PSW's database are used to highlight active substances or metabolites, which concentrations found in the various analysed catchments, are between 30 and 75% of the groundwater quality standard ($100\,\mathrm{ng\,l^{-1}}$). Mostly herbicides are found in waterbodies. To monitor groundwater, the Water Framework Directive networks (WFD) and a "supplementary" network include almost 1000 sites spread across Wallonia. We define a list of molecules, based on the frequency of detections, their concentrations and the agronomic pertinence. Based on this list, GW priority areas are studied and proposed. These areas, in which worrying detections have been confirmed, will allow us to identify plots of land. Then, advisors from the regional operator can prioritize the farms on which they need to focus their efforts. Finally, operational summaries will be produced to transfer knowledge. These summaries will include GW priority areas, as well as any other information deemed relevant (agricultural land parcels, statistics, hydrogeological information). The objective is to act more accurately and rapidly to support farmers in their efforts to reduce induced pressure on water resources.

Keywords: Water resource · PPP · spatialization

1 Introduction

Water is an essential resource for the Walloon Region; therefore, it needs to be protected. This resource is threatened by the use of plant protection products (PPPs). Due to the intensification of agriculture, the presence of contaminants has become ubiquitous in groundwater (GW) and surface water (SW). Contaminations have been observed, not only in Belgium but across the globe for more than three decades.

E. Lorencowicz et al. (Eds.): FMPMSA 2024, LNCE 609, pp. 52–61, 2024.
https://doi.org/10.1007/978-3-031-70955-5_6

In 2019, an article on the occurrence of PPPs in drinking water in Holland and Flanders reported the detection of the following herbicides: bentazone, bromacil, glyphosate, glufosinate-ammonium, the fungicide dimethomorph, and the insecticide oxamyl. Some of the PPPs were even found exceeding the drinking water quality standard of 100 ng l^{-1}: MCPA, mecoprop-P, terbuthylazine, isoproturon, etridiazole, and tolclofos-methyl. However, there is a significant difference in the concentrations detected between groundwater and surface water (Sjerps et al., 2019). Such contamination can have serious environmental and economic consequences.

Water pollution threatens aquatic environments. Indeed, PPPs can have deleterious consequences such as acute toxicity, chronic toxicity, and bioaccumulation, leading to habitat loss and biodiversity decline (Feckler et al., 2023). In 2013, a meta-analysis using data from Germany, France, and Australia demonstrated that PPPs can cause significant biodiversity loss of macroinvertebrates in stream, even at concentrations considered non-deleterious to the environment (Beketov et al., 2013). To avoid such issues, actions are taken across the EU.

In Wallonia (Belgium), the advisory service, called PROTECT'eau, has as main objective to assist and advise farmers and professional PPP users. It also conducts awareness-raising campaigns aimed at protecting groundwater and surface water bodies. Tools are developed to support the operator's teams and sometimes help them to expand their knowledge and strengthen their arguments. The PRIOR'eau project contributes to these efforts by identifying priority areas and enhancing the understanding of them by giving the operator relevant information (level of contamination, hydrogeologic context,...). It offers geomatics support and database management services. By addressing PPP detections before they reach an undesirable level, the project does not only expect to see local improvements but also aims to elevate the water body's status to 'good' as described in the WFD. This comprehensive approach includes monitoring all PPPs analyzed across the whole network of wells, not just those being part of the WFD network.

With the various outputs planned, including the operational summary and geomatics tool, the operator will possess the necessary tools to strengthens its arguments. These tools are designed to be didactic and easily accessible to field agents. Studied areas will include the hydrogeological context of specific regions or agricultural aspects such as land parcel analysis. This multifaceted approach was selected because it allows a comprehensive examination of the area from all angles, encouraging a deeper reflection and analysis rather than merely applying a potentially unsuitable model that might overlook important parameters or give wrong conclusions for an untrained eye. We have opted for a heuristic approach as first step, and we compare different method and their results, in order to develop a methodology for each area studied. This approach enhances our understanding of the area and sharpens our focus on evaluating and comparing model outputs. Another critical phase of the project involves continuously communicating our findings and insights to the operator throughout the project's duration, ensuring they also gain a thorough understanding of the studied areas. This will enable the operator to better assist farmers by guiding them toward the most effective solutions provided by the tools from other projects.

The PRIOR'eau project employs a collaborative approach, actively engaging various stakeholders including regional authorities (SPW Environment-DEE-DESo/DESu) and CDP'eau (Water Pesticides Diagnostic Unit), which aims to better understand the contamination sources of drinking water catchments by PPPs to target solutions at the field scale within the catchment area. By integrating diverse perspectives and expertise, we aim to enrich our analysis and maximize the impact of our actions.

There are numerous approaches to addressing groundwater contamination by active substances (a.s.) and their metabolites. Unfortunately, the solutions identified by scientific studies or regional projects are not always fully implemented in the field. This project aims to bridge the gap between research outputs and farmers, playing a role in the implementation of these solutions. The first step with our approach is to identify what parameters are usually used in models and what is already being done to protect GW.

Leaching potential as a function of soil type is studied as part of the European authorization process for the use of active substances. Models are employed to predict the concentration of an active substance at a depth of 1 m within the soil. The 'MetaPearl' model, used throughout Europe, assesses soil sensitivity to the transfer of active substances to groundwater. This model integrates various parameters including the PPP season of use (an empirical parameter), climate factors (rainfall and evapotranspiration), substance behavior (Kfoc and DT50 in soils), and soil characteristics (soil density, profile depth, textural class, etc.) (Bah, 2011). Another method, the 'DRASTIC index values,' primarily assesses groundwater vulnerability by assigning values based on its intrinsic characteristics (Yang and Wang, 2010). This assessment of intrinsic vulnerability is crucial for identifying areas most likely to be contaminated.

The integration of these different modelling techniques can produce risk analysis, which is often enhanced by using a Geographic Information System (GIS). GIS is instrumental in pointing out areas where interventions are necessary. This integration facilitates the development of decision-support tools, such as the 'pesticide fate tool' developed by Italian researchers (Bancheri et al., 2022). This graphical interface conducts a vulnerability analysis of groundwater, incorporating key physical parameters like hydrogeological properties, soil, active substances, climate, and legal restrictions, including Natura 2000 zones. These comprehensive considerations allow users to leverage the provided information to make informed decisions.

Due to the challenges of characterizing subsoils and water tables, spatial extrapolation techniques are sometimes employed. These methods range from simpler approaches, such as Inverse Distance Weighting (IDW), to more complex ones, like kriging. Recently, machine learning has emerged as a new type of extrapolation. These algorithms serve as alternatives to traditional geostatistical methods, accommodating a broader array of data and information types. However, they are more complex to implement. A notable advancement is the 'Smart-Map' module, developed by a Brazilian team to enhance the QGIS geographic information system using modern artificial intelligence tools. This tool also supports the generation of interpolated maps using algorithms such as Ordinary Kriging and Support Vector Machine (Pereira et al., 2022).

All these approaches, often highly theoretical, are sometimes complemented by in situ tests and monitoring to validate, refine, and add nuance to them. Creating one perfect model that captures all the subtleties of such complex problems is impossible. Indeed, hydrogeological contexts can vary greatly and are often difficult to precisely define. Therefore, it is essential to study these varied contexts in locations that are representative yet offer contrasting characteristics and are conducive to research.

2 Methodology

The groundwater WFD network in Wallonia consists of 550 sites. This network supports a monitoring program that covers all pollutants deemed relevant, including some PPPs, to assess the chemical status of GW bodies and track their changes over time. There is also an additional network, though its less frequent sampling results are not used to define the status of groundwater bodies. Results from nearly 1,000 structures across Wallonia are compiled into the database managed by the Public Services of Wallonia (PSW), which the PRIOR'eau project utilizes.

A large number of different molecules are analysed in water, necessitating a careful selection process for which ones the operator should focus on. Criteria have been established to guide this selection. The first criterion requires that the molecule be authorized for agricultural use. The second criterion considers a molecule if it is monitored by the WFD network and meets the first criterion. The third criterion is based on the detection level of the molecule; it is included in the list if detected at least once above 30% of the drinking water quality norm or its associated threshold. The final criterion includes all metabolites of the selected molecules, or the active substances (a.s.) of the metabolites that meet all previous criteria.

Around wells used for drinking water catchment, various zones are defined to protect the drinking water catchment area. These zones may be fixed and established by hydrogeologists, with their boundaries officially registered by the Walloon government. Alternatively, the zones can be determined based on the type of soil surrounding the well—such as sandy aquifers, gravelly aquifers, fissured aquifers, or karstic aquifers. This latter method is typically used when no study on the water transit time has yet been conducted.

Priority zones will be established around each structure for the active substances selected according to our criteria. The size of each zone is determined by rounding up the maximum possible dimension to the nearest and highest hundred—for example, 1035 m (1000 m + 35 m) would be rounded to 1100 m—where no specific dimensions have been predefined for the facility. If hydrogeological studies define perfectly the zones, they are used. We have opted to apply this zoning method even to structures not associated with drinking water extraction, to ensure that all potentially problematic structures are included within the priority zones. Such structures can include drains, wells, springs, and others. The issue of PPP contamination in groundwater is a complex phenomenon. The necessary knowledge to understand natural and specific vulnerabilities is not readily accessible to advisors from the regional operator. Since the goal includes transferring knowledge, effective communication is essential. By focusing on different aspects of the phenomenon, advisors are expected to strengthen their arguments for farmers. To achieve this goal, the concept of 'operational summaries' provides an excellent channel of communication. These summaries are usually presented in an easy-to-use A3 format, filled with information and interpretations tailored to the studied area and theme (contamination level, hydrogeology,…). Multiple summaries can cover various aspects of the phenomenon. This approach involves selecting relevant information that could help PROTECT'eau in understanding contaminations in an area and it also offers an opportunity to compare different model's output with the actual contamination situation (Fig. 1).

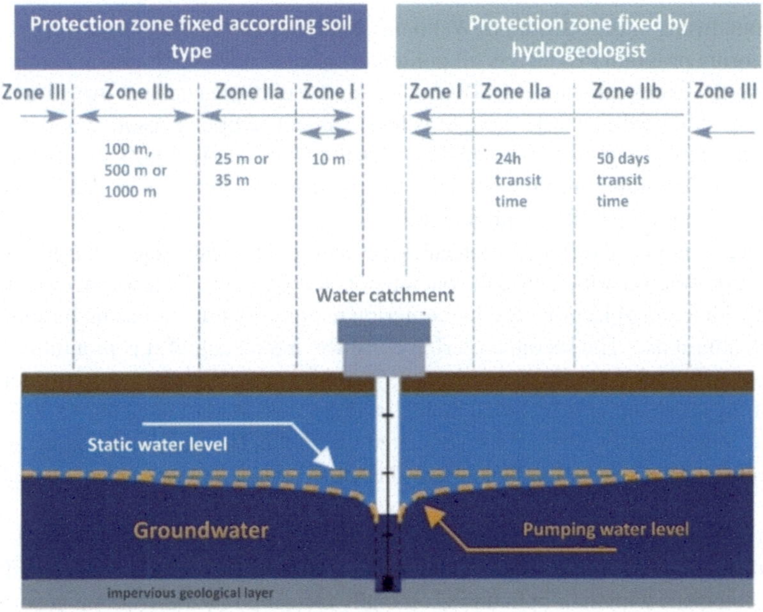

Fig. 1. Illustration of the zones around a drinking water catchment (source: CILE).

3 Results

3.1 List of the Active Substance

The analysis of groundwater from the PSW's database, using our criteria, identified a total of 30 active substances and metabolites. Depending on the molecule studied, either a norm or a threshold value is applied to identify the level of contamination. As outlined in the Water Framework Directive, the norm sets the concentration limit for individual pesticides in groundwater at 100 ng l^{-1}. For metabolites, the threshold values are defined at the regional level. Sometimes, there is no threshold value defined, in this case, values are chosen based on their associated active substances or other metabolites. It's indicated by an asterisk in the last column of the table below (Table 1).

Table 1. List of Active Substances and Metabolites in the PSW's Groundwater Contamination Database that meet the Criteria.

Criteria	Original Active substance	Active substance or metabolite	Quality standards/threshold value * no official threshold value
2		2,4-D	100 ng l^{-1}
2		bentazone	100 ng l^{-1}
2		chlorotoluron	100 ng l^{-1}
2		MCPA	100 ng l^{-1}
2	metazaclor	metabolite ESA	500 ng l^{-1}
4	metazaclor	metabolite 3 of metazachlor	500 ng l^{-1}*
4	metazaclor	metabolite 4 of metazachlor	500 ng l^{-1}*
4	metazaclor	metabolite 5 of metazachlor	500 ng l^{-1}*
4	metazaclor	metabolite 6 of metazachlor	500 ng l^{-1}*
4	metazaclor	metazachlor	100 ng l^{-1}
4	metazaclor	metazachlor OA	500 ng l^{-1}*
2	S-metolachlor	metolachlor ESA	1000 ng l^{-1}
4	S-metolachlor	S-metolachlor	100 ng l^{-1}
4	S-metolachlor	metolachlor OA	1000 ng l^{-1}*
2		metribuzin	100 ng l^{-1}
2	terbuthylazine	terbuthylazine	100 ng l^{-1}
4	terbuthylazine	Desethyl-terbuthylazine	100 ng l^{-1}*
3		dimethenamid-p	100 ng l^{-1}
3		ethofumesate	100 ng l^{-1}
3	flufenacet	flufenacet	100 ng l^{-1}
3	flufenacet	flufenacet ESA	100 ng l^{-1}*
3	flufenacet	flufenacet OA	100 ng l^{-1}*
3		mecoprop-P	100 ng l^{-1}
3	metamitron	desamino-metamitron	100 ng l^{-1}*
3	metamitron	metamitron	100 ng l^{-1}
3		lenacile	100 ng l^{-1}
3		metobromuron	100 ng l^{-1}

(continued)

Table 1. (*continued*)

Criteria	Original Active substance	Active substance or metabolite	Quality standards/threshold value * no official threshold value
3		prosulfocarb	100 ng l^{-1}
3		triclopyr	100 ng l^{-1}

3.2 Priority Zones

Priority zones based on the list have not been finalized yet, but the expected output should resemble the map shown below. It is important to note that the actual output encompasses more than just the map; it includes the entire shapefile, which enables the operator to dive deeply into the situation of contaminations—identifying which active substances are present, and at which level. With this tool, they will be able to select specific areas to begin engagements with farmers (Fig. 2).

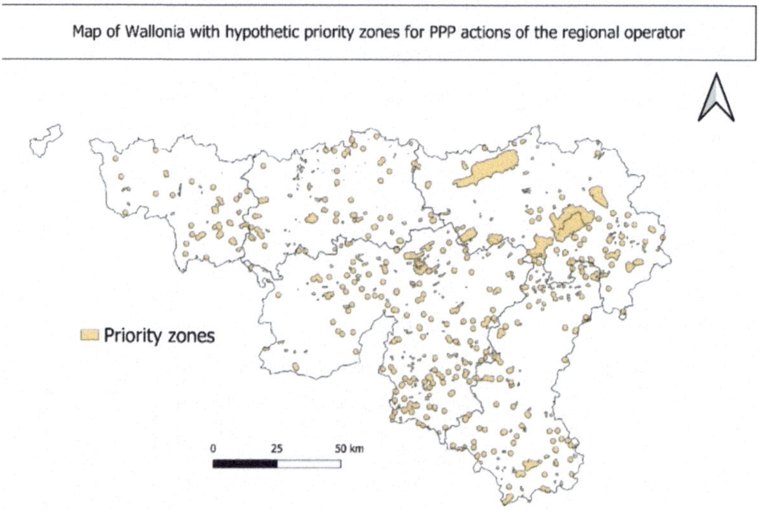

Fig. 2. Example of priority zones at Walloon level.

3.3 Operational Summaries

The operational summaries will include all information deemed necessary, including tables that display annual average concentrations for each well, a map illustrating the geographical location of the area studied, and detailed views of priority areas with information on the most problematic types of soil, the crop rotation, etc.

A simplified example of an operational summary is provided below. It depicts the situation of Dimethenamid-P at one of the operator's center, displaying a map with average and maximum concentrations from 2017 to 2022. The map highlights four structures where concentration peaks exceeded 30 ng l^{-1}, including one structure where concentrations ranged between 30 and 75 ng l^{-1}. On the left, a table displays the average annual concentrations by structure, while the left-hand side of the summary indicates the maximum concentrations recorded for each structure. Additionally, a detailed zoom on each of the catchment areas shows the referenced agricultural plots. These zooms can include various additional details such as soil type and the presence of carbonate rock. The table at the top right lists the crops authorized for Dimethenamid-P use in Belgium. Depending on specific needs, proposals, and possibilities, other information can also be incorporated into the operational summaries (Fig. 3).

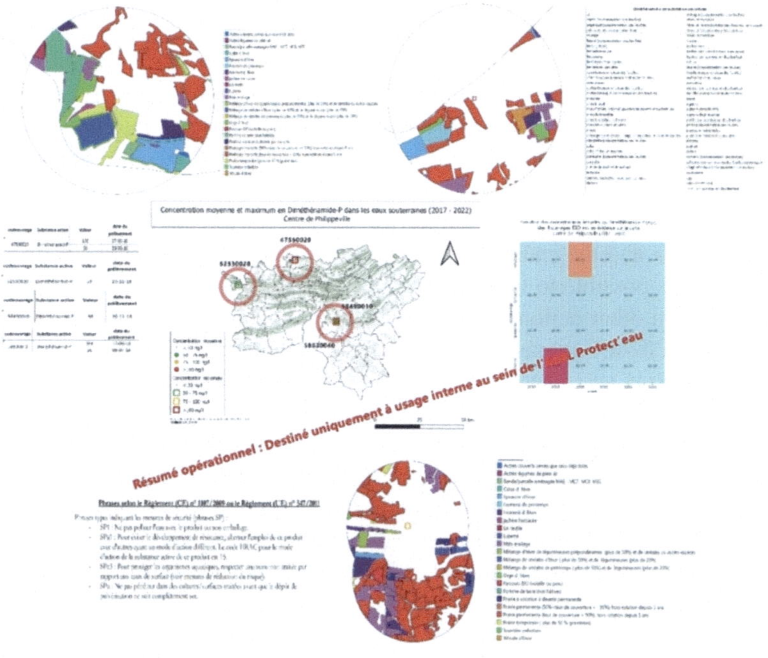

Fig. 3. Operational summary about Dimethenamid-P in Namur area (Belgium).

4 Discussion

The primary objective of the PRIOR'eau project is to extend its outreach to a broader farming audience. One approach is to proactively target areas at risk of contamination before problems occur (contamination above 100 ng l^{-1}). This strategy will not only allow us to raise awareness among a wider public but will also enable us to offer more targeted solutions to the concerned farmers.

This action is being carried out in partnership with the PSW, which manages databases of PPP detection and the associated report on the state of the Walloon environment. This collaboration ensures access to the most recent validated data.

These databases have revealed that many active substances and metabolites are not currently analyzed in groundwater. This oversight is critical; an active substance or metabolite could be present in groundwater without detection. Furthermore, the withdrawal of certain active substances may lead to the increased use of others, potentially substituting one contamination for another, especially if used in environmentally sensitive areas. Therefore, it is essential to maintain scientific vigilance and continually monitor molecules that are being researched in other countries.

The methodology used to target GW zones relies on the definition zones IIa, IIb, and zone III. This method allows for delineating a zone around any point—whether a well, an emergence or drain—and thereby targeting agricultural plots. The resulting shapefile helps visualize the affected areas and provides detailed information on the dynamics of contamination and detection, as well as the molecules involved. However, this approach does not always accurately pinpoint the hydrogeological origins of contamination due to significant regional variations. Initiated by the PSW in 2022, the integration of hydrogeological data into this methodology requires further refinement. To address these complexities, we propose incorporating hydrogeological insights into the 'operational summaries'. Understanding the specific hydrogeology of Wallonia is complex and will be further developed during the creation of these summaries. Over the long term, this enhanced knowledge will be used to improve the prioritization of GW areas.

Once the area and the causes (active substances and metabolites) have been identified, it becomes possible to target specific actions for implementation in the field, in coordination with other actions. For example, an automated tool can support farmers by calculating an ISAC (unbiased frequency treatment indicator) (Bergiers et al., 2023), which aids in raising awareness. If appropriate, alternatives or substitutions can be proposed. Additionally, with the help of operational summaries, advisors can effectively explain the hydrogeological context to farmers, supporting informed decision-making.

In this manner, the deliverables are designed to enable operators to target more effectively their field actions and tailor their arguments and action plans based on the specific conditions of the targeted area. Furthermore, these well-prepared arguments can be used to persuade farmers and other PPP users of the necessity and benefits of the proposed actions.

Funding. This research project was funded by the Service Public de Wallonie (SPW) and Société Publique de Gestion de l'Eau (SPGE).

References

Bah, B.: Valorisation de la Carte Numérique des Sols de Wallonie et d'une base de données disponible en analyse de sols, dans le cadre de l'évaluation du risque de pollution des eaux souterraines par les pesticides. Biotechnol. Agron. Soc. Environ. **15**, 709 (2011)

Bancheri, M., et al.: The pesticide fate tool for groundwater vulnerability assessment within the geospatial decision support system LandSupport. Sci. Total. Environ. **807**, 150793 (2022). https://doi.org/10.1016/j.scitotenv.2021.150793

Beketov, M.A., Kefford, B.J., Schäfer, R.B., Liess, M.: Pesticides reduce regional biodiversity of stream invertebrates. Proc. Natl. Acad. Sci. U.S.A. **110**, 11039–11043 (2013). https://doi.org/10.1073/pnas.1305618110

Bergiers, G., Durenne, B., Weickmans, B., Huyghebaert, B.: Development of a new unbiased impact index on water for pesticides use – a case study in Belgium for herbicides application on sugar beet crop. In: Pascuzzi, S., Santoro, F. (eds.) Farm Machinery and Processes Management in Sustainable Agriculture. Lecture Notes in Civil Engineering, pp. 357–364. Springer International Publishing, Cham (2023). https://doi.org/10.1007/978-3-031-13090-8_34

Feckler, A., Wolfram, J., Schulz, R., Bundschuh, M.: Reducing pollution to levels not harming biodiversity and ecosystem functions: a perspective on the post-2020 global biodiversity framework. Curr. Opin. Environ. Sci. Health **35**, 100495 (2023). https://doi.org/10.1016/j.coesh.2023.100495

Sjerps, R.M.A., Kooij, P.J.F., Van Loon, A., Van Wezel, A.P.: Occurrence of pesticides in Dutch drinking water sources. Chemosphere **235**, 510–518 (2019). https://doi.org/10.1016/j.chemosphere.2019.06.207

Yang, Y.S., Wang, L.: Catchment-scale vulnerability assessment of groundwater pollution from diffuse sources using the DRASTIC method: a case study. Hydrol. Sci. J. **55**, 1206–1216 (2010). https://doi.org/10.1080/02626667.2010.508872

Effectiveness of Biocoatings for Ammonia Emission from Manure

Rolandas Bleizgys(✉), Vilma Naujokienė, Arvydas Povilaitis, Juozas Pekarskas, and Ieva Knoknerienė

Vytautas Magnus University, K. Donelaičio 58, 44248 Kaunas, Lithuania
info@vdu.lt

Abstract. Ammonia is the main gas that acidifies precipitation and thus damages the ecosystem. These gases pollute the atmosphere and harm animals kept in barns and working people. The largest amount of dangerous gases enters the environment when urine bacteria break down proteins in cattle urine and manure. The change in gas emissions during manure storage is highly dependent on the biocoating used for manure. In order to evaluate the effectiveness of reducing emissions, studies were conducted with these biocoatings - chopped straw, sawdust, hemp leaves, peat. Ammonia (NH_3) gas concentration measurement analyzer GME700 (SICK MAIHAK GmbH, Germany) based on the principle of laser spectroscopy, using automatic mode (continuous or cyclic measurement with data accumulation). The obtained experimental research results were recalculated into conditional values: ammonia emission intensity per square meter of manure surface area, fixed ventilation intensity per square meter of manure surface area. Research has shown that the reduction of ammonia emissions from manure is greater when it is covered with bio-organic coatings with lower porosity and more liquid absorption. In order to achieve high efficiency in reducing ammonia emissions, it is recommended to cover with a layer of straw thicker than 10 cm, a layer of 8 cm of crushed hemp leaves, a layer of 5 cm of coniferous sawdust, a layer of 3 cm of wood peat. Scientific research was financed by the Lithuanian Science Council's Necessary Research Project (S-REP-22–5).

Keywords: Ammonia emission · Manure management · Animal husbandry

1 Introduction

The main source of ammonia is manure. More than 90% of the total amount of NH_3 is produced in ammonia animal husbandry [1], its evaporation occurs at all stages of manure formation, storage and spreading on the soil. Research on ammonia emission processes in animal husbandry has received more attention than other gases [2–5].

Changes in ammonia emissions are also greatly influenced by the crust that appears on the surface of the manure. Its formation is greatly influenced by the amount of straw, dry matter in the manure, and the environmental climatic conditions [6]. The most intense harmful gas evaporates during the first 8 days of the study, during which most of the ammonia formed during the entire study period enters the environment. Other

E. Lorencowicz et al. (Eds.): FMPMSA 2024, LNCE 609, pp. 62–68, 2024.
https://doi.org/10.1007/978-3-031-70955-5_7

researchers have found that NH_3 from uncovered manure piles is most volatilized at 10–14 days [7, 8]. Ammonia emissions from solid manure are high, so the manure needs to be covered. The authors claim that using appropriate manure application technologies can reduce NH_3 emissions to the environment by 40–85%. However, an improperly selected or applied ammonia reduction strategy can also increase emissions.

Emmerling [9] found that when using biocoatings to cover manure, it is very important to evaluate the properties of the selected material to form a non-porous layer and absorb liquids on the surface of the manure. Non-porous and hardly breathable manure coatings reduce ammonia emission into the environment by up to 68%. Scientific research was financed by the Lithuanian Science Council's Necessary Research Project (S-REP-22–5).

Ammonia emissions need to be reduced at all stages of manure management: barns, manure storage facilities and transport, and manure incorporation into the soil [10]. Straw or other litter absorbs moisture, the amount of liquid on the surface of the manure decreases and less ammonia evaporates. The highest emission of ammonia is from surfaces contaminated with liquid manure and where urine accumulates, and the lowest - over thick manure on which bedding is applied [11]. In order to reduce the emission of ammonia, it is possible to use various biocoatings for manure - chopped straw, fine sawdust of various wood, chopped hemp leaves, peat, oil, zeolite, straw pellets, etc. In order to evaluate the effectiveness of reducing emissions, studies were conducted with these biocoatings - chopped straw, sawdust, hemp leaves, peat.

2 Material and Methods

The influence of biocoatings on the evaporation of ammonia from manure was determined at the stand - in the manure storage chambers (Fig. 1). Various manures were used for the research. Before the tests, the manure is poured into a large container and mixed and homogenized. This ensures that when the manure is poured into two containers, the composition of the manure is the same: in one we pour manure and do not add anything else, and in the other we pour manure and cover it with a bio coating. Various bio-coatings are used - organic materials: straw, sawdust, peat, hemp leaves. Comparative studies of ammonia emission from both chambers are carried out. During the research, the same factors affecting the evaporation of ammonia from the manure are maintained in both chambers: temperature, air flow over the manure. Manure (1) is poured into sealed 20-L containers-manure chambers (10): the first container holds manure without additives - control, the second container holds manure that we cover with a biocoating. An equal amount of manure is added to both containers: 12–15 L each. The manure chamber is sealed with a lid. Two ducts are introduced into the chamber: clean ambient air is supplied to the manure chamber through one duct, and air is sucked from the chamber at a constant intensity of 6 l min-1 through the other duct (2). Air samples from the manure chambers are taken by probes (9) and supplied to the gas analyzer "lGME700" (5) by a heated hose (2). Air is continuously supplied to the analyzer by a pump (3) (Fig. 1).

Ammonia (NH_3) gas concentration is measured with a GME700 analyzer (SICK MAIHAK GmbH, Germany). Measuring limits from 0 ppm to 2000 ppm. The measuring principle of this device is laser spectroscopy, accuracy 2–4%. The measurement

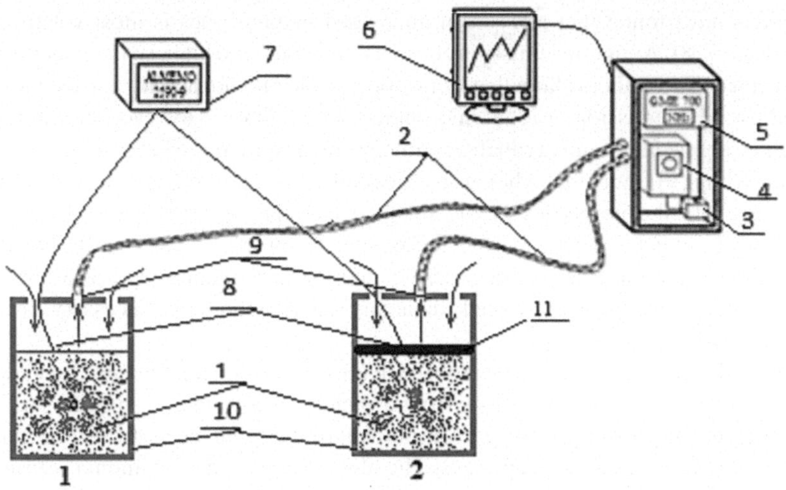

Fig. 1. Scheme of the research bench for ammonia emissions from manure: 1 – gas emission source - manure; 2 - heated air supply hose; 3 – membrane air pump; 4 – electrically heated three-channel dampers; 5 – laser gas analyzer "GME700"; 6 – computer (AMR program); 7 – meter-accumulator "Almemo 2590–9"; 8 – thermocouples; 9 - air sampling probes; 10 – manure chamber; 11 - bio coating.

of gas concentration in air, which depends on air flow and gas concentration, is less than 360 s. With an air flow of 6 l min^{-1} and an ammonia concentration in the air no higher than 30 ppm, the analyzer accurately measures the ammonia concentration in the air within 60–80 s, during which time the measured values stabilize. The analyzer is equipped with heating of the pumped gas, so the gas conditioning means ensuring that the cell is not contaminated, and condensate does not form in it. In order to ensure that the working temperature is higher than the sample dew point temperature, electrically heated three-channel dampers are used. Operating mode: automatic (continuous or cyclic measurement with data accumulation).

We recalculate the received results of experimental research into conditional values: ammonia emission intensity per square meter of manure surface area ($mg{\cdot}m^{-2}{\cdot}h^{-1}$); ventilation intensity per square meter of manure surface area ($m^3{\cdot}m^{-2}{\cdot}h^{-1}$).

3 Results and Discussion

The effect of bio-coatings on the intensity of ammonia evaporation from manure depends on the density, porosity, thickness of the coating layer and liquid absorption properties of the bio-coating. Therefore, the effect of biocoatings - organic materials used for research on gas emissions is significantly different. A greater difference is observed during a longer period of manure storage. After pouring small organic materials on the manure, the ammonia emission is significantly reduced, but due to the diffusion of the slurry to the surface of the coating, the gas emission quickly intensifies again. In order to effectively reduce ammonia emissions from manure, it should be covered with a layer of sawdust or other coatings so that the slurry cannot diffuse to the surface.

After pouring a 5 cm thick layer of chopped wheat straw on the manure, the average reduction of ammonia emission in one hour was up to 82 ± 5.8%, in 10 h. – up to 54 ± 5.3%, and within 20 h – up to 46 ± 3.9%. (Fig. 2). The most intense changes in ammonia emissions occur during 8 p.m. after covering the manure with bio-coating. Therefore, the following research periods were chosen. By increasing the thickness of the straw layer, the effect on ammonia emission increases during the short (1 h) and long (20 h) research period. The larger the straw layer, the more ammonia emissions are reduced over a longer period of time. With a small layer of straw (5 cm), the slurry diffuses intensively to the surface and the ammonia emission is significantly reduced only for a short period of time. When the straw layer is increased to a thickness of more than 10 cm, the slurry does not diffuse to the surface and the reduction of ammonia emission remains for a long time (up to 85% within 20 h). It is recommended to cover the manure with a layer of straw at least 10 cm thick. With such a layer of straw, the intensity of ammonia emission is lower by more than 60% for a long time (more than 20 h).

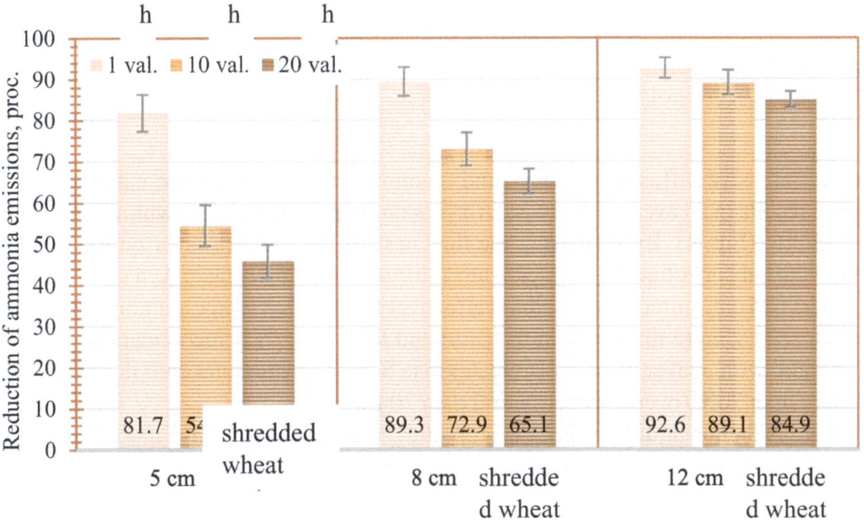

Fig. 2. Average reduction of ammonia emissions from manure over different time periods after adding a layer of shredded wheat straw to the manure.

The effect of a layer of shredded hemp leaves on top of manure is like that of straw on ammonia emissions. A small layer (2 cm) effectively reduces ammonia emissions from manure only for a short period of time (Fig. 3). Increasing the layer thickness to more than 6 cm effectively reduces gas emissions both in the short and long term. If a layer of crushed hemp leaves with a thickness of more than 10 cm is poured on the manure, the emission of ammonia is reduced by more than 91%. It is recommended to cover the manure with a layer of hemp leaves more than 8 cm thick, ammonia emission is reduced by more than 60% (over a longer period than 20 h).

Ammonia emissions from manure are effectively reduced by the added and not very large layer of softwood sawdust. When a very small (1 cm thick) layer of sawdust is

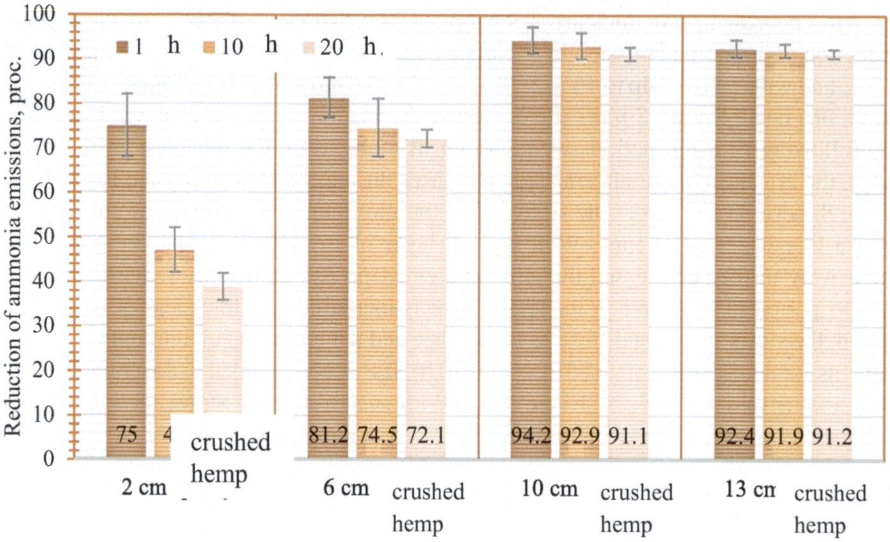

Fig. 3. Average reduction of ammonia emissions from manure over different time periods after adding a layer of crushed hemp leaves to the manure.

poured, intensive diffusion of slurry and ammonia emission from the manure to their surface results in an average decrease in 1 h. is only up to 64 ± 6.1%, within 20 h. - only 19 ± 5.3%. (Fig. 4). By increasing the thickness of the sawdust layer on the manure, the sawdust absorbs liquids, and less slurry diffuses to their surface. At a sawdust layer thickness of 5 cm, ammonia emission is effectively reduced. Average ammonia emissions are reduced by more than 90% in the short and long term. It is recommended to cover the manure with a layer of sawdust greater than 5 cm, the ammonia emission is reduced by more than 70% (over a longer period than 20 h).

Peat has better liquid absorption properties, it absorbs 2–3 times more liquid than straw. If peat is poured on the manure, less slurry diffuses from the manure to the surface, the peat also acidifies the manure, so even a small layer of peat effectively reduces ammonia emissions (Fig. 5). A peat layer only 5 mm thick reduces ammonia emissions to 88 ± 5.3%. In a short period of time (1 h). Such a small layer of peat is not enough to absorb all the slurry on the surface of the manure, it diffuses to the surface of the peat. Therefore, after pouring a 5 mm thick layer of peat on the manure, within 20 h the average reduction of ammonia emission is only 68 ± 3.1%. By increasing the thickness of the peat layer on the manure to more than 3 cm, the ammonia emission decreases very efficiently: more than 93%. It is recommended to cover the manure with a layer of peat more than 3 cm thick, the ammonia emission is reduced by more than 70% longer than 20 h for the period.

The most intense harmful gas evaporates during the first 8 days of the study, during which most of the ammonia formed during the entire study period enters the environment. Other researchers have found that NH_3 from uncovered manure piles is most volatilized in 10–14 days [7, 8]. Ammonia emissions from thick manure are high, so the manure needs to be covered. The authors state that with proper manure application technologies,

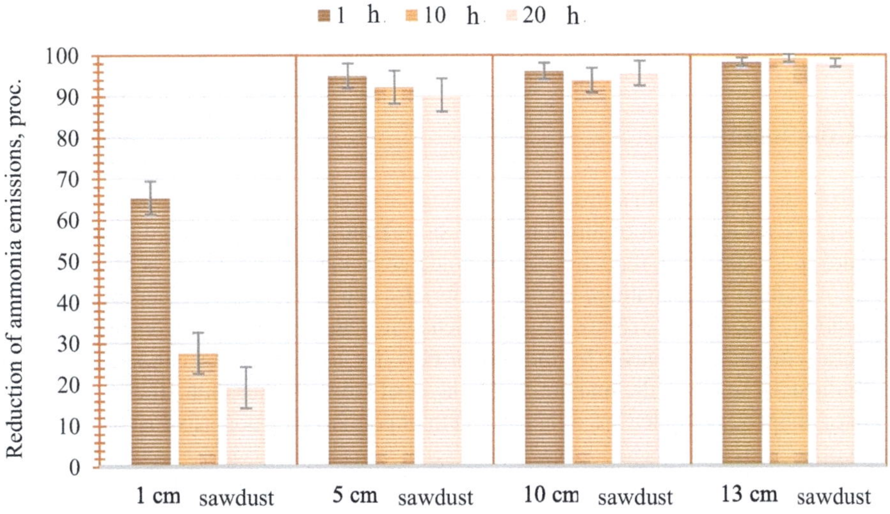

Fig. 4. Average reduction of ammonia emissions from manure over different periods after adding a layer of softwood fine sawdust to the manure.

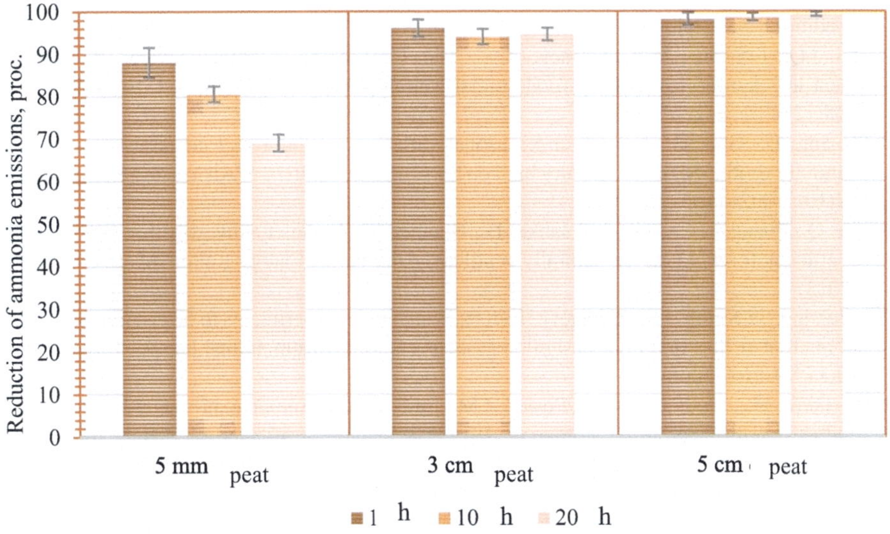

Fig. 5. Average reduction of ammonia emissions from manure over various periods after adding a layer of peat to the manure.

NH_3 emissions to the environment can be reduced by 40–85%, but an improperly selected or applied ammonia reduction strategy can increase emissions.

Emmerling [9] found that when using biocoatings to cover manure, it is very important to evaluate the properties of the selected material to form a non-porous layer and absorb liquids on the surface of the manure. Non-porous and hardly breathable manure coatings reduce ammonia emission into the environment by up to 68%.

4 Conclusions

Covering the manure with straw, hemp shavings, conifer wood sawdust, and peat results in a significant reduction of ammonia emissions from manure compared to uncovered manure ($p \leq 0.05$). The reduction of ammonia emissions from manure is greater when it is covered with biocoatings-organic materials with lower porosity and more liquid absorption. In order to achieve high efficiency in reducing ammonia emissions from manure (emission decreases by more than 60% for a period longer than 20 h), it is recommended to cover the manure with a layer of straw thicker than 10 cm, a layer of 8 cm of chopped hemp leaves, a layer of 5 cm of coniferous wood sawdust or a layer of 3 cm of peat.

References

1. Murawska, A., Prus, P.: The progress of sustainable management of ammonia emissions from agriculture in European union states including Poland—variation, trends, and economic conditions. Sustainability **13**, 1035 (2021)
2. Ni, J.Q., Erasmus, M.A., Croneyc, C.C., Lid, C., Lid, Y.: A critical review of advancement in scientific research on food animal welfare-related air pollution. J. Hazard. Mater. **408**(124468), 1–13 (2021)
3. Huang, D., Guo, H.: Diurnal and seasonal variations of Odor and gas emissions from a naturally ventilated free-stall dairy barn on the Canadian Prairies. J. Air Waste Manag. Assoc. **67**(10), 1092–1105 (2017)
4. Yi, Q., Zhang, G., König, M., Janke, D., Hempel, S., Amon, T.: Investigation of discharge coefficient for wind-driven naturally ventilated dairy barns. Energy Build. **165**(15), 132–140 (2018)
5. Morsing, S., Strøm, J.S., Zhang, G., Kai, P.: Scale model experiments to determine the effects of internal airflow and floor design on gaseous emissions from animal houses. Biosys. Eng. **99**(1), 99–104 (2008)
6. Aguerre, M.J., Wattiaux, M.A., Powell, J.M.: Emissions of ammonia, nitrous oxide, methane, and carbon dioxide during storage of dairy cow manure as affected by dietary forage–to–concentrate ratio and crust formation. J. Dairy Sci. **95**(12), 7409–7416 (2012)
7. Zhuang, M., Shan, N., Wang, Y., Caro, D., Fleming, R.M., Wang, L.: Different characteristics of greenhouse gases and ammonia emissions from conventional stored dairy cattle and swine manure in China. Sci. Total Environ. **722**, 137693 (2020)
8. Chadwick, D.R.: Emissions of ammonia, nitrous oxide and methane from cattle manure heaps: effect of compaction and covering. Atmos. Environ. **39**(4), 787–799 (2005)
9. Emmerling, C., Krein, A., Junk, J.: Meta-analysis of strategies to reduce NH_3 emissions from slurries in European agriculture and consequences for greenhouse gas emissions. Agronomy **10**(11), 1633 (2020)
10. Christoph, E., Krein, A., Junk, J.: Guidance document on preventing and abating ammonia emissions from agricultural sources. Agronomy **10**, 1633 (2014)
11. Bleizgys, R., Baležentienė, L.: Assessments of biogenic gas emission processes in cowsheds. Pol. J. Environ. Stud. **23**(4), 1107–1114 (2014)

Assessment of the Suitability of Selected Apple Cultivars for Production of Cloudy Juice

Agata Blicharz-Kania[1] and Anna Pecyna[2](✉)

[1] Department of Biological Bases of Food and Feed Technology, University of Life Sciences in Lublin, 20-612 Lublin, Poland

[2] Department of Technology Fundamentals, University of Life Sciences in Lublin, 20-612 Lublin, Poland

anna.pecyna@up.lublin.pl

Abstract. The aim of the study was to determine the physicochemical and sensory properties of cloudy apple juice made from Jonagored, Champion, Golden Delicious, Gala, and Gloster fruits. The research material was analysed for the content of total extract and sugar, juice density, pH, and total acidity. Additionally, sensory evaluation was carried out to determine the colour, smell, texture, and palatability of the juice. The product made from cv. Jonagored apples was characterised by the highest content of total extract and sugar accompanied by the lowest total acidity. Consumers assessing the sensory quality regarded this juice to be the most attractive product. Juice obtained from Golden Delicious and Gloster apples received relatively high rating as well. Juice produced from the Champion variety, which exhibited the lowest content of total extract and sugar, was regarded as the least attractive by the consumers.

Keywords: Cloudy apple juice · physicochemical properties · sensory properties

1 Introduction

Recently, increasing attention has been paid to the quality of consumed products. In their choices, consumers mainly focus on the nutritional properties of goods. Therefore, they are more willing to purchase natural products, which are usually unprocessed or slightly processed. Recently, naturally cloudy apple juices and nectars have appeared on the market. During production, the process of clarification and filtration is omitted; the product is naturally cloudy and contains substantially higher levels of bioactive compounds (Gerhauser, 2008). Incorporation of cloudy juices into the daily diet provides the organism with natural health-enhancing substances, including dietary fibre. They are therefore recommended as products with special functionality, in particular to subjects that intend to maintain normal body weight. There are investigations on the potential use of this type of products in the diabetic diet, as they are likely to delay intestinal sugar absorption (Johnston et al., 2002). Cloudy juices contain even a double level of phenolic compounds than clear juices. The compounds exhibit antioxidant and anti-inflammatory effects. Consumption of this type of products has a positive effect on the immune system and delays cell aging in the organism. Additionally, cloudy juices probably exert anticancer effects (Gerhauser, 2008).

E. Lorencowicz et al. (Eds.): FMPMSA 2024, LNCE 609, pp. 69–76, 2024.
https://doi.org/10.1007/978-3-031-70955-5_8

Cloudy apple juices are sometimes enriched with ascorbic acid to prolong their shelf life and appropriate colour. However, sugar, water, or other substances should not be added to such products. Hence, it is extremely important to choose good-quality raw material to meet consumer's expectations.

The aim of the study was to determine the physicochemical and sensory properties of cloudy apple juices obtained from fruits of different varieties.

2 Materials and Methods

2.1 Materials

The research material included apples of various varieties grown in Poland. The fruits were harvested in 2021. Five varieties of apples were used for processing: Jonagored, Champion, Golden Delicious, Gala, and Gloster. The cloudy juice was obtained using a Voran belt press (Fig. 1).

Fig. 1. Pressing process.

2.2 Determination of the Content of Total Extract

The content of the extract in the juices was determined with the use of an Abbe refractometer model G produced by Carl Zeiss JENA.

2.3 Determination of the Sugar Content

The sugar content was determined with the Lane-Eynon method. The analysed material was extracted and deproteinised. Inversion (with concentrated hydrochloric acid) was carried out to determine the total sugar content. The content of reducing sugars in the obtained liquid was determined by direct hot titration of copper salts with the analysed sugar solution (against methylene blue as an indicator of the end of the reaction).

2.4 Measurement of Density

The density of the juices was measured using a pycnometer in accordance with standard EN 1131:1994.

2.5 Determination of pH

The measurement was performed with the use of a Methorm pH meter composed of an electrode measuring the value of electromotive force, i.e. the ratio of the activity of hydrogen cations on both sides of the electrode.

2.6 Determination of Total Acidity

Total acidity was measured by titration of the analysed juice with a sodium hydroxide solution against phenolphthalein. The result was calculated with formula (1):

$$x = \frac{a \cdot n \cdot k \cdot 100}{b} \tag{1}$$

where:
x – total acidity [g/dm^3],
a –volume of 0.1 M sodium hydroxide solution used for the titration [cm^3],
n – normality of the sodium hydroxide solution.
k – malic acid conversion coefficient,
b – amount of the product in the solution [cm^3].

2.7 Sensory Analysis

The sensory analysis of the juice consisted in evaluation of their colour, consistency, odour - intensity and aroma (attractiveness), and palatability - intensity and flavour (attractiveness). The assessment was carried out in a scale from 1 to 5. A trained eight-person panel of males and females was selected to determine the sensory quality.

3 Results and Discussions

3.1 Physicochemical Properties

The total extract content in the juices ranged from 10.1 to 13.7% (Fig. 2). It should be noted that the highest value of this feature was determined for the Jonagored apple juice, whereas the juice from the Champion cultivar was characterised by the lowest value of this parameter. The total extract content in apple juice was determined by Nadulski et al. (2014) as well. The authors reported a high level of the parameter for juice produced from Jonagored fruits.

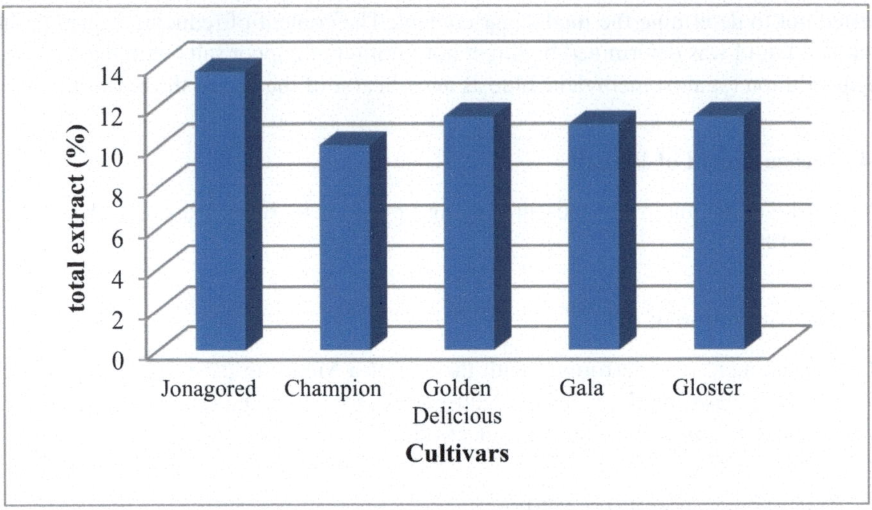

Fig. 2. The content of the extract in the apple juices produced from different cultivars.

As shown by the numerical data presented in Fig. 3, the amount of sugar in the analysed juices was similar to that of the total extract content. The highest content of sugar was detected in the Jonagored apple juice (13.4%), whereas the juice produced from the Champion cultivar had the lowest level of this parameter (9.1%). Significant differences in the sugar content in fruits of different varieties were reported by Nour et al. (2010). The authors analysed fifteen apple varieties and showed a sugar level in a range of 8.97–14.81%.

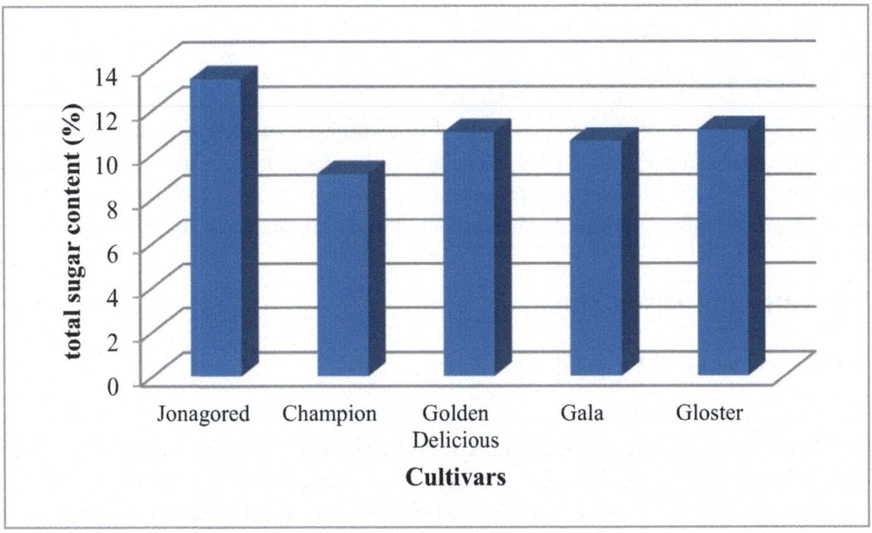

Fig. 3. The content of the total sugar in the apple juices produced from different cultivars.

The density of the apple juice produced from the different cultivars was in the range from 1.025 to 1.032 g/cm^3. There were no significant differences between the analysed samples. Investigations conducted by other authors indicate that the density of juice does not depend on the apple variety used. Greater changes in this parameter were observed for juice produced from fruits characterised by different levels of ripeness (Venkatachalam et al., 2018).

All the examined products had acidic reaction. A significantly higher pH value was recorded in the Gala apple juice (3.88), while the product from the Gloster cultivar was characterised by the highest acidity (Tab. 1). The other products had similar pH (3.56–3.58). As demonstrated by Eisele and Drake (2005), the apple variety used has a significant effect on the pH value of produced juice. Furthermore, the negative effect of a lower pH value on the stability of phenolic compounds has been proved (Zemel et al., 1990).

The total acidity of the juice varied depending on the apple cultivar used (Table 1). As in the case of pH, the juice made of Gala apples was characterised by the highest acidity of 3.35 g/dm^3.

Table 1. Physicochemical properties of the apple juices produced from different cultivars.

Cultivators	Jonagored	Champion	Golden Delicious	Gala	Gloster
Density (g/cm^3)	1,032 ± 0,01	1,025 ± 0,02	1,027 ± 0,03	1,03 ± 0,03	1,031 ± 0,01
pH	3,56 ± 0,02	3,58 ± 0,03	3,57 ± 0,02	3,88 ± 0,03	3,36 ± 0,04
Total acidity (g/dm^3)	2,01 ± 0,03	2,68 ± 0.03	3,02 ± 0,05	3,35 ± 0,05	3,02 ± 0,05

3.2 Sensory Analysis

The colour of the Golden Delicious apple juice was indicated by the consumers as the most desirable (Fig. 4). In their opinion, the product made from Gloster apples was characterised by an inadequate colour.

In terms of the consistency of the tested products, the panellists regarded the juice pressed from Jonagored and Champion apples as the most attractive. The use of the other apple varieties had a slightly negative impact on the attractiveness of this parameter. There were no significant differences in the evaluation of these products.

In the assessment of the odour of the juice, the consumers considered its intensity and aroma quality. The apple juice made from the Jonagored variety was rated as the most attractive (in both categories). The least perceptible and attractive scent was ascribed to the product from the Champion apples (Fig. 5).

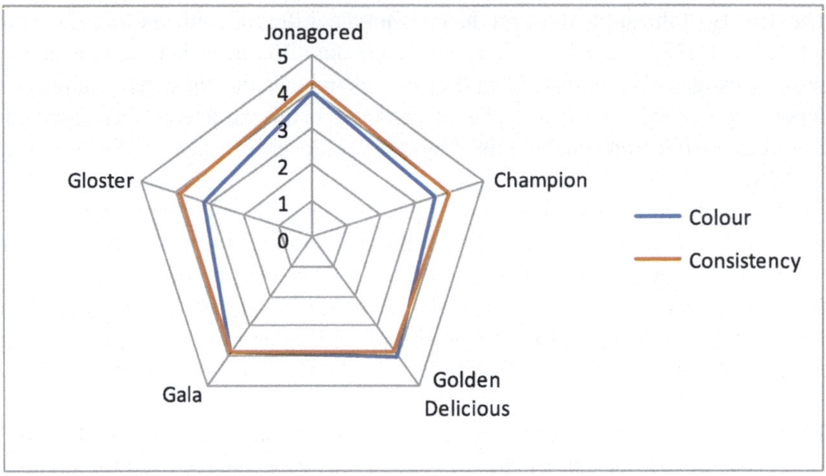

Fig. 4. Sensorial evaluation of colour and consistency by spider-plot.

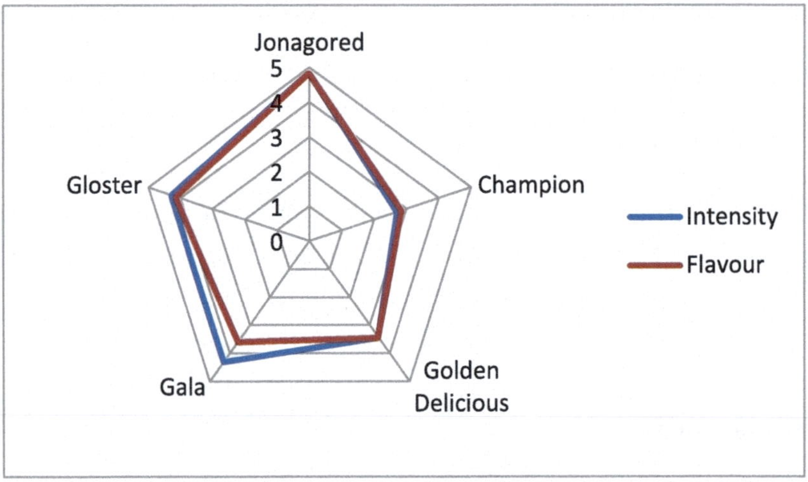

Fig. 5. Sensorial evaluation of odour by spider-plot.

Similar relationships were observed in the analysis of the juice palatability (Fig. 6). The consumers unanimously concluded that the Jonagored variety product was the most attractive in terms of the intensity and aroma. The use of Champion apples turned out to be the least advantageous for juice palatability.

As indicated by Zemel et al. (1990), the pH value has a negative effect on juice colour and palatability. Similar relationships were found in the material analysed in this study. The lowest pH was recorded for the juice made from Gloster apples. The colour of this product was also rated as the least desirable. However, no effect of pH on the juice palatability was demonstrated. This may be related to the method for modification of acidity applied by Zemel et al., who used hydrochloric acid for this purpose.

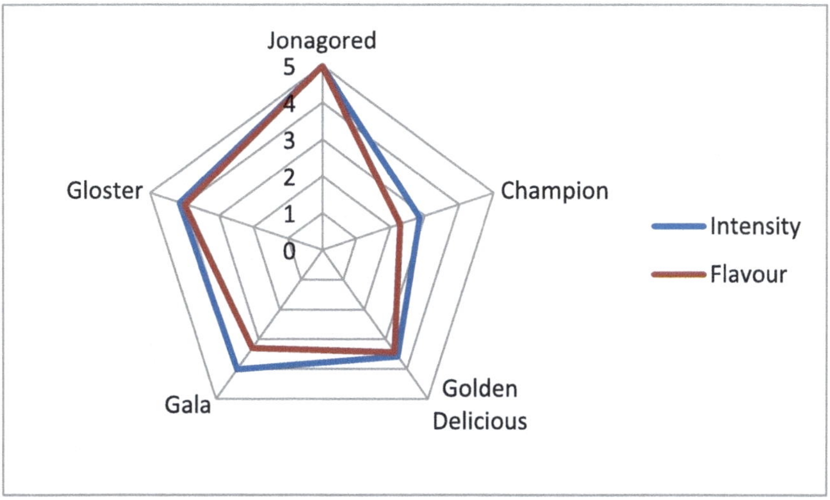

Fig. 6. Sensorial evaluation of taste by spider-plot.

The juice made from Jonagored apples was regarded as the most attractive in the overall assessment of the product (average rating of 5.00) (Fig. 7). The material produced from the Gloster and Golden Delicious cultivars was highly rated as well. In turn, as indicated by the consumers, the juices produced from Gala and Champion apples were not highly attractive (average rating for both raw materials of 3.14).

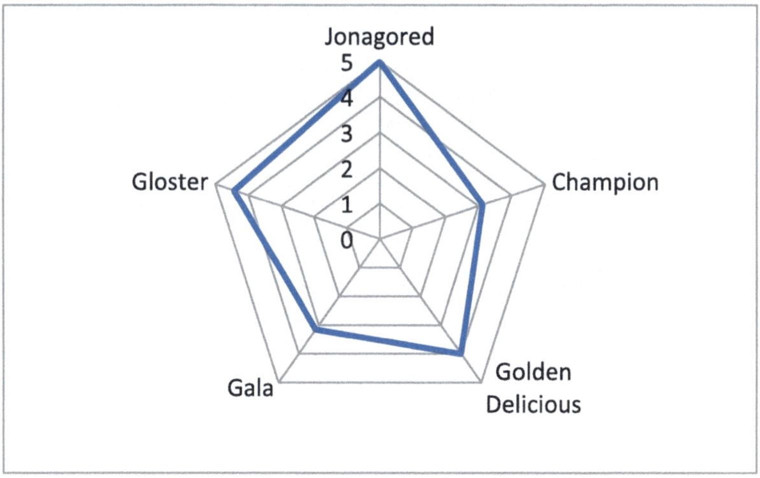

Fig. 7. Overall sensory impression by spider-plot.

4 Conclusions

In summary, we conclude that the analysed properties of cloudy juices (with the exception of density) vary depending on the apple variety used for production thereof. The results of the study clearly indicate that the juice pressed from the Jonagored variety was the most aromatic and the tastiest. It should also be underlined that this product was characterised by the highest content of total extract and sugar and the lowest total acidity. The apple juice made from the Golden Delicious and Gloster apples was highly rated as well. The juice from the Champion fruits was rated by the panellists as the least attractive product in terms of aroma and palatability; simultaneously, it had the lowest total extract and sugar content. As indicated by the analysis of the results, Jonagored, Golden Delicious, and Gloster apples can be successfully used as a raw material for production of cloudy juice.

References

1. Gerhauser, C.: Cancer chemopreventive potential of apples, apple juice, and apple components. Planta Med. **74**(13), 1608–1624 (2008)
2. EN 1131: 1994. Fruit and vegetable juices-Determination of the relative density (1994)
3. Johnston, K.L., Clifford, M.N., Morgan, L.M.: Possible role for apple juice phenolic compounds in the acute modification of glucose tolerance and gastrointestinal hormone secretion in humans. J. Sci. Food Agric. **82**(15), 1800–1805 (2002)
4. Nour, V., Trandafir, I., Ionica, M.: Compositional characteristics of fruits of several apple (Malus domestica Borkh.) cultivars. Notulae Botanicae Horti Agrobotanici Cluj-Napoca **38**(3), 228–233 (2010)
5. Nadulski, R., Kobus, Z., Guz, T., Strzałkowska, K., Kosik, D.: Przydatności wybranych odmian jabłek do tłoczenia w warunkach gospodarskich. Agric. Eng. **18**(1), 147–154 (2014)
6. Venkatachalam, K., Techakanon, C., Thitithanakul, S.: Impact of the ripening stage of wax apples on chemical profiles of juice and cider. ACS Omega **3**(6), 6710–6718 (2018)
7. Eisele, T.A., Drake, S.R.: The partial compositional characteristics of apple juice from 175 apple varieties. J. Food Compos. Anal. **18**(2–3), 213–221 (2005)
8. Zemel, G.P., Sims, C.A., Marshall, M.R., Balaban, M.: Low pH inactivation of polyphenoloxidase in apple juice. J. Food Sci. **55**(2), 562–563 (1990)

A Romanian Standpoint on Minimum Tillage Soil System and Prospects for an Sustainable Agriculture: A Review

Cătălin Bogdan, Ovidiu Ranta(✉), Alexandru Bogdan Ghețe, Ovidiu Marian, and Irimie Gheorghe Claudiu Andraș

Department of Technical Sciences and Soil Sciences, University of Agricultural Sciences and Veterinary Medicine Cluj-Napoca, Calea Mănăștur Street, no. 3-5, 400372 Cluj-Napoca, Romania

ovidiu.ranta@usamvcluj.ro

Abstract. The primary focus in nowadays discussions on agriculture is to strive for enhanced productivity that matches or surpasses conventional tillage systems in terms of energy efficiency, water conservation, soil preservation, and crop quality. Soil tillage plays a pivotal role in agricultural crop production, directly impacting soil stability and erosion susceptibility. Despite the prevalence of traditional tillage practices, various studies and practical applications in Romania have showcased the viability of minimum tillage systems. This alternative has become imperative in recent years due to the pedoclimatic conditions of Romanian agriculture and the overall food security of the country. With climate change manifesting prominently, the period between 2012 and 2022 witnessed 11 consecutive years of record-high temperatures, as reported by the National Meteorological Administration. Of the total arable land under cultivation of 8,005,889 hectares, only 3,165,767 hectares are currently irrigated or have the potential to be irrigated, according to INSSE data for the year 2022, with a significant portion facing challenging conditions. As we grapple with the impacts of a shifting climate and increasingly severe weather patterns, it becomes essential to restore and optimize the health and functionality of our soil resources. The adoption of minimum tillage practices can notably improve soil resilience, enhance crop yields and stability, reduce input costs, and promote biodiversity within ecosystems.

Keywords: Conservative tillage · conventional tillage · Romanian agriculture

1 Introduction

1.1 Soil Health

Soil health plays a vital role in order to have micro and macro view of agricultural practices, even if talking about *in situ* or *ex situ* conditions of growing. Soil is nothing than crucial living system within the confines of ecosystems and land-use parameters. It involves supporting plant and animal productivity, preserving or improving water and air quality, and fostering the health of plants and animals [1].

E. Lorencowicz et al. (Eds.): FMPMSA 2024, LNCE 609, pp. 77–86, 2024.
https://doi.org/10.1007/978-3-031-70955-5_9

It also stands as a critical position in providing essential ecosystem services, including the supply of food, feed, raw materials, and biofuels. Additionally, it serves vital regulatory functions such as carbon sequestration, water purification, contaminant reduction, pest control, climate regulation, nutrient cycling, and the maintenance of biological habitats [2].

1.2 Conventional Tillage

Conventional tillage pertains to the tillage procedure, specifically the spring and autumn plowing operations during which the entirety of the soil surface layer is disrupted, typically reaching depths of around 20 cm or deeper. Conventional tillage remains prevalent in sloping terrains, contributing to significant soil erosion and land deterioration. Soil compaction resulting from the passage of tractors and agricultural implements poses a significant threat to soil productivity, as well as its ecological and hydrological functions

The ongoing trend towards increasingly powerful and heavier agricultural vehicles exacerbates this problem, which is often inadequately diagnosed despite its severity. Tillage systems vary by location, thus their effectiveness is influenced by factors such as soil type, climate conditions, and management techniques [3, 4].

1.3 Conservation Tillage

Conservation tillage systems, in contrast to plow-based methods of seedbed preparation, offer the potential to promote sustainable utilization of soil resources. Minimum tillage soil systems, such as using cultivator, paraplow, rotary or disc harrow, present themselves as versatile alternatives for fundamental classic soil preparation, germination, bed preparation, and sowing purposes within agricultural contexts.

These minimum tillage systems play a critical role in ensuring an optimal aerial-hydrical regime essential for sustaining biological activity intensity and maintaining equilibrium in nutrient solubility. The residual vegetal matter left on the soil surface or superficially incorporated serves to enhance biological activity significantly, serving as a vital source of organic matter. Moreover, the minimum tillage systems play a pivotal role in restructuring the soil composition, thereby enhancing overall soil drainage efficiency, facilitating swift water infiltration into the soil matrix. Consequently, this leads to the development of a more fertile soil profile, fortified against potential wind and water erosion, and requiring reduced fuel consumption for the preparation of germination beds [6].

In the minimum tillage soil system, soil organisms serve as a tangible demonstration of the cause-and-effect relationship that connects land management choices to the overall productivity and health of plants and animals. Indicators used to determine soil health must be understandable and practical for farmers, who are responsible for maintaining soil quality and health. Traditionally, visible organisms like earthworms, insects, and molds have met these criteria and shows that the soil is healthy, facts that were less important for the romanian farmers or even not knew after decades of practicing conventional tillage systems as like ploughing and preparation with the disc harrow [7].

In terms of energy consumption, an research was conducted in 2004 with an experimental parcel of wheat winter where they managed to compare in terms of labor requirement the two tillage systems, revealing that conventional tillage necessitated 2.01 hours per hectare and 0.32 hours per ton, whereas conservation tillage required 1.54 hours per hectare and 0.25 hours per ton, indicating a reduction of 23.83% per hectare and 21.87% per hour in labor demand compared to the conventional tillage system, respectively. This analysis sheds light on the significant differences in labor efficiency between the two tillage systems, highlighting the potential for increased productivity and cost-effectiveness in agricultural practices [8].

2 Material and Method

A number of 19 of research articles published in the last 21 years meaning the interval between 2003–2024 in various internationally recognized open-access journals such as Springer, MDPI, as well as on research platforms like Google Scholar, Research Gate, and Web of Science were utilized for the purposes of this literature review.

The search process involved selecting the area and country of the study papers to be the Romanian region and identifying articles with titles containing key terms like *conservation agriculture, conservative tillage, conventional tillage, conservative agriculture, conventional tillage, minimum soil tillage systems, minimum till, minimum tillage, soil tillage, tillage, romanian agriculture.*

Numerous research studies showcase different approaches to implementing minimum tillage and the resulting outcomes observed over time. The results vary depending on the region of cultivation, considering factors such as climate conditions, soil type, and economic aspects. This study aims to present findings concerning energy consumption (such as labor hours and fuel usage, economic facts), crop health maintenance (including weed and disease control and pest management), crop yield (the quantity of harvest), and soil health management (addressing erosion).

3 Results and Discussions

3.1 Energy Consumption

Optimizing soil tillage and the quantity of mechanical operations can enhance the energy and economic efficiency in agricultural practices. This optimization leads to reduced fuel consumption (10–40%), decreased working time (over 50%), and lower equipment requirements, resulting in decreased production costs. Additionally, there are indirect benefits such as positive impacts on soil preservation and decreased greenhouse gas emissions. It is important to note that soil tillage plays a significant role in the direct energy consumption per unit area, accounting for approximately 52% in wheat cultivation and 60% in maize production [9].

An experimental investigation was carried out on a moderately fertile *Fluvisol* at the University of Agricultural Sciences and Veterinary Medicine in Cluj Napoca (north part of Romania –46°46′N, 26°36′E) to confirm the minimum tillage techniques aimed at reducing energy usage and promoting soil conservation in the elevated regions of Romania. The findings provide strong evidence that minimum tillage serves as a viable substitute for traditional soil tillage methods, owing to its positive impact on soil characteristics and crop yields. Specifically, the results indicate significant improvements in crop yields for the following harvests: maize (96–98.1%), soybean (102.9–111.9%), and wheat (93.4–96.8%) when compared to conventional tillage practices. Selecting the appropriate soil tillage approach for crop rotation plays a crucial role in reducing energy consumption, resulting in maize (97.3–97.9%), soybean (98.6–98.2%), and wheat (97.4–98%) yields surpassing those achieved through conventional methods [10].

The research was carried out at Agricultural Research and Development Station Turda (north parth of Romania, 46.579396327214155, 23.785998229460063) in the 2014 year focused on observing the behavior of four different varieties of afila peas under two distinct soil tillage systems, namely the classical method involving a mould plow, and the minimum tillage method known as the chisel grubber variant. It was determined that the implementation of the minimum tillage system led to a reduction in degradation caused by erosion, an increase in organic matter content, as well as the restoration of soil structure. Furthermore, by incorporating legumes into the crop rotation, this system ensured a higher nitrogen quantity for the crops in rotation. Notably, the utilization of the minimum tillage system during pea cultivation resulted in a significant enhancement of energy efficiency. This improvement translated into a remarkable 28.4% decrease in fuel costs compared to the traditional tillage approach. Moreover, the fuel consumption per ton of peas produced was notably lower in the minimum tillage system.

By adopting the minimum tillage system for pea cultivation, a substantial economic advantage of 28.4% in fuel costs equivalent to 337.2 RON (*166.7 € at 15.04.2024*) per hectare was observed in contrast to the classical tillage system application where there was 471 RON (*234.2 € at 15.04.2024*) per hectare. In the classical system, the production of one ton of peas necessitated a fuel consumption of 21.9 liters per ton and incurred fuel costs of 131.6 RON (*654 € at 15.04.2024*) per ton, which were higher when compared to the minimum tillage system. In the latter, the fuel consumption was reduced to 15.8 liters per ton, resulting in fuel costs of 94.9 RON (*472 € at 15.04.2024*) per ton, showcasing the economic and environmental benefits of such tillage practices. [11]

Reducing it nowadays should mean economical usage equivalent of 15.8 liters per ton in costs of 126.6 RON (*629.6 € at 15.04.2024*), higher with almost 33% than it was in 2014.

Another study of Agricultural Research and Development Station Turda (north parth of Romania, 46.579396327214155, 23.785998229460063) presents the findings of the investigation conducted in a *Phaeozem* soil located in the Transylvanian Plain, focusing on the impact of soil tillage methods (conventional system and minimum tillage-with chisel) and the economic dimensions of corn cultivation, from 2011 to 2014. The economic effectiveness of the minimum tillage approach is superior, attributed to the decrease in fuel usage to 84.4 liters per hectare in contrast to the conventional system, which requires 101.5 liters per hectare for all soil tillage activities. By deducting the total costs (materials and fuel) from the output value achieved within each technological approach, a profit of 112.9 RON (*561.5 € at 15.04.2024*) per hectare is obtained, representing a decrease of 6.6% in the context of the overall minimum tillage technology [12].

An experimental parcel was established on the chromic *Luvisol* at Moara Domneasca Teaching Farm, which is affiliated with the University of Agronomic Sciences and Veterinary Medicine of Bucharest. The experimental variants took place between 2014 – 2015 year in soil tillage included four variants of work; a1 - ploughing at a 20 cm depth (control - conventional system); a2 - chisel at 20 cm depth; a3 - chisel at 40 cm depth; a4 – harrow disk at 10 cm depth (minimum tillage system). Minimum tillage resulted in elevated energy indicators for wheat and slightly reduced indicators for maize in comparison to the conventional approach. In winter crops, minimum tillage led to similar or slightly higher yields than those achieved with traditional soil tillage methods; however, maize yield decreased by 5–11% due to less favorable growing conditions that year. Chiselling at a depth of 40 cm produced the most favorable outcomes for both crops, as well as for the energy indicators related to wheat and maize [13].

SCDA Drăgănești Vlașca (Teleorman, south part of Romania - 44.083896503718485, 25.55466315928514) conducted soil conventional works and minimum tillage on a 4 hectares parcel seeded with wheat and soy crops during the initial experimental year. Following an economic evaluation of the wheat crop, it was observed that minimum tillage led to a 52.97% reduction in costs compared to the conventional method. Similarly, for the soy crop, expenses decreased by 67.53% with the application of minimum tillage in contrast to the conventional approach [14].

Recently studies were made in 2023 when researchers from University of Life Sciences “King Mihai I” from Timisoara conducted an study in Oravița (Caraș-Severin , south vest part of Romania −45.036090250823484, 21.68689251885841) focusing on minimum tillage technology in wheat cultivation. The research highlights the negative impact of intensive soil operations, such as the conventional system, on soil erosion. The findings of the study indicated that the overall costs for implementing minimum soil tillage in wheat fields in Oravița amounted to 209 RON (*103.9 € at 15.04.2024*) per hectare. Specifically, direct costs totaling 175 RON (*870.3 € at 15.04.2024*) per hectare accounted for 80% of the total expenses. Fuel costs, at 117 RON (*581.9 € at 15.04.2024*) per hectare, made up 56% of the total expenses and 67% of the direct expenses associated with minimum tillage activities. From an economic perspective, this approach proves to be more cost-effective as it reduces fuel consumption and machinery operational time, underscoring the economic efficiency of employing modern mechanization technologies for various reasons [15].

3.2 Crop Health Maintenance

Another research made was investigating the impact of soil tillage and fertilization systems on corn cultivation was investigated in a three-year rotation cycle of soybean, wheat, and corn from 2011 to 2014. In the minimum tillage system, two perennial dicotyledonous weed species, namely Lepidium draba L. and Rorippa austriaca (Crantz) Besser, emerged in 2013, followed by the appearance of Matricaria inodora L. and Polygonum aviculare L. in 2014. The quantity and weight of weeds observed in the corn fields were higher under minimum tillage in the initial experimental years of 2012 and 2013. However, in 2014, the weed count reduced to 30 weeds m2, contrasting with the 33 weeds m2 found in the conventional tillage system [16].

A study was taken place between 2012 and 2017 with the aim of assessing the insect populations in conventional tillage (plowing) and minimum tillage (chisel) to determine if the abundance and dynamics of pests in maize crops are impacted by the tillage method. The investigation involved monitoring, abundance assessments, and dynamics evaluations of the most known pest species in maize crops, namely *Diabrotica virgifera virgifera LeConte, Autographa gamma L.,* and *Agrotis segetum Den. & Schiff*, using pheromone traps. By reducing tillage operations, there has been observed a rise in the pest population, notably in *Diabrotica virgifera virgifera LeConte*, in contrast to conventional tillage practices where the eggs or larvae become integrated into the soil and the population potentially decreases. This study potentially indicates the existence of adaptations of insects to changes to climate changes [17].

An interesting research was carried out between 2014–2016 at the Agricultural Research and Development Station Turda (north part of Romania, 46.579396327214155, 23.785998229460063). The research findings indicate that the utilization of a minimum tillage system may not adversely affect the overall isoflavone level in soybeans when compared to traditional soil tillage methods. The total isoflavone content is of paramount importance to both soybean growers and consumers, particularly in light of the pressing need for appropriate soil tillage practices in Europe, where arable land is limited. In conclusion, this research proposes that minimum tillage could serve as a suitable soil tillage approach for soybean cultivation without compromising the total isoflavone concentration. Nevertheless, it is worth noting that external factors like climatic conditions and soil tillage techniques can potentially influence their concentration [18].

The research took place in period of 2018–2020 years, at Agricultural Research and Development Station Turda (ARDS Turda), on a Faeoziom type soil wich looked out to the pest management of the maize field in different ways of working the soil, including minimum tillage and conventional tillage. The conclusion asserts that employing the practice of classic plowing is the most effective agrotechnical strategy for reducing the impact of diseases and pests. This technique involves the inclusion of vegetative remnants in tillage practices, which could contain spores that are resistant to pathogens [19].

Other researchers carried out a study between 2020–2022 on cambic *Chernozem* soil from Fundulea, in the non-irrigated area, in a stationary experiment. When minimum tillage system is used, the weeds from the maize crop with perennial dicotyledonous and monocotyledonous species are more usual to be seen growing than the annual ones because of the large reserve of seeds left in the soil and undisturbed over the years. Also, comparing the minimum tillage system to the control variant, there was a 26% increase in the number of weeds observed when utilizing the chisel plow and a 33% increase when using the disk [20].

3.3 Crop Yield

An experimental plot was established on mollic *Aluviosol* at the University of Agricultural Sciences and Veterinary Medicine in Cluj Napoca, located within the Research Center for Minimal Systems and Sustainable Agricultural Technologies. The impact of tillage systems on the productivity of cultivated species and consequently on crop yield is significant. Various crops in the rotation exhibited diverse responses to the implementation of minimum tillage systems. For example, wheat, soybean, and rape crops showed yields that were comparable between minimum tillage and conventional systems, with no statistically significant differences except for the rotary grape in the wheat crop. Yields ranged from 2747–2867 kg per hectare for soybean, 3282 kg – 3451 kg per hectare for wheat, and 1505 kg - 1588 kg per hectare for spring rape. Conversely, corn and particularly potato crops demonstrated lower yields under minimum tillage systems. In corn, there were notable yield variations between different tillage variants, with classic plough yielding 5857 kg per hectare and chisel and paraplow variants producing 5704 kg–5737 kg per hectare (97.3-97.9% of the control). The rotary grape variant had a significantly lower yield at 92.1% of the control. In potato crops, yields under minimum tillage systems ranged from 82.4–93.4% of the control (plough variant) [21].

A study made by an researcher comes with very powerful advices. The experimental field was situated in red *Preluvosoil* at an elevation of 130 meters in the Găvanu-Burdea Plain, specifically in the village of Scurtu Mare, Teleorman County. Regarding the climatic conditions, the crop year of 2011-2012 met dry weather, with irregular precipitation throughout the growth period. Precipitation of 374 mm was recorded from planting to harvesting, with 252 mm occurring at the conclusion of winter and throughout the growth phase. The field was cultivated with winter wheat, comprising 6 romanian varieties, 11 foreign varieties, and a hybrid wheat, each seeded using the minimum tillage (disk harrow) and traditional methods (plough, disk harrow, and combinator). The genotypes under examination were classified as very early to medium early varieties known for their strong resistance to drought, high temperatures, and winter conditions. Under identical climate conditions, the minimum tillage approach exhibited superiority over the conventional system; across all varieties, the minimum tillage method achieved an average yield of 5191 kg per hectare, whereas the conventional soil cultivation method yielded 3723 kg per hectare [22].

In 2018, a study was conducted at the Agricultural Research and Development Station Turda (north part of Romania, 46.579396327214155, 23.785998229460063) situated in the Transylvanian Plain. The research focused on a Faeoziom vertic soil characterized by a neutral pH, clay-loam texture, and a satisfactory supply of mobile phosphorus and potassium, with a medium humus content. The findings indicated that the optimal results for maize were achieved using the traditional soil cultivation approach, yielding 9566 kg per hectare. Conversely, employing minimal tillage methods such as chiseling or disc harrowing resulted in smaller yields, with differences of 94 kg per hectare and 93 kg per hectare respectively, which were not statistically significant. [23]

Researchers from University of Agronomic Sciences and Veterinary Medicine of Bucharest conducted a study in 2019 in the southern region of Dolj County aimed to investigate the potential impact of both the conventional system and the minimum tillage system on maize production and its characteristics. These experiments helped in establishing the most suitable technological system and optimal fertilizer levels for maize cultivation. The application of the minimum tillage system along with variant 4, which involved the addition of N150P50K50, resulted in the highest maize yield of 14 300 kg per hectare, surpassing the production of 13 970 kg per hectare achieved with the conventional system. Considering the pedoclimatic conditions, soil properties, and clay texture of the Chernozem in Southern Oltenia, Romania, it can be inferred that the minimum tillage system is a viable option for maize cultivation within an irrigated system [24].

3.4 Soil Health Management

The humus percent in soil tends to increase when minimum tillage systems are utilized. Soil infiltration assessments conducted at the soil surface using a field permeameter indicate a rate of 5.7 L per m^2 in a minute in the conventional tillage method and 6.5 L–7.9 L per m^2 in a minute for the minimum tillage approaches. In relation to this soil characteristic, the water retention in the 0–50 cm depth is higher by 1–32 m^3 per hectare in the minimum tillage systems [20].

A study was carried out at the Experimental Station of Ion Ionescu de la Brad University of Life Sciences in Iași between 2009 and 2010, focusing on the specific conditions of the forest steppe in Northeastern Romania. The objective of this study was to evaluate the impact of soil conservation practices through minimum tillage on soil biological activity in comparison to traditional tillage methods. The findings indicate that minimum tillage enhances soil biological activity when contrasted with conventional tillage. The results revealed increased biological activity levels across all experimental variations in comparison to the disk x2 method, particularly in the 0–15 cm depth, as this method only loosens the soil superficially at 5–7 cm from the 15 cm depth under investigation [25].

Some Romanian research institutions conducted a study in which they examined the evolution of humus content in both conventional and minimum tillage systems. The area under investigation is situated within the Teleorman Plain region, specifically within the Găvanu-Burdea Plain subunit, characterized by the presence of Haplic Phaeozem soil. The findings revealed that over a period of 54 months, there was an increase in soil humus content at a depth of 0–10 cm when the soil was managed using the minimum tillage system, reaching a value of 4.62%. In contrast, a lower value of 4.40% was observed in the conventional system. This phenomenon can be attributed primarily to the significant presence of plant residues on the soil surface, with a minimum of 30% in the minimum tillage systems, as well as the equilibrium between mineralization and humification processes [25].

One study was conducted in 2022 at the Chiscani Experimental Field located within A.R.D.S. Braila. The trial was implemented on a vermic, phreatic, wet, moderately carbonated *loess*, clay-sandy, arable soil. The soil management practices of conservation agriculture, specifically variant 3 involving scarification and variant 4 utilizing heavy-disk, played a role in the efficient utilization of water and nutrients from deeper soil layers. This in turn facilitated the optimal growth of maize crops and enhanced their ability to withstand the various stressors encountered during the experimental period. The mass of green matter per hectare, as determined from the analysis of soil tillage methods, ranged from 22.63 Mg ha^{-1} to 37.04 Mg ha^{-1}, with variant 3 and variant 4 ranking first, namely saying that this because the depth of the tillage that allowed the roots to go into the deep layers wich provided better water resources and nutrients [26].

4 Conclusion

In Romania, there has been significant interest in minimum tillage systems for many years, with Romanian experts anticipating climate change and implementing various practices through different research institutes. Trials and experiences have been conducted relatively uniformly across most agricultural areas of Romania. The main conclusions primarily alert us to the need for careful pest management in minimum tillage systems. Not all soils allow for minimum tillage, or rather, not all soils can be implemented with visible results in a short time, with clay content being the main obstacle. Crop rotation is mandatory to begin achieving conclusive results, implicitly promoting good agricultural practices advocated by the European Union. Predictions for sustainable agriculture in Romania are extremely positive and are taken into account by most farmers through platforms discussed by agricultural universities and beyond.

References

1. Doran, J., Zeiss, M.: Soil health and sustainability: managing the biotic component of soil quality. Appl. Soil Ecol. **15**, 3–11 (2000)
2. Duşa, E., Stan, V., Stanciu, A.M.: Evolution of soil ecosystem services under different cropping systems. Scientific Papers. Series A. Agronomy, LXV(2) (2022)
3. Qiong, P., Baoyuan, L., Yaxian, H., Aijuan, W., Qiankun, G., Bing, Y., Qi, C., Liang, H.: The role of conventional tillage in agricultural soil erosion. Agric. Ecosyst. Environ. **348**, 108407 (2023)
4. Six, J., Bossuyt, H., Degryze, S., Denef, K.: A history of research on the link between (micro) aggregates, soil biota, and soil organic matter dynamics. Elsevier, Soil and Tillage Research. 79(1): pp 7–31 (2004)
5. Rusu, T., Gus, P., Bogdan, İ, Moraru, P.İ, et al.: Effect of minimum tillage systems on the soil conservation and sustainability of agricultural production. Tarım Makinaları Bilimi Dergisi **5**(3), 241–245 (2009)
6. Hajabbasi, MA., Hemmat, A.: Tillage impacts on aggregate stability and crop productivity in a clay-loam soil in central Iran. Elsevier, Soil and Tillage Research (Vol. 56(3–4). pp 205–212) (2000)
7. Kosutic, S., Zimmer, R., Filipovic, D., Kovacev, I., Husnjak, S., Winter wheat production using conventional, conservation and no-till system. In: Kosutic, S.: 33rd international symposium on actual tasks of agricultural engineering, Actual Tasks on Agricultural Engineering, 33: 469–476 (2005)
8. Stănilă, S., Drocaş, I., Ranta, O., Molnar, A.: Nagy, Mihaela, Consideraţii privind consumul de energie în producţia agricolă şi la lucrările solului. Risoprint, ISSN **2247–7525**, 340–346 (2011)
9. Moraru, P.I., Rusu, T.: No-tillage and minimum tillage systems with reduced energy consumption and soil conservation in the hilly areas of Romania. WFL J. Food Agric. Environ. **11**(2), 1227–1231 (2013)
10. Şimon, A., Rusu, T., Cheţan, F., Cheţan, C.: Optimization the work of soil for the cultivation of afila peas in turda area. Bull. Univ. Agric. Sci. Veter. Med. Cluj-Napoca Agric. (2015). https://doi.org/10.15835/buasvmcn-agr:11535
11. Cheţan, F., Rusu, T., Cheţan, C., Moraru, P.I.: Influence of soil tillage upon weeds, production and economical efficiency of corn crop. AgroLife Scientific Journal Vol. 5, number 1 (2016)

12. Marin, I.D., Rusu, T., Mihalache, M., Ilie, L., Nistor, E., Bolohan, C.: Influence of soil tillage upon production and energy efficiency in wheat and maize crops. AgroLife Sci. J. **4**, 2 (2015)
13. Burtan, L., Vrinceanu, A., Sirbu, C., Cioroianu, T., Dumitrascu, M., Popovici, M.: Economic analysis between minimum tillage and conventional system. J. Environ. Protect. Ecol. **18**(3), 1289–1294 (2017)
14. Duma, C.A., Mihut, C., Mateoc-Sirb, N., Nita, L., Okros, A., Lato, I., Mateoc, T., Popa, D.: Studies on minimum wheat tilling scientific papers. Ser. Manag. Econ. Eng. Agric. Rural Develop. **23**(2), 217–222 (2023)
15. Cheţan, F., Rusu, T., Cheţan, C., Moraru, P.I., Influence of soil tillage upon weeds, production and Economical efficiency of corn crop. AgroLife Sci. J. 5(1) (2016)
16. Tărău, AD., Mureşanu, F., Vălean, A.M., Russu, F., Şopterean, L., Cheţan, F., Suciu, L., Urdă, C., influence of different tillage systems on abundance and dynamics of pests in maize crop. Nardi Fundulea, Romanian Agric. Res. 40, (2023)
17. Mureşan, L., Clapa, D., Borsai, O., Rusu, T., Wang, T.T.Y., Park, J.B.: Potential impacts of soil tillage system on isoflavone concentration of soybean as functional food ingredients. Land 2020, 9, 386 (2020)
18. Popa, A., Rusu, T., Şimon, A., Russu, F., Bărdaş, M., Oltean, V., Suciu, L., Tărău, A., Merca, N.C.: influence of biotic and abiotic factors on maize crop yield in transylvanian plain conditions, Scientific Papers. Series A. Agronomy. LXIV(2) (2021)
19. Partal, E., Oltenacu, C.V., Paraschivu, M., Cotuna, O., Dima, M., Contescu, E.L.: Effects of different soil tillage on soil moisture, weed control, yield and quality of maize (Zea mays L.). Nardi Fundulea, Roman. Agric. Res. **40**, 1045 (2023)
20. Rusu, T., et al.: Implications of minimum tillage systems on sustainability of agricultural production and soil conservation. J. Food Agric. Environ. **7**(2), 335–338 (2009)
21. Popa, M., Motcă, G., Genotype influence upon production and quality of wheat grown in găvanu-burdea plain, Scientific Papers. Ser. A. Agron. LVI, (2013)
22. Popa, A., Rusu, T., Coman, M., Şimon, A., Şopterean, L.: Influence of the tillage system and foliar fertilizations on yield and *fusarium* ear rot manifestation in maize crop in the transilvania plain, Scientific Papers. Ser. A Agron. LXII, No. 1 (2019)
23. Zafiu, C., Mihalache, M.: Research on the Influence of Technological Systems on Maize Cultivation in the South of Dolj County, Romania, Scientific Papers. Series A. Agronomy, Vol. LXIV, No. 1, (2021)
24. Bireescu, L., Bireescu, G., Sellitto, M.V.: Assessing the soil physiological potential using Pedo-biological diagnosis under minimum-tillage system and mineral fertilization. EQA Environ. Qual. **13**, 01–07 (2014)
25. Burtan, L., et al.: Various soil quality parameters and humus content evolution in conventional and minimum tillage systems. Roman. Agric. Res. **40**, 491–500 (2003)
26. Rîşnoveanu, L., Oprea, D., Joiţa-Păcureanu, M.: Preliminary results on maize biomass under the influence of tillage in the context of climate change, Scientific Papers. Series A. Agronomy, LXVI(1) (2023)

Theory of Operation of a Safety Finger Coupling Applied in the Drive of a Screw Conveyor for Bulk Agricultural Materials

Volodymyr Bulgakov[1], Simone Pascuzzi[2(✉)], Jüri Olt[3], Arlindo Almeida[4], Oleksandra Trokhaniak[1], Janusz Nowak[5], Yevhen Ihnatiev[3,6], Zbigniew Kiernicki[7], Francesco Paciolla[8,9], and Giacomo Scarascia Mugnozza[9]

[1] Department of Mechanics, National University of Life and Environmental Sciences of Ukraine, 15 Heroiv Oborony Str., Kyiv 03041, Ukraine
[2] Department of Soil, Plant and Food Science, University of Bari Aldo Moro, Via Amendola 165/A, 70126 Bari, Italy
simone.pascuzzi@uniba.it
[3] Institute of Technology, Estonian University of Life Sciences, 56 Kreutzwaldi Str., 51006 Tartu, EE, Estonia
[4] Polytechnic Institute of Bragança, Mountain Research Center, Bragança, Portugal
[5] Department Od Machine Operation and Production Processes Management, University of Life Sciences in Lublin, 13, Akademicka Str., 20-950 Lublin, Poland
[6] Dmytro Motornyi, Tavria State Agrotechnological University, 66 Zhukovsky Str., Zaporizhzhia 69600, UA, Ukraine
[7] Lublin University of Technology, Nadbystrzycka 38D, 20-618 Lublin, Poland
[8] PolySense Lab, Dipartimento Interateneo Di Fisica, Polytechnic and University of Bari, Via Amendola 173, 70126 Bari, Italy
[9] Polytechnic University of Bari, Via Edoardo Orabona 4, 70126 Bari, Italy

Abstract. The aim of the study is to enhance the operational characteristics of a screw conveyor during overloading of its working mechanism by substantiating the rational parameters of the safety coupling. Theoretical calculations and their graphical dependencies were performed using the Delphi programming environment. To increase the magnitude of axial displacement of the driven half-coupling with the jammed working element, while reducing dynamic loads during overloading, a design of a safety coupling has been developed. To determine the rational parameters of the safety coupling for the screw conveyor, the influence of dynamic loads on the operation of the safety coupling during overloading of the screw conveyor's working mechanism has been analyzed. From the analysis of the graphs, it has been determined that an increase in the rotation frequency n leads to an increase in the torque of the system. With a change in rotation frequency from 50 rpm to 200 rpm, the torque T_d increases by 64.8%, T_r rises by 65.2%, and T_m increases by 63.8%. It has also been determined that an increase in the mass m of moving links leads to an increase in loads on the drive. When the mass increases from 10 kg to 20 kg, the torque T_d increases by 15.8%, T_r rises by 12.5%, and T_m increases by 17.3%.

E. Lorencowicz et al. (Eds.): FMPMSA 2024, LNCE 609, pp. 87–97, 2024.
https://doi.org/10.1007/978-3-031-70955-5_10

Keywords: Safety Coupling · Half-Coupling · Groove · Rational Parameters · Dynamic Loading

1 Introduction

Technological processes in agricultural production involve many labor-intensive loading, unloading, and transportation operations [1–6]. Most commonly, these operations include the handling and transportation of grain crops, seed materials, granulated mineral fertilizers, and so on [7–10]. Screw auger conveyors, as a separate technical element of transportation mechanisms, have found widespread use in the layout schemes of machines for loading or moving bulk and lump materials due to their simple design, ease of maintenance, and the ability to load and unload materials at any point along the technological line [11–14]. During the transportation of such materials, due to the presence of clearances between the rotating surface of the screw and the inner surface of the guiding pipe, jamming of the screw working element is possible. To restore the functionality of the conveyor, it is necessary to disengage the jammed edge of the screw in the axial direction from contact with the material. Subsequently, after removing the overload, the drive elements should ensure the initial position of the screw working element for transporting the material to the unloading zone [15–18]. These issues can be addressed by axially disengaging the jammed screw working element using ball safety couplings with a profiled design of the recesses, both when disengaging and engaging, at the exit and entry points [19–21]. For the reliable operation of screw conveyors, both rigid [22] and flexible [23], specific designs of safety couplings have been developed for axial disengagement of a jammed screw, along with methodologies for conducting experimental research [24]. Considerable attention is given to analyzing the determination of dynamic loads in various types of drives with safety couplings, as well as their strength calculations [25–29]. Therefore, the task of developing new designs of safety couplings for screw conveyors to axially disengage the working element in case of overloading and ensure the restoration of its initial position is relevant. The aim of the research is to enhance the operational characteristics of the screw conveyor during overloading of its working mechanism by substantiating the rational parameters of the safety coupling.

2 Materials and Methods

To address this issue, a safety coupling with temporally separated slipping and axial displacement modes of the screw has been developed to restore the operational state of the screw conveyor. Its design and overall view are presented in Fig. 1. When transmitting torque, the cylindrical fingers (Fig. 1–4), which have a spherical shape on the working side and they are located in through axial openings in the anti-friction bushings (Fig. 1–5) of the driven half-coupling (Fig. 1–3), engage with the recesses (Fig. 1–7) of the driving half-coupling (Fig. 1–1). This ensures the rotation of the safety coupling and the screw working element (Fig. 1–3). The fingers (Fig. 1–4) and recesses (Fig. 1–7) are arranged at the same diameter, and on both sides of the recesses on the end surface of the driving half-coupling (Fig. 1), there are circular working (Fig. 1–10) and return (Fig. 1–11) grooves. The inclination angle of the working groove is significantly smaller than the inclination angle of the return groove. In the event of an overload, the driven half-coupling (Fig. 1–3) stops while the driving half-coupling (Fig. 1–1) continues to rotate, causing the cylindrical fingers (Fig. 1–4) to disengage from the recesses (Fig. 1–7). During this process, the cylindrical fingers (Fig. 1–4) move along the circular working grooves (Fig. 1–10). Due to the rotation of the driving half-coupling (Fig. 1–1), the cylindrical fingers (Fig. 1–4) return to their initial position, resulting in the restoration of the initial state. During this process, the cylindrical fingers (Fig. 1–4) move along the circular return grooves (Fig. 1–11).

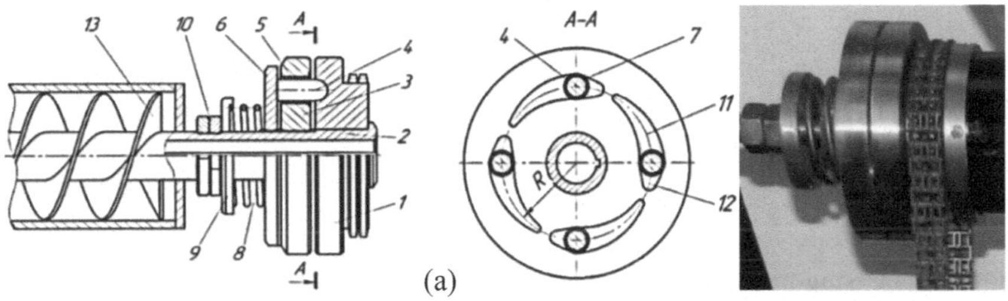

Fig. 1. Constructive diagram (a) and general view (b) of the safety coupling.

In determining the rational parameters of the screw conveyor's safety coupling, it is essential to consider the effect of dynamic loads, the absolute values of which will differ from static ones. To address this, a dynamic model of the screw conveyor with the safety coupling was examined, as shown in Fig. 2. It is described by a system of differential equations of motion (1):

$$\left.\begin{aligned} J_1 \cdot \ddot{\varphi}_1 &= T_d - c_1(\varphi_1 - \varphi_2), \\ J_{21} \cdot \ddot{\varphi}_2 &= c_1(\varphi_1 - \varphi_2) - T_m, \\ J_{22} \cdot \ddot{\varphi}_3 &= T_m - c_2(\varphi_3 - \varphi_4), \\ J_3 \cdot \ddot{\varphi}_4 &= c_2(\varphi_3 - \varphi_4) - T_r, \end{aligned}\right\} \quad (1)$$

where: c_1– equivalent torsional stiffness of the drive components between the engine and the driving half-coupling; c_2– equivalent torsional stiffness of the drive components between the driving half-coupling and the shaft of the working element; φ_1 – twist angle of the engine shaft; φ_2 – twist angle of the driving half-coupling; φ_3 – twist angle of the driven half-coupling; φ_4 – twist angle of the shaft of the working element; J_1 – consolidated moment of inertia of the drive; J_{21} – consolidated moment of inertia of the driving half-coupling; J_{22} – consolidated moment of inertia of the driven half-coupling; J_3– consolidated moment of inertia of the shaft of the working element; T_{d} – torque moment of the drive; T_{m} – torque moment of interaction between half-couplings; T_{r} – resistance moment on the shaft of the screw working element.

The drive torque moment T_d can be expressed as a constant moment:

$$T_d = T_{d0}, \quad (2)$$

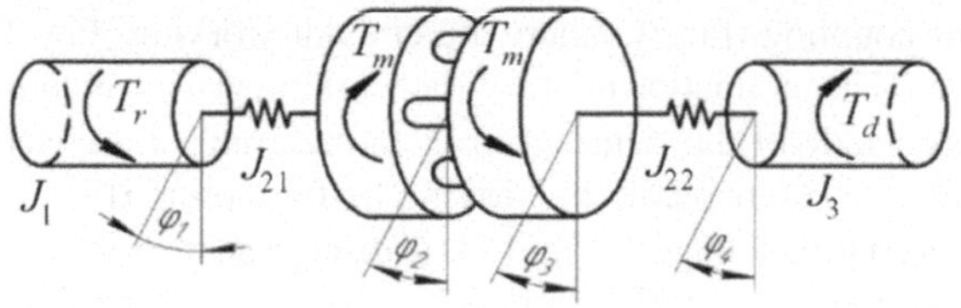

Fig. 2. Dynamic model of the screw conveyor with a safety coupling.

or as a linear characteristic of the electric drive in the form:

$$T_d = T_{d0}\left(1 - \frac{\dot{\varphi}_1}{\omega_0}\right), \tag{3}$$

where ω_0 is the angular velocity of rotation of all masses at the initial moment of time.

The resistance moment applied to the output link – the screw working element, T_r can be expressed as a constant moment:

$$T_r = T_{r0}, \tag{4}$$

or linearly increasing in the form:

$$T_r = T_{r0} + T_{r1} \cdot \frac{t}{\tau}. \tag{5}$$

Until the moment of time τ when complete stoppage of the screw occurs (its jamming).

The magnitude of the torque moment of interaction between half-couplings T_m is influenced by parameters such as the geometric dimensions of the half-coupling, the profiles of the recess and groove, the stiffness and initial tension of the compression spring, and the axial mass of the driven half-coupling together with the working element and a portion of the load. The influence of friction forces can be neglected since, with proper lubrication, it is sufficiently small.

Therefore, the moment T_m will be equal to the sum of two components – static, which depends on the deformation of the spring, and inertial, which is determined by the axial acceleration of the half-coupling.

The static component of the axial force during mutual displacement of half-couplings is equal to:

$$F_s = c(\delta_0 + \lambda), \tag{6}$$

where: δ_0 – initial tension of the spring; λ – current deformation of the spring, which is equal to the amount of the cylindrical finger exiting from recess; c – stiffness of the spring.

The dynamic component of the axial force was determined as the inertia force during the axial displacement of the driven half-coupling together with the working element and a portion of the load:

$$F_d = m\ddot{\lambda}, \tag{7}$$

where m is the mass of the driven links.

To determine the geometric relationships when the cylindrical finger exits from recess, a calculation scheme was constructed, as shown in Fig. 3.

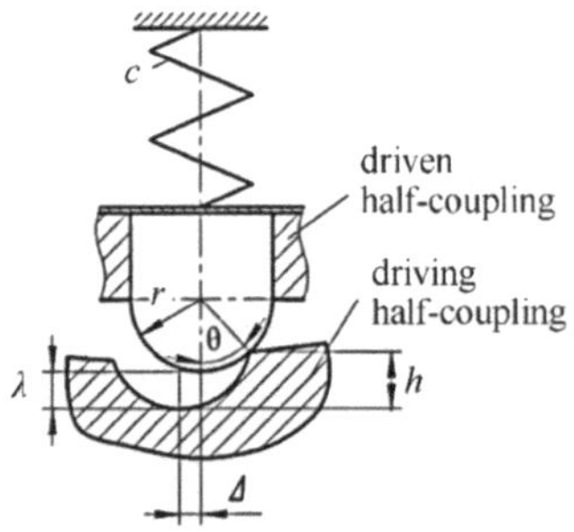

Fig. 3. Calculation scheme for determining the geometric relationships when the cylindrical finger exits from recess.

The tangential displacement of the cylindrical finger is equal to:

$$\Delta = R \cdot (\varphi - \varphi_3),$$

where R is the radius of the placement of the cylindrical fingers.

At the same time, the finger rises to a height λ, which is determined by the formula:

$$h - \lambda = r \cdot (1 - \cos\theta), \tag{8}$$

where: r is the radius of rounding of the cylindrical finger and θ is the current angle of the cylindrical finger pressing on the edge of the recess.

After substitutions and transformations, we obtain the expression:

$$\lambda = r \cdot (\cos\theta - \cos\theta_0), \tag{9}$$

where: θ_0 is the initial value of the angle of pressure of the cylindrical finger on the edge of the recess when the finger was completely in the recess.

The tangential displacement Δ of the cylindrical finger was determined through the angles of pressure:

$$\Delta = r \cdot (\sin\theta_0 - \sin\theta) = R \cdot (\varphi_2 - \varphi_3) = r \cdot (\sin\theta_0 - \sin\theta). \tag{10}$$

From Eqs. (9) and (10) we obtained:

$$\lambda = h - r + \sqrt{r^2 - [r \cdot \sin\theta_0 - R \cdot (\varphi_2 - \varphi_3)]^2}. \tag{11}$$

The current pressing angle of the finger is determined from (9) and (10), using (11), through the tangent of the angle θ:

$$\tan\theta = \frac{r \cdot \sin\theta_0 - R \cdot (\varphi_2 - \varphi_3)}{\lambda + r - h} = \frac{r \cdot \sin\theta_0 - R \cdot (\varphi_2 - \varphi_3)}{\sqrt{r^2 - [r \cdot \sin\theta_0 - R \cdot (\varphi_2 - \varphi_3)]^2}} \tag{12}$$

From here, the interaction torque of the half-couplings was determined by the equation:

$$T_m = R \cdot \tan\theta \cdot (F_s + F_d) = R\left[c(\lambda_0 + \lambda) + m_4 \cdot \ddot{\lambda}\right] \cdot \frac{r \cdot \sin\theta_0 - R \cdot (\varphi_2 - \varphi_3)}{\sqrt{r^2 - [r \cdot \sin\theta_0 - R \cdot (\varphi_2 - \varphi_3)]^2}} \tag{13}$$

By differentiating (11) with respect to time, the derivatives were determined $\dot{\lambda}$ and $\ddot{\lambda}$:

$$\begin{aligned} \dot{\lambda} &= \frac{[r \cdot \sin\theta_0 - R(\varphi_2 - \varphi_3)] \cdot R \cdot (\dot{\varphi}_2 - \dot{\varphi}_3)}{\sqrt{r^2 - [r \cdot \sin\theta_0 - R \cdot (\varphi_2 - \varphi_3)]^2}}; \\ \ddot{\lambda} &= \frac{[r \cdot \sin\theta_0 - R(\varphi_2 - \varphi_3)] \cdot R(\ddot{\varphi}_2 - \ddot{\varphi}_3) - (R)^2 \cdot (\dot{\varphi}_2 - \dot{\varphi}_3)^2}{\sqrt{r^2 - [r \cdot \sin\theta_0 - R \cdot (\varphi_2 - \varphi_3)]^2}} \\ &\quad - \frac{[r \cdot \sin\theta_0 - R \cdot (\varphi_2 - \varphi_3)]^2 \cdot (R)^2 \cdot (\dot{\varphi}_2 - \dot{\varphi}_3)^2}{\left(r^2 - [r \cdot \sin\theta_0 - R \cdot (\varphi_2 - \varphi_3)]^2\right)\sqrt{r^2 - [r \cdot \sin\theta_0 - R \cdot (\varphi_2 - \varphi_3)]^2}}. \end{aligned} \tag{14}$$

Substituting Eqs. (11) and (14) into the expression for the interaction torque of the half-couplings (13), we get:

$$\begin{aligned} T_m =& \left\{ R\left[c\left(\delta_0 + h - r + \sqrt{r^2 - [r \cdot \sin\theta_0 - R(\varphi_2 - \varphi_3)]^2}\right)\right] \right. \\ &+ mR\left\{ \frac{[r \cdot \sin\theta_0 - R \cdot (\varphi_2 - \varphi_3)]R \cdot (\ddot{\varphi}_2 - \ddot{\varphi}_3) - (R)^2 \cdot (\dot{\varphi}_2 - \dot{\varphi}_3)^2}{\sqrt{r^2 - [r \cdot \sin\theta_0 - R \cdot (\varphi_2 - \varphi_3)]^2}} \right. \\ &- \left.\left. \frac{[r \cdot \sin\theta_0 - R \cdot (\varphi_2 - \varphi_3)]^2 \cdot (R)^2 \cdot (\dot{\varphi}_2 - \dot{\varphi}_3)^2}{\left(r^2 - [r \cdot \sin\theta_0 - R \cdot (\varphi_2 - \varphi_3)]^2\right) \cdot \sqrt{r^2 - [r \cdot \sin\theta_0 - R \cdot (\varphi_2 - \varphi_3)]^2}} \right\}\right\} \\ &\times \frac{r \cdot \sin\theta_0 - R \cdot (\varphi_2 - \varphi_3)}{\sqrt{r^2 - [r \cdot \sin\theta_0 - R \cdot (\varphi_2 - \varphi_3)]^2}}. \end{aligned} \tag{15}$$

The torque depends on the structural, mass, and elastic characteristics of the coupling, as well as the difference in the angles of mutual rotation of the half-couplings and their derivatives. Then, the initial conditions for the system motion were considered.

The most significant dynamic loads will occur in the case of a sudden jamming of the system, which is associated with the impact mechanism of the drive operation. In the case of a soft jamming, the process occurs more slowly, and dynamic loads will accordingly be less significant.

In the worst-case scenario, i.e., during a sudden jamming, the system transforms into a two-mass one, as the rotational mass J_3 comes to a halt, and the angle $\varphi_4 = 0$.

The deformation of the elastic links at the initial moment in time is determined by the torque of resistance T_r preceding the jamming time:

$$\begin{aligned} &c_1(\varphi_{10} - \varphi_{20}) = T_r; \ c_1(\varphi_{30} - \varphi_{40}) = T_r; \\ &\varphi_{20} = \varphi_{30}; \ \varphi_{40} = 0. \end{aligned} \tag{16}$$

The third equation shows that the coupling rotates as a single unit.

Here, the subscript "0" denotes the initial value when the cylindrical finger was fully in the recess. The rotational velocities of all masses at the initial moment were the same and equal to ω_0.

$$\dot{\varphi}_{10} = \omega_0; \dot{\varphi}_{20} = \omega_0; \dot{\varphi}_{30} = \omega_0; \dot{\varphi}_{40} = 0. \tag{17}$$

So, Eqs. (1) and (15), along with the characteristics of the drive (2) or (3) and resistance (4) or (5), initial conditions (16) and (17), constitute the mathematical model of the motion of the screw conveyor's safety clutch system. The system is too complex for analytical solution, so numerical integration methods based on the well-known Runge-Kutta technique were applied.

To transform the system of second-order differential equations into a system of first-order differential equations, the following variable substitution was performed:

$$u = \dot{\varphi}_1; v = \dot{\varphi}_2; w = \dot{\varphi}_3. \tag{18}$$

Accordingly, after differentiating (18), we obtain:

$$\dot{u} = \ddot{\varphi}_1; \dot{v} = \ddot{\varphi}_2; \dot{w} = \ddot{\varphi}_3. \tag{19}$$

The expression for the interaction torque T_m contains the values of second derivatives of angles, which complicates the application of the numerical method. After substitutions and appropriate transformations, we obtain the final formula for determining the interaction torque T_m in the form:

$$T_m = \frac{m \cdot R^2\left(r^2 - k\right)\left[\frac{c_1(\varphi_2 - \varphi_3)}{J_{21}} + \frac{c_2(\varphi_2 - \varphi_3)}{J_{22}}\right] + m \cdot R^3 \cdot (\dot{\varphi}_2 - \dot{\varphi}_3)^2 \cdot \sqrt{r^2 - k}}{k + m \cdot R^2 \cdot \left(r^2 - k\right) \cdot \frac{J_{21} + J_{22}}{J_{21} \cdot J_{22}}} + \frac{mR^3 \cdot (\dot{\varphi}_2 - \dot{\varphi}_3) \cdot \sqrt{\left(r^2 - k\right)^3} + c \cdot R\left(\lambda_0 + h - r + \sqrt{k}\right) \cdot \sqrt{k^3 \cdot \left(r^2 - k\right)}}{k\left[k + m \cdot R^2 \cdot \left(r^2 - k\right) \cdot \frac{J_{21} + J_{22}}{J_{21} \cdot J_{22}}\right]}, \tag{20}$$

where $k = r^2 - [r \cdot \sin\theta_0 - R(\varphi_2 - \varphi_3)]^2$.

Using the substitutions and replacements mentioned above, we obtain the final system of first-order differential equations:

$$\begin{aligned} \dot{u} &= \frac{T_d - c_1(\varphi_1 - \varphi_2)}{J_1}; \dot{v} = \frac{c_1(\varphi_1 - \varphi_2) - T_m}{J_{21}}; \\ \dot{w} &= \frac{T_m - c_2(\varphi_3 - \varphi_4)}{J_{22}}; \dot{\varphi}_1 = u; \dot{\varphi}_2 = v; \dot{\varphi}_3 = w. \end{aligned} \tag{21}$$

Initial conditions, considering the transformations and substitutions, will take the form:

$$\varphi_{10} = \frac{T_r}{c_1} + \frac{T_r}{c_2}; \varphi_{20} = \frac{T_r}{c_2}; \varphi_{30} = \frac{T_r}{c_2}; u_0 = \omega_0; v_0 = \omega_0; w_0 = \omega_0. \tag{22}$$

3 Results and Discussions

Theoretical calculations of the obtained analytical dependencies and their graphical interpretation were performed on a PC using the Delphi programming environment.

To implement the numerical solution of the mathematical model described by Eqs. (20)–(22), a program was developed in the Pascal language within the Delphi visual programming environment. This program allows for the construction of graphical dependencies and the investigation of the influence of various system parameters on dynamic loads in the drive links of the screw conveyor.

The graphic dependencies illustrating the change in the torque magnitude during the slipping of the half-coupling are presented in Fig. 4.

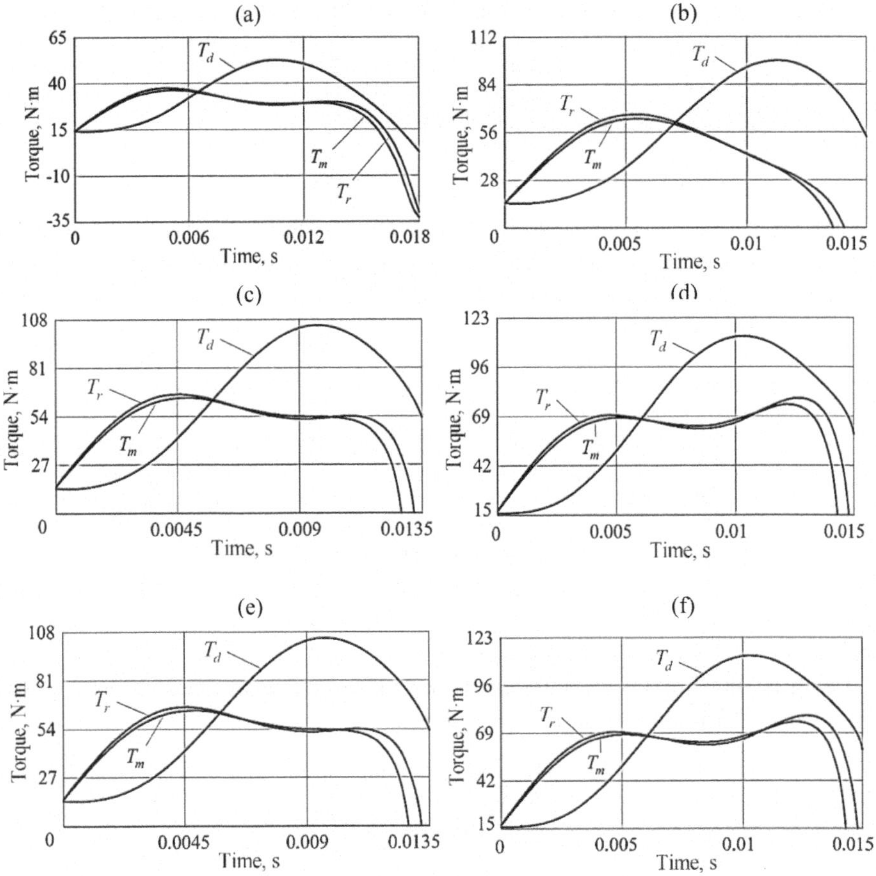

Fig. 4. Time-dependent plots of the torque moments T_d, T_r and T_m during the slip of the half-couplings: (a) – $m = 10$ kg, $n = 50$ rpm; (b) – $m = 10$ kg, $n = 100$ rpm; (c) – $m = 10$ kg, $n = 150$ rpm; (d) – $m = 10$ kg, $n = 200$ rpm; (e) – $m = 15$ kg, $n = 100$ rpm; (f) – $m = 20$ kg, $n = 100$ rpm.

The drive torque is equal to $T_d = c_1(\varphi_2 - \varphi_1)$, and the moment arising in the jammed driven half-coupling and screw working element is equal to $T_r = c_2\varphi_3$. In the numerical calculations, the following parameter values were adopted: $J_1 = 40$ kg·m^2; $J_{21} = 0.0157$ kg·m^2; $J_{22} = 0.00925$ kg·m^2; $c_1 = c_2 = 1600$ N·m·rad^{-1}; $c = 10000$ N·m^{-1}; $h = 4.6$ mm; $r = 12$ mm; $R = 60$ mm; $\delta_0 = 20$ mm. In this case, the rotation frequency n varied within the range of 50 to 200 revolutions per minute, and the mass of the driven links m ranged from 10 kg to 20 kg. Based on the conducted dynamic analysis of the safety coupling operation in the screw conveyor, it has been determined that an increase in the rotation frequency of the working mechanism leads to an increase in the torque moments of the system.

The graphs show that when the moment T_r arising in the jammed screw working element reaches the permissible drive moment T_d, i.e., the moment of interaction of the half-couplings T_m (the moment of the safety clutch operation provided by its design parameters), the smooth axial deflection of the screw occurs, returning it to its previous position. In this process, the cause of the working element jamming is eliminated, reducing the torque moments of the system, and the initial state is restored.

From the analysis of the graphs (Fig. 4, a-d), it has been determined that with a change in rotation frequency from 50 rpm to 200 rpm, the torque T_d increases by 64.8%, T_r rises by 65.2%, and T_m increases by 63.8%. Also, from the graphs (Fig. 4b, d, e), it has been established that an increase in mass m of moving links leads to an increase in loads on the drive. When the mass increases from 10 kg to 20 kg, the torque moment T_d increases by 15.8%, T_r rises by 12.5%, and T_m increases by 17.3%.

The obtained mathematical model allows determining the influence of key parameters on the dynamics of the safety coupling operation in the screw conveyor during overloading of the working mechanism. Thus, by applying the provided mathematical model, it is possible to select rational design and kinematic parameters for the developed safety coupling of the screw conveyor with respect to its engagement moment.

Subsequently, it is advisable to conduct research to determine the magnitude of contact stresses in the engagement elements during the operation of the safety coupling. This will enable the selection of material for the half-coupling of the safety coupling in the screw conveyor, ensuring strength conditions under contact stresses. The level of these stresses is a crucial factor influencing the wear intensity of contact surfaces.

4 Conclusions

Based on the dynamic model of the screw conveyor with the safety coupling, differential equations of motion for the system elements have been formulated.

Theoretical investigations have resulted in the construction of graphical dependencies illustrating changes in the torque magnitude during the slipping of the half-coupling of the safety coupling. These graphs serve to explore the influence of various system parameters on the dynamic loads in the drive links of the screw conveyor. From the analysis of the graphs, it has been determined that an increase in the rotation frequency n leads to an increase in the torque of the system.

With a change in rotation frequency from 50 rpm to 200 rpm, the torque T_d increases by 64.8%, T_r rises by 65.2%, and T_m increases by 63.8%. Also, from the graphs (Fig. 4b, d, e), it has been established that an increase in mass m of moving links leads to an increase in loads on the drive. When the mass increases from 10 kg to 20 kg, the torque moment T_d increases by 15.8%, T_r rises by 12.5%, and T_m increases by 17.3%.

Based on the conducted dynamic analysis of the safety coupling operation in the screw conveyor, it has been determined that an increase in the rotation frequency of the working mechanism leads to an increase in the torque within the system.

References

1. Bulgakov, V., Pascuzzi, S., Nadykto, V., Ivanovs, S., Adamchuk, V.: Experimental study of the implement-and-tractor aggregate used for laying tracks of permanent traffic lanes inside controlled traffic farming systems. Soil Tillage Res. **208**, 104895 (2021)
2. Bulgakov, V., Pascuzzi, S., Ivanovs, S., Nadykto, V., Nowak, J.: Kinematic discrepancy between driving wheels evaluated for a modular traction device. Biosys. Eng. **196**, 88–96 (2020)
3. Pascuzzi, S., Anifantis, A.S., Santoro, F.: The concept of a compact profile agricultural tractor suitable for use on specialised tree crops. Agriculture **10**, 123 (2020)
4. Bulgakov, V., Pascuzzi, S., Beloev, H., Ivanovs, S.: Theoretical investigations of the headland turning agility of a trailed asymmetric implement-and-tractor aggregate. Agriculture **9**, 224 (2019)
5. Bulgakov, V., Pascuzzi, S., Anifantis, A.S., Santoro, F.: Oscillations analysis of front-mounted beet topper machine for biomass harvesting. Energies **12**, 2774 (2019)
6. Bulgakov, V., Pascuzzi, S., Adamchuk, V., Kuvachov, V., Nozdrovicky, L.: Theoretical study of transverse offsets of wide span tractor working implements and their influence on damage to row crops. Agriculture **9**, 144 (2019)
7. Pascuzzi, S., Santoro, F., Manetto, G., Cerruto, E.: Study of the correlation between foliar and patternator deposits in a "Tendone" vineyard. Agric. Eng. Int. CIGR J. **20**(3), 97–107 (2018)
8. Pascuzzi, S., Blanco, I., Anifantis, A.S., Scarascia Mugnozza, G.: Electrolyzer performance analysis of an integrated hydrogen power system for greenhouse heating. A Case Study. Sustain. **8**, 629 (2016)
9. Pascuzzi, S.: Outcomes on the spray profiles produced by the feasible adjustments of commonly used sprayers in "Tendone" vineyards of Apulia (Southern Italy). Sustainability **8**(12), 1307 (2016)
10. Pascuzzi, S., Cerruto, E.: An innovative pneumatic electrostatic sprayer useful for tendone vineyards. J. Agric. Eng. **458**, 123–127 (2015)
11. Hevko, R.B., Klendiy, O.M.: The investigation of the process of a screw conveyer safety device actuation. INMATEH Agric. Eng. **42**(1), 55–60 (2014)
12. Pascuzzi, S., et al.: Theoretical study of the motion of a cut sugar beet tops particle along the inner surface of the conveying and unloading system of a topping machine. Agric. Eng. **6**, 409–422 (2024)
13. Hevko, R.B., Rozum, R.I., Klendiy, O.M.: Development of design and investigation of operation processes of loading pipes of screw conveyors. INMATEH Agric. Eng. **50**(3), 89–96 (2016)
14. Pascuzzi, S., Łyp-Wrońska, K., Gdowska, K., Paciolla, F.: Sustainability evaluation of hybrid agriculture-tractor powertrains. Sustainability **16**(3), 1184 (2024)
15. Hevko, R.B., Klendiy, M.B., Klendiy, O.M.: Investigation of a transfer branch of a flexible screw conveyer. INMATEH Agric. Eng. **48**(1), 29–34 (2016)
16. Kováčová, M., Matus, M., Krizan, P., Beniak, J.: Design theory for the pressing chamber in the solid biofuel production process. Acta Polytechnica **54**(1), 28–34 (2014)

17. Güneş, E.C., Çelik, I.: Use of disc springs in a pellet fuel machine. Czech Techn. Univ. Prague Acta Polyt.s **57**(2), 89–96 (2017)
18. Bulgakov, V., et al.: Justification of parameters for novel rotary potato harvesting machine. Agron. Res. **19**(2), 984–1007 (2021)
19. Trokhaniak, O.M., Hevko, R.B., Lyashuk, O.L., Dovbush, TA, Pohrishchuk, B.V., Dobizha, N.V.: Research of the of bulk material movement process in the inactive zone between screw sections. INMATEH Agric. Eng. **60**(1), 261–268 (2020)
20. Bulgakov, V., Holovach, I., Ruzhylo, Z., Fedosiy, I., Ihnatiev, Y., Olt, J.: Theory of oscillations performed by tools in spiral potato separator. Agron. Res. **18**(1), 38–52 (2020)
21. Adamchuk, V., Bulgakov, V., Ivanovs, S., Holovach, I., Ihnatiev, Y.: Theoretical study of pneumatic separation of grain mixtures in vortex flow. Eng. Rural Devel. **20**, 657–664 (2021)
22. Sokil, B., Lyashuk, O., Sokil, M., Popovich, P., Vovk, Y., Perenchuk, O.: Dynamic effect of cushion part of wheeled vehicles on their steerability. Int. J. Autom. Mech. Eng. **15**(1), 4880–4892 (2018)
23. Aulin, V.V., Pankov, A.O., Zamota, T.M., Lyashuk, O.L., Hrynkiv, A.V., Tykhyi, A.A., Kuzyk, A.V.: Development of mechatronic module for the seeding control system. Inmateh Agric. Eng. **59**(3), 1–8 (2019)
24. Lyashuk, O., Vovk, Y., Sokil, B., Klendii, V., Ivasechko, R., Dovbush, T.: Mathematical model of a dynamic process of transporting a bulk material by means of a tube scraping. Agric. Eng. Int. CIGR J. **21**(1), 74–81 (2019)
25. Gevko, R.B., Klendiy, O.M.: The investigation of the process of a screw conveyer safety device actuation INMATEH. Agric. Eng. **42**(1), 55–60 (2014)
26. Annunziato, A., Godfrey, M., Anelli, F., Dulieu-Barton, J., Holmes, C., Prudenzano, F.: Flexible photonic sensors: investigation of an approach based on ratiometric power in few-mode waveguides for bending measurement. IEEE Access (2024)
27. Merritt, A.S., Mair, R.J.: No access mechanics of tunnelling machine screw conveyors: model tests. Géotechnique **56**(9), 605–615 (2015)
28. Anelli, F., Annunziato, A., Erario, A., Holmes, C., Ciminelli, C., Prudenzano, F.: Design of microstructured flat optical fiber for multiaxial strain monitoring in composite materials. J. Lightwave Technol. **40**(17), 5986–5994 (2022)
29. Paul, Ph.J., Cleary, W.: Screw conveyor performance: comparison of discrete element modelling with laboratory experiments. Progr. Computat. Fluid Dyn. Int. J. **10**, 5–6 (2010)

Investigation of the Parameters of a Conveyor for Transportation of Bulk Agricultural Materials with a Bladed Working Body

Volodymyr Bulgakov[1], Simone Pascuzzi[2(✉)], Oleksandra Trokhaniak[1], Adolfs Rucins[3], Janusz Nowak[4], Mykola Klendii[5], Alessia Farella[2], and Tommaso Quartarella[2]

[1] Department of Mechanics, Faculty of Construction and Design, National University of Life and Environmental Sciences of Ukraine, 15 Heroiv Oborony Str., Kyiv 03041, Ukraine

[2] Department of Soil, Plant and Food Science, University of Bari Aldo Moro, Via Amendola 165/A, 70126 Bari, Italy

simone.pascuzzi@uniba.it

[3] Institute of Engineering and Energetics, Faculty of Engineering and Information Technologies, Latvia University of Life Sciences and Technologies, Cakstes Blvd. 5, Jelgava 3001, Latvia

[4] Department of Operation of Machines and Management of Production Processes, University of Life Sciences in Lublin, 13, Akademicka Str., 20-950 Lublin, Poland

[5] Separated Subdivision of National University of Life and Environmental Sciences of Ukraine, Berezhany Agrotechnical Institute, Berezhany, Ukraine

Abstract. The article presents a new design of a conveyor with a bladed working body, as well as a theoretical and experimental investigation of its operation, in particular its performance when transporting bulk agricultural materials. The article offers a new design of a conveyor with a bladed working body. An analytical relationship has been obtained to determine the productivity of a conveyor with a bladed working body. Based on the results of experimental studies, the corresponding regression equation and response surface were constructed to establish the impact of the controlled factors upon the performance value. Analysis of the above-mentioned regression equations shows that the main factor influencing the increase in productivity Q is the value of the factor x_3 (n) of the rotation frequency of the bladed working body. Increasing the conveyor inclination angle x_2 (β) leads to a decrease in productivity. An increase in the casing diameter x_1 (D) leads to an increase in conveyor productivity.

Keywords: transportation · grain material · conveyor · blade · rotation frequency · productivity

1 Introduction

Technological processes of production of agricultural products involve many labor-intensive loading, unloading and transport operations [1–10]. Screw conveyors are widely used during the transportation of various bulk agricultural materials, such as

E. Lorencowicz et al. (Eds.): FMPMSA 2024, LNCE 609, pp. 98–106, 2024.
https://doi.org/10.1007/978-3-031-70955-5_11

grains, seeds, granulated mineral fertilizers, etc. [11–13]. However, the existing designs of screw conveyors do not fully satisfy all the operational requirements. Their main disadvantages that arise when transporting bulk material by the screw conveyors are increased energy costs and limited functionality, which are associated exclusively with the process of moving the material [14–16]. The existing solid flexible spirals, when transporting materials along curved routes, are quickly destroyed because of significant cyclic loads, and the use of combined working bodies is characterized by high material consumption and the complexity of manufacturing screw working bodies of conveyors, especially with their significant overall dimensions. Also, when using screw conveyors, damage to the bulk material occurs.

From the analysis of research on the screw conveyors, we can conclude that the authors paid more attention to the derivation of analytical dependencies to determine the design and kinematic parameters of the screw working bodies and, to a lesser extent, to their energy-power parameters [17–20]. The known experimental studies are mainly aimed at determining such parameters of working bodies that would ensure their high functional and performance indicators with a minimum permissible mass of the working body, which will reduce the energy costs for the transportation process [21–23].

Therefore, the task of developing new designs of the transporting working bodies, in particular screw conveyors, remains relevant, ensuring an increase in their performance indicators while reducing material consumption and simplifying the design. In some publications, as an alternative to the screw body of a conventional design, a bladed organ in the form of flat blades, forming a helical surface, is considered [24–26].

The purpose of this investigation is to improve the performance of the conveyor with a bladed working body, when transporting bulk agricultural materials, by selecting their rational parameters.

2 Materials and Methods

To solve this problem, an experimental stand has been developed to study a conveyor with a bladed working body, the general view of which is shown in Fig. 1.

The use of a bladed working body when transporting bulk materials will reduce the material consumption and simplify the design of the screw because it is made in the form of a cylindrical shaft, on the surface of which inclined flat blades are sectionally fixed, limited by alternating half-ellipses, and they alternate by the sign. It is advisable to produce such blades by stamping sheet material with subsequent welding to a cylindrical shaft.

When conducting experimental research, wheat was used as a bulk material. To start the engine and adjust its rotation speed, frequency converter (Fig. 1 and 2) (Altivar 71) with Power Suite v.2.5.0 software was connected to the network and the computer.

When the required rotation frequency is set using a computer through the Altivar 71 system, a command is transmitted to the electric motor (Fig. 1–3), and it begins to rotate the shaft of the blade working body (Fig. 1–3) with the set parameters.

Previously, the bunker was completely loaded with the bulk material. Next, at certain speeds of the working parts and the angle of inclination of the conveyor body to the horizon, the material was unloaded into the measuring containers.

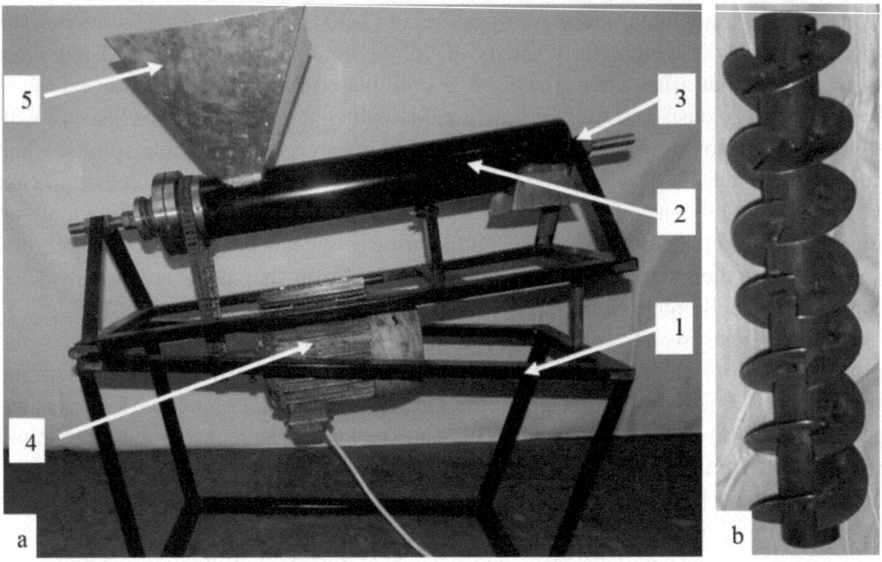

Fig. 1. General view of the conveyor (a) and a bladed working body (b): 1 – frame; 2 – cylindrical casing; 3 – shaft of the bladed working body; 4 – electric motor; 5 – bunker

To determine the kinematic parameters of a conveyor with a bladed working body, a theoretical investigation of transportation of a bulk material along the surfaces of the screw was carried out.

However, studying the entire flow for conveyors is a rather difficult task, since the interaction of particles with each other and with the working surfaces will be stochastic.

In this case it is advisable to describe the flow parameters using integral indicators.

In particular, the axial velocity component of the material flow, as it leaves the blade, is defined as the arithmetic mean of the velocity of each of the n-particles:

$$V_{Zn} = \frac{1}{n}\sum\nolimits_{i=1}^{n} R_i(\omega - \omega_i) \cdot \tan\gamma_0 \cdot \cos\varphi_i. \tag{1}$$

where R_i and ϕ_i are radial and angular parameters, respectively of the i-th particle in the flow of the material relative to the blade; ω_i is the angular velocity of an arbitrary particle; ω is the angular velocity of an arbitrary particle; γ_0 is the angle of placement of the blades on the working body.

Let us write dependence (1) in the form:

$$V_{Zn} = R \cdot \omega \cdot \tan\gamma_0 \cdot \sum\nolimits_{i=1}^{n} \frac{R_i}{R}\left(1 - \frac{\omega_i}{\omega}\right) \cdot \cos\varphi_i \tag{2}$$

According to the regularities of changes in the angular velocity of a particle, in the convergence zone the angular velocity of an arbitrary particle ω_i will depend little on its placement, therefore the speed coefficient of the conveyor $k_\omega = \frac{\omega_i}{\omega}$ will be characterized by a stable value, which depends only on the design of the working body and its angular velocity ω. For high-speed conveyors that use bladed working bodies with blade angles γ_0, close to 30º, $k_\omega = 0.13 \div 0.17$.

Since there is a gap between the outer radius of the blade R and the casing (the outer radius of the flow R_f is greater than R), it is advisable to take the average radius of the flow by analogy with the screw transportation, close to the radius R of the blade.

Then the coefficient of reduction of the radius of the flow of the bulk material $k_R = \frac{R_f}{R} = \frac{1}{n}\sum_{i=1}^{n}\frac{R_i}{R} = 0.95 \div 1$ is also a constant value, and, therefore, in the first approximate of dependence (2) will have the form:

$$V_{Zf} = k_R \cdot (1 - k_\omega) \cdot R \cdot \omega \cdot \tan\gamma_0 \cdot \sum_{i=1}^{n} \cos\varphi_i. \tag{3}$$

According to experimental observations, the value of the angle at which the material is leaving the blade is $\varphi_i = 45° \div 70°$, and the average value $k_\varphi = \sum_{i=1}^{n}\cos\varphi_i$ of the coefficient of leaving the blade by the material will be within the range $k_\varphi = 0.55 \div 0.65$.

Accordingly, the axial component of the flow velocity can be determined:

$$V_{Zf} = k_\varphi \cdot k_R \cdot (1 - k_\omega) \cdot R \cdot \omega \cdot \tan\gamma_0 = K_f \cdot R \cdot \omega \cdot \tan\gamma_0 \tag{4}$$

where $K_f = k_\varphi \cdot k_R \cdot (1 - k_\omega)$ – the parameter of the flow movement.

The angular component of the flow velocity is equal to:

$$V_\omega = R_f \cdot \omega_f = k_R \cdot k_\omega \cdot R \cdot \omega, \tag{5}$$

where ω_f – angular velocity of the flow.

The total flow rate according to the module will be equal to:

$$V_f = \sqrt{V_{Zf}^2 + R_f^2 \cdot \omega_f^2} = R \cdot \omega \cdot \sqrt{k_\varphi^2 \cdot \tan^2\gamma_0 + k_R^2 \cdot k_\omega^2}. \tag{6}$$

We determine the productivity of conveyor [10]:

$$Q = \pi \cdot \rho \cdot \varphi_q \cdot V_Z \cdot \left(R^2 - \mathrm{r}^2\right) = \pi \cdot \rho \cdot \varphi_q \cdot k_\varphi \cdot k_R \cdot (1 - k_\omega) \cdot R \cdot \omega \cdot \tan\gamma_0 \cdot \left(R^2 - \mathrm{r}^2\right), \tag{7}$$

where:φ_q – the conveyor fill factor;

ρ – the specific density of the material;

V_Z – axial velocity of the material flow;

R and r – radii of the working body and the shaft.

To determine the influence of the bulk material transportation parameters and the conveyor design parameters, i.e. of the blade working body *n*, rpm, upon productivity *Q*, a full-factorial experiment PFE - 3^3 was conducted.

Since during the experiments the variable independent factors are heterogeneous and have different units of measurement, and the numbers expressing the value of these factors are of different orders, they were brought to a unified calculation system by moving from actual values to coded ones, presented in Table 1.

Investigation of productivity *Q* on the drive shaft made it possible to determine the dependence of power on many factors characterizing the process, namely:

– of the internal diameter of casing *D*, m, which we coded with index x_1;

– of the conveyor inclination angle β, deg, coded by index x_2;

Table 1. Results of coding the factors and the levels of their variation in their performance research

Factors	Designation		Variation interval	Variation levels natur./coded		
	Code	Natur				
Innternal diameter of casing *D*, m	$X1$	x_1	0.02	0.06/–1	0.08/0	0.1/ + 1
Conveyor inclination angle β, deg	X_2	x_2	15	15/–1	30/0	45/ + 1
Rotation frequency of the blade workin body *n*, rpm	X_3	x_3	200	100/–1	300/0	500/ + 1

– of the rotation frequency of the bladed working body *n*, rpm, which was coded with by index x_3.

A study to determine the productivity Q of a conveyor with a bladed working body was carried out when transporting wheat with a volumetric mass of 760 kg·m^{-3} and with a humidity of W = 12 ÷ 15%, which made it possible to construct analytical regression equations.

Processing of the obtained data from the experimental array was performed according to well-known methods and methods of statistical processing, using well-known methods of the correlation and regression analysis to ultimately obtain empirical regression equations. To obtain a regression model for the optimization parameter, an appropriate full-factorial experimental design was selected, which was implemented in the following sequence.

The response function (optimization parameter), that is, productivity $Q = f(D, \beta, n)$, determined experimentally, is presented in the form of a mathematical model of a complete quadratic polynomial:

$$Q = b_0 + b_1x_1 + b_2x_2 + b_3x_3 + b_{12}x_1x_2 + b_{13}x_1x_3 + b_{23}x_2x_3 + b_{11}x_1^2 + b_{22}x_2^2 + b_{33}x_3^2 \quad (8)$$

where:$b_0, b_1, b_2, b_3, b_{12}, b_{23}, b_{13}, b_{11}, b_{22}, b_{33}$ – coefficients of the corresponding values x_i; x_1, x_2, x_3 – the corresponding factors to be coded.

The values of coefficients of the regression equations are presented in Table. 2.

Table 2. The significance of the coefficients of the regression equations

b_0	b_1	b_2	b_3	b_{12}	b_{13}	b_{23}	b_{11}	b_{22}	b_{33}
3.955	0.322	–0.232	0.803	–0.079	0.030	0.094	0	–0.131	–0.195

The regression equation for the determination of productivity when transporting wheat in an experimental installation has the form:

$$Q = 0.793 + 21.75D + 0.031\beta + 0.0008n - 0.265D\beta + 0.00003\beta n$$

$$+0.0075Dn - 0.000006n^2 - 0.0005\beta^2. \quad (9)$$

The resulting regression Eq. (9) may be used to determine productivity Q during transportation by a conveyor with a bladed working body, depending on the diameter of casing D, the angle of inclination β of the conveyor and the rotation frequency of the bladed working body n when transporting wheat within the following limits of change in the input factors: $0.06 \leq D \leq 0.1$ (m); $15 \leq \beta \leq 45$ (deg); $200 \leq n \leq 600$ (rpm).

3 Results and Discussions

In order to theoretically determine the productivity of the conveyor, the following initial data were selected: the conveyor filling factor $\varphi_q = 0.3 \div 0.7$, the inclination angle of the blade $\gamma_0 = 30°$, the specific mass of wheat $\rho = 760$ kg m^{-3}, the volumetric mass of barley $\rho = 610$ kg m^{-3}, the value of the coefficient of the material leaving the blades $k_\varphi = 0.55 \div 0.65$, coefficient of reduction of the radius of the flow of the bulk material $k_R = 0.9 \div 1.0$, the speed coefficient of the conveyor $k_\omega = 0.15$, outer radius of the blade working body $R = 0.049$ m, the radius of the shaft on which the blades are mounted $r = 0.023$ m, rotation speed of the blade working body n = $100 \div 500$ rpm.

The results of analytical determination of the productivity of the conveyor with the proposed blade working body are shown in Fig. 2.

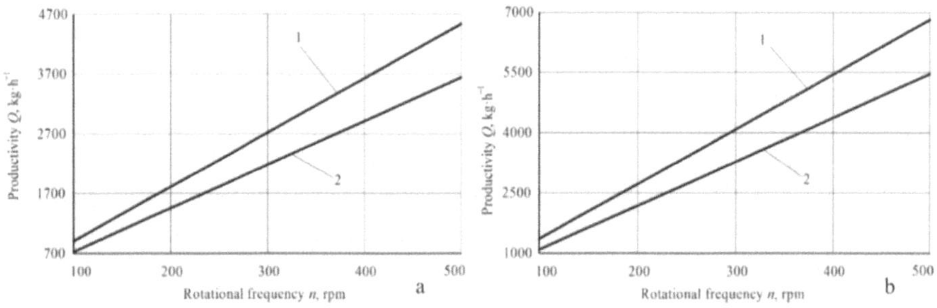

Fig. 2. Results of theoretical determination of the productivity of a conveyor with a bladed working body during transportation of 1 – wheat and 2 – barley, at the conveyor filling factor $\varphi_q = 0.4$ (a) and $\varphi_q = 0.6$ (b)

The analysis of dependencies (Fig. 3) indicates that the productivity of a conveyor with a bladed working body, when transporting wheat and barley within the limits of the rotation speed of the bladed working body $n = 100 \div 500$ rpm, grows linearly, in addition to this, productivity when transporting wheat and at the coefficient of the conveyor filling factor $\varphi_q = 0.4$, varies within the range of $Q = 908 \div 4540$ kg·h^{-1}, but at the conveyor filling factor $\varphi_q = 0.6$, productivity varies within the range of $Q = 1362 \div 6810$ kg·h^{-1}. Productivity, when transporting barley, at the conveyor filling factor $\varphi_q = 0.4$, varies within $Q = 729 \div 3645$ kg·h^{-1}, and in conveyor filling factor $\varphi_q = 0.6$, varies within $Q = 1094 \div 5468$ kg·h^{-1}.

Depending on the conveyor filling factor $\varphi_q = 0.4 \div 0.6$, productivity increases for both barley and wheat transportation by approximately 1.5 times.

Based on the results and laboratory experimental productivity research of the conveyor bladed working body, using the application program "STATISTICA 10", a graphical reproduction of intermediate general regression models was built in the form of quadratic response surfaces of productivity Q as a function of two variable factors $x_{i(1,2)}$ with a constant unchanged level of the corresponding third factor $x_{i(3)} = \text{const}$ (Fig. 3).

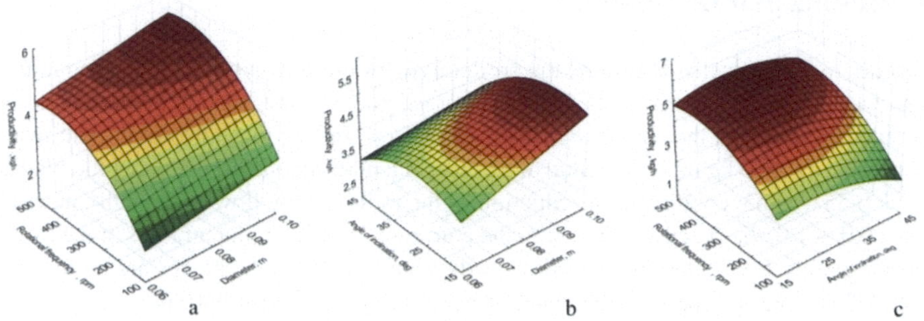

a b c

Fig. 3. Performance response surface of a conveyor with a bladed working body when transporting wheat, depending on: a) $Q = f\ (D;\ n)$; b) $Q = f\ (D;\ \beta)$; c) $Q = f\ (\beta;\ n)$

Analysis of the above regression equations shows that the main factor influencing the increase in productivity Q is the value of the factor x_3 (n) of the rotation frequency of the blade working body. Increasing the conveyor inclination angle x_2 (β) leads to a decrease in productivity. An increase in the casing diameter.

x_1 (D) leads to an increase in the conveyor productivity.

When diameter D changes within the range of $0.06 \div 0.10$ m, productivity Q increases by 14.7%, the conveyor inclination angles β from 15° to 45°, productivity Q decreases by 8.3%, but, when the rotation speed of the bladed working body is within the range of $100 \div 500$ rpm, productivity Q increases by 83%.

A comparison of the results of theoretical and experimental studies showed that at $D = 1.0$ m, $\beta = 45$ deg of different values of n within $100 \div 500$ rpm, the discrepancies in the obtained values of performance Q are equal to $4.6 \div 16.5\%$.

4 Conclusions

Based on the analysis of literary sources and a patent search, a new design of a screw conveyor with a bladed working body was developed and manufactured, and an experimental installation was also manufactured to conduct research of the operation of the conveyor, in particular its performance when transporting bulk agricultural materials.

An analytical relationship has been obtained to determine the productivity of a conveyor with a bladed working body. Analysis of the obtained dependence indicates that the productivity of a conveyor with a bladed working body, when transporting wheat and barley within the rotation speed of the bladed working body $n = 100 \div 500$ rpm, grows linearly, while the productivity, when transporting wheat and barley, with a conveyor filling factor $\varphi_q = 0.4$, varies within the range $Q = 908 \div 4540$ kg·h^{-1}, and with a conveyor filling factor $\varphi_q = 0.6$, productivity varies within the range of $Q = 1362 \div 6810$ kg·h^{-1}. Productivity, when transporting barley, at the conveyor filling factor φq = 0.4, varies within $Q = 729 \div 3645$ kg·h^{-1}, and at the conveyor filling factor φq = 0.6, it varies within $Q = 1094 \div 5468$ kg·h^{-1}. Depending on the conveyor filling factor φq = 0.4 ÷ 0.6, the productivity increases both for the barley and the wheat transportation by approximately 1.5 times.

Based on the results of experimental investigations, the corresponding regression equation and response surface were constructed to establish the impact of the controlled factors upon the performance value. Analysis of the above regression equations shows that the main factor, influencing the increase in productivity Q, is the value of the x_3 (n) frequency factor and the rotation of the bladed working body. Increasing the conveyor inclination angle x_2 (β) leads to a decrease in productivity. An increase in the casing diameter x_1 (D) leads to an increase in conveyor productivity.

When diameter D changes within the limits 0.06 ÷ 0.1 m, productivity Q increases by 14.7%, the conveyor inclination angles β from 15° to 45°, productivity Q decreases by 8.3%, and. When the rotation speed of the blade working body is wihin the range of 100 ÷ 500 rpm, productivity Q increases by 83%.

References

1. Bulgakov, V., Pascuzzi, S., Nadykto, V., Ivanovs, S., Adamchuk, V.: Experimental study of the implement-and-tractor aggregate used for laying tracks of permanent traffic lanes inside controlled traffic farming systems. Soil Tillage Res. **208**, 104895 (2021)
2. Bulgakov, V., Pascuzzi, S., Ivanovs, S., Nadykto, V., Nowak, J.: Kinematic discrepancy between driving wheels evaluated for a modular traction device. Biosys. Eng. **196**, 88–96 (2020)
3. Paciolla, F., Pace, N., Barile, G., Patimisco, P., Valenza, F., Pascuzzi, S.: Autonomous navigation simulation of an agricultural robot during soil fertilization in open fields. In: 2023 IEEE International Workshop on Metrology for Agriculture and Forestry (MetroAgriFor), pp 640–645
4. Pascuzzi, S., Anifantis, A.S., Santoro, F.: The concept of a compact profile agricultural tractor suitable for use on specialised tree crops. Agriculture **10**, 123 (2020)
5. Bulgakov, V., Pascuzzi, S., Beloev, H., Ivanovs, S.: Theoretical investigations of the headland turning agility of a trailed asymmetric implement-and-tractor aggregate. Agriculture **9**, 224 (2019)
6. Pascuzzi, S., Łyp-Wrońska, K., Gdowska, K., Paciolla, F.: Sustainability evaluation of hybrid agriculture-tractor powertrains. Sustainability **16**(3), 1184 (2024)
7. Quartarella, T., Farella, A., Paciolla, F., Pascuzzi, S. UVC Rays devices mounted on autonomous terrestrial rovers: A brief overview. In Farm Machinery and Processes Management in Sustainable Agriculture XI International Scientific Symposium (2024)
8. Bulgakov, V., Pascuzzi, S., Adamchuk, V., Kuvachov, V., Nozdrovicky, L.: Theoretical study of transverse offsets of wide span tractor working implements and their influence on damage to row crops. Agriculture **9**, 144 (2019)
9. Farella, A., Paciolla, F., Quartarella, T., Pascuzzi, S.: Agricultural Unmanned Ground Vehicle (UGV): A brief overview. In Farm Machinery and Processes Management in Sustainable Agriculture XI International Scientific Symposium (2024)

10. Paciolla, F., Patimisco. P., Farella, A., Quartarella, T., Pascuzzi, S.: Performance evaluation of hybrid-electric architectures for agricultural tractors. In: Farm Machinery and Processes Management in Sustainable Agriculture XII International Scientific Symposium (2024)
11. Pascuzzi, S., Santoro, F., Manetto, G., Cerruto, E.: Study of the correlation between foliar and patternator deposits in a "Tendone" vineyard. Agric. Eng. Int. CIGR J. **20**(3), 97–107 (2018)
12. Pascuzzi, S.: Outcomes on the spray profiles produced by the feasible adjustments of commonly used sprayers in "Tendone" vineyards of apulia (Southern Italy). Sustainability **8**(12), 1307 (2016)
13. Pascuzzi, S., Cerruto, E.: An innovative pneumatic electrostatic sprayer useful for tendone vineyards. J. Agric. Eng. **458**, 123–127 (2015)
14. Pascuzzi, S., et al.: Theoretical study of the motion of a cut sugar beet tops particle along the inner surface of the conveying and unloading system of a top-ping machine. Agric. Eng. **6**, 409–422 (2024)
15. Anelli, F., Annunziato, A., Erario, A., Holmes, C., Ciminelli, C., Prudenzano, F.: Design of microstructured flat optical fiber for multiaxial strain monitoring in composite materials. J. Lightwave Technol. **40**(17), 5986–5994 (2022)
16. Bulgakov, V., Pascuzzi, S., Anifantis, A.S., Santoro, F.: Oscillations analysis of front-mounted beet topper machine for biomass harvesting. Energies **12**, 2774 (2019)
17. Lech, M.: Mass flow rate measurement in vertical pneumatic conveying of solid. Powder Technol. **114**(1–3), 55–58 (2001)
18. Lyashuk, OL., et al.: Mathematical model of bending vibrations of a horizontal feeder-mixer along the flow of grain mixture. INMATEH Agric. Eng. **55**(2), 35–44 (2018)
19. Mondal, D.: Study on filling factor of short length screw conveyor with flood-feeding condition. Mater. Today Proc. **5**, 1286–1291 (2018)
20. Hevko, R.B., Rozum, R.I., Klendiy, O.M.: Development of design and investigation of operation processes of loading pipes of screw conveyors. INMATEH Agric. Eng. **50**(3), 89–96 (2016)
21. Hevko, R.B., Klendiy, M.B., Klendii, O.M.: Investigation of a transfer branch of a flexible screw conveyer. INMATEH Agric. Eng. **48**(1), 29–34 (2016)
22. Buteler, M., Gitto, J.G., Stadler, T.: Enhancing the potential use of microparticulate insecticides through removal of particles from raw grain. J. Stored Prod. Res. **89**, 101707 (2020)
23. Olt, J., et al.: Harrow with screw-type operating tools: optimisation of design and process parameters. Agron. Res. **20**(4), 751–763 (2022)
24. Tkachenko, I., Hevko, R., Gandziuk, M., Synii, S., Trokhaniak, O.: Substantiation of the parameters of a horizontal conveyer-cleaner of root crops. Bull. Transilv. Univ. Brasov Ser. II Fores. Wood Ind. Agric. Food Eng. **14**(63), 213–222 (2021)
25. Bulgakov, V., et al.: Research of a contact stresses in swivel elements of flexible shaft in screw conveyor for transportation of agricultural materials. Estonian Acad. Agric. Soc. **33**(1), 67–73 (2022)
26. Гевко, Б.М., Рогатынский, Р.М. Винтовые подающие механизмы сельскохозяйственных машин. /Screw feeders for agricultural machines/ Львов. Вища школа, (1989) (In Russian)

A System for Precise Diagnosis of Diseases, Pests and Fertilization Needs in Horticultural Production

Michał Cupiał[1,3(✉)], Bogdan Kulig[2,3], Mirosław Maziarka[4], Anna Szeląg-Sikora[1,5], and Aneta Oleksy-Gębczyk[6]

[1] Faculty of Production Engineering and Energetics, University of Agriculture in Krakow, Krakow, Poland
michal.cupial@urk.edu.pl
[2] Faculty of Agriculture and Economics, University of Agriculture in Krakow, Krakow, Poland
[3] Research and Development Center for Agriculture 4.0, Krakow, Poland
[4] Agro Smart Lab, Niegardów, Poland
[5] Cavalry Captain Witold Pilecki State University of Małopolska in Oświęcim, Oświęcim, Poland
[6] University of Applied Sciences in Nowy Sącz, Nowy Sącz, Poland

Abstract. Management of modern horticultural production, in addition to possessing technical facilities, requires specialist knowledge and advanced IT tools to support this activity. Such software should enable production planning (including fertilization and chemical plant protection), ongoing crop supervision, prevention of adverse phenomena (e.g. pests threatening crops) and prediction of possible threats.

Such solution is the FarmSmart system, provided by Agro Smart Lab, and de-veloped as part of the "Precision Horticulture" project in cooperation with the University of Agriculture in Krakow and the Agrosan Company. The system includes many elements that are designed to support primarily horticultural crops. The software can also be used on agricultural farms and fruit farms.

The system, based on meteorological stations located throughout the country, pest traps and pathogen development models, generates alerts on the occurrence of threats to crops in a given area.

Other elements of the system are a field manager and specialized calculators de-signed to calculate the necessary fertilization and greenhouse gas emissions.

The system has been implemented and made available to interested farmers. It is still being developed and enriched with new features.

Keywords: Precision Agriculture · Agriculture 4.0 · Fertilization · Horticultural Production

1 Introduction

The management of modern horticultural production, in addition to having technical facilities, requires specialist knowledge and advanced IT tools to support this activity. Such software should be able to plan production. Planning should include, m.in, fertilization [1] and chemical plant protection [2]), ongoing supervision of crops, prevention

E. Lorencowicz et al. (Eds.): FMPMSA 2024, LNCE 609, pp. 107–115, 2024.
https://doi.org/10.1007/978-3-031-70955-5_12

of adverse phenomena (e.g. pests threatening crops) and anticipation of possible threats. The introduction of precision farming systems on farms requires farmers to have a lot of knowledge, systematically expand their skills and learn from their mistakes [3]. Implementing modern technologies without appropriate knowledge and qualifications does not bring the expected results [4]. Studies have shown that the purchase of new agricultural machinery, despite the benefits of its use, increased fuel consumption on farms, which generated high costs [5].

Progress in the use of modern technologies and the digitization of this field are conducive to the transition to the era of Agriculture 4.0. This fourth revolution is now seen as an enabling solution for agricultural development, ensuring that the future needs of the world's population are met in a sustainable way [6]. Agriculture 4.0 sets four essential requirements: increasing productivity, allocating resources wisely, adapting to climate change and avoiding food waste. The implementation of advanced ICT systems enables the collection and processing of vast amounts of data, such as meteorological information, satellite data, soil conditions, marketing requirements and land use [7, 8]. These solutions help farmers make the right decisions, obtain higher quality crops and, consequently, obtain higher profits [9].

2 Materials and Methods

The FarmSmart system is an advisory system that has been developed for years and provides farmers with advice solutions in the field of pest control that threatens crops. The system contains many elements that are designed to support horticultural crops in particular. The software can also be used on agricultural farms and fruit farms.

The 'Precision Horticulture' project has developed new solutions for the system. The project, led by Agro Smart Lab, was carried out in a consortium with the University of Agriculture in Krakow and Agrosan. The scope of work included, among others, the development of algorithms that are the basis for the fertilizer calculator and the greenhouse gas emissions calculator, the development of a field manager, as well as the performance of field tests, the results of which were used to improve the pest threat prediction system. The obtained results have been implemented and applied in the FarmSmart system. The system is offered in a basic version (free) and extended version (paid). Farmers can use the field manager with fertilizer and greenhouse gas emission calculators free of charge.

3 Result and Discussions

3.1 The Concept of the FarmSmart System

The system, based on meteorological stations located throughout the country, pest traps and pathogen development models, generates alerts about the occurrence of threats to crops in a given area. Thanks to weather stations and camera traps located throughout the country (Fig. 1), the service can be provided in different regions of Poland (or other given country). An extensive network of stations enables precise delivery of information based on data obtained in a short distance from the farm. The density of stations in each region depends on the number of system users in a given area.

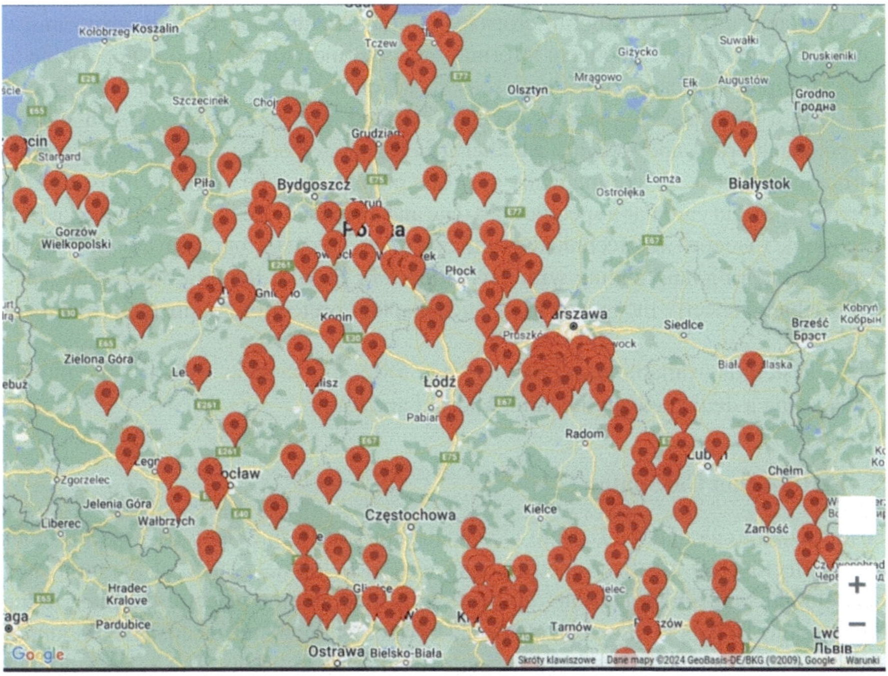

Fig. 1. Location of regions covered by the FarmSmart service (weather stations and camera traps).

The system uses more than 70 disease models to predict threats (data from 2024-04-17). Models based among others on weather data, allow predicting the level of threat to crops caused by individual pathogens. This makes it possible to apply appropriate protective treatments at the right time, which results in their high effectiveness. In addition to the economic effects, it also gives measurable environmental benefits, eliminating unnecessary, usually late, and therefore ineffective, spraying of plants. The number of crops, diseases and pests monitored is presented in Table 1.

Table 1. List of monitored crops, diseases and pests.

Group of plants	Number of plants	Number of diseases	Number of pests
fruit	13	32	5
vegetables	22	33	7
agricultural plants	9*	25	
ornamental plants	1*	12	

Source: https://farmsmart.pl (data from 2024-04-17)
*The number of plants, in the case of agricultural plants, may mean similar species (this applies to e.g. cereals, ornamental plants - many species of such plants)

The system allows you to define crops in individual fields and record all agrotechnical treatments. The fields tab collects information about individual crops. It is possible to define multiple field cards in a single season, which is helpful when you have several crops with a short growing season in the same field. For each crop, you can specify its type, catch crops and indicate whether it is organic or conventional. Based on this information, the system automatically suggests plant protection products registered for this crop. The field card has a specific start date and an expected end date, based on which it is possible to record agrotechnical treatments in the field, starting from cultivation and sowing or planting, and ending with harvesting. For each of the described treatments, it is possible to specify the means used, such as plant protection products, fertilizers or seeds.

The system is available for computers, via a web browser, as well as from mobile devices using a dedicated application (Android and iOS).

3.2 Fertilization Calculator

The fertilization calculator takes into account the nutrient requirements of crops and other factors affecting the fertilizer needs, i.e. species, development stage of the cultivated plant, climatic and soil conditions, cultivation method and the place of the plant in the crop rotation. The application calculates the balance of the plant's needs, the amount of available nutrients and determines the needs for missing nutrients. The balance applies to both organic and mineral fertilizers, for conventional and organic farming systems. The calculated fertilizer requirement determines the amount of the component for N, P_2O_5, K_20, Mg, S, Ca, respectively. The result is presented in kg/ha, and these values can be converted into the amount of a specific fertilizer that should be applied to the crop.

Figure 2 shows a simplified scheme of fertilizer rate calculations for the selected crop.

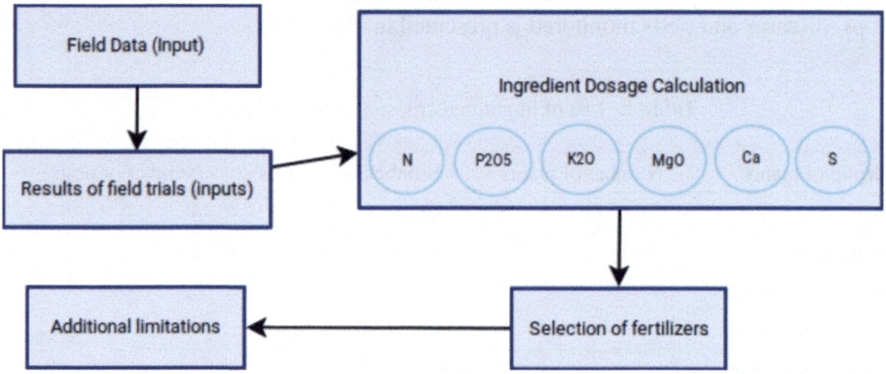

Fig. 2. Simplified method of calculating fertilizer doses.

The calculator contains five modules that carry out the subsequent stages of the calculation, from downloading the input data to obtaining the results. Three types of calculators have been designed for: agriculture, horticulture and fruit growing. The methods of calculation in these calculators differ from each other, and the input data is also different (e.g. the parameters are given in different units of measurement). Table 2 shows selected input data and calculated indicators for calculations intended for agricultural and horticultural crops.

Table 2. Selected input data and calculation results.

type of calculator	module	fertilizer ingredient	program variables
Horticulture	Field data		crop, field area, forecrop, catch crop, crop volume, organic fertilization, method of managing plant remains, soil type, amount of rainfall
	Result of field trials		organic matter content in the soil, mineral nitrogen, phosphorus, potassium, magnesium in soil, soil reaction
	Ingredient dose calculation	Nitrogen	plant needs, fertilizers not used by the forecrop, nitrogen released from the soil, nitrogen from legumes, manure, crop residues, fertilizer requirements (balance)
		Phosphorus, Potassium and Magnesium	optimal content for the expected yield, the amount of the ingredient in the soil, fertilizer requirements (balance)

(*continued*)

Table 2. (*continued*)

type of calculator	module	fertilizer ingredient	program variables
Agriculture	Field data		crop, field area, forecrop, catch crop, crop volume, organic fertilization, method of managing plant remains, soil type, amount of rainfall
	Result of field trials		organic matter content in the soil, mineral nitrogen, phosphorus, potassium, magnesium in soil, soil reaction
	Ingredient dose calculation	Nitrogen	plant needs, fertilizers not used by the forecrop, nitrogen released from the soil, nitrogen from legumes, manure, crop residues, fertilizer requirements (balance)
		Phosphorus, Potassium and Magnesium	plant needs, amount of manure ingredient, from crop residues, the amount of the ingredient in the soil, fertilizer requirements (balance)

The calculated doses of nutrients can be used to select the right amount of specific mineral fertilizers. In addition, the program takes into account various restrictions and recommendations regarding the use of specific fertilizers for specific groups of plants. The program is based on data available in the literature [10–15] as well as the project's own research. The calculation methodology is in line with the applicable legislation.

In order to check the methods used, a test version of the calculator was made (Fig. 3). After testing, the calculation algorithms were transferred to the FarmSmart system.

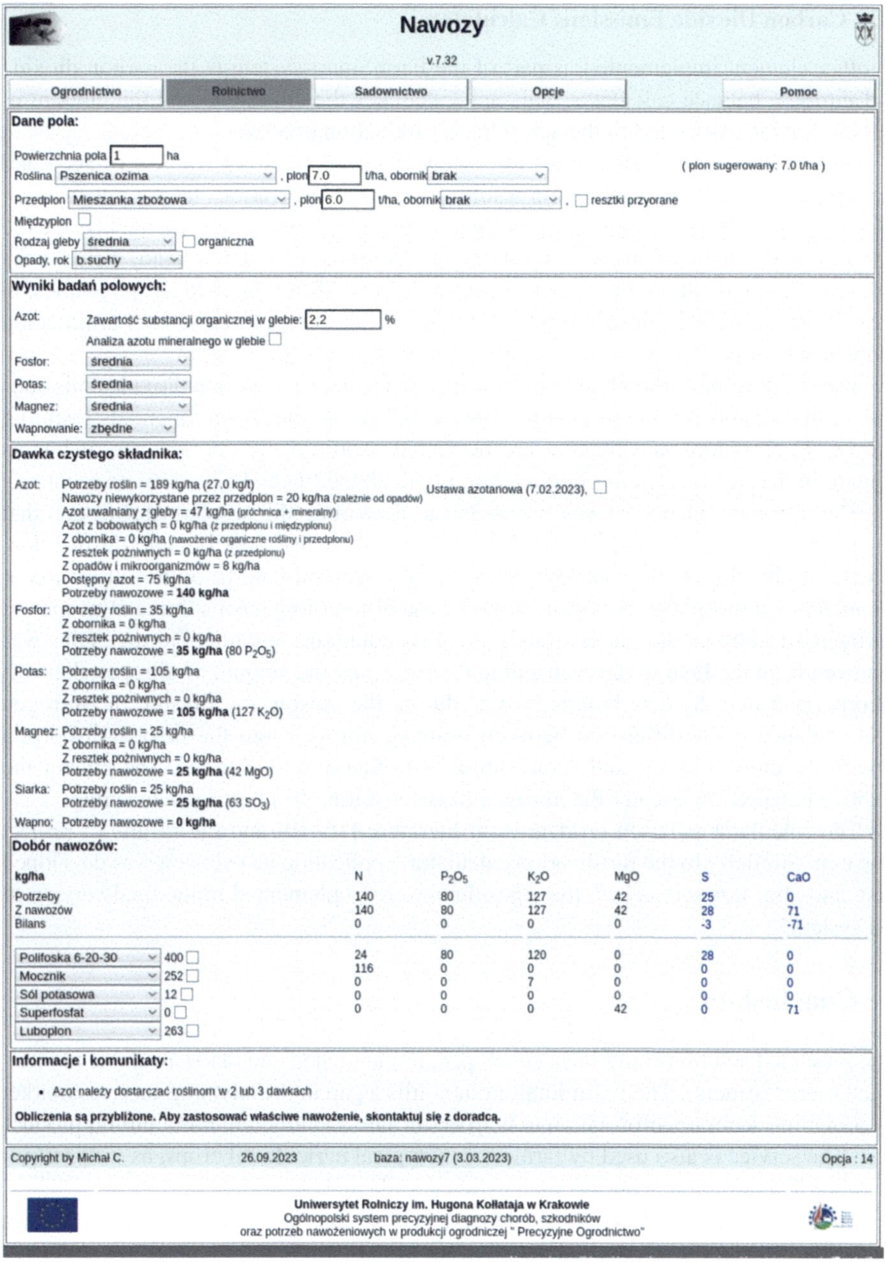

Fig. 3. Screenshot of the test version of the fertilizer calculator.

3.3 Carbon Dioxide Emissions Calculator

Another element implemented as part of the FarmSmart system is the carbon dioxide and nitrogen balance calculator. This app calculates the emissions and the amount of carbon dioxide associated in the agricultural production process.

Carbon dioxide emissions take into account factors such as: the use of seed and planting material, fuel consumption resulting from field operations taking into account the type of soil, the use of fertilizers, pesticides, biostimulants, biological agents, and the transport and storage of crops. Carbon dioxide fixation, on the other hand, includes the amount of carbon bound: in the crop, in straw removed from the field, crop residues, in the soil (increase in humus in the soil), as a result of catch crop, as a result of humification of the catch crop.

The result is the carbon balance as the difference between carbon dioxide emissions and carbon fixation from cultivation. The overall result and the components resulting from CO2 emissions and fixation are presented. Additionally, the balance of humus content in the soil as a result of cultivation is calculated (it can be positive or negative).

The nitrogen balance makes it possible to determine the amount of nitrogen that has been brought to the field as a result of: mineral fertilization, organic fertilization, release of nitrogen from humus during the season, with rainfall, from the cultivation of plants fixing atmospheric nitrogen, from the use of microorganisms fixing atmospheric nitrogen. In addition, the calculations took into account: the amount of nitrogen that was removed from the field in the crop and in the straw and the amount of nitrogen that was incorporated into the new humus formed during the season. As a result, the nitrogen cycle balance is the difference between bringing nitrogen into the field, transporting it with the crop or straw, and "embedding" nitrogen into the humus. The greater the positive balance, the greater the nitrogen losses resulting from nitrogen leaching.

The calculator is based on data from literature [16–19] and the results of project research. Similarly to the fertilization calculator, application test version was developed first, and after it was checked, the algorithms were implemented in the final version of the system.

4 Conclusions

The presented solutions have been developed, implemented and made available to interested users (farmers). The FarmSmart system fills a gap in the advising services market by providing accurate information to help to run safe and efficient horticultural production. The service is also used by farmers with typical agricultural crops, as well as fruit growers. This solution is unique on a national scale and beyond.

Thanks to the use of scientific methods, the system is highly effective and gives the opportunity to introduce modern solutions of agriculture 4.0 to farms, regardless of their size (including small-scale farms). The system is constantly updated and further developed. The number of plants, diseases and pests monitored is increasing. It is also planned to introduce additional functionalities and increase the accuracy of forecasts, among other by expanding the network of measuring stations. Cooperation with the University of Agriculture in Krakow and other research units means that Agro Smart Lab can provide farmers with a high-quality tool.

Acknowledgement. The development of the system was supported by the "Precision Horticulture" project, co-financed by the European Agricultural Fund for Rural Development and national public funds. The publication is financed from the subsidy of the Ministry of Education and Science for the Agricultural University in Kraków.

References

1. Cupiał, M., Kowalczyk, Z.: Computer-aided fertilisation using the Nawozy-5 (Fertiliser-5) software. BIO Web Conf. **10**, 02002 (2018). https://doi.org/10.1051/bioconf/20181002002
2. Cupial, M.: Concept of a system of plant protection support based on the ERP system. (2017). https://doi.org/10.24326/fmpmsa.2017.15
3. Borusiewicz, A., Kapela, K.: Implementation of precise farming technique on the example of a multi-surface agricultural farm. Agric. Eng. **18**, 14 (2014)
4. Borusiewicz, A., Kapela, K., Drożyner, P., Marczuk, T.: Application of precision agriculture technology in podlaskie voivodeship. Agric. Eng. **20**, 5–11 (2016). https://doi.org/10.1515/agriceng-2016-0001
5. Kuboń, M., Cupiał, M., Szeląg-Sikora, A., Kobuszewski, M.: The impact of purchasing new agricultural machinery on fuel consumption on farms. Sustainability **16**, 52 (2023)
6. Araújo, S.O., Peres, R.S., Barata, J., Lidon, F., Ramalho, J.C.: Characterising the agriculture 4.0 landscape—emerging trends, challenges and opportunities. Agronomy. **11**, 667 (2021)
7. Alimohammadi, F., et al.: Hyperspectral imaging coupled with multivariate analysis and artificial intelligence to the classification of maize kernels. Int. Agrophys. **36**, 83–91 (2022)
8. Findura, P., et al.: Comparison of tillage technologies in relation to changes in soil properties using soil biostimulation. Mechan. Agric. Conserv. Resour. **67**, 102–104 (2023)
9. Zhai, Z., Martínez, J.F., Beltran, V., Martínez, N.L.: Decision support systems for agriculture 4.0: survey and challenges. Comput. Electron. Agric. **170**, 105256 (2020)
10. Dyśko, J., Kaniszewski, K., Kowalczyk, W., Nowak, J., Wójcik, P.: Zrównoważone nawożenie roślin ogrodniczych. Instytut Ogrodnictwa, Skierniewice (2014)
11. Grzebisz, W.: Nawożenie roślin uprawnych: Podstawy nawożenia. Państwowe Wydawnictwo Rolnicze i Leśne (2008)
12. Grzebisz, W.: Nawożenie roślin uprawnych: Nawozy i systemy nawożenia. Państwowe Wydawnictwo Rolnicze i Leśne (2009)
13. Grześkowiak, A.: Vademecum nawożenia. Police: Grupa Azoty Zakłady Chemiczne Police SA. (2016)
14. Jadczyszyn, T., Kowalczyk, J., Lipiński, W.: Zalecenia nawozowe dla roślin uprawy polowej i trwałych użytków zielonych. Wydawnictwo IUNG-PIB (2010)
15. Schweder, P., Kape, E., Neubauer, W.: Düngung 1998, Hinweise und Richtwerte für die landwirtschaftliche Praxis–Leitfaden zur Umsetzung der Düngeverordnung. Schwerin: Min. für Landwirtsch. u. Naturschutz, Mecklenburg-Vorpommern. (1998)
16. Eggleston, H.S., Buendia, L., Miwa, K., Ngara, T., Tanabe, K.: 2006 IPCC guidelines for national greenhouse gas inventories. (2006)
17. Berdechowski, K.: Analiza składowych emisji GHG z upraw rzepaku wykorzystywanego do produkcji estrów metylowych kwasów tłuszczowych. Nafta-Gaz **73**(5), 360–364 (2017). https://doi.org/10.18668/NG.2017.05.08
18. Brwinów, C.D.R.: Normatywy produkcji rolniczej (database). (2020)
19. Faber, A., Jarosz, Z., Borek, R., Borzęcka-Walker, M., Syp, A., Pudełko, R.: Poziom emisji gazów cieplarnianych (CO2, N2O i CH4) dla upraw pszenicy, pszenżyta, kukurydzy i żyta przeznaczonych do produkcji bioetanolu oraz upraw rzepaku przeznaczonych do produkcji biodiesla. Ekspertyza wykonana na zlecenie Ministerstwa Rolnictwa i Rozwoju Wsi. 91 (2011)

Automatic Classification of Farmer's Weather Station Siting Based on Geodata

Sébastien Dandrifosse(✉), Alban Jago, Valéry Michaud, Jean Pierre Huart, Viviane Planchon, and Damien Rosillon

Agriculture, Territory and Technlogies Integration, Walloon Agricultural Research Centre, 9 Rue de Liroux, Gembloux, Belgium
s.dandrifosse@cra.wallonie.be

Abstract. More and more farmers are equipped with private weather stations which transmit local weather data in real-time. However, those weather stations are sometimes installed near obstacles such as farm buildings or trees, which influence the measurements. Those biased data have consequences when included in a wider weather monitoring network or in decision support systems, but the experts' managing networks of hundreds of stations cannot afford to visit all the stations to check the siting.

The goal of this study was to develop an algorithm to automatically assess the siting quality of weather stations based on their geographic coordinates.

The algorithm relied on existing public high-resolution geodata. A land cover map was used to identify heat sources such as buildings or artificial surfaces, which can bias temperature measurements. Digital terrain and surface models were used to identify obstacles located too close to the station compared to their height. Ultimately, one of the following siting classes was assigned to each station: excellent, good, poor or bad.

The siting class was computed for 30 stations from a public network. In 73.3% of cases, the siting class determined by the algorithm aligned with the field experts' assessment. The algorithm never diverged from the experts when they announced a class "excellent" or "bad". The method was then used to assess the siting of 108 farmers' stations. More than 65% of these stations had poor or a bad siting. This proves the importance of assessing the station siting.

Keywords: land cover · digital surface model · geospatial · air temperature · rain

1 Introduction

Farmers can now deploy affordable automated stations to monitor weather in real-time for their plots. In the context of precision agriculture, such local weather observations are more useful than ever. They feed decision-making tools that help the farmers to manage water resources or to assess the risks of diseases. However, the farmers operating their own stations often lack experience in meteorological measurement [1]. They may especially not install the station in a good location. The station should not be too close to obstacles such as buildings or trees that may intercept the rain, the wind or the

E. Lorencowicz et al. (Eds.): FMPMSA 2024, LNCE 609, pp. 116–130, 2024.
https://doi.org/10.1007/978-3-031-70955-5_13

direct sunlight. Nearby sources or sinks of heat should also be avoided. This concerns artificial elements such as asphalt roads [2], buildings or air-conditioning systems. These recommendations may seem obvious, but from our experience they are often overlooked due to logistic siting constraints. The farmers seek to locate the weather station close to their farmhouse to prevent theft and perform maintenance operations. If the station is located further from the farmhouse, it must be placed outside of a cultivated plot, in a strip of land that may contain hedges and trees. These obstacles cause incorrect weather measurements that could lead to wrong agricultural decisions, and ultimately to environmental pollution and economic losses for the farmer. Poor siting of a station also poses problems if its data is shared in a station network. The biased data could be wrongly used for spatial interpolation, to compute regional statistics or for the quality control of other stations.

The World Meteorological Organization (WMO) has developed a classification to report the environmental conditions of weather observation sites, based on the work of [3] and [4]. This classification has been published by ISO as an international ISO standard: ISO 19289:2014(E). It is also an appendix of each recent edition of the Guide to Instruments and Methods of Observation - Volume 1, published by the WMO. In the 2021 edition [5], it constitutes Appendix 1.D. Numerous national weather services use this classification to document the siting of their stations. In Belgium, it was implemented by the Royal Meteorological Institute (RMI) [6]. Multiple criteria are used to determine a class from 1 to 5 for each weather variable. The lower the class, the higher the representativeness of the measurement for a large area. As an example, the criteria for a class 1 site for air temperature are i) a flat, horizontal land, surrounded by an open space, with a slope less than 19°, ii) a ground cover with natural vegetation < 10 cm, iii) measurement point situated at more than 100 m from heat sources or reflective surfaces, at more than 100 m from an expanse of water, and away from all projected shade when the sun is higher than 5°. In practice, class 1 criteria are rarely met for all the measured weather variables, even for the reference stations of the national weather services. This demanding classification may not be adapted to farmers' stations. The fact that the classification differs for each weather variable complicates the communication of siting advice. Moreover, most farmers may not find sites with low class values (good representativeness) on their lands. They could end up with only class 4 or class 5 sites because of some specific criteria such as the height of the grass or some occasional projected shade. The higher classes of the WMO system lack nuances for on-farm agrometeorological measurements, for which the most obvious obstacles should be avoided, but without the need to satisfy the strictest siting requirements of national networks.

There is a high interest among agrometeorological institutes to evaluate the siting of farmers' weather stations that share their data in a network. Given the potentially high number of farmers' stations, an automatic siting method is necessary. Such a method could determine the siting conditions, and then warn the farmer in case of improper siting and exclude the station from agrometeorological computations. During studies about citizens' stations in urbanised areas, some automatic checks have been designed to identify indoor stations [7–9], which is the worst siting scenario. Those checks are based on the data recorded by the station. Basically, it is possible to determine if a station stays indoors if the minimum temperatures are too high or if temperature variations are limited compared to reference outdoor stations. Such a method would be interesting to identify farmers' stations kept indoors. Indeed, farmers sometimes bring in their station during the winter. Nevertheless, the outdoor environment of a station cannot be characterised based on the recorded data.

The utilisation of stations coordinates and geodata should permit to automatically assess the siting of a weather station. Geodata were already used to plan the installation of stations at the regional scale in Greece [10], Turkey [11], Philippines [12, 13] and Peru [14]. The typical geodata leveraged in these studies had spatial resolutions of 25 or 30 m. To our knowledge, the master's thesis of [15] was the only attempt to develop an automatic method to characterise station siting from geodata with high spatial resolution, accounting for obstacles such as small trees and buildings. Based on land cover, Digital Surface Model (DSM) and solar angle calculations, the author was able to compute all the criteria of the WMO classification for air temperature. The research highlighted some weaknesses of the WMO guidelines for an automatic siting computation based on geodata: some criteria are poorly specified, and the shading criteria should be adapted. Nevertheless, the results were promising. Further research needs to be done in that direction. Advances in remote sensing now allow open access to highly accurate and high resolution geodata.

This paper presents an automatic method to classify the siting of agrometeorological weather stations. Criteria are inspired from the WMO guidelines but have been adapted to allow a simple automatic approach, which only requires high resolution DSM, Digital Terrain Model (DTM) and land cover geodata. The algorithm yields a single class that translates the overall representativeness of the siting, considering the measurement of all the usual weather variables. The aim is to attribute one of the following siting class values – excellent, acceptable, poor or bad –, that can easily be communicated to the farmer.

2 Material and Methods

2.1 Weather Stations

Two groups of weather stations were studied: 30 reference stations from the Pameseb network of the Walloon Agricultural Research Centre (CRA-W) and 108 farmers' stations. All the considered farmers' station shared their data with the CRA-W, either via a direct sharing agreement with the CRA-W, an agreement via the non-profit WalDigiFarm organisation, or an agreement in the frame of the Vigimap warning system. The geographic distribution of the reference and farmers' stations is depicted on the map of Wallonia (Belgium) in Fig. 1. To determine the siting, the necessary data included the coordinates of the stations, the height of the sensors and the geodata. The coordinates of the stations used were those of 2 April 2024. All the chosen stations were active at this date.

2.2 Geodata

The geodata were provided by the Public Service of Wallonia via the Géoportail: https://geoportail.wallonie.be. Three data sources were used for obstacle detection: the 2021–2022 DSM, the 2021–2022 DTM and the WALOUS 2018 land cover map. The DSM

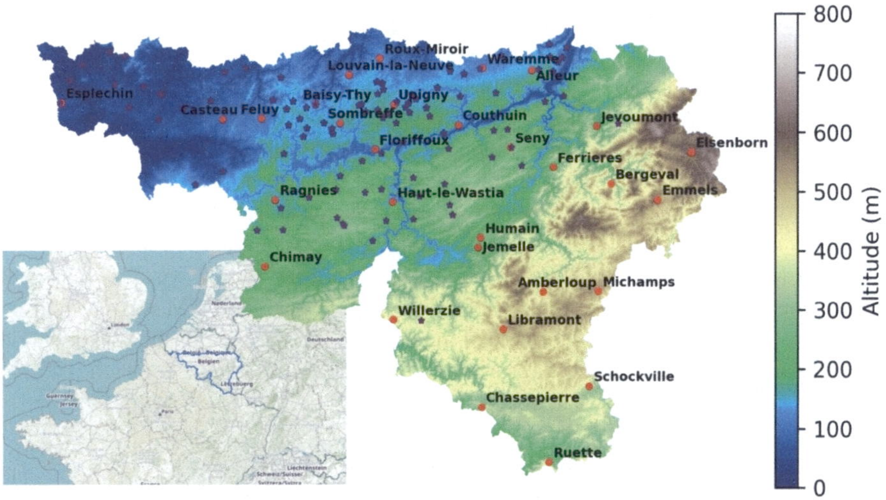

Fig. 1. Positions of reference and farmers' stations on the topographic map of Wallonia, Belgium. Reference stations are indicated by a red point, with their name written next to it. Farmer's stations are indicated only by a purple star, in order to guarantee anonymity. The location of Wallonia in Europe is indicated by a blue shape on the map in the lower left corner. The map of Europe comes from OpenStreetMap. The topographic map comes from EuroGeographics 2022 and is available for free at http://www.mapsforeurope.org/licence.

was a raster layer describing the elevation of the terrain and all the features on its surface (buildings, trees, etc.) while the DTM was a raster describing only the elevation of the terrain. Both models had a spatial resolution of 0.5 m and were derived from Lidar acquisitions. The WALOUS map was a raster layer with a spatial resolution of 1 m describing land cover using 12 different classes: water, bare soil, artificial ground cover, the rail network, buildings, greenhouses, herbaceous cover in rotation during the year, year-round herbaceous cover, broadleaf trees > 3 m, broadleaf trees ≤ 3 m, coniferous trees > 3 m and coniferous trees ≤ 3 m. A fourth layer of geospatial data was the 2020 orthophotoplan of Wallonia, having a spatial resolution of 0.25 m. This layer was not directly used to detect obstacles but rather to visualise the surroundings of the weather stations and to manually verify the results of obstacle detection.

2.3 Automatic Detection of Obstacles Around Weather Stations

An obstacle was defined as any element likely to significantly affect the measurement of one of the meteorological variables. The variables concerned were those measured by the reference agrometeorological stations, i.e. air temperature at a height of 1.5 m, ground temperature at 20 cm, air temperature in a 10 cm grass canopy, relative air humidity, precipitation, wind speed at a height of 2 m, wind direction, global radiation and leaf wetness duration. For the sake of simplicity and flexibility, it was decided not to define the obstacles separately for each weather variable, but rather to define the obstacles in such a way as to best integrate the constraints linked to the different variables. The chosen criteria were strict enough to ensure the correct measurement of all the weather variables, though even stricter criteria could enhance the quality of wind speed measurement. A single sensor height of 1.5 m was considered for both the reference and the farmer's stations. It corresponded to the height of measurement of air temperature.

The first category of obstacles included objects that were too high with respect to their distance from the station, which can skew the measurements by stopping the wind or rain, or by casting shadows on the station. For each station, obstacles were identified within a 200 m radius. This distance was chosen as a compromise between the possibility of taking distant obstacles into account and the computation time. An object was considered an obstacle as soon as its distance from the station was less than four times its height with respect to the sensors. To compute this, the DSM was normalised to the elevation of the sensors: the value of the DTM at the station plus the sensor height (1.5 m) was subtracted from all the surrounding DSM values. Then, for each 0.25 m^2 pixel, an obstacle weight was calculated (Eq. 1).

$$Weight = MIN\left(MAX\left(\frac{4DSM_{norm} - d}{3.5DSM_{norm}}, 0\right), 1\right) \tag{1}$$

where DSM_{norm} is the normalised DSM value for this pixel (in meters), i.e. the height of potential obstacles with respect to the sensors, and d is the distance (in meters) between this potential obstacle pixel and the pixel representing the station. The MIN and MAX conditions simply imply that the obstacle weight was bounded between 0 and 1. If the distance was greater than four times the height of the obstacle, then there was no siting problem, and the obstacle weight was 0. On the other hand, if the distance was half the height of the obstacle or less, then the obstacle weight was 1. For distances between half and four times the height of the obstacle, the obstacle weight varied between 0 and 1. In short, the smaller the distance with respect to the height of the obstacle, the higher the obstacle weight. The following score was then calculated by summing the weights of all the k pixels within the 200 m radius divided by their distance (Eq. 2).

$$Score_{height} = \sum_{i=1}^{k}\left(\frac{Weight_i}{d_i}\right) \tag{2}$$

This score represented the influence on the station of the obstacles derived from the normalised DSM. It considered the obstacle weight for each pixel, as well as the distance of the pixel from the station. For the same weight, an obstacle close to the station was considered more problematic because it intercepts direct radiation or rain in a greater azimuth.

The second category of obstacles included heat source or heat sink obstacles. The pixels considered as obstacles were those belonging to the following WALOUS land cover classes: water, artificial ground cover, the rail network, buildings and greenhouses. These pixels also had to be located within a 30 m radius of the station. Considering all these m heat source/sink pixels, the following score was calculated (Eq. 3).

$$Score_{heat} = \sum_{i=1}^{m}\left(\frac{1}{d_i}\right) \tag{3}$$

This score represented the influence of heat source obstacles on the station. It considered both the number of heat source pixels and their distance to the station. More distant heat source pixels were considered to have a lesser effect.

A representation of the obstacles is shown in Fig. 2 for the reference station in Ruette. The pixels identified as obstacles are shown alongside the photo of the area, the WALOUS land cover map and the normalised DSM. Colours are used to distinguish three classes of obstacles: heat sources, obstacles higher than four times the distance, and obstacles higher than twice the distance.

2.4 Agrometeorological Siting Classes

The obstacles scores (Sect. 2.3) were combined to provide a single value assessing the station siting. Each score was discretised into four siting classes: 1 = excellent, 2 = acceptable, 3 = poor, 4 = bad. The thresholds used are shown in Table 1. A human operator, based on the scores and obstacle maps of the 30 reference stations, determined these thresholds. The operator had never seen the stations in the field, with the exception of Libramont (Belgium). In this way, the determination of the thresholds was based solely on the algorithm's outputs (siting scores and obstacle maps), and did not interfere with comparison of the algorithm's output with field observations, which is described in Sect. 2.5.

Table 1. Thresholds used to convert each obstacle score into one of the four siting classes.

	Class 1: Excellent	Class 2: Acceptable	Class 3: Poor	Class 4: Bad
$Score_{height}$	$<=0.1$	>0.1	>1	>30
$Score_{heat}$	$<=7.5$	>7.5	>20	>70

The final station siting class was set to the most unfavourable siting class. For example, for a $Score_{height}$ of 3.1 and a $Score_{heat}$ of 5.2, the classes are class 3 and class 1 respectively. This indicates issues with surrounding obstacles being too high but none regarding heat sources. The final siting class is 3.

2.5 Comparison of the Algorithm's Outputs with Field Observations

The automatic classification was tested on the reference stations, for which the experts of the CRA-W knew the terrain inside out. Three field experts were brought together. Each was asked to assign one of the four siting classes to each station. They could not consult each other or see the results of the algorithm. They were then shown the algorithm outputs and were asked to work together to reach a consensus between their ratings and the algorithm result. This would lead to the final siting class, which may or may not be the same as the one determined by the algorithm.

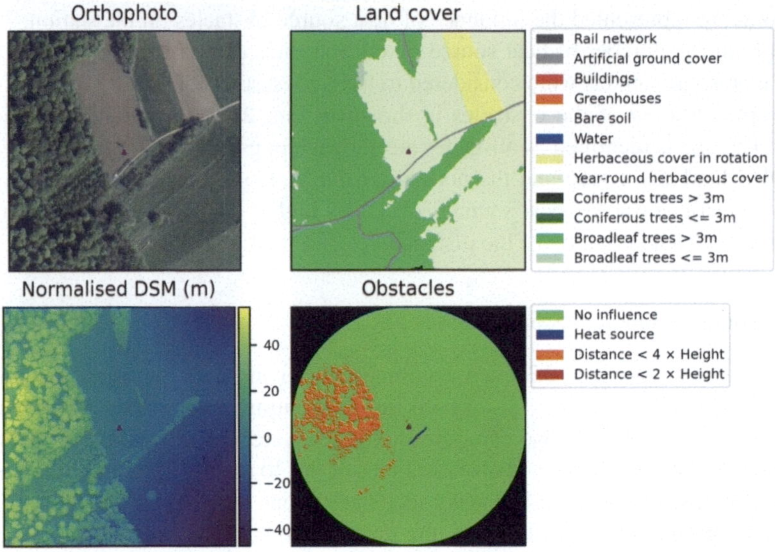

Fig. 2. Orthophoto, land cover, normalised digital surface model (DSM) and detected obstacles in a square zone of 401 × 401 m around the reference station of Ruette. The station is indicated by a purple triangle at the centre of the four sub-images. The normalised DSM is the difference between the DSM and the elevation of the sensors.

2.6 Comparison of the Agrometeorological and the WMO Siting Classes

For the purpose of comparison with agrometeorological siting classes, the WMO siting classes were established for the 30 reference stations, following the international ISO standard: ISO 19289:2014(E). This classification was performed in the field from June to August 2023 for the stations of Chassepierre, Chimay, Feluy, Ferrières, Humain, Jevoumont, Libramont, Roux-Miroir, Ragnies and Upigny (Belgium). For the other stations, the classification was last performed in 2021.

3 Results

3.1 Comparison of the Algorithm's Outputs with Field Observations

The comparison between the agrometeorological siting classes of the 30 reference stations determined by the experts and the automatic method is detailed in Table 2. In 73.3% of cases, the siting class determined by the expert consensus and the algorithm was the same. For the other stations, the difference was never more than one class. In 62.5% of the cases with a difference, the algorithm produced a higher class than the experts, and in 37.5% a lower class. There was no systematic overestimation or underestimation. Moreover, the algorithm never diverged from the experts when they announced a class 1 or a class 4, the most desirable cases to identify. Finally, the experts did not always agree with each other when classifying their cases. In 53.3% of cases, one of the experts gave a different class from the other two. However, it never happened that the three experts gave different classes, or that one expert gave more than one class of difference from the others. Considering all these factors, it can be concluded that the automatic method performed well. The objectivity provided by this method is also of great interest. For several stations, experts have changed their classification at the time of the consensus on the basis of the information and maps provided by the automatic method, recognising that they had incorrectly assessed certain distances or taken certain obstacles into account.

The two next sections delve into the cases where the experts reached a consensus that differed from the result provided by the algorithm.

3.2 Cases Where the Algorithm has Calculated a More Unfavourable Siting Than the Experts

These cases include the Amberloup and Ruette stations (Belgium). In these two cases, the algorithm assigned class 4 because of the presence of large clumps of tall trees, many of which were detected as obstacles. The experts were a little more lenient, particularly as the problematic trees at Amberloup were only located to the north of the station. The algorithm did not consider the direction in which the obstacles were located. Nor did it base its verdict on the slope of the terrain, which was one of the criteria used by the experts to judge the siting of the Amberloup station as poor.

Other stations for which the algorithm assigned a more unfavourable class were Roux-Miroir and Esplechin (Belgium). In these two cases, the algorithm gave class 3 and the experts class 2. There were no problematic high obstacles, but a wide road very close to the station. The experts were more lenient than the algorithm about this potential heat source.

For the Chassepierre station, the algorithm assigned a class 3 rating because there were several trees very close to the station. Humans assigned class 2 because the trees had been cut down in the meantime.

3.3 Cases Where the Algorithm has Calculated a More Favourable Siting Than the Experts

The algorithm assigned class 1 to Floriffoux because there were no obstacles according to its criteria, but the experts were stricter due to the location of the station in a stretch of grass in the centre of a village.

The Michamps station was located near tall hedges. The algorithm gave a class 1 rating because it considered that the hedges were far enough, but the experts judged that the station did not deserve a first-class rating since such obstacles were present.

The Waremme station was given a class 1 by the algorithm. However, it was right in the middle of crops, very close to the plots. This was not a problem at the time the DSM was carried out, but in practice maize crop can have an impact on weather observations at certain periods. This is why the experts gave a class 2.

3.4 Comparison with the WMO Classes

In Table 3, for the 30 reference stations, the agrometeorological siting classes resulting from the consensus between the experts and the algorithm are compared to the WMO

Table 2. Agrometeorological siting classes of the reference stations, established by the algorithm and three field experts, called Expert A, B and C. The consensus is the final expert decision, based on the algorithm's output and their three ratings. A colour code is used to ease the comparisons: green means that the expert consensus validated the algorithm output, and orange means that the consensus differed from the algorithm output. Remark: Sometimes, there is no data for Expert A, as he felt he did not know the station well enough to make a decision.

Station	$Score_{Heat}$	$Score_{Height}$	Agrometeorological siting classes: from 1 to 4		
			Experts A, B, C	Algorithm	Consensus
Alleur	27.24	4.57	2, 2, 3	3	3
Amberloup	0	111.45	3, 3, 3	4	3
Baisy-Thy	0	0	1, 1, 1	1	1
Bergeval	24.98	10.37	3, 3, 4	3	3
Casteau	2.93	0	2, 1, 2	1	1
Chassepierre	2.02	10.97	2, 2, 2	3	2
Chimay	10.20	0.04	3, 3, 2	2	2
Couthuin	0	0	1, 2, 1	1	1
Elsenborn	9.57	1.69	/, 3, 2	3	3
Emmels	0	0	/, 1, 1	1	1
Esplechin	32.44	0	2, 1, 1	3	2
Feluy	105.18	0.51	4, 3, 4	4	4
Ferrières	0	0	1, 1, 2	1	1
Floriffoux	0	0	/, 2, 3	1	2
Haut-le-Wastia	0	0	1, 1, 2	1	1
Humain	12.35	0.64	1, 1, 2	2	2
Jemelle	0	22.05	2, 3, 3	3	3
Jevoumont	0	0	2, 1, 1	1	1
Libramont	0	0.66	1, 2, 2	2	2
Louvain-la-Neuve	0	0	1, 1, 1	1	1
Michamps	0	0	/, 2, 2	1	2
Ragnies	0	0	1, 1, 1	1	1
Roux-Miroir	35.21	0	3, 2, 2	3	2
Ruette	5.32	77.11	/, 3, 3	4	3
Shockville	57.34	9.40	/, 3, 3	3	3
Seny	5.54	0.04	2, 2, 2	1	1
Sombreffe	30.87	0.01	3, 3, 3	3	3
Upigny	0	0	1, 1, 1	1	1
Waremme	0	0	/, 2, 2	1	2
Willerzie	0	185.82	4, 4, 4	4	4

siting classes. For air temperature and humidity, the WMO class was often 4 or 5 because the vegetation sometimes rose to more than 25 cm, despite frequent mowing maintenance. This simple criterion was very punitive. For wind speed, the WMO class was always 4 or 5, because the anemometer was installed at a height of 2 m, which is the standard for agrometeorological purposes but differs from the 10 m height required to obtain a smaller class. Comparing the agrometeorological classes and the WMO classes for rain, 66.7% of the stations had a low value (1 or 2) in both classifications and 23.3% had a high value (3, 4 or 5) in both classifications. Comparing the agrometeorological classes and the WMO classes for global radiation, 50% of the stations had a low value (1 or 2) in both classifications and 33.3% had a high value (3, 4 or 5) in both classifications.

3.5 Automatic Classification of Farmers' Stations

Among the 108 farmers' stations, the proportion of stations classified class 1, 2, 3 and 4 were respectively 25.0%, 9.3%, 41.7%, and 24.0%. In other words, more than 65% of the farmers' stations were poorly located. This shows the importance of assessing the station siting. The poorly located stations included stations in the garden too close to the houses, many stations on artificial ground in the farmyards, stations too close to roads, stations in grass strips too close to trees and stations indoors.

3.6 A Siting Map for the Whole Region

In addition to the evaluation of the siting of existing weather stations, a siting map was built for the whole region (Wallonia). This siting map intended to help the farmers to position their new stations or to find a proper location to move the badly positioned ones. The construction of the siting map was computationally intensive. It consisted of assigning a siting class to each 1 m^2 pixel in the region. Firstly, each pixel was classified as obstacle or possible location. A pixel was classified as possible location only if the height of the object on this pixel (DSM minus DTM) was less than 3 m and if the WALOUS land cover was one of the followings: bare soil, herbaceous cover in rotation or year-round herbaceous cover. For each pixel classified as possible location, the process of detecting nearby obstacles (Sect. 2.3) and siting classification (Sect. 2.4) was repeated. In the end, each pixel of the map was assigned one of the following siting classes: 1 = excellent, 2 = acceptable, 3 = poor, 4 = bad, 5 = obstacle to positioning. An example of siting map is presented in Fig. 3 for a square area of 401 $\times$ 401 m.

Table 3. Comparison of the agrometeorological siting classes and the WMO siting classes. The agrometeorological siting class is the consensus of the experts and the algorithm. The WMO siting class is different depending on the weather variable.

Station	Agrometeo siting class: from 1 to 4	WMO siting classes: from 1 to 5			
		Temperature and humidity	Rain	Global radiation	Wind speed
Alleur	3	4	2	4	4
Amberloup	3	4	3	4	5
Baisy-Thy	1	4	2	1	4
Bergeval	3	5	3	4	5
Casteau	1	3	1	3	4
Chassepierre	2	4	1	2	4
Chimay	2	4	2	4	4
Couthuin	1	4	2	1	4
Elsenborn	3	4	2	4	4
Emmels	1	4	2	2	4
Esplechin	2	4	2	1	4
Feluy	4	4	4	4	4
Ferrières	1	2	1	2	4
Floriffoux	2	3	2	2	4
Haut-le-Wastia	1	4	2	1	4
Humain	2	2	1	1	4
Jemelle	3	4	2	4	5
Jevoumont	1	3	1	2	4
Libramont	2	4	2	4	4
Louvain-la-Neuve	1	4	2	1	4
Michamps	2	4	2	4	4
Ragnies	1	1	1	1	4
Roux-Miroir	2	3	2	2	4
Ruette	3	5	2	4	5
Shockville	3	5	3	4	4
Seny	1	4	2	4	4
Sombreffe	3	5	3	4	4
Upigny	1	1	1	1	4
Waremme	2	1	1	1	4
Willerzie	4	5	3	5	5

Note: Temp. is an abbreviation for air temperature.

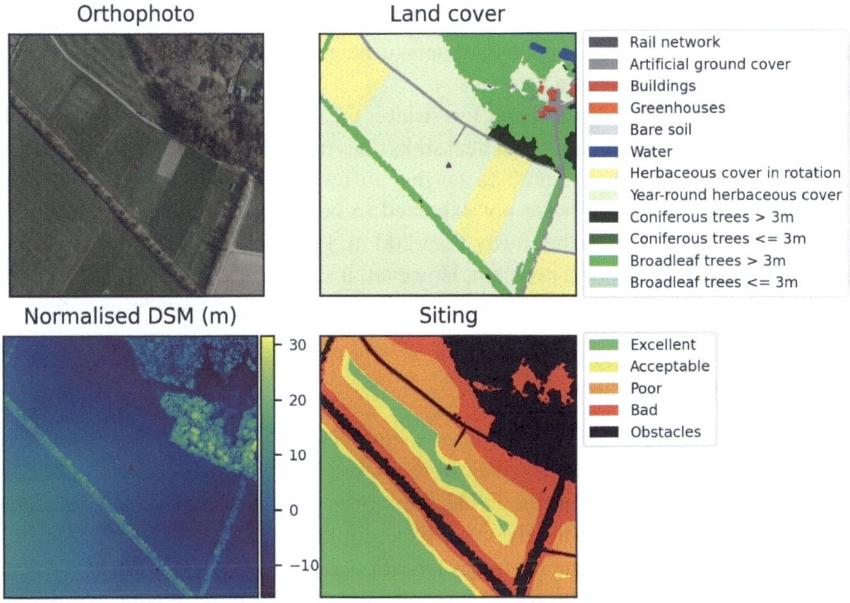

Fig. 3. Orthophotoplan, land cover, normalised DSM and siting map in a zone of 401 × 401 m. The normalised DSM is the difference between the DSM and the elevation of the sensors.

4 Discussion

4.1 The Measurement of Each Weather Variable has Different Siting Requirements

The proposed approach assigned a siting class for a given station position without using different parameters or criteria depending on the weather variable of interest. The aim was to propose locations that would be suitable for observing all the classical agrometeorological variables, so that a single siting map could be communicated to the farmers. A number of fairly strict criteria were used, and it could be argued that the siting class was determined in each case by the most limiting weather variable. This is not totally true, and it is interesting to see how this overall classification can be interpreted for each variable considered individually.

Air temperature can be influenced by both the height of surrounding obstacles (through shading) and the presence of heat sources or sinks. For agrometeorological measurements, the height of the sensor is usually 1.5 m, which is the value used to normalise the DSM relative to the measurement height. The classification is considered well adapted for the measurement of air temperature.

Global radiation is influenced by both the height of surrounding obstacles and the presence of reflective surfaces. The classification of the WMO [5] mentions that an obstacle is considered as reflecting if its albedo is greater than 0.5. Such reflective surfaces may be artificial surfaces or water bodies under certain incidence angles, and thus they are included in the land cover classes used to compute the $Score_{heat}$ criterion. For global radiation, the siting requirements based on geodata were thus considered similar to those for temperature. In the reference network, radiation sensors were installed at a height of 2 m while the algorithm used a standard sensor height of 1.5 m. This second value is slightly stricter, but this is not a problem.

The algorithm criteria are however too strict if the only sensor of the station is a rain gauge. The presence of heat sources or heat sinks does not influence the rain observations, and it is not necessary to consider obstacles that are too far away. Wind speed is the only variable for which the criteria are not expected to be strict enough. The height score should consider obstacles further away than 200 m. For example, a wide strip of trees beyond this limit may still be a problem. However, it should be noted that wind speed is measured at 2 m in the agricultural context, and not at 10 m as for synoptic stations.

4.2 Limits of the Automatic Method

The automatic method was not capable of seeing certain realities on the ground. For the station of Feluy, high voltage cables were considered by the algorithm as obstacles due to their height, recorded in the DSM. Moreover, the geodata were not frequently updated and the algorithm had to rely on geodata acquired several years before the date of interest. In the meantime, some obstacles have appeared, and others have disappeared. The vegetation may especially grow or be cut down. In Haut-le-Wastia, it was once observed that a huge pile of manure was temporarily stored very close to the station.

The geodata were not able to account for the presence of animals (cows, horses, ...) in the grassland of the station. In presence of animals, the station risks to be destroyed if it is not well protected. In the past, one of the stations of the reference network was ransacked by cows.

The algorithm did not account for the proximity of the farm when suggesting adequate zones to position the station. Yet, the proximity of the farm is very important to encourage the farmer to frequently inspect and clean the station. The algorithm also did not account for the proximity of axes of human passage. Yet, farmers are reluctant to expose their station too much, because of the risk of theft.

Finally, it can happen that the coordinates of a station used by the algorithm do not correspond to the actual position of the station. This could be due to the farmer moving the station without switching it off, while new coordinates are only recorded when the station is switched off and on again.

4.3 Avenues for Improving the Algorithm

The algorithm could be refined by incorporating the following elements: shading calculations using the heights of obstacles and the position of the sun over the course of the year [15], considering the direction of obstacles, the slope of the terrain and the 'roughness' of the terrain (in the context of wind speed observation). These additional factors were not initially included for the sake of simplicity and robustness of the method.

In Wallonia, the land cover and DSM geodata are updated only after several years. However, the orthophotos are updated more often. They could be used to detect the appearance or disappearance of obstacles around the weather stations.

4.4 The Notion of Good or Bad Station

The WMO specifies that its classes are not intended to be considered 'from best to worst', although the numerical ordering may suggest this. The classification of the WMO does not define good and bad stations in an absolute way, simply the stations that are most representative of a larger area, minimising the influence of obstacles in the local environment. The agrometeorological siting classes introduced here work in a similar way. The measurements from stations with high class values can still be very useful for certain applications of weather observations. For example, the Pameseb station in Willerzie was deliberately placed in a forest environment.

5 Conclusions

A simple automatic method for classifying the siting of weather stations in agricultural environments was developed. The method assigned a class from 1 to 4, the higher the number, the more the observations may be influenced by surrounding obstacles, and therefore the less representative the station is of a wider area. The classification considered the observation of all the weather variables of interest together: rain, temperature, humidity, wind speed and global radiation. A site classified 1 was therefore suitable for measuring all these variables.

The classes computed by the algorithm were compared with classes calculated by three field experts for 30 well-known stations from a reference network. In 73.3% of cases, the siting classes determined by the experts' consensus and the algorithm was the same. For the other stations, there was never more than one class difference. It should also be noted that the divergences between the experts themselves were of the same order as those between the expert consensus and the algorithm, and that the algorithm never diverged from the experts when they announced a class 1 or a class 4, which were the most desirable cases to identify.

For the reference network, automatic classification identified 46.7% of stations as class 1, 10.0% as class 2, 30.0% as class 3 and 13.3% as class 4. The farmers' stations studied were less well located, with 25.0% class 1, 9.3% class 2, 41.7% class 3 and 24.0% class 4. These numbers highlight the importance of studying the siting of agricultural weather stations, indicating that there is still room for improvement in both the public reference network and the farmers' network.

Acknowledgment and funding. The authors are grateful to Rodolphe Geradin, Arnaud Verlinden, Sébastien Weykmans, Michel Journée, Patrick Bogaert and Yannick Curnel. The authors also thank the Walloon Public Service, UCLouvain, ULB and ISSeP for creating and providing the geodata. This work was supported by the Wallonia recovery plan: project 142, Agromet II, led by the Agriculture, Natural Resources, Environment section of the Public Service of Wallonia and the Agence du Numérique (AdN).

References

1. Gleason, M.L., Duttweiler, K.B., Batzer, J.C., Taylor, S.E., Sentelhas, P.C., Monteiro, J.E.B.A., Gillespie, T.J.: Obtaining weather data for input to crop disease-warning systems: leaf wetness duration as a case study. Sci. agric. (Piracicaba, Braz.) **65**, 76–87 (2008)

2. Kumamoto, M., et al.: Field experiment on the effects of a nearby asphalt road on temperature measurement. SOLA **9**, 56–59 (2013)
3. Leroy, M.: Meteorological measurement representativity, nearby obstacles influence. In: Papers Presented at the WMO Technical Conference on Meteorological and Environmental Instruments and Methods of Observation (TECO–1998), Instruments and Observing Methods Report No. 70 (1998)
4. Leroy, M.: Sites classification. In: Papers Presented at the WMO Technical Conference on Meteorological and Environmental Instruments and Methods of Observation (TECO–2010), Instruments and Observing Methods Report No. 104 (2010)
5. World Meteorological Organization: Guide to Instruments and Methods of Observation - Volume I: Measurement of Meteorological Variables. WMO, Geneva (2021)
6. Sotelino, L.G., Coster, N.D., Beirinckx, P., Peeters, P.: Environmental classification of RMI automatic weather station network. In: Presented at the WMO Technical Conference on Meteorological and Environmental Instruments and Methods of Observation (2016)
7. Meier, F., Fenner, D., Grassmann, T., Otto, M., Scherer, D.: Crowdsourcing air temperature from citizen weather stations for urban climate research. Urban Clim. **19**, 170–191 (2017)
8. Napoly, A., Grassmann, T., Meier, F., Fenner, D.: Development and application of a statistically-based quality control for crowdsourced air temperature data. Front. Earth Sci. **6** (2018)
9. Fenner, D., Bechtel, B., Demuzere, M., Kittner, J., Meier, F.: CrowdQC+—A quality-control for crowdsourced air-temperature observations enabling world-wide urban climate applications. Front. Environ. Sci. **9** (2021)
10. Baltas, E.A., Mimikou, M.A.: GIS-based optimisation of the hydrometeorological network in Greece. Int. J. Digit. Earth. **2**, 171–185 (2009)
11. Yildirim, V., Nisanci, R., Colak, E.H., Yildiz, O.: A GIS-based siting technique for automatic weather stations in Trabzon. Turkey. Weather. **71**, 43–49 (2016)
12. Alejo, L.: Suitability analysis for optimum network of agrometeorological stations: a case study of Visayas region, Philippines. J. Agrometeorol. **20** (2018)
13. Jalil, A., Caguiat, L.S., Khaing, K.T., Alos, B.M., Moreno, N.A.: Augmentation of agrometeorological stations network in Southern Luzon, Philippines. EPRA Int. J. Multidiscip. Res. **7**, 187–197 (2021)
14. Rojas Briceño, N.B., et al.: Site selection for a network of weather stations using AHP and near analysis in a GIS environment in Amazonas. NW Peru. Clim. **9**, 169 (2021)
15. Stuurman, J.: Automating air temperature Siting Classification of meteorological stations according to the World Meteorological Organization guidelines. In: Master's thesis, Royal Netherlands Meteorological Institute (2019)

Evaluating Heavy Metal Contamination in the Rehova Cu-Mine Area for Sustainable Soil Management

Arta Dollani[1], Enkeleda Shkurta[1(✉)], Fatbardh Sallaku[2], and Seit Shallari[2]

[1] National Environmental Agency, Rruga Sami Frasheri nr. 4, Tirana, Albania
{arta.dollani,enkeleda.shkurta}@akm.gov.al, ledi.mera@yahoo.com
[2] Agricultural University of Tirana, Rruga Paisi Vodica 1025, Tirana, Albania
{fsallaku,sshallari}@ubt.edu.al

Abstract. This study investigates the extent of heavy metal contamination in soils near Rehove Cu-Mine site in the Korca District. Covering about 1 km^2, this site includes a former mine, an enrichment factory, and waste dumpsites, presenting notable hurdles for environmental and agricultural sustainability. Despite the cessation of copper mining operations in 1990, residual heavy metal contamination, particularly Pb, Cd, Cr, Cu, Zn, and Ni, persists in the soil from the enrichment processing residues. Employing systematic soil sampling techniques, including natural soil, dumpsites, and sediment areas, this study evaluates the distribution of heavy metal concentrations across various distances and depths from the mine site. The findings reveal significant variations in heavy metal concentrations across the sampled sites and depths: Cd from 3.048 mg/kg to 6.425 mg/kg, Cr from 34.615 mg/kg to 932.096 mg/kg, Cu from 67.562 mg/kg to 249.199 mg/kg, Ni from 39.88 mg/kg to 1571 mg/kg, Pb from 1.95 mg/kg to 492.80 mg/kg, and Zn from 3.363 mg/kg to 10.957 mg/kg. Notably, samples collected near the mine site exhibited elevated concentrations of heavy metals, highlighting the urgent need for remediation efforts to ensure environmental and agricultural sustainability. By integrating these findings into sustainable soil management practices, stakeholders can develop strategies to mitigate soil pollution and promote the long-term health of agricultural ecosystems in the affected regions.

Keywords: Heavy metals · Rehove Cu-Mine site · sustainable soil management · soil pollution · agricultural sustainability

1 Introduction

The Rehove copper deposit is located at an altitude of 950–1200 m, while the Geshtenje hill is at an altitude of 1050–1150 m. The Rehova copper enrichment plant was designed and built with machinery and equipment mainly produced in the country where, according to the technological scheme, the enrichment of copper ores is carried out by the method of selective flotation of copper ore. Copper concentrate with about 18% Cu and 10–12% moisture was obtained there, which was sent to Laç metallurgical plants. The tailings that remain after the separation the copper ore have been deposited in the dam

E. Lorencowicz et al. (Eds.): FMPMSA 2024, LNCE 609, pp. 131–136, 2024.
https://doi.org/10.1007/978-3-031-70955-5_14

erected for this purpose near the factory which has worked for about 11 years, during which time it has processed 1,117,728 tons of ore producing 678,930 tons of ore and about 47,000 tons of copper concentrate 17.5% Cu.

Contamination of soils with heavy metals may pose risks and hazards to humans and the ecosystem [1]. Due to their very material composition, the large amount of material deposited in them as well as their very fine granulometry, the roofs constitute a very large potential risk for the surrounding environment. Solid and liquid substances with the composition of heavy metal ions are discharged from the dams, which constantly dissolve and pollute streams and rivers.

Heavy metal contamination has adverse effects on the agro-ecosystems, such as loss of high-quality farmland and pollution of soil and groundwater, enhanced demand for clean water, contamination of urban areas and increased public health problems, low grazing quality and reduced crop yields and livestock production [2]. Recent studies by several investigators have shown that atmospheric fallout from smelters can contribute significantly to soil contamination with heavy metals [3]. Metal deposition patterns, though depending to a considerable extent on climatic conditions (i.e. wind and rainfall distribution) [4], generally decline exponentially with distance from the smelter. Consequently, metal concentrations in soils near the smelter may be elevated significantly and may pose a significant health hazard. The total soil metal concentration can be used to estimate the degree of soil exposure to heavy metal contamination, but they are not generally [5].

Damba of Rehova has 285714 m3 of sterile factory composed of: volcanic and subvolcanic rocks such as Calcic basalts, spilites, Keratophyre, pyritized and altered calcidacites, as well as low content of Cu minerals (0.19%), chalcopyrite, chalcosine, coveline, pyrite etc. The amount of waste in the damba is 873600 tons, where 0.25% of it is copper or a total of 2184 tons of copper metal. [10].

This is a difficult problem to prevent, but one that must be studied in depth to know the influence that these ions have on the lives of living things that live in terrestrial and aquatic environments polluted by them. The purpose of this study is to evaluate the pollution of the surrounding area with heavy metals and to propose a methodology for the rehabilitation of that area.

2 Materials and Methods

Sampling site: Sampling was carried out at the location where the enrichment factory was as a starting point, moving towards the dam through the national road that leads to Korça from 0 m, 1 m, 10 m, 100 m, 1k m-10 km. This dam is located near the enrichment factory, in the valley between two hills. The dam is located under the quota of the enrichment factory, which is why the waste has been sent to the dam with free flowing.

The Rehove mine site covers an area of approximatively 1 km^2, including mine area, the remaining of a former enrichment factory and dumpsites of mine waste near tributary of river Osumi.

Fig. 1. Located area of Rehova copper mine

Soil Sampling, Preparation and Analyses: Analyzes carried out in dam's waters, many years ago, showed that the presence of SO4 ions is present everywhere, its content ranges from 300–1340 mg/l. The pH value in these waters ranges from 5.15–2.86 [10].

In July 2023, 16 surface soil samples (0–30 cm) were collected from the different locations according to the distance (0 m, 10 m, 100 m and 1, 2, 3, 4, 5, 6, 7, 8, 9, 10 km) from the first sample (Fig. 1). Subsamples of each soil were air-dried and ground to pass through a 2-mm stainless-steel sieve and are analyzed with METHOD 3051A [6]. The water content (w %), is calculated as the ratio of the weight of water to the weight of the solids in a given mass of soil and organic C by elementary analysis. Total contents of heavy metals were determined by the mineralization of soil samples in the solution of aqua regia (HCl and HNO3) in the ratio (3:1). The metal concentrations in the extracts were measured by Atomic Absorption Spectroscopy.

3 Results and Discussion

The chemical composition of the water that drains from the dams depends on the mineral composition of the material deposited in these dams. We have emphasized that the minerals that take part in the composition of copper ores are generally sulphide minerals such as: Chalcopyrite CuFeS2, Pyrite FeS2, Sphalerite ZnS, Arseno-pyrite FeAsS, etc. These minerals, under the action of oxidizing agents, such as air oxygen and water, are oxidized giving water-soluble products. The phenomenon of oxidation and dissolution of sulphide minerals found in waste is continuous, the dissolution occurs in natural conditions, they constantly pollute the environment with the ions of their metals, such as copper, zinc, iron, arsenic, lead, sulphur, etc. Figure 2 gives the chemical composition of the sterile materials at the Rehova dams [10].

These metals are generally found in the form of sulfides that favor oxidation reactions. During the development of pyrite mineral oxidation reactions, in the presence of air and water oxygen, it breaks down into ferrous sulfate and sulfuric acid as follows:

$$2FeS_2 + 7H_2O + 7O_2 = 2FeSO_4 + 2H_2SO_4 \quad (3.1)$$

$$2FeS_2 + 2H_2SO_4 + O_2 = 2FeSO_4 + 2H_2O + 4S \quad (3.2)$$

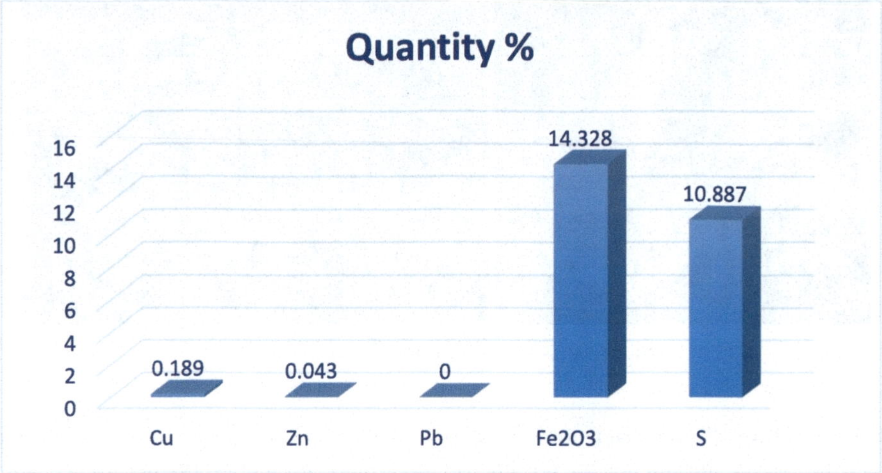

Fig. 2. Chemical compositions of sterile materials in the Rehove dams.

These reactions are dictated not only by the presence of SO4 ions in the discharge waters of the dams, but also by the presence of free sulfur that separates in the path that these waters pass through. These reactions do not always occur as above, they are also in the presence of some microorganisms that have the ability to oxidize the iron and sulfur of minerals, they can also be developed as below, up to the separation of ferric sulfate and sulfuric acid.

$$2\,FeSO_4 + 2H_2SO_4 + Bacter = Fe_2(SO_4)_3 + H_2O \tag{3.3}$$

$$2\,S + 3O_2 + Bacter = 2H_2SO_4 \tag{3.4}$$

Pyrite oxidation continues

$$FeS_2 + Fe_2(SO_4)_3 + Bacter = 3FeSO_4 + 2S \tag{3.5}$$

The total contents of heavy metals of the studied samples in 2023, are presented in Fig. 3. We compared the total content of heavy metals in the soil samples with international and German threshold values. The mean contents of Cd, Cr, Cu, and Ni were above the permissible limits of the 86/278/EEC Directive [7] (CEC 1986, BBodSchV (1999) [8] and Average Earth Crust* [9]. The mean contents of Pb were within values of international and German threshold values and the mean contents of Zn were above of permissible limits of Average Earth Crust*.

We also emphasize that the amount of Ni and Cr have very high values in areas far from the surface of the mine, while Cd has very high values in all samples.

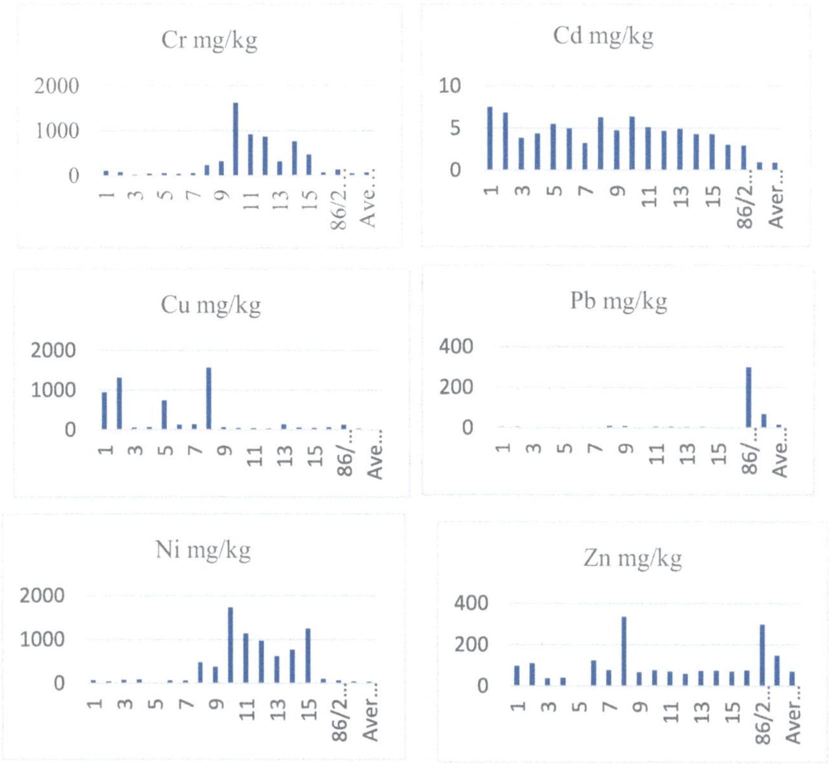

Fig. 3. Mean, international thresholds values, average earth crust (mg/kg) of heavy metals.

4 Conclusion

We note that the amount of water is not the same in time, it varies widely and depends on the weather, in the dry summer season, the amount of water coming out of the dams decreases a lot, in some cases it becomes immeasurable or is completely interrupted, in winter, especially during the rainy season, the amount of water increases significantly.

What should be emphasized is that, regardless of the amount of water that flows, their character is always acid, which has also affected the amount of heavy metals in those areas.

References

1. Wuana, R.A., Okieimen, F.E.: Heavy metals in contaminated soils: a review of sources, chemistry, risks and best available strategies for remediation. Int. Sch. Res. Netw. ISRN Ecol. **2011**, 402647 (2011). https://doi.org/10.5402/2011/402647
2. Shallari, S., Schwartz, C., Hasko, A., Morel, J.L.: Heavy metals in soils and plants of serpentine and industrial sites of Albania. Sci. Total. Environ. **209**, 133–142 (1998)
3. Paces, T.: Critical loads of trace metals in soils: A method of calculation. Water Air Soil Pollut. **105**, 451–458 (1998)

4. Elsokkary, H., Lag, J.: Distribution of different fractions of Cd, Pb, Zn and Cu in industrially polluted and nonpolluted soils of Odda Region. Norway. Acta Agric. Scand. **28**, 262–268 (1978)
5. Sallaku, F., Fortuzi, S., Tota, O., Huqi, B., Chachalis, D., Darawsheh, M.: Heavy metal soil contamination around the metallurgical plant of Elbasani in Albania. J.: Food, Agric. Environ. (JFAE) **7**(3&4), 878–881 (2009)
6. METHOD 3051A Microwave Assisted Acid Digestion Of Sediments, Sludges, Soils, And Oils. https://www.epa.gov/sites/default/files/2015-12/documents/3051a.pdf
7. CEC: Council Directive of 12 June 1986 on the protection of the environment, and in particular of the soil, when sewage sludge is used in agricultural soil. Official Journal of the European Communities L181, pp. 6-12 (1986)
8. BBodSch, V.: Federal soil protection and contaminated sites ordinance. Fed. Law Gaz. **I**, 1554 (1999)
9. Rudnick, R.L., Gao, S.: Composition of the continental crust. In: Celo, V. (ed.) The Crust, pp. 1–64. Elsevier, Amsterdam
10. National Agency of Natural Resources in Albania
11. https://www.scribd.com/doc/284311355/Industria-Minerare?doc_id=284311355&order=637047563

Agricultural Unmanned Ground Vehicle (UGV): A Brief Overview

Alessia Farella[1(✉)], Francesco Paciolla[2,3], Tommaso Quartarella[1], and Simone Pascuzzi[1]

[1] Department of Soil, Plant and Food Science, University of Bari Aldo Moro, Via Amendola 165/A, 70126 Bari, Italy
alessia.farella@uniba.com

[2] PolySense Lab-Dipartimento Interateneo di Fisica, Polytechnic and University of Bari, Via Amendola 173, 70126 Bari, Italy

[3] Dipartimento di Ingegneria Elettrica e dell'Informazione, Polytechnic University of Bari, Via Edoardo Orabona 4, 70126 Bari, Italy

Abstract. Agricultural Unmanned Ground Vehicle are the center of the new green revolution that is characterizing agriculture, contributing to a digital and sustainable future. Agricultural robots improve working conditions and production efficiency, address labor shortages, taking care of the environment. This study provides a brief review of robotics applications in agriculture, exposing the key features of the employment of this technology. Through a detailed analysis, the principal tasks performed by agricultural robots have been identified. In addition, a classification of prototypes and currently available on the market agricultural robots, has been introduced. Starting from planting, current technologies cover most agricultural tasks such as monitoring, sprayers, harvesting, and transportation. The objective of this overview is to provide a summary of the state of the art in the field of agricultural robots to highlight the potential of such devices that contribute to a greater efficiency in agricultural processes. The study can be also useful to improve the design of these devices, which currently still have some limitations to their application.

Keywords: Agricultural robotics · Unmanned Ground Vehicle (UGV) · Field operations · Automation · Precision farming · Farm management

1 Introduction

1.1 An Agricultural Revolution

Agriculture is a major contributor to global economic output. A few problems are putting this sector at risk: unpredictable climate change, rapid population growth, declining labor supply, and dwindling primary resources are putting it at risk. These are the challenges that agriculture and farmers face every day, but fortunately numerous solutions are being experimented with to ensure adequate food production [1]. In addition, today's consumer seeks sustainably produced food. Meeting the demand for this type of product requires

E. Lorencowicz et al. (Eds.): FMPMSA 2024, LNCE 609, pp. 137–146, 2024.
https://doi.org/10.1007/978-3-031-70955-5_15

providing farmers with new plant varieties and new technologies. The development of alternative agricultural technologies is a great opportunity for the advancement of agriculture and the future of the planet.

According to FAO data - to meet food demand - agricultural production must increase by 60 percent globally and nearly 100 percent in developing countries. This increase must take place by 2040, but the chances are very low. The 33% of soil is highly degraded due to erosion, nutrient loss, acidification, salinization, compaction, and chemical pollution. More efficient use of water, reduced use of pesticides and improved soil health can increase average agricultural yield by 79%. Responding to this requires continued innovation, transferring the best of scientific research to this field.

The green revolution that occurred between the 1940s and 1970s had many positive effects pe our lives but also created new problems, especially environmental ones. Agriculture 4.0 is the new vision of agriculture designed to solve these problems. This agriculture is the fusion of precision agronomic practices with computer engineering and allows the use of different technologies for the improvement of crop yield and sustainability. High production quality and better working conditions for farmers are ensured [1].

One of the main features of agriculture 4.0-also called digital agriculture, -is the ability to use real-time information through the development of sensors. Sensors make it possible to improve techniques in the field and optimize production processes by taking advantage of multidisciplinary services (agronomic, information technology, meteorological, etc.). In this context, we find modern agricultural machines that become technical means that perform operations and elements of production data acquisition referenced in time and space. These data are indispensable for the analysis of more efficient and sustainable operational strategies [2].

This paper is presented as a review that will cover the use of Unmanned Ground Vehicle (UGV) in agriculture to perform different agricultural operations.

1.2 Characteristics of Unmanned Agricultural Ground Vehicle (UGV)

An unmanned ground vehicle (UGV) is a vehicle that operates while in contact with the ground and without an onboard human presence. The UGV is a remotely controlled robot and is the land-based counterpart to unmanned aerial vehicles. It is an intelligent system that integrates environmental perception, location, navigation, route planning, decision making and motion control. It combines high technologies including computer science, data fusion, computer vision, deep learning, etc., to meet real needs to achieve predetermined goals [3].

The development of robotic technologies that can be used in field operations has become one of the hot research topics of the past 20 years. These innovative technologies are being developed, and some are already successfully commercialized and adopted by farmers [4].

Currently, most agricultural activities are practiced by traditional means. One of the biggest challenges facing the agricultural industry in many countries today is the shortage of human labor to do strenuous work in the fields, and the trend is expected to continue and worsen. One solution to address this challenge is to adopt robotic technologies in agricultural production in the open field.

Traditional agriculture has several problems [5]:

- It requires more time and a lot of effort.
- Involves more human resources to manage the various agricultural processes.
- Lacks accurate information on weather, soil conditions and fertilizer use.
- It requires manual monitoring of crop health.
- Identifying diseases requires more time and effort.
- Traditional pesticide spraying affects farmers' health and reduces crop productivity.
- Old methods of crop cutting, and segmentation of healthy crops and fruits are tedious tasks.
- Tractor rollovers represent a primary cause of death or serious injury in agriculture.
- Many younger generations are breaking away from agricultural activities causing problems in food production.

The agricultural technology revolution has occurred to overcome all these problems. The traditional agricultural system is replaced with a more advanced system based on artificial intelligence in which the machine itself makes decisions to solve problems in real time. At present, young engineers and scientists are working hard to make the agricultural process a simple, intelligent, cost-effective, highly productive, time-efficient, sustainable, healthy, and rich society. Artificial intelligence-based systems include sensor technology, IoT, data management technology, intelligent decision-making algorithm, robotics and advanced mechanisms [5].

The essential features required by autonomous agricultural vehicles and field-operated systems tend to fall into five categories [6]:

- Navigation: the vehicle must move through the field independently, selecting a predetermined path, following the trajectory dictated by the operator.
- Sensing: the vehicle using various sensors must be able to detect, measure and sample anything that might be relevant in planning crop and soil management operations. The robot, through knowledge of the environment and/or using on-board sensors, must recognize where obstacles are and avoid them in order to navigate safely (without collisions) within the workspace.
- Mapping: surveying activities generate a large amount of data, which are generally processed by Geographic Information Systems (GIS), which are maps that collect all the essential features of the field.
- Processing: the robot processes and analyzes the data received from its sensors to gene-rate reasoning and a perception of the environment, task, or state at some level of situational awareness.
- Action: the robot generates an operational plan based on its perception of the environment and its state and the objectives of the task. It will then be able to perform all the actions required in the operational plan.

Environmental perception is an extremely important technology for UGVs, including perception of the external environment and estimation of the state of the vehicle itself. A high-precision environmental perception system is the basis for UGVs to drive safely and perform their tasks efficiently. Environmental perception for UGVs requires various sensors such as Lidar, monocular camera, and millimeter-wave radar to collect environmental information as input for planning, decision making and motion control system [7].

At present, agricultural robots can replace humans in simple agricultural production, but they still fail to meet the requirements when dealing with complex operations with complex agronomic processes. With the development of robotics and the continuous improvement of image processing technology, research on agricultural robots working in complex and unstructured farmland has become a current research hot spot.

Most agricultural operations occur in unstructured environments characterized by rapid changes in time and space. Soil, vegetation, lighting, and other weather conditions are constantly changing. The inherent uncertainty of these environments generates unpredictable and dynamic situations. Complexity increases when dealing with natural objects such as fruits and leaves, due to the high variability in shape, texture, color, size, orientation, and position that cannot be determined empirically [8].

The requirements for a robot to operate in unstructured environments complicates robotic application and results in a system that is difficult and expensive to develop. Research on self-driving vehicles kicks off in the 1960s, and at present, technological progress is ripe to ensure the development of robots even in unstructured environments such as agriculture [9].

Robots are perceptual machines on which it is possible to program, perform specific tasks, make decisions and act in real time. Agricultural robots include a moving frame, robotic arms, end effectors, and environmental sensing systems [10]. Their structure varies according to the operation for which they are designed.

2 Methodology

For the development of agricultural robots, it is necessary to start by researching what is currently on the market and what is currently being developed. With agricultural robots, a new frontier of mechanization is opening (or will open in the very near future) that demands interdisciplinary approaches between different worlds. Therefore, a review of the state of knowledge of robots used in agriculture is important to identify future research priorities in this area.

The research was divided into several phases. In the first phase, several documents were retrieved and identified using the following keywords: "farming robots", "agricultural robots", "robots" and "field operation robots". In the second stage to refine the search, keywords concerning the agricultural operation that was to be investigated were used. Therefore, the keywords "monitoring" "seeding" "weeding" "spraying" were added. The Boolean operator AND was used to limit the about to robots used in agriculture. The main databases used for the search were Google Scholar, Science direct, Springer Link, ProQuest, IEE explore, Scopus, ResearchGate and MDPI. Non-academic or scientific commercial material was also retrieved from the Google search engine because commercial companies and corporations often prefer to publicize products instead of writing scientific papers leaving the patent to maintain exclusivity on the patented product. Since this is a new field of research, the presence of materials concerning robots is limited so we used other sources of information such as YouTube and websites of commercial agricultural robotics companies.

The research was filtered by limiting the information to the last 3 years from 2022 to dwell on the current state of the art. It was worked by classes divided by crop operation. Each of the retrieved items was analyzed and placed in the relevant class. Then the items were evaluated and the main differences between the systems were extracted.

3 Robotic Applications in Agriculture

The various applications of robots in agriculture have been divided into subsections: transport, planting, spraying, monitoring, harvesting. Each subsection groups the main technologies used to perform that agricultural task.

3.1 Seeding Robotic Systems

Seeding is the primary process in field production. Therefore, seeding robots facilitate the sowing of seeds in exact locations, saving time and costs for farmers.

To date, numerous functional seeding robots have been invented and put into practice. FarmDroid F20 (see Fig. 1a) is a robot designed by FarmDroid ApS [11]. This robot autonomously handles seeding and mechanical weeding of crops. It is equipped with 4 solar panels to ensure entirely independent power supply and can be used on multiple crops. After configuring operational settings, adjusting the working depth and seed spacing, the robot begins sowing autonomously thanks to GPS-RTK technology. For high-precision seeding, it can mount from 4 to 12 containers on rows and different seeding discs depending on the crop used. Thanks to GPS technology, it is possible to know the position of each individual seed planted and this allows the robot to perform precision weeding before the crop emerges thus reducing weed competition. The precision of these operations not only increases productivity but also reduces the need for pesticides and fertilizers, thus reducing the environmental impact. A precision wheat seeding robot was designed with four wheels, servo motors and stepper motors. According to the test results, the seeding rate exceeded 93% with a typical seeding speed [12]. Kumar et al. [13] developed an intelligent planting robot controlled by an IoT system that can achieve complete automation of planting. After conducting the comprehensive tests with various types of seeds, the seeding yield rate was higher than that of the conventional seeder.

3.2 Weeding Robotic Systems

Mechanical weeding destroys weeds by uprooting them, burning them, or generating high-voltage current. Aside from add-on devices for conventional tractors, a generation of robots designed to perform these functions has been developed over the past decade. An early example of a robot designed for weed removal is BoniRob (see Fig. 1b) from Deepfield a start-up company owned by Bosch. Locating is done through GNSS and LiDAR measurements. Detection is via cameras and ultrasonic sensors while weed control is done with a manipulator that strikes weeds in the ground not only using mechanical tools but also by spraying herbicides. The effectiveness is estimated at 80 percent and has the capacity to remove two weeds per second. The database of the weeds to be treated is made from a series of images collected by farmers and then fed, in digital form, into the robot's artificial intelligence system to train it [14].

Vitirover (see Fig. 1c) is a technology for cutting all types of grass. During operation, this robot moves autonomously in an area while cutting the lawn using rotating blades. Its navigation strategy makes use of a GPS and two separate IMUs, one located on the main electronic board and the other on the tool holder. Its structure integrates two solar cells connected to batteries that give it the ability to self-charge. This robot is particularly suitable for maintaining grass height in vineyard rows, tree rows, and grassy playgrounds [15]. In Europe, France is a pioneer in the development of automated systems and offers various automated solutions already on the market. The company Naio Technology has contributed greatly to the spread of robots on the market. In fact, it has developed numerous robots to meet the needs of various sectors. Oz (see Fig. 1d) aims at weed control and hoeing work in all types of plants. This robot offers three different operating modes: manual, track-based, and autonomous. In manual mode, the robot is teleoperated. In track mode, it follows the user's movement along the field. In autonomous mode, Oz uses a LiDAR and some preliminary field information (e.g., length, width and number of crop rows) to navigate between crop rows. Its structure integrates four wheels, a navigation system, sensors, and special tools for weeding and hoeing. In the case of Dino (see Fig. 1e) and Ted (see Fig. 1f) robots, their goal is weed control in vineyards and organic farms, respectively. Their mechanisms have four omnidirectional wheels and tools for weeding crop paths. Bob (see Fig. 1g), on the other hand, is a redesigned version of Oz with more range and tracks to deal with slopes and muddy soils. It was developed to perform tasks such as weeding and soil work in vineyards and nurseries [15].

3.3 Harvesting Robotic Systems

Among the robots designed for crop harvesting, we can mention the Agrobot E-series (see Fig. 1h), a patented robot developed by the Spanish company Agrobot. It is one of the first robotic pickers designed for strawberry harvesting. It is entirely made of stainless steel and military-grade aluminum to withstand various weather conditions in the field. It has 24 independent robotic arms that work on four rows of strawberry plants simultaneously. Infrared and color cameras are integrated for fruit identification, while cutting-edge graphic processing units assist in assessing fruit ripeness. The robotic arms carefully pick the strawberries from the plants by grasping and cutting them from their stems using two sharp blades. This method prevents pressure on the fruit that could damage it, especially when it is overripe. After harvesting, the fruit is transported to containers or conveyor belts for further packaging [16].

Mechanical abrasions and bruises on the fruit can accelerate water loss, increasing the possibility of decaying fungi and bacteria entering the fruit and causing rapid decay. Therefore, for some crops it is difficult to talk about harvesting robots as a mature technology.

VineScout (see Fig. 1i) is an agricultural robot being industrialized, developed by the Agricultural Robotics Laboratory (ARL) of the Universidad Politécnica de Valencia (UPV). This robot is designed to improve the quality of wine production by collecting big data about the vineyard and then analyzing it in the field. When VineScout explores an area, the state of the plants is classified and displayed in georeferenced maps with high spatial resolution (color-coded). This enables selective harvesting that leads to production of normal and premium quality wine from the same field [17].

3.4 Multi-purpose Robotic Systems

To extend the functionality and make it compatible with various operations and crops, the agricultural robot Trektor (see Fig. 1j) was developed by Sitia, a French engineering company. It is a versatile and autonomous tractor equipped with a three-point hitch that allows to attach all tools owned by the user, provided they are suitable for this type of attachment, and it can automatically activate and deactivate them by programming the vehicle. This autonomous hybrid vehicle is equipped with both a diesel and an electric motor, allowing users to configure missions by deciding the percentage of electric and diesel usage based on their needs. Trektor is capable of working autonomously in narrow and wide-row vineyards, horticulture, and tree farming by varying both wheelbase (adjustable at a 530 mm pitch) and the width and height of the vehicle (motorized with a 400 mm pitch) [18].

Agri.Q, is a rover developed specifically for precision agriculture applications in vineyards and can operate in an unstructured environment on uneven terrain, cooperating with drones if necessary. It is equipped with various tools and sensors to perform specific tasks, such as field mapping and crop monitoring. With a redundant seven-degrees-of-freedom (DOF) collaborative robotic arm, it is also possible for the rover to interact with the environment, such as performing soil, leaf and grape sample collections [19].

3.5 Agricultural Robots for Plant Protection

Ara is a robot for precise application of herbicides. It integrates a navigation system, sensors, a processor, a color camera, two arms, four wheels, solar cells, and self-charging batteries. During operation, the control algorithm identifies an invasive plant in real time through captured images. Then, the processor estimates a position just above the plant and moves one of its arms to this position to spray a microdose of herbicide. This technology has the advantage of minimizing herbicide waste, reducing costs and environmental impact [15]. To promote more sustainable agriculture, the world's first autonomous robot with hybrid UV-C radiation treatment technology for vineyards was introduced in 2021. This is Icaro X4 (see Fig. 1k), an Italian robot designed by the startup Free Green Nature of Colle Umberto (TV). Its purpose is to protect grapevines from damage caused by downy mildew and powdery mildew using two foldable and adaptable side panels equipped with UV-C radiation emitters. These radiations can devitalize microorganisms on the surface of leaves or clusters and stimulate the plant's natural defenses by inducing it to produce fungicides and fungistatic substances that can block the development of pathogenic fungi. UV radiation is 100% effective against powdery mildew as this fungus develops on the leaf's surface. As for downy mildew, which penetrates through the stomata and settles deep in the tissues, the effectiveness is more limited, but the use of UV rays is still considered essential because it significantly reduces the need for treatments, relying on the plant's own defenses stimulated by UV radiation [20].

A promising future lies ahead with the spread of these machines that can be successfully integrated into both conventional and organic defense strategies to reduce the number of treatments and minimize the use of copper and sulfur in defense. Thorvald (see Fig. 1l) began as a research project at the Norwegian University of Life Sciences. It is a modular platform, available in three different versions: normal, compact and research. The specific sensor and capabilities of the robot are adjustable according to the customer's needs and budget. It can be delivered with four-wheel drive and steering and is equipped with IMU or AHRS units, LiDAR, and cameras. For specific tasks that require better accuracy in pose estimation, it comes with a GNSS RTK that provides centimeter-level accuracy. In this case, preloaded maps are used for robot navigation. The robot is responsible for applying UV-B light to the strawberries [21]. Thorvald is a multi-functional robot capable of controlling powdery mildew through UV-C treatment. It can also be used for other autonomous tasks around the farm or vineyard [22, 23].

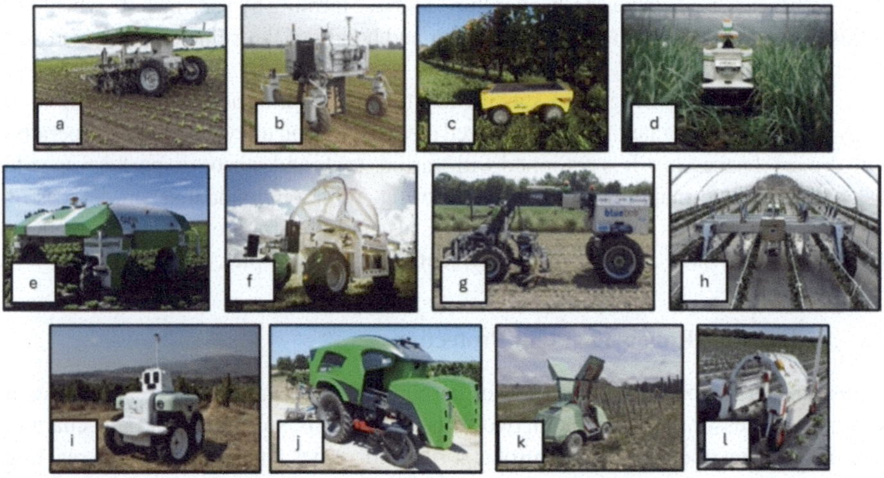

Fig. 1. Robotic applications in agriculture. a. FarmDroid F20, b. BoniRob, c. Vitirover, d. Oz, e. Dino, f. Ted, g. Bob, h. Agrobot E-series, i. VineScout, j. Trektor, k. Icaro X4, l. Thorvald.

4 Results

The intensification of research in robotic equipment for agriculture over the last five years has demonstrated a clear commitment to seeking innovative solutions to improve the efficiency and sustainability of the agricultural sector. However, to ensure effective dissemination of such systems and obtain farmers' approval, it is essential to develop robust and uncertainty-free solutions. The delay in commercialization can be understood considering the thorough studies conducted to ensure proper market diffusion. The prototype simulation phase, preceding the implementation phase, plays a crucial role in ensuring that the equipment meets the practical needs of farmers [24]. Furthermore, the focus on the possibility of hybridizing these systems to reduce environmental impact demonstrates a commitment to sustainability [25, 26]. Although it is still premature to speak of a full transition to robotic agriculture, it is evident that research is progressing in the right direction, offering promising opportunities for the future of the agricultural sector.

5 Conclusions

Robotics is a solution to agricultural challenges. Agriculture is facing growing challenges, including the increasing demand for food, fluctuations in commodity prices, and climate uncertainty. Agricultural robotics offers a promising solution to address these challenges. With further investments in research and development, we could see a future in which western agriculture will be able to produce food more efficiently, sustainably, and affordably.

In the end, robotics is transforming agriculture in the western world. This revolution offers concrete hope for addressing global food challenges and reducing the environmental impact of agriculture. With constant technological progress, we can look to the future of agriculture with optimism, knowing that we are not only growing crops, but also a better future.

References

1. Mandrioli, M.: Nove miliardi a tavola. Droni, big data e genomica per l'agricoltura 4.0. Zanichelli (2020)
2. Karunathilake, E.M.B.M., Le, A.T., Heo, S., Chung, Y.S., Mansoor, S.: The path to smart farming: innovations and opportunities in precision agriculture. Agriculture **13**, 1593 (2023)
3. Liu, B.: Recent advancements in autonomous robots and their technical analysis. Math. Probl. Eng. **2021**, 6634773 (2021)
4. Yang, Q., Du, X., Wang, Z., Meng, Z., Ma, Z., Zhang, Q.: A review of core agricultural robot technologies for crop productions. Comput. Electron. Agric. **206**, 107701 (2023)
5. Wakchaure, M., Patle, B.K., Mahindrakar, A.K.: Application of AI techniques and robotics in agriculture: a review. Artif. Intell. Life Sci. **3**, 100057 (2023)
6. Bagyaveereswaran, V., Ghorui, A., Anitha, R.: Automation of Agricultural Tasks with Robotics- Agrobot, Innovations in Power and Advanced Computing Technologies (i-PACT), pp. 1–7. Vellore, India (2019)
7. Liu, Q., Li, Z., Yuan, S., Zhu, Y., Li, X.: Review on vehicle detection technology for unmanned ground vehicles. Sensors **21**, 1354 (2021)
8. Bechar, A., Vigneault, C.: Agricultural robots for field operations: concepts and components. Biosys. Eng. **149**, 94–111 (2016)
9. Bechar, A., Vigneault, C.: Agricultural robots for field operations. Part 2: operations and sys-tems. Biosyst. Eng. **153**, 110–128 (2017)
10. Xie, D., Chen, L., Liu, L., Chen, L., Wang, H.: Actuators and sensors for application in agricultural robots: a review. Machines **10**, 913 (2022)
11. FARMDROID Homepage. https://farmdroid.com/. Accessed 14 Apr 2024
12. Cheng, C., Fu, J., Su, H., Ren, L.: Recent advancements in agriculture robots: benefits and challenges. Machines **11**, 48 (2023)
13. Kumar, P., Ashok, G.: Design, and fabrication of smart seed sowing robot. Mater. Today: Proc. **39**, 354–358 (2021)

14. Oliveira, L.F.P., Moreira, A.P., Silva, M.F.: Advances in agriculture robotics: a state-of-the-art review and challenges ahead. Robotics **10**, 52 (2021)
15. Gil, G., Casagrande, D.E., Cortés, L.P., Verschae, R.: Why the low adoption of robotics in the farms? Challenges for the establishment of commercial agricultural robots. Smart Agric. Technol. **3**, 100069, ISSN 2772–3755 (2023)
16. Zhou, H., Wang, X., Au, W., et al.: Intelligent robots for fruit harvesting: recent developments and future challenges. Precision Agric. **23**, 1856–1907 (2022)
17. Saiz-Rubio, V., Rovira-Más, F.: From smart farming towards agriculture 5.0: a review on crop data management. Agronomy **10**(2), 207 (2020)
18. SITIA Homepage. https://www.sitia.fr/innovation-2/trektor/. Accessed 24 Mar 2024
19. Botta, A., Cavallone, P., Baglieri, L., Colucci, G., Tagliavini, L., Quaglia, G.: A review of robots, perception, and tasks in precision agriculture. Appl. Mech. **3**, 830–854 (2022)
20. Quartarella, T., Farella, A., Paciolla, F., Pascuzzi, S.: UVC Rays devices mounted on autonomous terrestrial rovers: a brief overview. In: Farm Machinery and Processes Management in Sustainable Agriculture XII International Scientific Symposium (2024)
21. Grimstad, L., From, P.J.: The Thorvald II agricultural robotic system. Robotics **6**, 24 (2017)
22. Pearson, S., et al.: 2022 Robotics and autonomous systems for net zero agriculture. Current Robot. Rep. **3**, 57–64 (2022)
23. Bulgakov, V., et al.: Investigation of the parameters of a conveyor for transportation of bulk agricultural materials with a bladed working body. In: Farm Machinery and Processes Management in Sustainable Agriculture XII International Scientific Symposium (2024)
24. Paciolla, F., Pace, N., Barile, G., Patimisco, P., Valenza, F., Pascuzzi, S.: Autonomous navigation simulation of an agricultural robot during soil fertilization in open fields. IEEE Int. Workshop Metrol. Agric. For. (MetroAgriFor) **2023**, 640–645 (2023)
25. Paciolla, F., Patimisco, P., Farella, A., Quartarella, T., Pascuzzi, S.: Performance evaluation of hybrid-electric architectures for agricultural tractors. In: Farm Machinery and Processes Management in Sustainable Agriculture XII International Scientific Symposium (2024)
26. Pascuzzi, S., Łyp-Wrońska, K., Gdowska, K., Paciolla, F.: Sustainability evaluation of hybrid agriculture-tractor powertrains. Sustainability **16**(3), 1184 (2024)

Concentration of Pollutants in the Air of a Cattle Farm

Mateusz Gancarz, Maciej Wilk, Sebastian Jaguszewski, Katarzyna Karpińska(✉), and Bożena Nowakowicz-Dębek

University of Life Sciences in Lublin, Lublin, Poland
katarzyna.karpinska@up.lublin.pl

Abstract. Livestock farms are recognized as contributing to environmental degradation, especially climate change. Therefore, it is necessary to control the concentration of pollutants around these facilities. In order to determine the concentration of gaseous pollutants and dust in the air, tests were carried out on a cattle farm. The research was performed in the autumn and winter period on a farm with approximately 150 dairy cows. The monitoring covered pollutants as NO, NO_2, NH_3, VOCs, H_2S, CO_2, O_3, as well as PM1, PM2.5, PM10, respirable dust and total dust. The highest concentrations among the identified pollutants were obtained for volatile organic compounds (4.57 ppm) and O_3 (1.03 ppm). The average concentration of total dust on the farm was 0.76 mg/m^3, and of suspended dust PM10 and PM2.5: 0.48 mg/m^3 and 0.34 mg/m^3, respectively. The monitored conditions on the farm did not pose a threat to animals or employees handling them.

Keywords: Pollutants · Cattle farm · Environmental

1 Introduction

In light of legal requirements, gaseous pollutants in the air of farms are an increasing environmental problem (Hu et al., 2017). The effects of pollution can be analyzed in terms of environmental changes, but also the impact on people working on farms. Breeders and farm workers are exposed to dust, biological aerosol, NH_3, and many other pollutants suspended in the air of livestock buildings (Ji et al. 2017; Ossowski et al., 2018). Monitoring these substances is crucial for human health and animal welfare. The share of individual farms in the emission of these pollutants varies and depends, among others, on: the size of the farm, the species of animals kept, the animal housing, and the feeding system. The research has been conducted for better understanding using the models, as well as estimating the concentrations of these pollutants inside breeding facilities (Borghi et al., 2023; Gieseke et al., 2018; Takai et al., 1998; Stefański et al., 2020, Nowakowicz-Dębek et al., 2019). Determining the exposure of workers to pollutants released on cattle farms, as well as the animals themselves, is a complex problem that requires commitment. The use of new technologies in animal production is important in managing this environment and improving production in accordance with the principles of sustainable development (Teeter et al., 2021; Wlazło et al., 2019; Nowakowicz-Dębek

E. Lorencowicz et al. (Eds.): FMPMSA 2024, LNCE 609, pp. 147–152, 2024.
https://doi.org/10.1007/978-3-031-70955-5_16

et al., 2019; Katie et al., 2022). However, the implementation of selected forms of action requires monitoring the conditions in farm facilities each time. Therefore, research was carried out to monitor the amount of deposited gaseous pollutants on a dairy cattle farm.

2 Material and Methods

The research was conducted on one of the dairy farms in the southeastern part of the country. All aspirators were calibrated, before starting the study. Measurements were carried out three times, i.e. at the end of summer, during autumn, and winter, at 10 points. 10 measurement points covering individual buildings were selected throughout the facility:

I - cowshed 1 point 1; II - cowshed 2 point 1; III - cowshed 2 point 2; IV - cowshed 3 point 1; V- cowshed 3 point 2; VI - cowshed 4 point 1; VII - manure pile 1; VIII - manure pile 2; IX - cowshed 4 point 2.

The portable aerosol analyzer DustTrak™ II Aerosol Monitor 8532 TSI was used for dust testing. Multi-gas analyzers (MultiRAE LITE; MX6 iBrid) equipped with NO, NO_2, volatile organic compounds (VOCs), H_2S, NH_3, and CO_2 sensors were used to measure gaseous pollutants, while the O_3 concentration was tested with a single-gas analyzer (Watch Gas Uni MP100). At the same time, monitoring of microclimatic parameters was carried out using a thermo-hygrometer (Abatronik AB-3321) and an anemometer (LT Lutron Am-4204).

3 Results and Discussion

The results of the tests are presented in Tables 1 and 2 and Figs. 1 and 2. The presence of NO and H_2S was not identified during the tests and the O_2 level remained constant at 20.9%. The concentration of VOCs was the highest in the first sampling period (4.56 ppm), while the lowest in the second testing period (0.99 ppm). The average NO_2 and CO_2 levels were 0.07 – 0.17 ppm and 0.01 – 0.03 ppm, respectively (Table 1). Throughout the entire study period, the average NH_3 level was 0.70 ppm (Fig. 1). The concentration of O_3 was the highest at the 9th measurement point, 3.36 ppm. This magnitude exceeded the permissible levels. However, the average O_3 concentration in the studied period was low at 0.37 ppm (Table 1, Fig. 1).

The concentration of total dust (Total) at individual measurement points ranged from 0.26 mg/m^3 to 0.36 mg/m^3, with an average level of 0.30 mg/m^3 (Table 2, Fig. 2). The highest concentrations of suspended PM10 (0.85 mg/m^3), PM2.5 (0.49 mg/m^3), PM1 (0.49 mg/m^3) and respirable dust fractions were obtained at the Study Point I. The highest levels of dust of all fractions were recorded during the first material collection (in the summer), while the lowest during the second collection (Fig. 2). Low levels of particulate matter (PM) correspond to low levels of NH_3, as this pollutant is considered a precursor of secondary pollutants (Teeter et al., 2021, Katie et al., 2022).

Microclimatic conditions on the farm were monitored during the research. The results are presented as average values in individual study periods (Fig. 3). The highest temperatures were recorded in first measurement period (18.38 °C), while the lowest were recorded in third period. The average for the entire period was 12.37 °C (Fig. 3). However,

Table 1. The average concentration of pollutants in the studied periods at individual points (ppm).

Parameter	Place of sampling									
	I	II	III	IV	V	VI	VII	VIII	IX	X
NO	0.00	0.00	0.00	0.00	0.00	0.00	0.00	0.00	0.00	0.00
NO_2	0.07	0.17	0.17	0.17	0.17	0.17	0.13	0.17	0.17	0.17
VOCs	3.30	3.27	2.93	2.77	2.67	2.43	2.17	2.00	1.93	1.73
H_2S	0.00	0.00	0.00	0.00	0.00	0.00	0.00	0.00	0.00	0.00
CO_2	0.04	0.04	0.03	0.03	0.02	0.01	0.01	0.01	0.02	0.02
NH_3	0.67	0.33	0.33	1.00	1.00	1.00	1.00	1.00	0.33	0.33
O_3	0.15	0.02	0.02	0.02	0.02	0.03	0.06	0.02	3.36	0.05

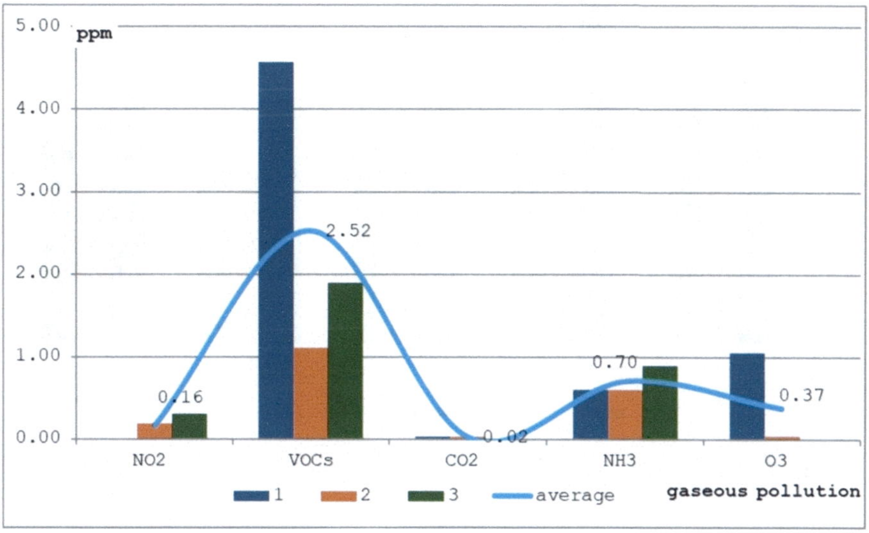

Fig. 1. Concentration of pollutants in individual samples on a cattle farm (ppm).

the relative humidity ranged from 56.66% to 75.31%, creating an average of 63.58%. The air movement was determined to be 0.74 m/s throughout the entire study period (Fig. 3).

The volatility of market conditions affects the dynamics of animal production development. The cattle population according to the Central Statistical Office (GUS, 2022) in December 2022 numbered 6,448.3 head and was 1.1% higher than last year. The increase in cattle numbers results from the increase in the calf population and the number population of cattle aged 2 years and older. In contrast, the cow population decreased by 5.1% compared to 2021. However, production concentration is observed, which results from the economic conditions of production. To meet the requirements, breeders strive

Table 2. Average dust concentration at individual points in the examined period (mg/m^3).

Parameter	Place of sampling									
	I	II	III	IV	V	VI	VII	VIII	IX	X
Total	0.29	0.33	0.26	0.27	0.29	0.31	0.28	0.36	0.29	0.36
PM10	0.85	0.28	0.25	0.25	0.27	0.27	0.26	0.29	0.26	0.30
PM2.5	0.49	0.26	0.24	0.24	0.25	0.25	0.25	0.27	0.25	0.27
PM1	0.49	0.26	0.24	0.24	0.25	0.25	0.25	0.27	0.25	0.27
Resp.	0.53	0.27	0.25	0.24	0.25	0.25	0.25	0.27	0.25	0.27

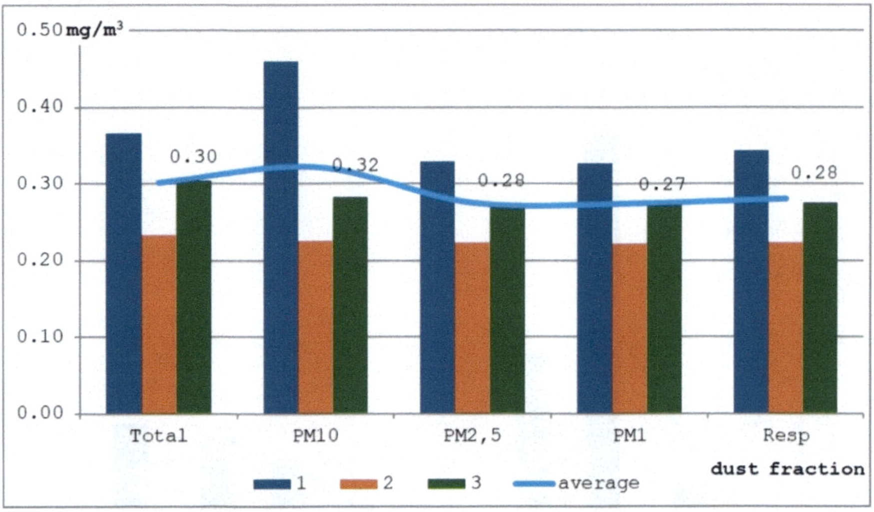

Fig. 2. Average dust concentration in individual samples on a cattle farm (mg/m^3).

to maintain a high level of welfare, which translates into the quality of the raw material and optimal living conditions. Especially since the Food and Drug Administration (FDA), out of concern for the environment and human health, strives to reduce the concentration of contaminants in food production (Teeter et al., 2021). Various feed and bedding additives are being developed to minimize greenhouse gas (GHG) and PM emissions (Chmielowiec-Korzeniowska et al., 2022; Ossowski et al., 2018; 2022; Teeter et al., 2021; Lee et al., 2012; Stefański et al. 2020; Katie et al., 2022). The results of these studies indicate that it is possible to effectively reduce emissions and reduce the environmental load, thereby reducing work exposure on farms.

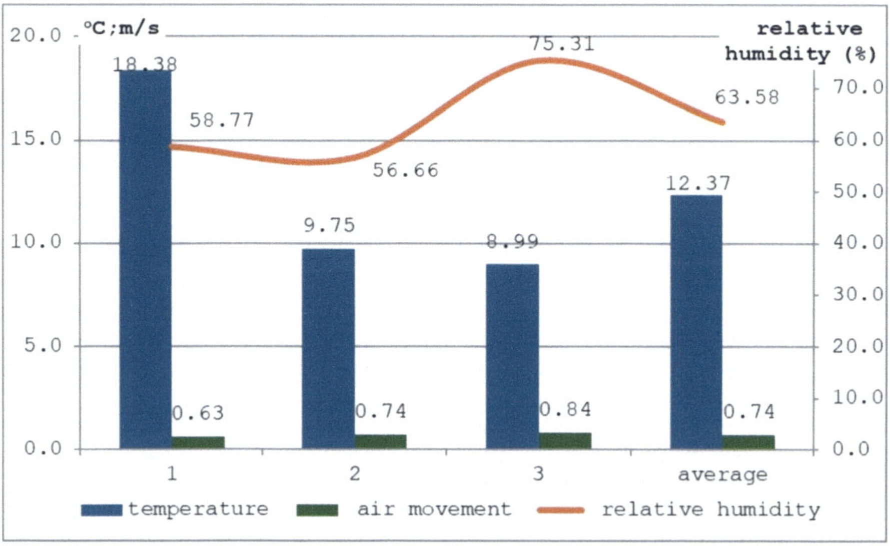

Fig. 3. Microclimatic parameters in individual samples during the studied period.

4 Conclusion

Cattle farms are perceived as emitters of many pollutants, especially GHGs. The levels of NO, NO_2, VOC, H_2S, NH_3 and O_3 were monitored during the research. With the test method used, NO and H_2S were not detected, and the levels of other identified pollutants did not exceed the permissible values.

There were also no exceedances of the recommended values for total dust and PM in the air of the studied farm. Therefore, the tested work environment can be considered safe.

Acknowledgement. The research was funded by the Ministry of Science and Higher Education as part of the implementation of the programme 'Student Scientific Circles Create Innovations' (No. SSKN/SP/569532/2023).

Minister of Science Republic of Poland

Studenckie Koła Naukowe Tworzą Innowacje

References

Gieseke, D., Lambertz, C., Gauly, M.: Relationship between herd size and measures of animal welfare on dairy cattle farms with freestall housing in Germany. J. Dairy Sci. **101**(8), 7397–7411 (2018)

Takai, H., et al.: Concentrations and emissions of airborne dust in livestock buildings in Northern Europe. J. Agric. Eng. Res. **70**(1), 59–77 (1998)

Ji, B., Banhazi, T., Wang, C., Li, B.: A review: the influence and mitigation of ammonia and dust in modern animal buildings. In: International Symposium on Animal Environment & Welfare, pp. 1–10. China Agriculture Press, Chongqing (2017)

Hu, Y., Cheng, H., Tao, S.: Environmental and human health challenges of industrial livestock and poultry farming in China and their mitigation. Environ. Int. **107**, 111–130 (2017)

Potseluev, A.A., Nazarov, I.V., Tolstoukhova, T.N., Kostenko, M.V.: Ecological safety of the internal space of the cattle-breeding facility (Cowshed). In: IOP Conference Series: Earth and Environmental Science, pp. 012005. IOP Publishing, Yurga (2017) (2018)

Borghi, F., et al.: Studies on air pollution and air quality in rural and agricultural environments: a systematic review. Environments **10**(12), 208 (2023)

Wlazło, Ł, Ossowski, M., Nowakowicz-Dębek, B., Krzaczek, P., Kułażyński, M., Maj, G.: Ocena emisji amoniaku z fermy trzody chlewnej z wykorzystaniem sieci pomiarowej (Evaluation of ammonia emission from a swine farm with the use of the measurement network). Przemysł Chemiczny **98**(7), 1075–1077 (2019)

Nowakowicz-Dębek, B., Ossowski, M., Wlazło, Ł, Krzaczek, P., Kułażyński, M.: Wykorzystanie sieci pomiarowej do określenia emisji metanu z fermy trzody chlewnej (The use of a measuring network to determine methane emissions from a pig farm). Przemysł Chemiczny **98**(10), 1606–1607 (2019)

Wlazło, Ł, Nowakowicz-Dębek, B., Kułażyński, M., Wnuk, W., Ossowski, M.: Modelowanie rozprzestrzeniania się amoniaku w powietrzu atmosferycznym wokół fermy drobiu (Modeling the spread of ammonia in atmospheric air around a poultry farm). Przemysł Chemiczny **97**(4), 645–647 (2018)

Nowakowicz-Dębek, B., et al.: Estimating methane emissions from a dairy farm using a computer program. Atmosphere **11**(8), 803 (2020)

Wlazło, Ł, Nowakowicz-Dębek, B., Ossowski, M., Stasińska, B., Kułażyński, M.: Estimation of ammonia emissions from a dairy farm using a computer program. Carbon Manag. **11**(2), 195–201 (2020)

Ossowski, M., Wlazlo, Ł., Nowakowicz-Debek, B., Krukowski, H., Bis-Wencel, H., Martyna, J.: Problem dermatoz u rolników i zwierząt hodowlanych (Dermatoses problem in farmers and farm animals). J. Anim. Sci., Biol. Bioecon. **36**(2), 27–37 (2018)

GUS (Statistics Poland), The cattle population according to the Statistics Poland (2022)

Teeter, J.S., et al.: Effects of feeding lubabegron on gas emissions, growth performance, and carcass characteristics of beef cattle housed in small-pen environmentally monitored enclosures during the last 3 mo of the finishing period. J. Anim. Sci. **99**(12), skab338 (2021)

Lee, C., Hristov, A.N., Dell, C.J., Feyereisen, G.W., Kaye, J., Beegle, D.: Effect of dietary protein concentration on ammonia and greenhouse gas emitting potential of dairy manure. J. Dairy Sci. **95**(4), 1930–1941 (2012)

Stefański, T., Ahvenjärvi, S., Vanhatalo, A., Huhtanen, P.: Ruminal metabolism of ammonia N and rapeseed meal soluble N fraction. J. Dairy Sci. **103**(8), 7081–7093 (2020)

Wyer, K.E., Kelleghan, D.B., Blanes-Vidal, V., Schauberger, G., Curran, T.P.: Ammonia emissions from agriculture and their contribution to fine particulate matter: a review of implications for human health. J. Environ. Manage. **323**, 116285 (2022)

Chmielowiec-Korzeniowska, A., Tymczyna, L., Wlazło, Ł., Trawińska, B., Ossowski, M.: Emissions of gaseous pollutants from pig farms and methods for their reduction–a review. Ann. Anim. Sci. **22**(1), 89–107 (2022)

Ossowski, M., Wlazlo, L., Bis-Wencel, H., Krzaczek, P., Nowakowicz-Debek, B.: Zastosowanie naturalnych sorbentów w żywieniu świń jako metoda zmniejszania zanieczyszczeń z pomieszczeń hodowlanych (The use of natural sorbents in the diet of pigs as a method for reducing gaseous pollutants and manure nutrients from livestock housing). Przemysł Chemiczny **101**(5), 297–303 (2022)

Impact of Soil Erosion on Environmental and Agricultural Sustainability in Bovilla Watershed (Tirana, Albania)

Fran Gjoka(✉), Valmir Baloshi, and Elvin Toromani

Agricultural University of Tirana, Rruga Paisi Vodica 1025, Tirana, Albania
fgjoka@ubt.edu.al

Abstract. This study investigates soil erosion within Tirana's Bovilla Watershed, Albania, and its implications for environmental and agricultural sustainability. Through thorough analysis of erosion rates and runoff volumes across different land slope categories, the complex relationship between topography, land use, and erosion processes is clarified. In the 0–20% slope category, pastureland exhibits lower runoff volumes (201.81 L/m^2) and soil loss (0.28 t/ha) compared to agricultural land (256.47 L/m^2 and 9.43 t/ha, respectively). Conversely, in the 21–40% slope category, both pastureland and agricultural land experience elevated runoff volumes and soil loss, highlighting the vulnerability of steeper slopes to erosion processes. The findings underscore the importance of integrated land management strategies to sustainably manage soil erosion in agricultural landscapes. Further investigation into factors influencing erosion rates and the effectiveness of erosion control measures is essential for tailored conservation approaches. Additionally, data on nutrient loss and economic impact resulting from soil erosion are presented. In agricultural land, the loss of organic matter is 0.55 t/ha, with total N loss at 0.035 t/ha, P loss at 0.011 t/ha, and K loss at 0.066 t/ha. Pastureland experiences lower losses, with organic matter loss at 0.1 t/ha, total N loss at 0.008 t/ha, P loss at 0.002 t/ha, and K loss at 0.01 t/ha. Economically, soil erosion in agricultural land results in a total loss of 177 Euro/ha, with nitrogen accounting for 89 Euro/ha, phosphorus for 25 Euro/ha, and potassium for 63 Euro/ha. These findings emphasize the need for sustainable soil management practices to mitigate erosion's environmental and economic impacts.

Keywords: Erosion Plots · Replacement Cost Method · Soil Nutrients · Economic Losses · Agricultural Sustainability

1 Introduction

Erosion, defined as the rapid loss of soil due to human actions, exceeds the natural rate at which soil forms [6], representing the most significant threat to soil quality. Soil erosion is the most common type of soil damage [7]. As the global population expands, the intensification of agricultural production to meet food demands has intensified pressure on soil re-sources, leading to elevated erosion rates and endangering long-term agricultural sustainability [8].

E. Lorencowicz et al. (Eds.): FMPMSA 2024, LNCE 609, pp. 153–158, 2024.
https://doi.org/10.1007/978-3-031-70955-5_17

Soil erosion presents a significant challenge throughout Albania, particularly evident in regions like the Bovilla watershed in Tirana, where flysch formations predominate. Recognizing its essential role in maintaining ecological balance, agricultural productivity, and water availability, effective conservation and sustainable management of the Bovilla watershed are imperative. The economic losses assessed from soil erosion underscore the viability of soil conservation efforts in Albania [3]. Furthermore, erosion not only damages the soil and ecosystems within the Bovilla watershed but also contributes to non-point pollution of the water in the Bovilla reservoir. This reservoir serves as the primary source of drinking water for Tirana city, Albania's capital, potentially introducing heavy metals like Ni and Cr [4], as well as other chemicals from agricultural inputs [5], and organic pollutants such as sewage.

In this study, our objective is to assess and compare soil erosion rates in arable and pastureland areas within the Bovilla watershed. By analyzing erosion levels in these distinct land use types, we aim to understand their varying impacts on agricultural sustainability and environmental well-being. Through this comparative analysis, we strive to identify customized conservation and management strategies to mitigate soil erosion and uphold the ecological balance and productivity of the Bovilla watershed.

2 Materials and Methods

The study concentrated on the Bovilla watershed, spanning from 41°30' - 41°15' N latitudes and 19°50' - 20°05' E longitudes in northeastern Tirana, encompassing a 95 km2 area. Its climate resembles the Mediterranean, with cold, wet winters and warm, dry summers. The region presents diverse topography and soil types. For a more detailed overview of the study area, refer to [1]. Additionally, it features a water reservoir crucial for providing drinking water to Tirana city.

Erosion caused by water runoff was quantified using erosion plots, each with an area of approximately 100 m^2. The distribution of erosion plots differed between agricultural and pastureland areas. Specifically, in agricultural land, there were three plots on slopes of 0–20% and one plot on slopes of 21–40%. Conversely, in pastureland, there was one plot on slopes of 0–20% and three plots on slopes of 21–40%. These plots varied in shape, including both regular and irregular shapes.

Following each erosion event, water flow volume and sediment were measured using 200-L barrels installed in each experimental plot. Data was recorded and contents of each barrel were mixed until homogeneous. Three 0.5-L bottles were filled from each barrel, labeled, and left for 1–2 weeks to settle out the sediment. Sediment was then quantitatively transferred to a 1000 ml container for monthly collection. After one year, sediment was transferred to the same container for further analysis. Samples were sent to the Soil Laboratory of the Agricultural University of Tirana for dry weight determination and analysis of humus and NPK contents using standard methods. Additional information can be found in the publication by [2].

Economic loss resulting from soil erosion was calculated using the replacement cost method [9]. Data on runoff and soil loss were collected over the period from June 2017 to May 2018. Specifically, data of soil loss and runoff for arable land (cultivated with maize and plantation) were compared with corresponding data for pastureland with similar land slope categories (0–20%). The economic loss (EL) is calculated by the formula:

$$EL = NL * NP \tag{1}$$

where:

NL - Loss of Nutrients – NPK (t).

NP- Market price for plant nutrients (Euro/t).

$$NP = FP / NF \tag{2}$$

where:

FP = Market price for fertilizers (Euro/t); NF = Nutrient content in 1 ton of fertilizer (t).

3 Results and Discussion

The erosion rates and runoff volumes varied considerably across different categories of land slope, reflecting the complex interplay between topography, land use, and erosion processes (Table 1).

Table 1. Runoff volumes and soil loss across different categories of land slope

Category of land slope. %	Pastureland			Agriculture land		
	Erosion plot code	Runoff, L/m^2	Soil loss, t/ha	Erosion plot code	Runoff, L/m^2	Soil loss, t/ha
0–20	P_1	201,81	0,28	A_1	158,15	6,09
				A_2	257,25	8,56
				A_3	354,02	13,65
Mean (0–20%)	P_1	201,81	0,28	A_1-A_3	256,47	9,43
21–40	P_2	200,55	2,17	A_4	421,48	15,87
	P_3	288,7	0,93			
	P_4	269,43	0,98			
Mean (21–40%)	P_2-P_4	252,89	1,36	A_4	421,48	15,87

In the 0–20% slope category, pastureland demonstrated a lower runoff volume (201.81 L/m2) compared to agricultural land (256.47 L/m2). Similarly, the soil loss in pastureland was considerably lower (0.28 t/ha) compared to agricultural land (9.43 t/ha). These findings suggest that lower slopes are generally associated with reduced erosion rates and runoff volumes, highlighting the importance of land slope in erosion dynamics. Conversely, in the 21–40% slope category, both pastureland and agricultural land exhibited higher runoff volumes and soil loss. Pastureland had a runoff volume of 252.89 L/m^2 and a soil loss of 1.36 t/ha, whereas agricultural land showed significantly higher values, with a runoff volume of 421.48 L/m^2 and a soil loss of 15.87 t/ha. The substantial increase in erosion rates on steeper slopes underscores the vulnerability of such terrain to erosion processes, necessitating targeted conservation efforts to mitigate soil loss and preserve soil fertility.

These results underscore the complex relationship between land slope, land use, and erosion dynamics, highlighting the need for integrated land management strategies to sustainably manage soil erosion in agricultural landscapes. Further research into the specific factors influencing erosion rates on different slopes, as well as the effectiveness of erosion control measures, is essential for devising tailored conservation approaches to protect soil resources and maintain ecosystem health.

The data presented in following figures reveals the extent of nutrient loss and economic impact resulting from soil erosion in both agricultural land and pastureland. Firstly, in agricultural land, the loss of organic matter was 0.55 t/ha, with total N loss at 0.035 t/ha, P loss at 0.011 t/ha, and K loss at 0.066 t/ha. In comparison, pastureland experienced lower losses, with organic matter loss at 0.1 t/ha, total N loss at 0.008 t/ha, P loss at 0.002 t/ha, and K loss at 0.01 t/ha.

Further analysis of nutrient loss reveals that the total N loss was 0.063 t/ha (0.035 t/ha direct + 0.028 t/ha through humus) in agricultural land and 0.013 t/ha 0.013 (0.008 t/ha direct + 0.005 t/ha through humus) in pastureland.

The economic impact of soil erosion in agricultural land is substantial, amounting to a total loss of 177 Euro/ha. This loss is distributed with 89 Euro/ha attributed to nitrogen (N), 25 Euro/ha to phosphorus (P), and 63 Euro/ha to potassium (K). These figures underscore the significant financial ramifications of soil erosion on agricultural productivity and sustainability.

In comparison, pastureland also experiences economic losses, albeit to a lesser extent. The total loss from pastureland amounts to 12.4 Euro/ha, with breakdowns of 7.12 Euro/ha from N, 1.5 Euro/ha from P, and 3.8 Euro/ha from K. Despite being lower than agricultural land, these losses still emphasize the economic challenges posed by soil erosion in maintaining the sustainability of pastureland ecosystems (Figs. 1, 2).

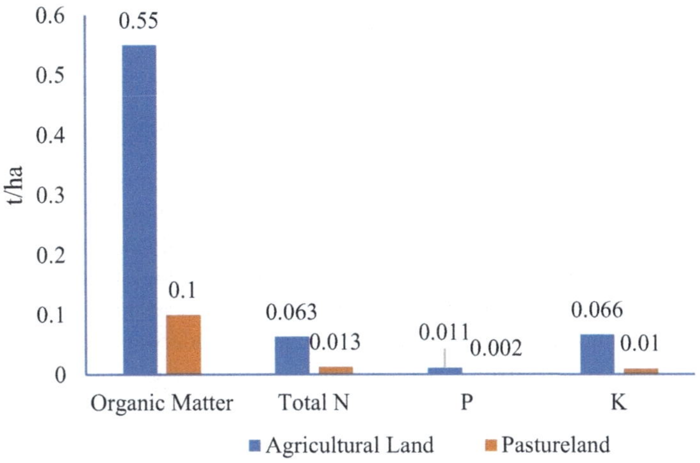

Fig. 1. Humus and nutrient loss from soil erosion in agricultural and pastureland

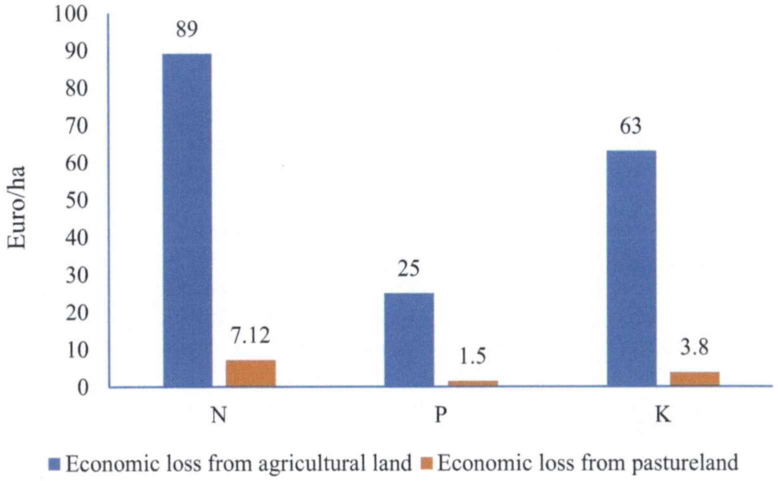

Fig. 2. Economic losses for two types of land use

4 Conclusion

The study revealed significant variations in erosion rates and runoff volumes across different categories of land slope within the Bovilla watershed. These variations underscore the complex interplay between topography, land use, and erosion processes.

Lower slopes (0–20%) generally exhibited lower erosion rates and runoff volumes compared to steeper slopes (21–40%). Pastureland showed lower runoff volumes and soil loss in both slope categories, highlighting the importance of land slope in erosion dynamics.

Steeper slopes (21–40%) demonstrated higher erosion rates and runoff volumes, indicating their vulnerability to erosion processes. This emphasizes the need for targeted conservation efforts to mitigate soil loss and preserve soil fertility on such terrain.

The complex relationship between land slope, land use, and erosion dynamics underscores the necessity for integrated land management strategies. Tailored conservation approaches are essential to sustainably manage soil erosion in agricultural landscapes.

The study revealed significant nutrient loss and economic impact resulting from soil erosion in both agricultural land and pastureland. Agricultural land experienced higher nutrient loss and economic loss compared to pastureland, highlighting the financial ramifications of soil erosion on agricultural productivity and sustainability.

Overall, these findings emphasize the importance of understanding erosion dynamics and implementing effective conservation measures to protect soil resources, maintain ecosystem health, and ensure long-term agricultural sustainability in the Bovilla watershed.

References

1. Baloshi, V., Gjoka, F., Colllaku, N., Toromani, E., Iaroslav, L.K.: Relationship between surface water quality and land use in bovilla watershed, Tirana region (Albania). Fresenius Environ. Bull. **28**(6), 4435–4441 (2019)
2. Baloshi, V., Gjoka, F., Colllaku, N., Toromani, E.: Determination of soil loss by erosion in different land covers categories and slope classes in bovilla watershed, Tirana, Albania. World Academy of Science, Engineering and Technology. Int. J. Environ. Ecol. Eng. **13**(2), 57–61 (2019)
3. Binaj, A., Veizi, P., Beqiraj, E., Gjoka, F., Kasa, E.: Economic losses from soil degradation in agricultural area in Albania. Agric. Econ. – Czech **60**(6), 287–293 (2014)
4. Gjoka, F., Duering, R.A., Siemens, J.: Background concentrations and spatial distribution of heavy metals in Albania's soils. Environ. Monit. Assess. **194**, 115 (2022)
5. Gjoka, F., Miho, L., Vata, N., et al.: Temporal variability of environmental indicators for agriculture in Albania. Environ. Monit. Assess. **193**, 829 (2021)
6. Huber, S., et al.: Environmental assessment of soil for monitoring volume i: indicators and criteria. EUR 23490 EN/1. Luxembourg (Luxembourg): OPOCE. JRC47184 (2008)
7. Lal, R.: Soil erosion and the global carbon budget. Environ. Int. **29**(4), 437–450 (2003)
8. Montgomery, D.R.: Soil erosion and agricultural sustainability. Proc. Natl. Acad. Sci. **104**(33), 13268–13272 (2007)
9. Eswaran, H., Lal, R., Reich, P.F.: Land degradation: an overview. In: Bridges, E.M., Hannam, I.D., Oldeman, L.R., Pening de Vries, F.W.T., Scherr, S.J., Sompatpanit, S. (eds.) Responses to Land Degradation. Proceeding 2nd International Conference on Land Degradation and Desertification, Khon Kaen, Thailand. Oxford Press, New Delhi (2001)

Soil Compaction and Maize Yield in Various Plowing Systems from a Sustainability Perspective

Fran Gjoka[1(✉)], Enkeleda Shkurta[2], and Elian Kasa[1]

[1] Agricultural University of Tirana, Rruga Paisi Vodica 1025, Tirana, Albania
fgjoka@ubt.edu.al

[2] National Environmental Agency, Nr. 4, Rruga Sami Frashëri, Tirana, Albania

Abstract. This paper explores the impact of different plowing systems on soil compaction and maize yield from a sustainability perspective, drawing on data from the ex-experimental base in Mjede, Shkoder, from 1988 to 1989. Conducted in Alluvial soil or Fluvisols (WRB, 2014) with Rozafa 133 hybrid maize, the study employs a randomized design with 6 tillage depth variants (12, 20, 25, 30, 35, 40 cm) and 4 replicates. Soil bulk density (SBD) is measured using the core method. The assessment of maize yield loss due to soil compaction utilizes the linear relationship between depth of tillage and plant yield and soil bulk density (SBD), which calculates the relative production impacted by soil compaction. The results of this study indicate a clear relationship between soil tillage depth and bulk density ($y = -0.0018x + 1.317$; $R2 = 0.926$), with deeper tillage depths generally associated with lower bulk densities, suggesting a potential method to mitigate soil compaction. Additionally, the findings reveal a significant positive correlation between maize yield (%) and bulk density ($y = -497.36x + 719.63$, $R2 = 0.858$), emphasizing the importance of deeper tillage practices for optimizing agricultural productivity and sustainability. The economic loss from maize yield loss due to soil compaction is 325 Euro/ha.

Keywords: Sustainability · Yield Loss · Soil Compaction · Economic Loss · Tillage Depth · Bulk density

1 Introduction

Soil compaction is described as the densification and distortion of soil, leading to the reduction of total and air-filled porosity, and causing deterioration or loss of one or more soil functions (Van den Akker 2008 quoted by [8]). It represents a significant challenge in modern agriculture, impacting agricultural productivity and sustainability worldwide. Soil compaction can be estimated using soil density as an indicator and measure of its severity [2].

Despite extensive research on tillage systems' effects on soil properties, wheat yield, and associated costs [4, 5, 7], the specific impact of these practices on soil compaction remains understudied. This research gap, emphasized by [1], underscores the necessity

E. Lorencowicz et al. (Eds.): FMPMSA 2024, LNCE 609, pp. 159–164, 2024.
https://doi.org/10.1007/978-3-031-70955-5_18

for comprehensive investigations into the relationship between tillage practices and soil compaction. The practice of tillage is a significant component of plant cultivation technology, which directly impacts soil properties and crop production. Understanding this complex relationship is crucial for mitigating adverse effects on crop yield and economic sustainability.

Therefore, the primary objective of this study is to examine how soil compaction resulting from tillage activities influences maize yield reduction and subsequent economic losses. Through the analysis of multi-year experimental data obtained from soils suitable for maize cultivation, our aim is to provide insights that can inform more effective soil management strategies and enhance agricultural sustainability despite soil compaction challenges.

2 Materials and Methods

The study was conducted at the ex-experimental base in Mjede, Shkoder, situated in the northwestern region of Albania. This experimental site is characterized by Alluvial soil or Fluvisols (WRB, 2014), known for its high fertility and suitability for agriculture. However, these soils typically exhibit low structural stability, primarily attributed to their high silt content and limited humus levels. Detailed morphological and physicochemical characteristics of these alluvial soils are provided by [3]. These soils in Albania occupy an area of about 130,000 ha and are used for the cultivation of cereals and especially maize.

The three-year experiment, conducted from 1988 to 1990, employed a randomized complete block design with six tillage depth variants: 12 cm, 20 cm, 25 cm, 30 cm, 35 cm, and 40 cm. Each tillage depth variant was replicated four times to enhance reliability. Rozafa 133 hybrid maize (Zea mays L.) was chosen as the test crop, with maize seeds planted between April and May, and harvests carried out from September to October, depending on the experimental year.

Soil bulk density (SBD) was assessed utilizing the core method employing a volumetric cylinder. Soil samples were gathered at two specific depths: 5–15 cm and 25–35 cm, representative of the maize plant root zone. Bulk density measurements were conducted both after maize sowing and before harvest to monitor soil compaction alterations throughout the growing period.

Maize yield (q/ha) was determined through the harvesting of maize plants at maturity from each experimental plot. The yield data underwent analysis to assess the impact of different tillage depths on maize productivity. The effect of soil compaction on crop yield was evaluated via the calculation of relative yield of maize (RYM) using the adopted model proposed by [6].

$$\mathrm{RYM} = 100 - \mathrm{g}(\mathrm{SBD} - \mathrm{f}),\ \text{for SBD} \geq \mathrm{f} \tag{1}$$

where, g is the slope of the compaction response curve, f is the bulk density threshold value at which the plant yield begins to be affected, and SBD is a common measure of the bulk density of soil.

All statistical analyses were carried out using SPSS 23.0, with a significance level set at $\alpha = 0.05$.

3 Results and Discussion

Table 1 illustrates the impact of various tillage treatments on soil bulk density at different depths (5–15 cm, 25–35 cm, and overall, 5–35 cm) alongside maize yield. An increase in tillage depth from 12 cm to 40 cm generally correlates with a reduction in soil bulk density across all depths (5–15 cm, 25–35 cm, and overall, 5–35 cm). At the 5–15 cm depth, soil bulk density ranges from 1.21 g/cm^3 for a tillage depth of 40 cm to 1.27 g/cm^3 for a tillage depth of 12 cm, indicating that deeper tillage results in lower soil bulk density in this layer.

Similarly, at the 25–35 cm depth, soil bulk density ranges from 1.28 g/cm^3 for a tillage depth of 12 cm to 1.32 g/cm^3 for a tillage depth of 12 cm, highlighting the trend of deeper tillage leading to lower soil bulk density. The overall soil bulk density (averaged across 5–35 cm depth) follows a similar pattern, with lower values observed for deeper tillage depths (Fig. 1).

Descriptive statistics reveal variations in maize yield across different tillage depth variants, ranging from 85.84 q/ha in the 12 cm variant to 118.31 q/ha in the 40 cm variant, with standard deviations ranging from 0.466 to 2.703, indicating the degree of variability within each treatment group.

Table 1. Effect of tillage treatment on soil bulk density (g/cm^3) and maize yield (q/ha)

Depth of tillage, cm	Average soil bulk density (5–15 cm)	Average soil bulk density (25–35 cm)	Average soil bulk density (5–35 cm)	Yield, q/ha
12	1.27	1.32	1.29	85.84
20	1.27	1.3	1.28	102.47
25	1.23	1.31	1.27	107.1
30	1.25	1.3	1.27	105.16
35	1.23	1.29	1.26	111.58
40	1.21	1.28	1.24	118.31
Significance	–	–	–	F = 0.012

The ANOVA analysis presented in Table 2 reveals a statistically significant difference between the means of repetitions ($p = 0.012$), indicating variability among the experimental replicates.

Tukey's post-hoc test identified significant differences in maize yield among all variants ($p < 0.05$), except for the 25 cm and 30 cm variants ($p = 0.558$), indicating similar yields for these two depths. However, specific data from this analysis is not provided here.

To quantify the reduction in crop yield, we calculate the relative yield of maize by determining two key statistical parameters from the previously mentioned model: the slope of the yield response to soil compaction (g) and the threshold value of bulk density (f).

Table 2. Results of ANOVA analysis

	Sum of Squares	df	Mean Square	F	Sig.
Between Groups	4,468	3	1,489	,012	,998
Within Groups	2434,475	20	121,724		
Total	2438,943	23			

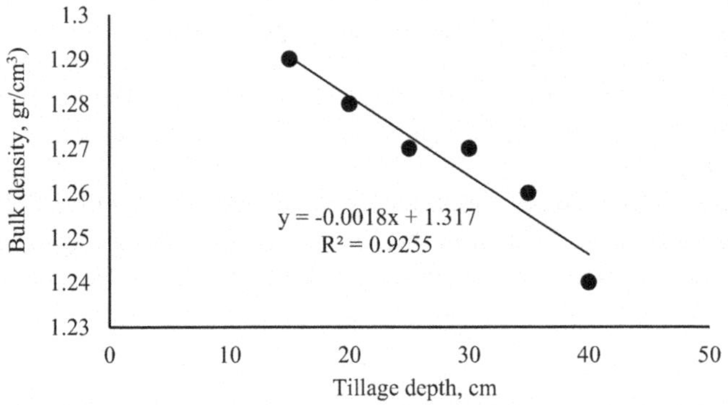

Fig. 1. Relationship between bulk density (5–35 cm) of soil and tillage depth

This process involves solving the linear function ($y = -497.36x + 719.63$) illustrated in Fig. 2, which outlines the relationship between relative maize yield and soil bulk density.

The slope parameter (represented by "g") is determined to be −497.36. To compute the bulk density threshold ("f"), Eq. (2) is utilized:

$$f = (b - 100)/a \tag{2}$$

where, b represents 719.63, representing the intercept of the linear function. Consequently, for a relative yield of 100%, the bulk density threshold ("f") is calculated to be 1.25 g/cm^3. BD = 1.28 g/cm^3, which corresponds to a relative yield of 87% (data is sourced from Gjoka 1994).

For a bulk density threshold value of 1.25 g/cm^3, the relative maize yield was found to be 85%. This relative yield is then utilized to determine the potential yield of maize (PYM) by the formula:

$$PYM = AYM/RYM \tag{3}$$

where, AYM represents the actual maize yield (ton/ha) obtained from official statistics.

Subsequently, the loss in maize yield due to soil compaction (LYM) is calculated using the formula:

$$LYM = PYM - AYM \tag{4}$$

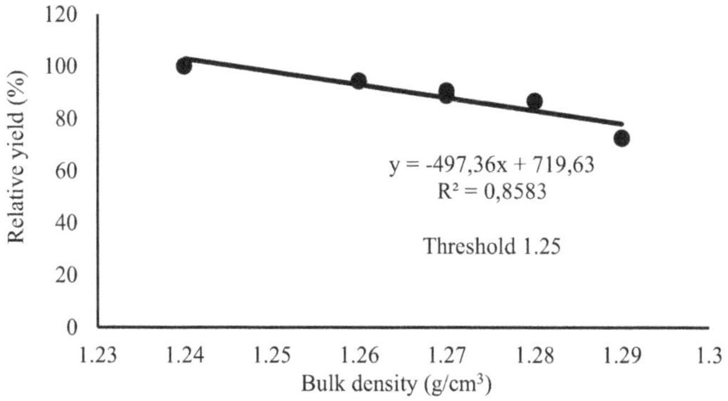

Fig. 2. Correlation between soil compaction and relative maize yield

Finally, the loss in maize yield is converted into economic loss (ELM, in Euro) by using the formula:

$$ELM = LYM * \mathrm{AM} \times PM \tag{5}$$

where, AM represents the cultivated area (ha) and PM is the market price for maize (Euro).

Using official data, the economic loss from maize yield reduction due to soil compaction is calculated to be 325 Euro/ha. This finding underscores the critical importance of mitigating soil compaction through appropriate tillage practices to optimize crop productivity and minimize economic losses for farmers.

These results have significant implications for sustainable agriculture, emphasizing the crucial role of soil management practices in enhancing crop productivity and ensuring long-term environmental sustainability. By adopting appropriate tillage depths and strategies to minimize soil compaction, farmers can improve soil health, water infiltration, and nutrient availability, thereby contributing to sustainable agricultural systems and long-term food security.

4 Conclusion

The findings of this three-year investigation into the impact of tillage depth on soil compaction and subsequent crop yield loss offer valuable insights for predicting production losses in maize cultivated in similar alluvial soils.

The study highlights the significant influence of tillage systems on soil compaction and maize productivity. Deeper tillage effectively reduces soil bulk density, particularly at depths of 5–15 cm and 25–35 cm, suggesting the efficacy of appropriate practices in mitigating compaction.

Variations in maize yield across tillage depths underscore the importance of depth in maximizing productivity. Relative yield analysis indicates an 85% yield at a bulk density threshold of 1.25 g/cm^3, highlighting the detrimental impact of compaction.

Economic analysis estimates a loss of 325 Euro/ha from compaction-induced yield reduction, underscoring the advantages of sustainable soil management.

Overall, this study emphasizes the critical role of tillage practices in enhancing crop productivity and ensuring sustainability.

References

1. Binaj, A., Veizi, P., Beqiraj, E., Gjoka, F., Kasa, E.: Economic losses from soil degradation in agricultural area in Albania. Agric. Econ. – Czech **60**(6), 287–293 (2014)
2. Filipovic, D., Husnjak, S., Kosutic, S., Gospodaric, Z.: Effects of tillage systems on compaction and crop yield of Albic Luvisol in Croatia. J. Terramechanics. **43**(2), 177–189 (2006)
3. Gjoka, F., Miho, L., Spaholli, A., Kasa, E.: Weathering patterns in alluvial soils under Mediterranean climatic conditions in Albania and implications for soil fertility. Carpath. J. Earth Environ. Sci. **18**(2), 461–474 (2023)
4. Gjoka, F.: Problems of genetic and production characterization and of improvement of the alluvial soil fertility. Ph.D.Thesis. Agricultural University of Tirana, Tirana (1994)
5. Gjongecaj, B., Veizi, P.: The impact of various types of tillage on the soil water availability. Albanian J. Agric. Sci. **13**(2), 29–40 (2014)
6. Houk, E., Frasier, M., Schuck, E.: The regional effects of waterlogging and soil salinization on a rural county in the Arkansas river basin of Colorado. In: Western Agricultural Economics Association, 2004 Annual Meeting, June 30–July 2, 2004, Honolulu, Hawaii, No. 36229
7. Sallaku, F.: The plowing depth, agronomical and deriving economical effects. Bull. Agric. Sci. **1**, 31–34 (1996)
8. Schjønning, P., et al.: Soil compaction. In: Soil Threats in Europe - Status, Methods, Drivers and Effects on Ecosystem Services, EU Joint Research Centre: Petten, The Netherlands, Chapter 6 (2016)

Evaluation of Engine Oil Degradation: Enhancing Cost-Efficiency in Vehicle Fleet Maintenance

Wojciech Gołębiowski[1](✉), Artur Wolak[2], and Grzegorz Zając[1]

[1] Department of Power Engineering and Transportation, Faculty of Production Engineering, University of Life Sciences in Lublin, Głeboka 28, 20-612 Lublin, Poland
wojciech.golebiowski@up.lublin.pl

[2] Department of Quality and Safety of Industrial Products, Cracow University of Economics, Sienkiewicza 4, 30-033 Kraków, Poland

Abstract. Vehicle fleet operators face many challenges. The most important of them are maintaining low operating costs, optimizing energy consumption, managing the risk of failure, maintaining high fleet efficiency, and meeting increasingly stringent environmental standards. An important role is to strive to eliminate undesirable losses of energy and materials resulting from improper lubrication of the combustion engine. This task can be made easier by regularly monitoring the physicochemical parameters of engine oil. The article aims to answer the research question regarding the degree of differentiation in the degradation of engine oil of the same viscosity and quality class, used in different types of vehicles and engine designs. The research material was engine oil in the SAE 10W40 viscosity class. The oil was used in four buses of the Municipal Transport Company. The buses differ in type (12 and 18 m) and engine design (2 different types). As part of the experimental work, the kinematic viscosity at 40 °C and 100 °C and the viscosity index were examined using the Stabinger SVM 3001 viscometer. Oil parameters were determined using the FTIR method, such as the degree of oxidation, nitrooxidation and sulfonation, soot and fuel content, degradation of additives, and acid and base number. Moreover, to examine changes in the content of elements responsible for oil additives, elemental analysis based on the HDXRF method was used. The test results showed that vehicle type and engine design influence oil degradation, considering these factors is crucial when choosing optimal oil service strategies.

Keywords: cost-efficiency · oil analysis · degradation · viscosity · FTIR · elemental analysis · additive depletions

1 Introduction

Vehicle fleet operators face several technical, economic, and environmental challenges. The key issues are maintaining low operating costs, optimizing energy consumption, ensuring high fleet reliability, and compliance with increasingly stringent emission standards. In this context, the condition of the engine oil plays an important role, directly

E. Lorencowicz et al. (Eds.): FMPMSA 2024, LNCE 609, pp. 165–178, 2024.
https://doi.org/10.1007/978-3-031-70955-5_19

affecting the performance and durability of vehicles. Proper maintenance and use of recommended supplies can help avoid unexpected machine failures [1, 2]. After use, engine oil becomes a valuable source of diagnostic information, and the benefits of regular oil analysis include reducing operating costs, optimizing fuel consumption, managing failure risk, maintaining high fleet efficiency, and meeting environmental standards. Additionally, regularly monitoring your engine oil can help extend the life of other engine components, such as the turbocharger and catalytic converter, and make servicing easier [3, 4].

Early detection of signs of oil wear and degradation allows for its replacement at the optimal time. This prevents engine damage and extends its lifespan, which translates into significant savings on repair and replacement part costs.

A study by Raposo et al. [5] found that regularly monitoring oil in city buses extended oil change intervals by 25%. This resulted in savings on engine oil purchase costs and labor costs associated with oil changes. The methodology proposed by the authors can be applied to many other machines, vehicles, and devices.

The selection of appropriate lubricants is paramount to optimizing vehicular performance and extending the service life of components. Synthetic lubricants, in particular, exhibit superior friction reduction and wear mitigation capabilities [6, 7]. On the other hand, the employment of an unsuitable lubricant can have a detrimental impact on vehicle performance, resulting in elevated friction, wear, and heat generation, which, in turn, leads to diminished efficiency and a shortened lifespan of engine components.

A well-chosen and regularly monitored engine oil ensures optimal lubrication, leading to reduced internal friction within the engine. This, in turn, translates into lower fuel consumption and emissions. Macian et al. [8] demonstrated that using a low-viscosity oil in conjunction with regular condition monitoring can reduce fuel consumption in city buses by up to 4.5% without observing any significant adverse effects on engine wear. A well-maintained engine operates more efficiently, delivering better performance and lower fuel consumption, which translates into reduced emissions of pollutants such as nitrogen oxides (NOx), carbon dioxide (CO2), and particulate matter (PM). Meeting emission standards is becoming increasingly important due to the growing emphasis on environmental protection and public health [9].

Regular oil analysis enables the early detection of potential issues, such as excessive oil consumption, oil contamination, loss of lubricating properties, and the presence of water or fuel in the oil. Early identification of these problems allows for preventive maintenance actions and averts major engine failures. For instance, detecting excessive oil consumption could indicate an engine leak [10]. Prompt repair will prevent further engine damage and costly breakdowns.

Conducting oil testing under real-world conditions poses challenges in achieving reproducible results [11]. There is still a scarcity of research presenting long-term real-world testing of engine oil degradation [12].

Inefficient engine lubrication leads to undesirable energy and material losses. Reducing costs and energy expenditures are primary objectives for fleet operators. As part of experimental research, the viscosity of four samples of used lubricating oils was investigated. These oils were obtained from two different engine designs installed in buses belonging to a municipal public transportation company.

This article delves into the topic of monitoring the physicochemical parameters of engine oil as a tool for optimizing fleet operations [13–17].

Among the numerous parameters that characterize engine oil quality, viscosity is widely recognized as one of the most critical indicators of lubricating oil performance. Subsequently, properties such as the degree of oxidation, nitrooxidation, and sulfonation, the content of contaminants in the form of soot and fuel, the degradation of additives, and the acid and base numbers are also considered to be significant measures of engine oil quality. All of these parameters were taken into account in the presented investigation. Additionally, elemental analysis was employed to determine changes in the elements associated with oil additives. The primary objective of this study was to conduct a comprehensive analysis and comparison of the changes occurring in engine oil operated in different bus types and engine designs.

2 Materials and methods

2.1 Materials

Four diesel-powered urban buses were subjected to comprehensive monitoring under real-world driving conditions. All buses were lubricated with a semi-synthetic SAE 10W40 UHPDO (Ultra High-Performance Diesel Oil) lubricant specifically formulated for high-performance diesel engines. The detailed specifications and performance characteristics of this oil are presented in Table 1. Engine oil samples were collected during scheduled maintenance intervals by the manufacturer's recommendations, corresponding to approximately 60,000 km of vehicle operation. A comprehensive breakdown of the mileage data for each engine oil is provided in Table 2.

Table 1. Characteristics of the new engine oil

Parameters of the new engine oil	Value
SAE grade	10W40
ACEA category	E7, E9
OEM specification	MB228.5
Kinematic Viscosity @ 40 °C [mm^2/s]*	95.96
Kinematic Viscosity @ 100 °C [mm^2/s]*	14.41
Viscosity Index*	155.30
TBN [mgKOH/g]*	10.57
Ca (Calcium) [ppm]*	4305
P (Phosphor) [ppm]*	960
S (Sulfur) [ppm]*	3806
Zn (Zinc) [ppm]*	1376

* based on own research

A comprehensive analysis of four samples of used engine oil was conducted to evaluate the alterations in their physicochemical properties and additive consumption over a single oil change cycle. The cycle duration, expressed in terms of the kilometers driven by each bus, was determined based on the similar operating conditions of each vehicle, with the flexibility to extend it up to 4000 km in certain instances. The two concluding columns of Table 2 present data on oil top-ups (the addition of fresh oil) and fuel consumption per 100 km of travel.

Table 2. Characteristics of evaluated engine oil samples

No.	Sample code	Bus model	Bus mileage [km]	Mileage since last oil change [km]	Amount of the oil top-ups [l]	Fuel consumption per 100 km [l]
1	2334	Mercedes Benz 628 Conecto G	678,258	64,162	30	55.55
2	2335	Mercedes Benz 628 Conecto G	667,290	64,683	27	57.72
3	2340	Mercedes Benz 628 Conecto G	732,678	63,347	25	56.32
4	2328	Mercedes Benz 628 Conecto LF	846,313	61,131	29	39.04

2.2 Research Objects

Four buses served as the subjects of this research study. Specifically, three 18-m (articulated) Mercedes Benz (MB) 628 Conecto G buses and one 12-m MB 628 Conecto LF bus were utilized, each operating under typical urban conditions. Table 3 provides a detailed overview of the specifications and engine type for each bus. Notably, the study encompasses two distinct vehicle designs (12-m and 18-m) and two engine types. This experimental setup was meticulously crafted to address a crucial question: will the same oil undergo similar degradation in different vehicles?

The OM457 and OM 926 engines boast a wide range of applications, including trucks, marine vessels, military vehicles, municipal vehicles, buses, agricultural machinery, and stationary engines. These engines are water-cooled and feature a cast-iron cylinder block with replaceable wet cylinder liners. They utilize diesel fuel delivered through a direct injection system with individual electronically controlled unit pumps driven by the camshaft. The cylinder heads are separate assemblies for each cylinder. The OM 926 LA engine is also employed in grain harvesters. The "LA" designation signifies the

Table 3. Engine specifications of the selected buses

Parameter	Mercedes Benz 628 Conecto LF	Mercedes Benz 628 Conecto G
Model year	2009	2009
Length of bus [m]	12	18
Engine model	OM 926 LA	OM 457 LA.V
Engine displacement [cm^3]	7201	11,967
Cylinders	6	6
Engine power	205 kW at 2500 rpm	260 kW at 2000 rpm
Maximum torque	1250 Nm at 1300 rpm	1600 Nm at 1100 rpm
Emission class	EURO V	EURO V
Oil sump volume [l]	29 l	35 l

turbocharged version of the engine. For instance, the OM 926 LA engine powers the CLAAS Tucano grain harvester.

2.3 Methods

The kinematic viscosity of the used oil samples at 40 °C and 100 °C, as well as the viscosity index, were measured using an SVM Stabinger model 3001 (Anton Paar) viscometer in accordance with ASTM D7042.

Fourier transform infrared (FTIR) spectroscopy, employing a Thermo Nicolett IS5 (Thermo Fisher Scientific) instrument, was employed to evaluate the degree of oxidation, nitrooxidation, sulfonation, and soot and fuel content. The analysis involved comparing the FTIR spectra of fresh and used oil, which were obtained from difference spectra (spectra of used oil minus fresh oil). Additionally, the total base number (TBN) and total acid number (TAN) were determined using an ERASPEC Oil (Eralytics) instrument in accordance with ASTM E2412.

The trace element content in the oils was determined using an HD Maxine (XOS) multi-element analyzer employing high-resolution X-ray fluorescence (HDXRF). Before conducting the measurements, the instrument was subjected to calibration. Subsequently, the oil samples were mixed, heated to 40 °C, and loaded into a measuring cup utilizing propylene foil. A 1 ml aliquot of the test material was withdrawn using an automated pipette. The measurement involved three replicates, and the final result was reported as the average value.

The dispersant properties of the oil were evaluated using paper chromatography. A representative oil sample was applied to chromatographic paper (model SG), and the appearance of the spot was observed after drying (24 h). The evaluation focused on the spot's core (inner circle) and the dispersion zone (second circle).

3 Results and Discussion

Table 4 presents the results of kinematic viscosity changes at 40 °C and 100 °C for the examined engine oil samples. Table 4 also includes the percentage changes in viscosity parameters, calculated based on the values assigned to the new oil (Table 1). The study established that a change exceeding 15% is classified as a critical change (red color), a change exceeding 10–15% as a warning change (orange color), and a change below 10% is considered an acceptable change (green color).

Table 4. Relative kinematic viscosity changes at 40 °C and 100 °C of the examined engine oils

Sample code	Bus model	Mileage since oil change [km]	KV @ 40°C [mm²/s]	KV @ 100°C [mm²/s]	VI	Percentage changes [%] KV @ 40°C	KV @ 100°C	VI
2334	Conecto G	64,162	78.32	12.42	156.6	**-18.4%**	-13.8%	**0.8%**
2335	Conecto G	64,683	78.35	12.45	157.1	**-18.4%**	-13.6%	**1.1%**
2340	Conecto G	63,347	79.17	12.63	158.7	**-17.5%**	-12.3%	**2.2%**
2328	Conecto LF	61,131	85.39	13.54	161.5	-11.0%	**-6.0%**	**4.0%**

* where KV is kinematic viscosity and VI is viscosity index

As shown in Table 4, the kinematic viscosity of all oil samples decreased. Notably, the oil samples with codes 2334 and 2335 from the Conecto G articulated buses displayed a similar change in viscosity parameters at both 100 °C and 40 °C. This could indicate the consistency of the results, which could potentially be extrapolated to other similar vehicles with analogous engines in the future. It's worth noting that the kinematic viscosity at 40 °C for all oils from the Conecto G buses exceeded the established critical limit of ±15% change from the viscosity of the fresh oil. A warning value was observed for this parameter for the Conecto LF model. On the other hand, for kinematic viscosity measured at 100 °C, warning ranges were only observed for oils from the Conecto G articulated buses. For the oil from the Conecto LF model, the 6% change is not a cause for concern. The decrease in oil viscosity was mainly attributed to the presence of unburnt fuel, as observed in the FTIR spectrum. Sejkorova et al. [18] conducted a study on a public transport vehicle fleet and presented similar observations regarding changes in these parameters. On the other hand, Macian et al. [8] suggest that the primary factor contributing to the viscosity decrease is the shearing of viscosity index improvers (VIIs), additives that enhance oil viscosity. Similar conclusions were made by Honig et al. [19] about possible causes of decrease in kinematic viscosity. The smaller percentage changes in viscosity for the Conecto LF model could potentially be linked to the highest soot content among all tested oils. For oils from the Conecto G articulated bus models, the average percentage change in kinematic viscosity decrease at 40 °C was 18.1%. The reduced viscosity change in the oil may be attributed to the use of top-up oils during operation. The Conecto G bus with code 2340 required an additional 25 l of oil to be added to its 35-l oil pan. This resulted in a kinematic viscosity decrease of 17.5% at 40 °C and 12.3% at 100 °C. In contrast, for the oil from the Conecto LF bus with code 2328, 29 l were added to a 29-l sump. This implies that all the oil was replaced, which could contribute to the minimal viscosity change for this model.

The analysis of soot content using the infrared spectrometer involved measurements at two key wavenumbers: ~2000 cm^{-1} and ~3980 cm^{-1}. Figure 1 presents the slope of the curve between these points for the examined oils, which confirms the high soot content for the oil sample from the Conecto LF bus.

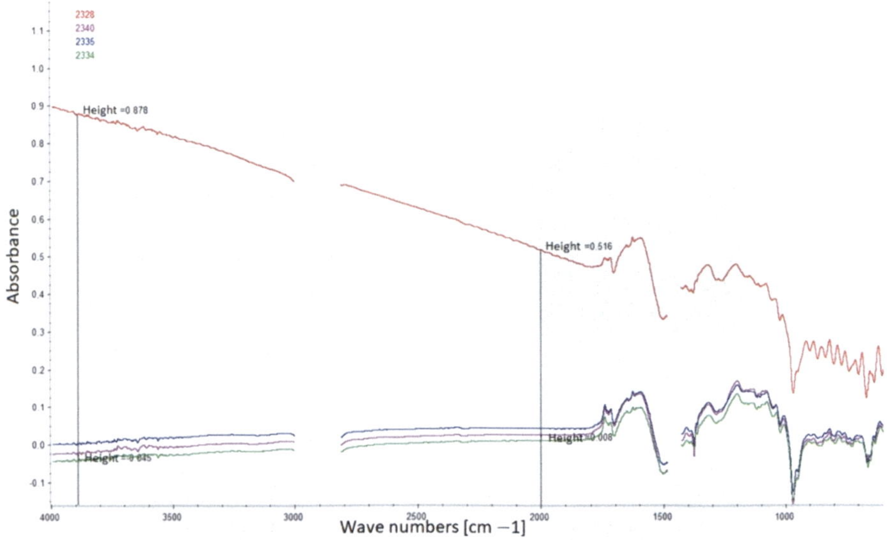

Fig. 1. FTIR spectrum - Soot content in engine oil samples

No significant soot accumulation was observed in the Conecto G articulated buses. Soot is a byproduct of incomplete combustion, and its formation can be influenced by various factors, including fuel type (sulfur content), engine operating conditions, and combustion efficiency. The difference between the two analyzed engine types is substantial and warrants clear emphasis. It is challenging to pinpoint the exact cause of the high soot content in the oil from the Conecto LF bus. Notably, while a decrease in sulfur content was observed at the end of the operating period for the Conecto G articulated buses, an increase was recorded for the Conecto LF bus. However, both vehicles were fueled with the same low-sulfur fuel (maximum 10 ppm). Additionally, the experimental conditions assumed similar engine operating conditions under typical urban driving conditions. Therefore, the only potential explanation for the significant differences in soot levels in the examined samples lies in the structural differences in the combustion chamber or more precise injection equipment enabling a more efficient combustion process.

Table 5 presents the results of paper chromatography, which corroborate the FTIR findings. The Conecto LF bus with code 2328, which had the lowest oil mileage, exhibited the darkest spot, confirming the high soot content. In contrast, the Conecto G buses, despite their higher mileage, displayed significantly lighter spots, suggesting better preservation of the oil's dispersant properties. The paper spot test further revealed that the oils from the Conecto G buses (2334 and 2340) retained the highest dispersant-detergent properties, as the spot image allows for the differentiation of the dispersion zone from the spot's core.

Table 5. Characteristics of changes in dispersing properties in two different bus engines based on blotter spot test

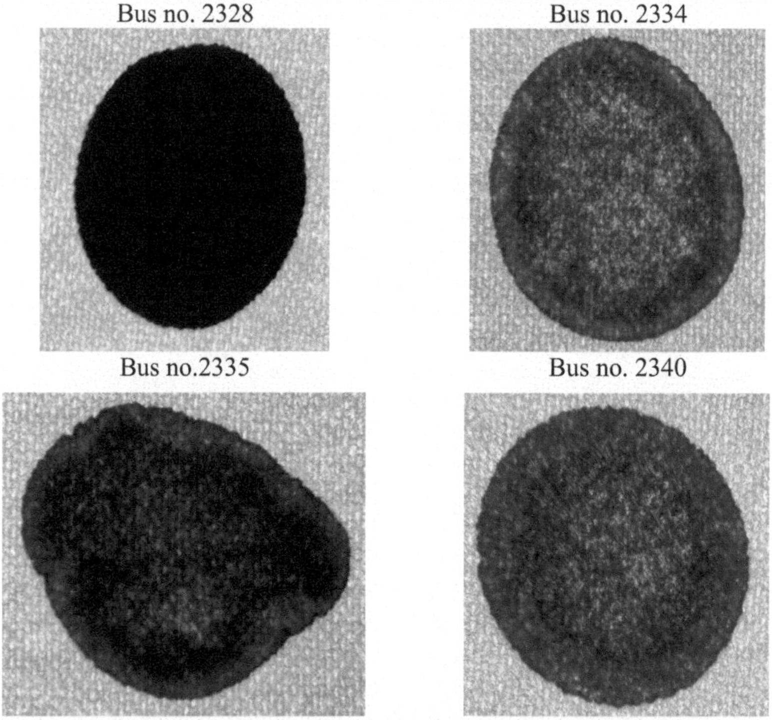

Figure 2 presents the FTIR spectra in the region related to nitrooxidation (~1630 cm^{-1}), oxidation (~1720 cm^{-1}), and the presence of fuel in the oil (~1747 cm^{-1}).

It is noteworthy that the results correspond well with the viscosity measurements. The peak around ~1747 cm^{-1}, indicating the presence of fuel (ON), is most pronounced in the spectra of oils from the Conecto G articulated buses, while the oil sample from the Conecto LF bus model exhibits the weakest signal in this region.

The best results in terms of oxidation levels (0.036 [Abs/0.1 mm]) were obtained for the oil from the Conecto LF bus, which had the lowest operating mileage. This is an interesting observation in the studied sample set, as the oil with the issue of high soot content also exhibits one of the better percentage changes in viscosity parameters. The weakest result in the analyzed set (0.061 [Abs/0.1 mm]) was observed for the oil from the Conecto G articulated bus with code 2340. The oil sample from this bus was characterized by the highest oxidation degree value – 41% higher than the best result.

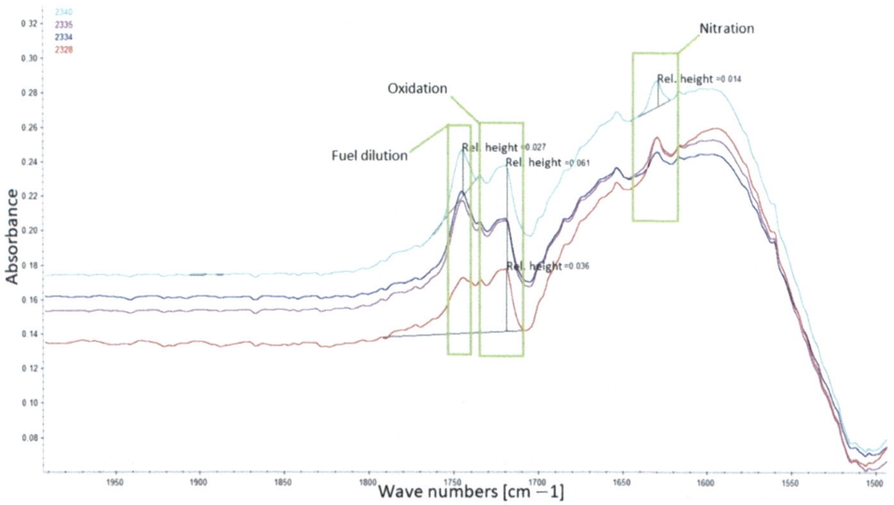

Fig. 2. FTIR spectrum – Degree of oxidation and nitration and fuel dilution in engine oil samples

When considering the degree of nitrooxidation (~1630 cm^{-1} band), the highest degree (0.014 [Abs/0.1 mm]) was also observed in the Conecto G articulated bus with code 2340.

Figure 3 presents the FTIR spectrum for the ~970 cm^{-1} region, corresponding to additive degradation.

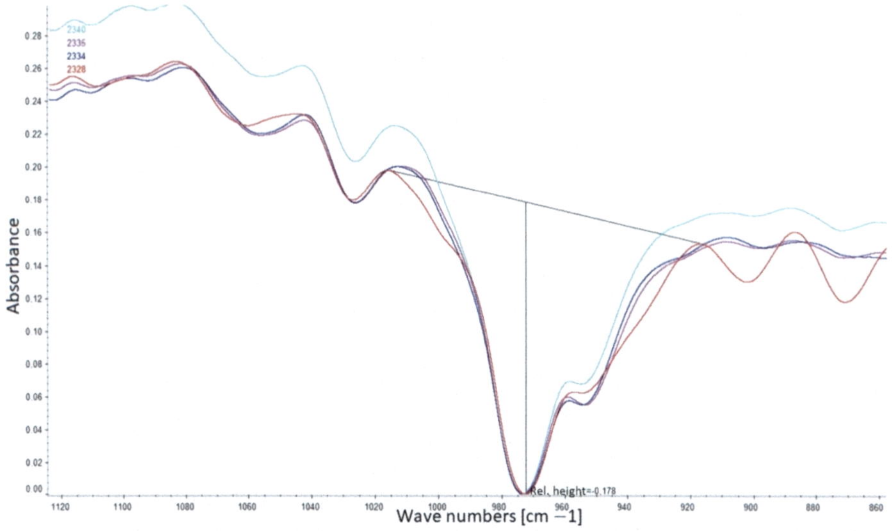

Fig. 3. FTIR spectrum - Additives depletion in engine oil samples

The degradation of extreme pressure (EP) and anti-wear (AW) additives, as read from the FTIR spectrum, was at a similar level for all analyzed oil samples. Nevertheless, a noticeably higher additive degradation was observed for the oil sample with code 2340 from the Conecto G articulated bus. This situation may have a direct connection with the previously observed highest values for the degree of oxidation and nitrooxidation of the studied oil. Similar relationships and influences were shown in a study [18], in which the authors performed a correlation analysis between these parameters and the degradation of anti-wear and extreme pressure additives in oils from city buses. A potential cause of the reduced level of additive depletion may be the amount of oil top-ups used during operation. The least degree of additive degradation was observed in buses with the most oil top-ups. Both in the Conecto G articulated bus with code 2334 (30 l top-up in a 35-l oil sump) and in the Conecto LF bus with code 2328 (29 l top-up in a 29-l oil sump), this resulted in a decrease in additive degradation to 0.179 [Abs./0.1 mm] for bus 2334 and to 0.177 [Abs./0.1 mm] for bus 2328. On the other hand, in the case of a 25-l oil top-up in a 35-l oil sump in bus 2340, a decrease to 0.205 was observed, which represents a degradation 13% higher than that of the other oils.

Table 6 shows the total acid and base number results.

Table 6. Changes in the Total Base Number and Total Acid Number in the analyzed engine oil samples

Sample data		Parameters		Percentage changes
Sample code	Bus model	TBN [mg KOH/g]	TAN [mg KOH/g]	TBN [%]
		$\overline{x}$±SD	$\overline{x}$±SD	$\overline{x}$±SD
2334	Conecto G	8.63 ±0.1	0.73 ±0.1	−18.6% ±0.01
2335	Conecto G	8.60 ±0.2	0.83 ±0.1	−18.9% ±0.01
2340	Conecto G	8.07 ±0.1	1.13 ±0.1	−23.9% ±0.02
2328	Conecto LF	7.77 ±0.1	1.40 ±0.1	−26.7% ±0.01

$\overline{x}$ - arithmetic average, SD – Standard deviation

Examining the TAN and TBN parameters for the Conecto G articulated buses, the oil with code 2334 retained the largest alkaline reserve of 8.63 [mg KOH/g] (representing a decrease of 18.6% compared to the fresh oil). In contrast, the oil with code 2340 exhibited a decrease of 23.9% compared to the fresh oil. A potential reason for the preservation of better oil neutralization properties may be the increased amount of fresh oil topped up during operation. In bus 2334, the amount of topped-up oil was 30 l, while in bus 2340, 25 l of oil were topped up. A similar relationship is evident in the degree of oil acidity; the fewer top-ups applied, the higher the TAN parameter obtained.

Table 7 presents the elemental analysis of the examined oil samples.

Table 7. Relative additives depletion changes in analyzed engine oil samples

Sample data	Additives [ppm]					Percentage changes [%]				
Sample code	Ca	Mo	P	S	Zn	Ca	Mo	P	S	Zn
	$\bar{x}\pm$SD	$\bar{x}\pm$SD	$\bar{x}\pm$SD	$\bar{x}\pm$SD	$\bar{x}\pm$SD	$\bar{x}\pm$SD	$\bar{x}\pm$SD	$\bar{x}\pm$SD	$\bar{x}\pm$SD	$\bar{x}\pm$SD
2334	4577 ±31	358 ±9	937 ±25	3645 ±30	1383 ±1	6.3% ±0.7	−5.5% ±2.5	−2.4% ±2.6	−4.2% ±0.8	0.5% ±0.04
2335	4528 ±9	356 ±15	940 ±13	3678 ±23	1390 ±2	5.2% ±0.2	−6.1% ±2.5	−2% ±1.4	−3.4% ±0.6	1% ±0.2
2340	4607 ±12	356 ±5	966 ±18	3588 ±17	1407 ±1	7% ±0.3	−6.1% ±1.5	0.7% ±1.8	−5.7% ±0.4	2.2% ±0.1
2328	4819 ±28	354 ±19	1003 ±42	3836 ±4	1444 ±6	11.9% ±0.7	−6.6% ±4.9	4.5% ±4.4	0.8% ±0.1	4.9% ±0.4

$\bar{x}$ - arithmetic average, SD – Standard deviation

Examining the changes in the content of elements corresponding to the additives' presence in the oil, it can be observed that the oil with code 2340 (Conecto G) experienced the largest decrease (5.7%) in the sulfur content responsible for antioxidant additives, which corresponded well with obtaining the lowest oxidation degree result among all tested oils. Additionally, a 2.2% increase in zinc, which is responsible for anti-wear additives in the oil, was observed, suggesting an engine wear process rather than additive depletion. The Conecto LF bus is an example of oil in poor condition. In this case, the most visible is the highest increase in calcium content (11.9%) among all analyzed oils, which is responsible for detergent-dispersant additives in the oil and resistance to overheating. Similar conclusions were presented in a paper [20] in which the authors, based on ICP analysis of soot-contaminated oil, showed a decrease in the concentration of additive elements from anti-wear and detergent-dispersant additives. Additionally, calcium is also a component of coolants, which in this case may indicate coolant leaks and poor engine condition. The increase in sulfur content (0.8%) indicates a high degree of oil acidity, which is confirmed by the highest TAN value among all analyzed oils. The observed highest phosphorus content (4.5%), which is present as an anti-abrasive additive in engine oil, also confirms the high TAN value. The amount of fresh oil replenishment during operation may be a potential reason for the slowdown in oil degradation. In the case of Conecto G articulated buses, the largest oil top-ups of 30 l - in the bus with code 2334 resulted in the smallest decrease in molybdenum content (5.5%), which is responsible for anti-wear additives. Comparing this to bus 2340, where 25 l of oil were topped up, the content of this additive was 6.1%.

4 Conclusions

Monitoring the physicochemical parameters of engine oil is a crucial element of an effective vehicle fleet management strategy. It enables the reduction of operating costs, optimization of energy consumption, minimization of the risk of breakdowns, maintenance of high fleet efficiency, and compliance with environmental standards.

The conducted rheological studies revealed that the oil used in different types and designs of buses operating in typical urban conditions degrades in varying manners. The differences were observed in the kinematic viscosity parameter of the oil.

FTIR analysis also revealed differences in the degradation patterns of oil used in different vehicle types. Variations were observed in the degree of oxidation, nitrooxidation, and soot content. The high soot content in the Conecto LF bus correlated well with both the paper chromatography method [21] and the observations of the degree of oxidation and nitrooxidation in the corresponding FTIR spectral bands. Furthermore, a concordance was observed between the decrease in refining additives responsible for antioxidant additives, evident as sulfur content in the elemental analysis, and the simultaneous increase in the degree of oxidation visible in the FTIR spectrum. This demonstrates the good complementarity of both research methods. In the four analyzed cases, the depletion of oil refining additives was observed at a similar level. Similar conclusions regarding complementarity were presented in a study [19] where the authors focused on engine oils obtained from agricultural tractors.

From the perspective of viscosity parameters and maintaining adequate oil lubricity, the oil life recommended by vehicle manufacturers/service providers was found to be excessive. The potentially better kinematic viscosity parameters in the Conecto LF bus did not indicate better oil condition but rather a high soot content that counteracted the effect of fuel dilution of the oil. Based on these findings, it is not recommended to generalize the results of engine oil property tests for resource optimization across an entire vehicle fleet. Instead, it is advisable to focus on the data obtained from the analysis of used oils from the specific engine type and vehicle design used in a given fleet to develop customized maintenance models that will enhance lubricant management efficiency. Additionally, it is crucial not to make oil change decisions based on a single parameter alone. Only a comprehensive assessment of oil degradation allows for rational decision-making regarding oil interval optimization.

It is important to emphasize that this work is merely an introduction to the topic and requires further research to thoroughly analyze the impact of oil monitoring on various aspects of vehicle fleet operations, such as public transportation or agricultural tractors.

References

1. Zhu, X., Zhong, C., Zhe, J.: Lubricating oil conditioning sensors for online machine health monitoring – a review. Tribol. Int. **109**, 473–484 (2017). https://doi.org/10.1016/j.triboint.2017.01.015
2. Ľubomír, H., et al.: Design of laboratory test equipment for automotive oil filters to evaluate the technical life of engine oil. Appl. Sci. **11**, 483 (2021). https://doi.org/10.3390/app11020483

3. Wakiru, J.M., Pintelon, L., Muchiri, P.N., Chemweno, P.K.: A review on lubricant condition monitoring information analysis for maintenance decision support. Mech. Syst. Signal Process. **118**, 108–132 (2019). https://doi.org/10.1016/j.ymssp.2018.08.039
4. Hamasha, M.M., et al.: Strategical selection of maintenance type under different conditions. Sci. Rep. **13**, 15560 (2023). https://doi.org/10.1038/s41598-023-42751-5
5. Raposo, H., Farinha, J., Fonseca, I., Ferreira, L.: Condition monitoring with prediction based on diesel engine oil analysis: a case study for urban buses. Actuators **8**, 14 (2019). https://doi.org/10.3390/act8010014
6. Dörr, N., et al.: Correlation between engine oil degradation, tribochemistry, and tribological behavior with focus on ZDDP deterioration. Tribol. Lett. **67**, 1–17 (2019). https://doi.org/10.1007/s11249-019-1176-5
7. Agocs, A., Besser, C., Brenner, J., Budnyk, S., Frauscher, M., Dörr, N.: Engine oils in the field: a comprehensive tribological assessment of engine oil degradation in a passenger car. Tribol. Lett. **70**, 1–18 (2022)
8. Macián, V., Tormos, B., Bastidas, S., Pérez, T.: Improved fleet operation and maintenance through the use of low viscosity engine oils: fuel economy and oil performance. Eksploatacja i Niezawodność – Maint. Reliab. **22**, 201–211 (2020). https://doi.org/10.17531/ein.2020.2.3
9. Šarkan, B., Loman, M., Harantová, V.: Identification of places with deteriorated air quality in city of Žilina in relation to road transport. Arch. Autom. Eng. Archiwum Motoryzacji **102**, 68–90 (2023). https://doi.org/10.14669/AM/176958
10. Gołębiowski, W., Zając, G.: Impact assessment of farm tractor engine fuel injector damage on engine oil properties. Agric. Eng. **24**, 65–75 (2020). https://doi.org/10.1515/agriceng-2020-0018
11. Abdulqadir, L.B.: Investigation of the effects of soot on the wear of automotive engine components. Energy **253**, 124150 (2017). https://doi.org/10.1016/j.energy.2022.124150
12. Chourasia, S., Patel, P.D., Lakdawala, A., Patel, R.N.: Study on tribological behavior of biodiesel – Diethyl ether (B20A4) blend for long run test on compression ignition engine. Fuel **230**, 64–77 (2018). https://doi.org/10.1016/j.fuel.2018.05.055
13. Wolak, A.: TBN performance study on a test fleet in real-world driving conditions using present-day engine oils. Measurement **114**, 322–331 (2018). https://doi.org/10.1016/j.measurement.2017.09.044
14. Zhang, R., et al.: A study into the impact of engine oil on gasoline particulate filter performance through a real-world fleet test. Presented at the WCX SAE World Congress Experience, 2 April 2019.https://doi.org/10.4271/2019-01-0299
15. Nagy, A.L., et al.: Rapid fleet condition analysis through correlating basic vehicle tracking data with engine oil FT-IR spectra. Lubricants **9**, 114 (2021). https://doi.org/10.3390/lubricants9120114
16. Kučera, M., Kopčanová, S., Sejkorová, M.: Lubricant analysis as the most useful tool in the proactive maintenance philosophies of machinery and its components. Manag. Syst. Prod. Eng. **28**, 196–201 (2020). https://doi.org/10.2478/mspe-2020-0029
17. de Jonge, B., Scarf, P.A.: A review on maintenance optimization. Eur. J. Oper. Res. **285**, 805–824 (2020). https://doi.org/10.1016/j.ejor.2019.09.047
18. Sejkorová, M., Hurtová, I., Jilek, P., Novák, M., Voltr, O.: Study of the effect of physicochemical degradation and contamination of motor oils on their lubricity. Coatings **11**, 60 (2021). https://doi.org/10.3390/coatings11010060
19. Hönig, V., et al.: Determination of tractor engine oil change interval based on material properties. Materials **13**, 5403 (2020). https://doi.org/10.3390/ma13235403

20. Al Sheikh Omar, A., Motamen Salehi, F., Farooq, U., Morina, A., Neville, A.: Chemical and physical assessment of engine oils degradation and additive depletion by soot. Tribol. Int. **160**, 107054 (2021). https://doi.org/10.1016/j.triboint.2021.107054
21. Gołębiowski, W., Wolak, A., Šarkan, B.: Engine oil degradation in the real-world bus fleet test based on two consecutive operational intervals. Lubricants **12**, 101 (2024). https://doi.org/10.3390/lubricants12030101

The Course of Changes in the Starch Index of Apples Calculated Using Selected Optical Instruments

Tomasz Guz[1] (✉) and Małgorzata Szczepanik[2]

[1] Department of Food Engineering and Machinery, University of Life Sciences in Lublin, Lublin, Poland
tomek.guz@up.lublin.pl

[2] Department of Applied Mathematics and Computer Science, University of Life Sciences, Lublin, Poland

Abstract. The aim of the study was to check the usefulness of computer image analysis methods in calculating the starch index (S) value. Starch index determined properly facilitates the harvesting of fruit at the optimal time for a given variety. Starch tests were performed on apple fruit of two varieties: Gloster and Jonagold, with circular and radial starch disintegration on the surface of their cross-section. Images were recorded using selected digital image recording methods. During the four-year experiment cycle, they were assessed in terms of the course of starch breakdown using visual assessment (VA) and three optical techniques (BW – black and white images of slices (CCD camera), DC – colour images from a digital camera, SC – images from scanner). Relationships were determined between simple regression equations illustrating the course of changes in S as a function of time. According to the methodology proposed by Janishidian et al. (2005), the values of the probability of compliance of the course of the compared straight pairs were calculated. It was shown that the course of straight lines obtained using the program, which is a component of the Supervist system software (used only in the BW method), was characterized by the highest compliance.

Keywords: apples · starch index · computer image analysis · optimization · cost reduction

1 Introduction

1.1 Apples Maturity at Harvest

High demands and expectations of consumers regarding the quality of food force the necessity of objective and fast control of raw materials and products in the food industry. The quality assessment performed with the help of the human senses is subjected to subjectivism and is also prone to a series of errors caused by fatigue and monotony of the activities performed. That is why there has been a need to adapt to the research and application aims of the image analysis system known since the 1960's (Brosnan, Sun

E. Lorencowicz et al. (Eds.): FMPMSA 2024, LNCE 609, pp. 179–189, 2024.
https://doi.org/10.1007/978-3-031-70955-5_20

2004). Fruits, which are climacteric organisms, show an increase in ethylene production and starch decomposition during development (Vanoli 1995; Song, Bangerth 1996), which is associated with increased respiration (Tomala 2003; Tomala 2004; Toivonen 2008). There is a close relationship between fruit maturity and starch breakdown rate (Fan et al. 2011; Lohner et al. 2021). The harvest maturity of apples expressed by the starch index is one of the most accurate and simplest methods of assessing the physiological condition of these fruits.

1.2 Starch Index and Its Interpretation

Despite the simplicity of the determination of the starch content by a color reaction with iodine, there is another, significant problem consisting in the correct interpretation of the images of the "starch pattern" obtained in this way. According to researchers dealing with this issue (Peirs et al. 2002), apple maturity grading using comparative tables (benchmarks), caused by the subjectivity of perception by the assessors are burdened with a large error, which, committed by the evaluators, reach up to 60%. The average error values for estimation of the starch index values are about 50% for varieties with a radial starch type decay, and about 43% for varieties with a circular disintegration.

The largest dispersion of the results was reported for images where the starch index was 5, 6 or 7 on a 10-point scale. This is the starch disintegration phase, which for many apple varieties occurs in a group maturity state, whose accurate indication is the basis for the proper storage of harvested fruits (Watkins 2008; Brookfield 1997; De Long et al. 1999; Sjöstrand et al. 2024). A similar discrepancy in the interpretation of starch index images was noted by Guz (2009, 2010), who compared the values of this indicator using the traditional method performed by the evaluation panel with the values obtained by computer image analysis methods.

2 Materials and Methods

The paper attempts to develop a methodology for precise and objective calculation of the starch index value S. This important indicator is estimated so far by visual comparison of fruit samples with reference tables being photographs of fruit sections stained with iodine-starch complex.

The purpose of the work was to check the suitability of computer image analysis methods in calculating the starch index value. As part of the research, starch tests of apple fruits with a circular and radial type of starch disintegration on their cross-section area were performed in a wide range of harvest dates. The images were recorded using selected methods of digital image recording. In the four-year cycle of the experiment, three optical tools were used to record images of stained fruit slices:

- CCD camera registering black and white images of patches (BW),
- a digital camera that records color images (DC),
- scanner (SC).

In the four-year research cycle (A (2019) – D (2022)), the samples for testing consisted of two varieties of apples: Gloster and Jonagold, were picked up from the Experimental Orchard of the University of Life Sciences in Lublin, located in the Felin district.

Harvesting was carried out between 253 and 285 days of the year at 4-day intervals. Fruits were collected from selected trees of the given variety, located in the same quarters. Four fruits were collected from each tree, growing at a height of 1.5–2 m above ground level. The stained fruit slices were laid on the scanner plate (Epson 600) and then scanned at 300 dpi. Stained fruit slices were also recorded using a digital camera (DC method) and CCD camera (BW method), which was mounted on a column for reproduction placed in a light-tight, ventilated chamber.

The received images were saved in the computer's memory and transferred for further analysis. Images stored in the computer's memory were displayed to five assessors who compared them with the reference tables, assigning them values from 1 to 10, in accordance with the valid rating scale. Photos taken using the scanner and digital camera were analysed in Photoshop CE 4.0. The histogram function was used as the basic tool for determining the surface areas of slices, areas of surface occupied by starch and background surface of images. For this purpose, it was necessary to determine the brightness ranges of pixels in the image that formed the area necessary for the analysis.

On this basis, the object distribution threshold was determined in two parts for a whole series of photos taken during one session. From the histogram showing the distribution of brightness of the points of the analysed image, the percentage of the surface area of each image field was read. The area calculated in the program was used as follows (Fig. 1):

$$S = 10 \times (1 - \frac{p_w}{p_c}) \tag{1}$$

where:

S – starch index value,
p_c – the whole slice area,
p_w – the area occupied by starch pattern.

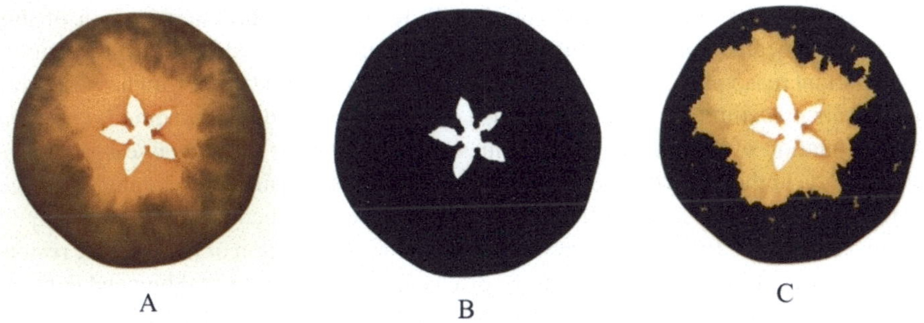

Fig. 1. Image transformation stages: A – slice image with starch pattern (primary image), B – image of the whole slice area (Pc), C – image of starch occupied area during measurement procedure (Pw)

3 Results and Discussion

The starch index value, for Gloster apples cultivar is between 2 and 3 during harvest. In this experiment the fruit harvest should take place between 269 and 281 days of the year (Fig. 2).

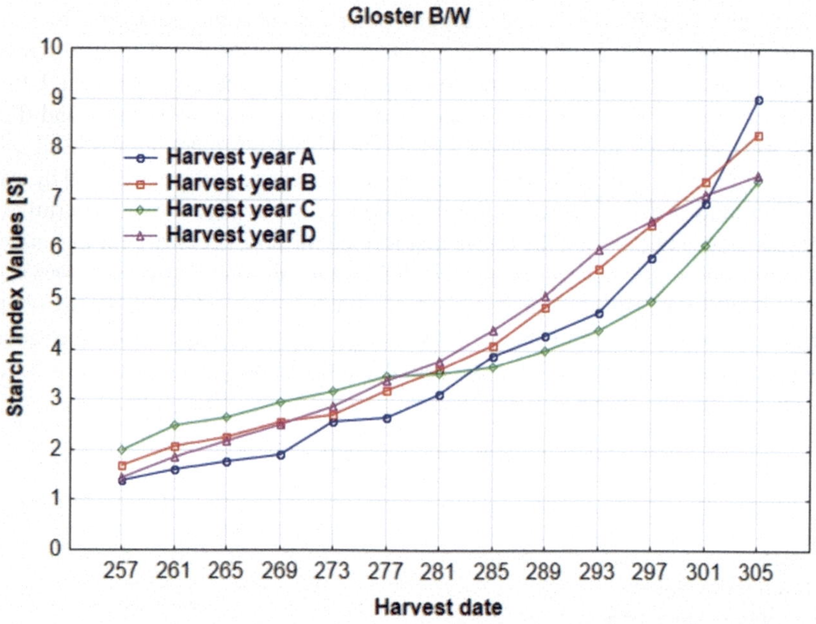

Fig. 2. The course of changes in the S value in the fruits of the Gloster variety

This interval (so-called "harvest window") results from the interpretation of this indicator by means of the starch image registration methods used on the fruit sections. An important reason for choosing such a range was also observation of fruit harvest dates in the years preceding the experiment. Visual assessment (traditional method) based on the presentation of images recorded by means of the scanner 5 to the assessors shows significantly higher starch index values (S) than those obtained by computer image analysis (CIA) methods. According to the indications of this method, the harvest should take place between 261 and 273 days of the year. The acceleration of harvest date results in higher susceptibility of fruit to storage diseases, mainly surface scalding and damage caused by high concentration of CO_2 in storage chambers (Tomala 1995, Łysiak 2011; Thewes et al. 2023). Another, important effect of the early harvest is poor fruit growth and their smaller mass and dimensions (Rutkowski et al., 2003). In the final dates of the set (289–305), S values were underestimated by artifacts not belonging to the area occupied by the "starchy pattern" using the SK method. They appeared where scanned objects were thick and not planar surfaces. This was the case when illuminating a slice of fruit with a thickness of 5 mm. There was then a shadow whose brightness was within the range of the iodine complex with starch and was identified as its "additional" surface. This introduced a systematic measurement error that appeared when the space occupied by the complex was small. At high S values in the final phase of the experiment, the presence of such artifacts significantly interfered with the accuracy of the measurement. At low values of the starch index, no significant influence of this factor on the interpretation of the value of this index was noted. Due to the conditions of image production using the scanner (SC), the effect of "lowering" the starch index value by the presence of a patch shadow was already visible in the considered range of harvest dates.

The Jonagold variety fruits represent a "radial" starch disappearance on the cross-sectional area. The breakdown of starch was observed on the whole surface, which was manifested by a gradual disappearance along the radius of the circular cross-section of the fruit, in the form of bright areas of the flesh, which occupied an increasing surface area. The fruit harvest of this variety, which is intended for long-term storage, is recommended to be carried out when the starch index value (S) is in the range 5 to 7. This means that 50 to 70% of the cross-section area of the fruit no longer shows a dark blue iodine complex with starch. In the experiment, the values of this index were interpreted by calculating them on the basis of computer image analysis using, which were then compared with the values obtained using the traditional method (VA). Detailed analyses were conducted for the results obtained in the collection days from 265 to 281 days of the year (Fig. 3). This range was established on the basis of long-term observation of the dates of fruit harvesting of this variety. It was also confirmed in the experiment.

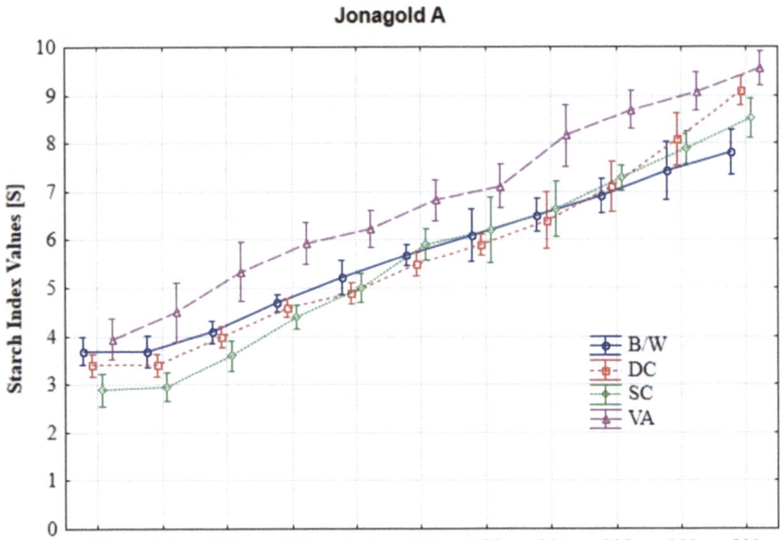

Fig. 3. An example of course of changes in the S value in the fruits of the Jonagold variety

The S values obtained with the VA method were higher by 1 degree on average compared to those obtained using the CIA methods, which means that the proportion of the surface devoid of color was estimated according to the VA method by 10% higher. Considering the complicated distribution of the "starchy pattern" in Jonagold fruits, which belong to the group of varieties with a radial starch decomposition model, such an error was plausible (Peirs et al. 2002).

Tables 1 and 2 show the relationships between simple regression equations illustrating the course of S changes as a function of time. According to the methodology proposed by Janishidian (2005), the probability values were determined (Tables 3 and 4) in the course of the compared straight pairs. These coefficients, for simple imaging diagrams of S changes in two collective seasons, determined by the same method, indicate a large variation in the similarities between simple regression equations. Analysing data from Table 3, it can be noticed that the course of simple lines obtained with the help of the program, which is a component of the Supervist system software (used only in the BW method), was characterized by the highest compatibility. This was demonstrated by summing the coefficients of probability of conformity of the course of straight pairs, at which the comparisons of simple for this method showed the highest values.

Table 1. Regression equations of Starch index during Gloster apples maturation

Harvest year	method	Regression equation	R^2
A	BW	$S = 0{,}140t - 35{,}570$	0,946
	DC	$S = 0{,}144t - 36{,}409$	0,958
	SC	$S = 0{,}119t - 30{,}104$	0,979
	VA	$S = 0{,}158t - 39{,}320$	0,981
B	BW	$S = 0{,}134t - 33{,}478$	0,970
	DC	$S = 0{,}136t - 34{,}480$	0,971
	SC	$S = 0{,}108t - 27{,}246$	0,942
	VA	$S = 0{,}157t - 38{,}811$	0,984
C	BW	$S = 0{,}134t - 33{,}024$	0,940
	DC	$S = 0{,}125t - 31{,}983$	0,930
	SC	$S = 0{,}106t - 26{,}568$	0,964
	VA	$S = 0{,}146t - 36{,}231$	0,979
D	BW	$S = 0{,}132t - 32{,}799$	0,990
	DC	$S = 0{,}154t - 39{,}481$	0,960
	SC	$S = 0{,}114t - 28{,}329$	0,972
	VA	$S = 0{,}144t - 34{,}905$	0,991

The coefficient of compliance values lower than in the BW method were observed for the DC and SC assessment methods. The traditional method of assessing the starch index (VA), due to errors resulting from the subjective assessment of people who determine its value, obtained the lowest sum of probabilities. Among the six comparison pairs, the course of simple one pair differed significantly when the objects evaluated were the cross sections of the Gloster fruit. When the Jonagold fruit was evaluated, only the simple equations of one pair were not significantly different.

Table 2. Regression equations of Starch index during Jonagold apples maturation

Harvest year	method	Regression equation	R^2
A	BW	S = 0,110t – 24,288	0,996
	DC	S = 0,138t – 32,042	0,984
	SC	S = 0,149t – 35,014	0,988
	VA	S = 0,142t – 31,995	0,996
B	BW	S = 0,163t – 39,223	0,987
	DC	S = 0,124t – 28,748	0,984
	SC	S = 0,124t – 28,493	0,974
	VA	S = 0,130t – 29,179	0,980
C	BW	S = 0,117t – 26,416	0,983
	DC	S = 0,127t – 29,674	0,982
	SC	S = 0,109t – 25,586	0,949
	VA	S = 0,151t – 35,350	0,993
D	BW	S = 0,127t – 28,956	0,994
	DC	S = 0,138t – 31,622	0,987
	SC	S = 0,121t – 27,566	0,991
	VA	S = 0,131t – 28,623	0,993

The results presented in Table 4 indicate a smaller convergence of simple runs expressing the maturation of the Jonagold fruit. It can be assumed that the fruits of this variety are more sensitive to changes in climate parameters during maturation or, due to the starch decomposition method in fruit and the complex form of the "starchy pattern", determination of the S value is more difficult both by traditional method and CIA methods.

According to Table 2, in which the regression equations and straight-line (determination coefficients have been placed, they are high (0.94–0.99). The use of equations in this form is in this case a good approximation of the actual course of the experiment. Linear models of changes in the starch index were confirmed by previous studies (Reid 1982, Peirs et al 2002).

Apple fruit varieties Jonagold are those in which the breakdown of starch proceeds according to the scheme characterized by the radial disappearance of this polysaccharide, which is visible in the form of disappearing dark blue spots on the cross-section of the fruit. In this case, the length of the line delimiting the fields occupied by the starch from areas where it is no longer detected is long, which may affect the accuracy of the determination of the value of S, even with precise determination of the filtration threshold value. In this case, the identification of the object, which consists in calculating its surface area as a percentage of the whole image, depends on the error that was made when determining the filtration threshold.

Table 3. The compatibility comparison of *S* values obtained from linear regression models (Gloster variety)

Years of comparison	method			
	DC	BW	VA	SC
A/B	0,82	0,32	0,78	0,39
A/C	0,01	0,04	0,20	0,64
A/D	0,82	0,33	0,19	0,31
B/C	0,04	0,52	0,06	0,74
B/D	0,59	0,99	0,42	0,03
C/D	0,01	0,49	0,02	0,15
Total	2,31	2,69	1,67	2,26

Table 4. The compatibility comparison of S values obtained from linear regression models (Jonagold variety)

Years of comparison	method			
	DC	BW	VA	SC
A/B	0,00	0,00	0,00	0,09
A/C	0,00	0,67	0,00	0,00
A/D	0,01	0,10	0,15	0,02
B/C	0,43	0,00	0,00	0,00
B/D	0,00	0,00	0,00	0,64
C/D	0,00	0,29	0,00	0,00
Total	0,44	1,06	0,15	0,75

4 Conclusions

Attempts to use optical tools to assess the quality of apples have been made in numerous publications (Peirs et al. 2000; Peirs et al., 2002; Peirs et al., 2005; Quing et al. 2007; Solovchenko et al. 2010; Solovchenko et al. 2011; Linker and others 2012; Hengari et al. 2014; Jarolmasjed et al. 2016 Cardenas-Perez et al. 2017). Maturity of apples has been evaluated for a long time with a small but reliable group of tests that accurately describe the changes during their maturation (Zhang et al. 2020). The most important among them include the content of ethylene in seed nests, the content of starch, the firmness of the pulp and the content of the extract. (Łysiak 1998). Among them there are those in which the researchers tried to establish a relationship between the quality changes occurring in the fruit flesh and the state of their surface (Zhang et al. 2021).

According to Szalay et al. (2013) the starch content, expressed in the starch index, is the most reliable indicator of the ripening process of apples. However, the results of observation of their maturation, despite the abundant research material, carried out only in one season, should be treated with caution according to the authors. This is particularly true for the determination of the optimal harvesting date and the length of the "harvest window" interval. The results of the research included in the work included 7 storage seasons, including 4 seasons, treated as the first, preliminary stage of research. The purpose of this step was to pre-assess the suitability of the optical tools used to calculate the starch index. In the initial phase of experiments, the comparison of the starch index value determined by various methods was made, starting from traditional (VA), using the scanner (SC), black and white CCD camera (BW) and digital camera in the RGB (DC) system. Regression equations illustrating S changes in 4 consecutive seasons of the set were determined using the significance analysis of differences of these equations for individual methods. In the summary of the experiment, the following conclusions can be made:

- The highest compliance rates of the regression equations, illustrating changes in the S index, were achieved by black and white photo analysis (BW), regardless of the breakdown of starch in the fruit.
- For fruits representing the circular type of starch decay (Gloster), the coefficients of conformity of the course of regression equations were assumed higher than for fruits with radial starch distribution (Jonagold).
- The lowest values of total compatibility factors characterizing the course of changes in the S index in four consecutive years of the set were achieved by comparing the regression equations obtained with the VA method. The course of changes in this indicator obtained with the use of reference tables (VA) was burdened with errors resulting from the individual perception of the assessors.
- Starch index values obtained using the traditional method (VA) were significantly higher than those obtained by CIA methods. This concerned the assessment of the S-value of the Jonagold and Gloster varieties, which would suggest an earlier harvest when using this method of assessment in comparison with the CIA methods.
- The starch index did not change significantly when its values were calculated using image analysis of black and white (VA) images taken with a digital camera (DC) and scanner (SC). This relation was observed in the range of probable "harvest window". The exception here was the variant of the SC method, where for some seasons of the set (Gloster 2020, 2022, Jonagold 2021) the S values calculated with this method were significantly lower in the considered range of the set.

References

Brookfield, P., Murphy, P., Harker, R., MacRae, E.: Starch degradation and starch pattern indices; interpretation and relationship to maturity. Postharvest Biol. Technol. **11**, 23–30 (1997)

Brosnan, T., Sun, D.W.: Improving quality inspection of food products by computer vision—a review. J. Food Eng. **61**, 3–16 (2004)

Cardenas-Perez, S., et al.: Evaluation of the ripening stages of apple (Golden Delicious) by means of computer vision system. Biosys. Eng. **159**, 46–58 (2017)

De Long, J.M., Prange, R.K., Harrison, P.A.: Using the Streif Index as a final harvest window for controlled-atmosphere storage of apples. Hort Sci. **34**(7), 1251–1255 (1999)

Fan, L., Song, J., Forney, C.F., Jordan, M.A.: Fruit maturity affects the response of apples to heat stress. Postharvest Biol. Technol. **62**, 35–42 (2011)

Guz, T.: The use of image analysis to estimate harvest ripeness of apples TEKA Kom. Mpt Energ. Roln OL PAN **9**, 61–68 (2009)

Guz, T.: Harvest date determined by different methods of starch index readings and texture changes during apple storage. TEKA Kom. Mpt Energ. Roln OL PAN, Vol **10**, 74–81Author, F.: Contribution title. In: 9th International Proceedings on Proceedings, pp. 1–2. Publisher, Location (2010)

Hengari, S., Theron, K.I., Midgley, S.J.E., Steyn, W.J.: The effect of high UV-B dosage on apple fruit photosystems at different fruit maturity stages. Sci. Hortic. **170**, 103–114 (2014)

Jamishidian, F., Zhang, Y., Liu, W., Jamishidian, M.: A software including some new developments in multiple comparison and simultaneous confidence bands for linear regression models. J. Stat. Softw. **12**(2), 1–23 (2005)

Jarolmasjed, S., Espinoza, C.Z., Sankaran, S., Khot, L.R.: Postharvest bitter pit detection and progression evaluation in Honeycrisp apples using computer tomography analysis. Postharvest Biol. Technol. **118**, 35–42 (2016)

Linker, R., Cohen, O., Naor, A.: Determination of the number of green apples in RGB images recorded in orchards. Comput. Electron. Agric. **81**, 45–57 (2012)

Lohner, S.A., Biegert, K., Nothelfer, S., Hohmann, A., Mccormick, R., Kienle, A.: Determining the optical properties of apple tissue and their dependence on physiological and morphological characteristics during maturation. Postharvest Biol. Technol. **181**, 111647 (2021)

Łysiak G., 1998. Ocena metod określania dojrzałosci zbiorczej jabłek. Zeszyty naukowe AR w Krakowie. Sesja Naukowa z. 57 s. 509–513

Łysiak, G.: The determination of harvest index of 'Šampion' apples intended for long storage Acta Sci. Pol. Hortorum Cultus **10**(3), 273–282 (2011)

Peirs, A., Lammertyn, J., Ooms, K., Nicolaı, B.M.: Prediction of the optimal picking date of different apple cultivars by means of VIS:NIR-spectroscopy. Postharvest Biol. Technol. **21**, 189–199 (2000)

Peirs, A., Scheerlinck, N., Perez, A.B., Jancsók, P., Nicolai, B.M.: Uncertainty analysis and modelling of the starch index during apple fruit maturation. Postharvest Biol. Technol. **26**, 199–207 (2002)

Peirs, A., Schenk, A., Nicolaı, B.M.: Effect of natural variability among apples on the accuracy of VIS-NIR calibration models for optimal harvest date predictions. Postharvest Biol. Technol. **35**, 1–13 (2005)

Qing, Z., Ji, B., Zude, M.: Predicting soluble solid content and firmness in apple fruit by means of laser light backscattering image analysis. J. Food Eng. **82**, 58–67 (2007)

Reid, M.S., Padfield, C.A.S., Watkins, C.B., Harman, J.E.: Starch iodine pattern as a maturity index for Granny Smith apples. NZ J. Agric. Res. **25**, 229–237 (1982)

Rutkowski, K.P., Miszczak, A., Płocharski, W.: The influence of storage conditions and harvest date on quality of 'Elstar' apples. Acta Hort. **600**, 809–812 (2003)

Sjöstrand, J., Tahir, I., Helena Hovmalm, H.P., Garkava-Gustavsson, L., Stridh, H., Olsson, M.E.: Comparison between IAD and other maturity indices in nine commercially grown apple cultivars. Scientia Horticulturae **324**, 112559 (2024)

Solovchenko, A.E., Merzlyak, M.N., Pogosyan, S.I.: Light-induced decrease of reflectance provides an insight in the photoprotective mechanisms of ripening apple fruit. Plant Sci. **178**, 281–288 (2010)

Solovchenko, A., Kozhina, L., Nazarov, Y., Gudkovsky, V.: Relationships between internal ethylene and optical reflectance in ripening Antonovka apples grown under sunlit and shaded conditions. Postharvest Biol. Technol. **59**, 206–209 (2011)

Song, J., Bangerth, F.: The effect of harvest date on aroma compound production from'Golden Delicious' apple fruit and relationship to respiration and ethylene production. Postharvest Biol. Technol. **8**, 259–269 (1996)

Szalay, L., Ordidge, M., Ficzek, G., Hadley, P., Tóth, M., Battey, N.H.: Grouping of 24 apple cultivars on the basis of starch degradation rate and their fruit pattern. Hort. Sci. Prague **40**(3), 93–101 (2013)

Thewes, F.R., Argenta, L.C., Anese, R.O., Stanger, M.C., de Freitas, S.T.: The response of 'Monalisa' apples to high CO_2 storage conditions, harvest maturity and 1-MCP treatment. Sci. Hortic. **317**, 112038 (2023)

Tomala, K.: Prognozowanie zdolności przechowalniczej i określenie terminu zbioru jabłek. Fundacja Rozwój SGGW (1995)

Tomala, K.: Zanim zaczniesz zbiór owoców oceń stan fizjologiczny jabłek. Sad Nowoczesny 9/2003 (2003)

Tomala, K.: Jabłka zbieramy na podstawie oceny stanu fizjologicznego. Owoce, Warzywa, Kwiaty 16/2004 (2004)

Vanoli, M., Visai, C., Rizzolo, A.: The influence of harvest date on the volatile composition of'Starkspur Golden' apples. Postharvest Biol. Technol. **6**, 225–234 (1995)

Watkins, C.B.: Postharvest ripening regulation and innovation in storage technology. Acta Hort. (ISHS) **796**, 51–58 (2008)

Zhang, M., et al.: Determination of bagged 'Fuji ' apple maturity by visible and near-infrared spectroscopy combined with a machine learning algorithm. Infrared Phys. Technol. **111**, 103529 (2020)

Zhang, B., et al.: Quality monitoring method for apples of different maturity under long-term cold storage. Infrared Phys. Technol. **112**, 103580 (2021)

The Market for Renting Tractors and Agricultural Machinery as an Element of Sustainable Agriculture

Sławomir Juściński(✉)

Department of Power Engineering and Transportation, Faculty of Production Engineering, University of Life Sciences in Lublin, Głęboka 28, 20-612 Lublin, Poland
slawomir.juscinski@up.lublin.pl

Abstract. The concept of sustainable development in agricultural production has undergone significant modifications over the last decade. The aspect of economic security of farms in the long term began to be treated as one of the priorities. Sustainable agricultural production can only be achieved if the farm's financial stability and its long-term development potential are ensured. Hence, the logistics of agricultural vehicles and machine distribution has significantly expanded its rental segment, whose global value in 2022 was USD 52.7 billion. The need to reduce the financial burden on farms after the COVID-19 pandemic, with rising prices of equipment, fertilisers and plant protection products, has strengthened the development of the agricultural equipment rental market. Its systematic growth is forecast for the next decade with its value doubling by 2032. This article presents the characteristics of the global and domestic agricultural vehicle and machinery rental market in terms of rules and requirements. The national pricing structure for individual rental options at various levels of agricultural tractor engine power is discussed, as well as the scope of fees for operating hours above a certain limit. The article concludes with a presentation of trends in the development and principles of operation of the agricultural tractor rental market in Poland.

Keywords: tractor and agricultural machinery · rental market · sustainable agriculture · agricultural tractor rental options · rental costs

1 Introduction

The World Commission on Environment and Development (WCED) presented the first report in 1987 which defined the main rules of sustainable development. The document titled „Our Common Future" is often called the „Brundtland Report" coming from the name of the chairman of the Commission, Gro Harlem Brundtland. At the end of the twentieth century, global problems concerning protecting the environment were an effect of two phenomena. The first one was the poverty of the Southern Hemisphere countries, while the second was large-scale production and consumption using excessive amounts of natural resources in the countries of the Northern Hemisphere. The proposed term

E. Lorencowicz et al. (Eds.): FMPMSA 2024, LNCE 609, pp. 190–202, 2024.
https://doi.org/10.1007/978-3-031-70955-5_21

"sustainable development" called for creating a strategy combining economic development with protection of the environment. A definition of the action was: „Sustainable development is development that meets the needs of the present without compromising the ability of future generations to meet their own needs". In 1989 UN General Assembly organised a conference concerning Environment and Development [2, 20, 29].

Sustainable development is a strategy comprised of three main areas [14, 30]:

- economic, which should ensure the current and future profitability of economic endeavours,
- environmental, which should lead to decreasing the amount of waste and carbon footprint to reduce pollution and global warming,
- social, which is focused on creating healthy and safe conditions of life and work.

The paradigm of the world's sustainable development assumes the necessity of integrating efforts in all three areas – economic, environmental, and social. Thus, the agriculture sector, the second biggest production branch after industry, is crucial to achieving that goal. However, it is necessary to emphasise that agriculture and its success depend on natural resources, especially soil quality. Furthermore, the method of cultivation and level of mechanisation also plays a crucial role. In order to be effective there is a need for modern and efficient, thus costly, agricultural tractors and other machinery [3–6, 11] (Fig. 1).

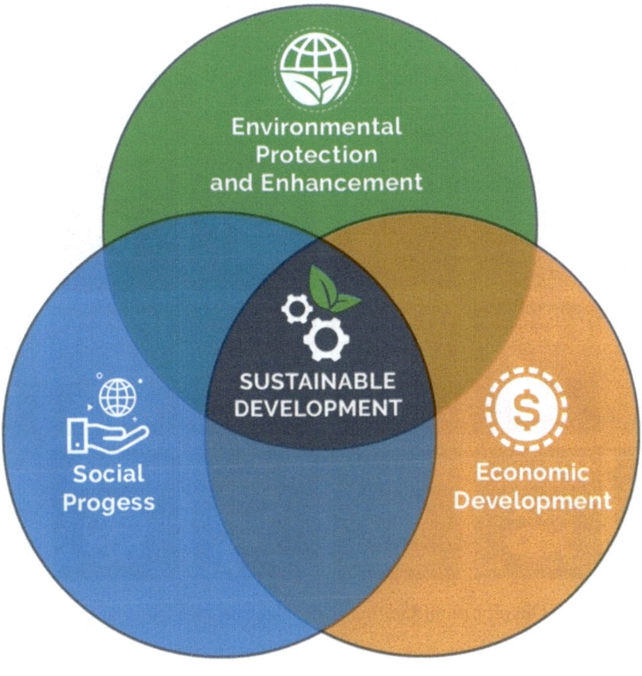

Fig. 1. Three areas sustainable development. Source: [29]

The structure of farms in the European Union was characterised during the last decade by two mutually dependent trends. Firstly, the number of farms was decreasing. In 2010–2020 three million farms ceased to exist, meaning that from the starting point of twelve million farms, only nine million were left. The process of liquidation affected mostly farms with the lowest acreage. Secondly, the average farm acreage in the European Union was rising. An average farm in the European Union in 2022 had an acreage of 17.4 ha [8, 24]. There is a wide diversity in the size of farms between countries of the EU since an average farm in France and Germany has over 60 ha, in Denmark almost 75 ha, in Czechia 120 ha, while in Italy and Poland only 11 ha. Generally, in countries leading in agricultural production, large farms are more common: in the US and Canada, an average farm has 118 ha, while in Ukraine it is 132 ha. Furthermore, over 80% of the biggest farms in Ukraine have over 500 ha [1, 9, 12, 19].

In family farming, which is the main type of farming in many countries, development is dependent on social resources. Thus, the biggest problem is the ageing of residents of rural areas [7, 10, 13]. Another problem is the migration of young people from rural areas caused by decreasing profitability of agricultural production and continually increasing costs of mechanisation, fertilisers, and plant protection products. In the long term, low agricultural produce prices preclude securing the accounting liquidity of the farm. The quality of the farmers' lives depends on the farms' profitability. Sustainable development is only possible with a stable economic situation which enables farmers to invest and pursue chosen business models [15, 21]. Agricultural issues are affected by regional conditions, however, some of them are universal [22].

Fig. 2. Sustainable Development Goals, as a collection of 17 global goals. Source: [36]

During the United Nations Organization summit, which took place in New York on 25–27 September 2015, a document titled „Transforming Our World: The 2030 Agenda for Sustainable Development" was signed. In response to new problems, leaders of the

UNO countries are obliged to work to realise 17 Sustainable Development Goals. The document contains a detailed description of 169 actions contained in 17 goals which are to be fulfilled by all interested parties: governments, international organisations, non-governmental agencies, the world of science and business, as well as all citizens (Fig. 2). These actions are focused on five areas: People, Planet, Prosperity, Peace, and Partnership [36].

Goal number 8 pertaining to economic growth, stable employment, and decent working conditions is particularly important. The Association for Sustainable Agriculture & Food in Poland puts ensuring the financial stability of farms as the first and foremost requirement of sustainable agricultural production. Profitability of agricultural production is necessary to ensure long-term growth and effective implementation of business plans [32].

2 Subject of Research

Various international crises during the last few years caused phenomena which reduced economic growth. Unfavourable relation between decreasing income, caused by the decrease in price of grain, and increased cost of labour affected all farms. The start of the war in Ukraine on 24th February 2022 caused a destabilisation in the industry production market. Prices of energy and steel skyrocketed and the supply chains of parts for tractors and machines were broken. The result was an increase in tractor and machine prices as well as delays in their manufacture. The unstable situation on the market, high fuel prices and increased inflation stopped investments. A notable decrease in the registration of new tractors when compared to the previous year was observed. These factors naturally influenced tractors and agricultural machine rental dealers to offer a wide range of options.

Acumen Research and Consulting (ARC) being the main provider of market research and consulting services published a report titled "Farm Equipment Rental Market Size – Global Industry, Share, Analysis, Trends and Forecast 2023 – 2032" in February of 2024. According to the report the value of the global agriculture equipment rental market reached 52.7 bln USD. Acumen Research and Consulting predicts it will reach 104.1 bln USD by 2032 based on a Compound Annual Growth Rate (CAGR) of 7,2% in 2023–2032. CAGR indicates how quickly a specific market's value grows in a year [27]. In comparison, Allied Market Research in a report titled „Agriculture Equipment Market Size, Share, Competitive Landscape and Trend Analysis Report by Type, by Automation, by Business: Global Opportunity Analysis and Industry Forecast, 2023–2032" estimates the value of global agricultural equipment market in 2022 as 121,2 bln USD. They forecast that by 2032, it will reach 192,5 bln USD, given its CAGR will average at 4,6% in 2023–2032 [28].

Doubling the value of the farm equipment rental market will be caused by farmers' goal of lowering maintenance costs. Simultaneously, it is expected to increase farm productivity due to using modern, thus effective, equipment. Rental enables farmers to access more sophisticated and technologically advanced equipment which usually is more expensive. Thus, according to the prognosis, the farm equipment rental market will grow quicker by 50% more than the farm equipment sales market [34] (Fig. 3).

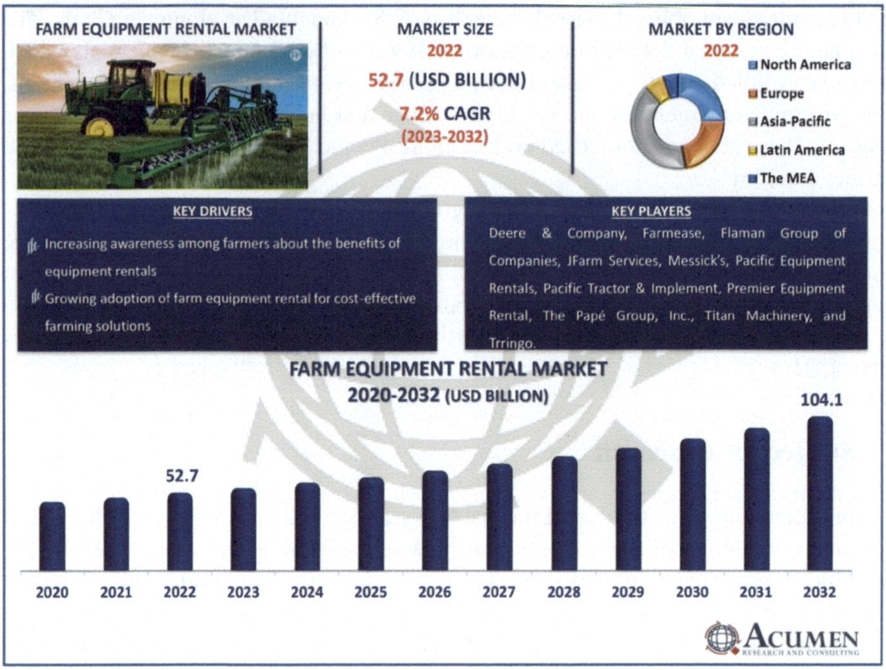

Fig. 3. Global Farm Equipment Rental Market 2023–2032. Source: [26]

The world's most prominent companies that offer farm equipment rental are Deere & Company (USA), FarmEase (USA), CNH Industrial (UK), Flaman Group of Companies (Canada), JFarm Services (India), AGCO Corporation (USA), Messick's (USA), Pacific Tractor & Implement (USA), Titan Machinery (USA), Kubota Corporation (Japan), Mahindra & Mahindra (India), JCB (UK), Escorts Ltd (India), Tractors and farm equipment's Ltd (India).

The agricultural machinery rental market in the world in 2022 was divided into the following groups: tractors, harvesters, sprayers, balers and other agricultural machinery. Tractors generated over 21.1 bln USD, which is 40% of the agricultural machinery rental market value. The most rented pieces of equipment were four-wheel drive tractors which account for 75% of rentals in their group. An analysis of the size of the agricultural machinery rental market in 2022 confirmed dominance (41% of the market) of the APAC (Asia-Pacific) region. This region contains continental Asia, as well as countries and territories counted as a part of the Pacific area. Agriculture is the crucial sector of the economy of that region which is caused by the high demand for food needed by its dynamically growing population. Hence, some governments in the region created subsidiary initiatives for food producers [27].

The Polish agricultural machinery rental market in the twenty-first century is growing steadily. The range of rentable vehicles and machines was limited in the first part of the century. A system of subsidies and grants offered to Polish farmers after Poland joined the European Union in 2004 increased the demand for agricultural equipment causing the growth of the domestic market. Grant aids from the EU reinforced the trend of owning agricultural equipment by each, even a small, farm. For over a decade, it led to poor investments. Many small farms purchased tractors and other machinery of high utility parameters. Thus, they were used to a minimal degree. The termination of subsidies and grants was one of the causes of the growth of the agricultural equipment rental market.

The machines purchased individually by the farmers are usually treated as an investment, which is judged based on their complementarity with already owned machinery. In the case of renting vehicles and/or machines, these activities are treated as a short-term substitute for the agricultural labour needed during e.g. harvest or the work in the fields. The cost level of renting agricultural equipment, used to mechanise the farm work, compared to the cost of buying such equipment is becoming an increasingly attractive business model. The nature of these processes is universal in different geographical areas of the world [16–18].

Purchasing costs, followed by the operational, maintenance and repairs costs of modern equipment, are disproportionately high for a small farm with regard to the short usage time for own needs during the harvest season or in the autumn. Owning and using agricultural equipment can be characterised in relation to the cultivated area as three business models [33]:

- In the largest farms, area-wise, it is considered as a long-term investment of the capital held. Thus, while the selling of tractors has decreased overall between 10% and 20%, year-over-year, over the last two years, the selling of tractors with engines over 250 HP is still on the rise.
- The owners of the medium-sized farms treat renting as a way of temporarily increasing the already had engine power (in the case of the tractors) or supplementing the technical infrastructure with the missing specialised machine.
- As for small farms, renting tractors and agricultural machinery is, in practice, the only way for further production.

High inflation rates and fuel prices have caused a stop in investments, particularly among the farms with a small cultivated area. Such a tendency can be confirmed with data concerning the average engine power of the registered tractors. During the second decade of the century, the segment of tractors with engine power up to 100 HP was purchased most often. When it comes to the tractor purchase decision, the most deciding factor was and still is the price.

It is important to note that there has been considerable technological progress considering tractor construction and its basic and additional equipment. Thus, particular current tractor offers are priced radically higher. By referencing the price of the engine power, it can be stated that during the last 25 years, the price has risen tenfold [35, 37]:

- 1 HP at the end of the 1990s cost around 500 PLN.
- 1 HP in current offers costs even 5,000 PLN.

The rental market of tractors and agricultural machines is operated in Poland by authorised dealers of specific brands or independent providers. Overall, it is commonly accepted in the rental market that a farmer should have at their disposal a fully functioning tractor while the provider does its maintenance.

Machinery rental is based on different periods, including [36]:

- 24 h,
- a week or its multiple,
- a month or its multiple.

Common breakdown in the market offer includes three-time ranges:

- 24 h,
- short-term (up to 12 months),
- long-term (up to two years).

The rental agreement contains a number of conditions, detailing e.g. [36]:

- hours of operation limit,
- additional fee (rate) for exceeding the limit of hours of operations stipulated in the agreement,
- tractor parameters and its equipment,
- possibilities and terms concerning rental extension,
- tractor storage conditions and training concerning the usage of a chosen model,
- tractor return (technical acceptance and inspection, e.g. amount of fuel in the tank),
- guarantee of the replacement machine being delivered within 48 h after receiving a malfunction notification,
- amount of additional payment for transporting a tractor and/or a machine from the provider's service point to the customer's farm.

CLAAS has launched a tractor rental programme on their company website. Presented rental conditions (Table 1) constitute commercial information, specifying initial net rates for one hour of tractor operation. Proposed base periods for renting vary from one week to six months with a flexible extension option. In the beginning, the FIRST CLAAS RENTAL programme registered six service points run by Authorised CLAAS Dealers in north-western Poland.

Rental end conditions depend on tractor class, equipment, and the time it is to be rented. The initial offer contains a price reduction of several per cent for every additional longer period of rental for specified engine power. Every one of the five suggested periods of rental has an assigned minimal working time, from 40 h of operation in one week to 500 h of operation in six months. The programme stipulates that such a number of hours will be billed as agreed, even if the user does not use up all of the limit. A dealer guarantees that a chosen tractor is ready at the agreed point of service during the 48 h since the conclusion of the agreement. Every CLAAS tractor is equipped with an automated steering system GPS PILOT CEMIS 1200, which allows for a precision $\pm$ 2–3 cm through the use of a mobile RTK NET network. Rented tractors are insured with liability insurance and comprehensive cover and periodically go through technical reviews (Table 1) [31].

In 2010, a company called Porszak started to operate as a professional service point for renting agricultural machines. In the beginning, the offer was addressed only to clients from the Kujawsko-Pomorskie, Mazowieckie and Wielkopolskie voivodships but was later extended across Poland and other EU countries. Machinery offer is systematically supplemented with new tractors and restored, and the company offers vehicles with engine power ranging from 80 HP to 500 HP belonging to brands like Fendt, John Deere, Valtra, Massey Ferguson i Case [25].

Table 1. Tractor rental offer under the FIRST CLAAS RENTAL programme

Rental cost per 1 h of operation net					
Rental duration (working limit)	1 week (40 h of operation) [PLN]	1 month (120 h of operation) [PLN]	2 months (210 h of operation) [PLN]	3 months (500 h of operation) [PLN]	6 months (500 h of operation) [PLN]
CLAAS ARION (160–235 km)	145	136	132	128	126
CLAAS AXION (220–250 km)	160	152	148	145	140
CLAAS AXION (260–300 km)	180	171	166	163	158
CLAAS AXION (350–400 km)	215	205	199	195	188
CLAAS AXION (450 km)	270	257	249	244	237
CLAAS XERION (500 km)	290	276	268	262	254

Source: Own elaboration based on [31]

Company employees offer help in choosing an appropriate vehicle for the area and type of work, e.g. basic arable farming, sowing, and transport. The company also ensures transportation of the rented machine to the indicated address and the support of a technical department in case of a machinery malfunction. The base offer for particular tractors sets the price of rental for one week with a limit of 40 h of operation and one month with a limit of 100 or 130 h of operation. The price for the 1-week rental period falls within the range of 2800 PLN (for 86 HP) to 12,000 PLN (for 500 HP). Respectively, for the period of one month, the price is set between 7,000 PLN and 39,000 PLN. A longer period of rental results in a price reduction of 20% for the individual tractors. Reductions at the 25–40% levels are applied for tractors with higher engine power and in cases where additional fees are added for use over the limit set in the agreement (Table 2) [25].

AgroSharing is an example of a modern company providing agricultural machinery rental and forestry and construction equipment rental services in central and north-western Poland. The offer includes John Deere tractors with engine power from 172 HP to 458 HP priced for 24 h of rental, although the base rental period is one month (Table 3) [23].

Table 2. Tractor rental offer from Porszak

Tractor brand	Horsepower (HP)	1-week rental with 40 h of operation limit Net price [PLN]/	1-month rental with 130 h of operation limit Net price [PLN]/	Rate for every hour above the hours of operation limit Net price [PLN]/
MF 5425	86	2 800	7 000 (limit 100 mth)	70
MF 5445	110	3 000	7 500 (limit 100 mth)	75
John Deere 6330	120	3 200	8 000 (limit 100 mth)	80
Fendt 714	140	3 400	11 050	85
Case Puma 145	145	3 400	11 050	85
John Deere 6150R	150	3 400	11 050	85
Fendt 716	160	3 400	11 050	85
Fendt 718	180	3 600	11 700	90
Fendt 920	200	4 000	13 000	100
Fendt 820	200	4 000	13 000	100
Fendt 722	220	4 600	14 950	115
Fendt 824	240	5 000	16 250	125
Fendt 924	240	5 000	16 250	125
Fendt 927	270	5 600	18 200	140
Fendt 930	300	6 000	19 500	150
Fendt 936	360	6 400	20 800	160
Fendt 939	390	7 400	24 050	185
Fendt 1050	500	12 000	39 000	300

Source: Own elaboration based on [25].

Available to rent are also sets containing a tractor and a machine, including spreaders, disc harrows, cultivators, subsoilers, reversible mounted ploughs, grain trailers or trailers. The machine rental offer contains a detailed parameters description and technical data. In order to choose a tractor or a machine, a client fills in an internet form on the company website and then is contacted by an advisor. As a result, a client books a date, discusses the price, signs an agreement and indicates a pick-up location, one of fourteen available. Typically, a client receives an inspected tractor and/or a machine with valid insurance and service support in case of a malfunction. As for fuel used by a working machine, it is a cost added to the rental price and it is borne by a client.

Table 3. Tractor rental offer from AgroSharing

Tractor brand	Tractor engine horsepower {HP]	24-h rental [HP]Net price [PLN/24 h]*
John Deere 6155M	172	from 700
John Deere 6175M	194	from 900
John Deere 6195M	216	from 1 100
John Deere 6R 250	301	from 1 670
John Deere 7R 310	352	from 1 770
John Deere 7R 350	385	from 1 770
John Deere 8R 410	458	from 2 150
John Deere 250 + HAWE spreader DST16	275	from 3 000
John Deere 6195M + Kverneland disc harrow Qualidisc Farmer 4001F	216	from 2 200
John Deere 6195M + Enduro cultivator Pro 3M	216	from 2 200
John Deere 6155M + Kverneland subsoiler CLG	172	from 1 800
John Deere 6155M + Kverneland reversible mounted plough 150S 4K	172	from 1 800
John Deere 8R 410 + grain trailer ULW 3000	443	from 2 810
HAWE trailer ULW 3000	Capacity 38 m3	from 660
HAWE trailer ULW 2000	Capacity 30 m3	from 600

* The price pertains to the rental period of over 1 month.
Source: Own elaboration based on [23].

3 Conclusion

The Polish agricultural machinery rental market in the twenty-first century is growing steadily. Rentals are offered by dealers authorised by large concerns producing tractors and agricultural machinery as well as business entities which engage only in renting. An advantage of using rental equipment is the ability to use it only in a specified time necessary for completing the farm work. This usage model excludes costs of purchase, service, repair, and insurance of machinery. The client pays only for rental and fuel. The price of rental depends mostly on the type of equipment rented and the time of rental. The shortest but the most expensive type of rental is renting for one day. All rental companies prefer to rent equipment for longer periods – weeks or even months.

Price lists inform that a discount of up to several dozen per cent per hour of operation is offered for long-term agricultural machinery rental.

The characteristics of the Polish agricultural machinery rental market:

- The application of the laws of supply and demand means that the negotiated prices depend on the seasonal farm work.
- Large dealerships offer a wider range of tractors and machinery, including the newest products of the producer they represent.
- The falling profitability of agricultural production forced farmers to reduce the mechanisation costs by refraining from buying new tractors and agricultural machinery.
- Polish farmers' mentality has to change to make agricultural machinery rental common in Poland.
- Using the newest technical solutions in tractors makes them more effective, simultaneously increasing their price.
- It is necessary to raise awareness among farmers that it is possible to rent machinery which is highly effective and otherwise expensive to buy.

Reducing the costs while increasing productivity is the way to improve and stabilise the economic situation of agricultural producers, which is necessary for the development of sustainable agriculture.

References

1. Baer-Nawrocka, A., Kiryluk-Dryjska, E.: The Influence of the Integration with the European Union on Central and Eastern European Countries' Agricultural Productivity. In: Piotrowska, M., Kurowski, L. (red.), Global Challenges and Politics of the European Union – Consequences for the New Member States, Research Papers of Wrocław University of Economics, No. 59, pp. 13–21 (2009)
2. Bardy, R., Rubens, A., Massaro, M.: The systemic dimension of sustainable development in developing countries. J. Organisational Transformation Soc. Change **12**(1), 22–41 (2015)
3. Brodzińska, K.: Program rolnośrodowiskowy i jego zakres realizacji w aspekcie rozwoju rolnictwa i ochrony środowiska. Zeszyty Naukowe SGGW w Warszawie, Problemy Rolnictwa Światowego **IV**(19), 83–91 (2008)
4. Brodzińska, K.: Świadomość ekologiczna rolników a praktyka gospodarowania. Roczniki Naukowe Stowarzyszenia Ekonomistów Rolnictwa i Agrobiznesu **XIV**(5), 34–38 (2012)
5. Czyżewski, B., Matuszczak, A.: Wspólna polityka rolna w kształtowaniu zrównoważonego rozwoju rolnictwa w aspekcie dostosowań regionalnych. Roczniki Naukowe Stowarzyszenia Ekonomistów Rolnictwa i Agrobiznesu **XV**(3), 229–236 (2013)
6. Czyżewski, B.: Produktywność zasobów w rolnictwie w Polsce wobec paradygmatu zrównoważonego rozwoju. Studia ekonomiczne **2**(LXXIII), 165–188 (2012)
7. Eizenberg, E., Jabareen, Y.: Social sustainability: a new conceptual framework. Sustainability **9**(1), 68 (2017)
8. Foster, A.D., Rosenzweig, M.R.: Are there too many farms in the world? labor market transaction costs, machine capacities, and optimal farm size. J. Polit. Econ. **130**(3), 636–680 (2022)
9. Gargano, G., Licciardo, F., Verrascina, M., Zanetti, B.: The agroecological approach as a model for multifunctional agriculture and farming towards the european green deal 2030—some evidence from the italian experience. Sustainability **13**, 2215 (2021). https://doi.org/10.3390/su13042215

10. Glauben, T., Petrick, M., Tietje, H., Weiss, C.: Probability and timing of succession or closure in family firms: a switching regression analysis of farm households in Germany. Appl. Econ. **41**, 45–54 (2009)
11. Knickel, K., Ashkenazy, A., Chebach, T.C., Parrot, N.: Agricultural modernization and sustainable agriculture: contradictions and complementarities. Int. J. Agric. Sustain. **15**(5), 575–592 (2017). https://doi.org/10.1080/14735903.2017.1373464
12. Machnacki, M.: Wpływ dzierżaw na zrównoważenie gospodarstw typu uprawy polowe regionu Mazowsze i Podlasie. Roczniki Ekonomii i Zarządzania **10**(4), 123–136 (2018)
13. Migliorini, P., Galioto, F., Chiorri, M., Vazzana, C.: An integrated sustainability score based on agroecological and socioeconomic indicators. A case study of stockless organic farming in Italy. Agroecol. Sustain. Food Syst. **42**, 859–884 (2018)
14. Murawska, A.: Tendencje zmian wzorców konsumpcji w krajach Unii Europejskiej w kierunku zrównoważonego rozwoju. J. Agribus. Rural Dev. **3**(37), 477–485 (2015). https://doi.org/10.17306/JARD.2015.50
15. O'Donoghue, C., Devisme, S., Ryan, M., Conneely, R., Gillespie, P., Vrolijk, H.: Farm economic sustainability in the European Union: a pilot study. Stud. Agricul. Econ. **118**(3), 163–171 (2016)
16. Qian, L., Lu, H., Gao, Q., Lu, H.: Household-owned farm machinery vs. outsourced machinery services: the impact of agricultural mechanization on the land leasing behavior of relatively large-scale farmers in China. Land Use Policy **115**, 106008 (2022). https://doi.org/10.1016/j.landusepol.2022.106008
17. Qiu, T., Luo, B.: Do small farms prefer agricultural mechanization services? Evidence from wheat production in China. Appl. Econ. **53**(26), 2962–2973 (2021). https://doi.org/10.1080/00036846.2020.1870656
18. Qiu, T., Choy, S.T.B., Luo, B.: Is small beautiful? Links between agricultural mechanization services and the productivity of different-sized farms. Appl. Econ. **54**(4), 430–442 (2022). https://doi.org/10.1080/00036846.2021.1963411
19. Salvioni, C., Ascione, E., Henke, R.: Structural and economic dynamics in diversified Italian farms. Bio-Based Appl. Econ. **2**, 257–275 (2013)
20. Velten, S., Leventon, J., Jager, N., Newig, J.: What is sustainable agriculture? A systematic review. Sustainability **7**(6), 7833–7865 (2015)
21. Wezel, A., Herren, B.G., Kerr, R.B., Barrios, E., Gonçalves, A.L.R., Sinclair, F.: Agroecological principles and elements and their implications for transitioning to sustainable food systems. a review. Agron. Sustain. Dev. **2020**(40), 1–13 (2020)
22. Wojewódzka-Wiewiórska, A., Kłoczko-Gajewska, A., Sulewski, P.: Between the social and economic dimensions of sustainability in rural areas—in search of farmers' quality of life. Sustainability **12**, 148 (2020). https://doi.org/10.3390/su12010148
23. https://agrosharing.pl/. Last accessed 28 April 2024
24. https://fpg24.pl/czy-polskie-gospodarstwa-rolne-sa-zbyt-male/. Last accessed 30 Apr 2024
25. https://porszak.pl/. Last accessed 29 Apr 2024
26. https://www.acumenresearchandconsulting.com/farm-equipment-rental-market. Last accessed 30 Apr 2024
27. https://www.alliedmarketresearch.com/agriculture-equipment-market. Last accessed 27 Apr 2024
28. https://www.are.admin.ch/are/en/home/media/publications/sustainable-development/brundtland-report.html. Last accessed 29 Apr 2024
29. https://www.arenasolutions.com/resources/glossary/sustainable-development/. Last accessed 30 Apr 2024
30. https://www.claas.pl/zakup-finansowanie/maszyny-uzywane/first-claas-rental. Last accessed 27 Apr 2024

31. https://www.farmer.pl/produkcja-roslinna/ochrona-roslin/rolnictwo-zrownowazone-co-to-wlasciwie-oznacza,99948.html. Last accessed 28 Apr 2024
32. https://www.farmer.pl/technika-rolnicza/maszyny-rolnicze/niewielka-sprzedaz-ciagnikow-a-faktycznie-jeszcze-mniejsza,145747.html. Last accessed 30 Apr 2024
33. https://www.topagrar.pl/articles/aktualnosci-branzowe-technika/jak-sie-zmieni-swiatowy-rynek-sprzetu-rolniczego-do-2027-roku-2500211. Last accessed 29 Apr 2024
34. https://www.topagrar.pl/articles/ciagniki/sprzedaz-ciagnikow-w-2022-roku-seria-twoja-opinia-o-ciagniku-2455070, Last accessed 28 Apr 2024
35. https://www.tygodnik-rolniczy.pl/maszyny-rolnicze/ciagniki-rolnicze/ciagnik-rolniczy-na-wynajem-ile-kosztuje-za-dobe-tydzien-albo-miesiac-mf-fendt-john-deree-case-2486797. Last accessed 27 Apr 2024
36. https://www.un.org/sustainabledevelopment/sustainable-development-goals/. Last accessed 30 Apr 2024
37. https://www.wrp.pl/ciagnik-kupic-czy-wynajac-ile-kosztuje-wynajecie-ciagnika/. Last accessed 29 Apr 2024

The Effect of Digestate on the Antioxidant Properties of Hemp Leaves (*Cannabis sativa* L)

Magdalena Kachel(✉), Milan Koszel, and Karolina Sokal

Department of Machine Operation and Production Processes Management, Faculty of Production Engineering, University of Life Sciences in Lublin, 28 Głęboka Street, 20-612 Lublin, Poland
magdalena.kachel@up.lublin.pl

Abstract. Agriculture has an enormous environmental footprint. Sustainable agriculture consists of environmentally friendly methods of farming that allow the production of crops or livestock without damage to human or natural systems. Implementation of modern technological solutions in agricultural can have a beneficial effect on environmental protection, mainly due to the reduction of waste, including hazardous substances. Post-digestion matter rich in nutrients NPK can be used as a fertilizer in agriculture either in a liquid or solid form. That is why we used it in the cultivation of hemp (*Cannabis sativa* L.), that is a plant rich in numerous compounds such as Cannabinoids and phenolic compounds (PC) and a low content of psychoactive cannabinoids. In humans, polyphenols can display health-promoting showing cardio or chemoprotective activity. In our case, this fertilizer in various concentrations (20; 30; 50 L/ha) was used for the cultivation of industrial hemp (*C. sativa*), where is used in industry there are of leaves, seeds and oil.

In our research, we focused on the analysis of leaves in terms of the content of flavonoids and polyphenols depending on the digestate concentration used.

The content of polyphenols and flavonoids in hemp leaves decreased as the plants grew in the control sample. The obtained polyphenol values for the control sample were lower at 15.06, respectively; 9.76 and 19.87%. A downward trend was also observed when a dose of 50 L/ha of digestate was used.

Keywords: Sustainable agriculture · Hemp leaves · Phenolic compounds · Biological activities · Mechanism · Post-digestate fertilizers

1 Introduction

Cannabis sativa L. is a well-known plant throughout Europe due to its widespread cultivation worldwide. It is an environmentally friendly plant that complements a sustainable growth system and can be grown under various agroecological conditions without herbicides, fungicides, or pesticides [1, 2]. It is reported that one hectare of hemp is assumed to consume 9 to 15 tons of CO_2, which is as much as a young forest consumes and takes only four to five months to grow [2, 3]. In pursuit of the goals of sustainable agriculture with an environmental impact, enabling more efficient and environmentally friendly use

E. Lorencowicz et al. (Eds.): FMPMSA 2024, LNCE 609, pp. 203–210, 2024.
https://doi.org/10.1007/978-3-031-70955-5_22

of resources, it is appropriate to use the cultivation of these plants with fertilizers of natural origin (digestate) experimentally. Digestate is a well-recognized valuable organic fertilizer, effectively reducing nitrogen losses compared to mineral fertilizers using the exact amounts of this component [4]. Restoring the soil with nutrients accumulated in fertilizers from biogas plants is necessary not only from an economic standpoint but also to maintain and restore the ecological balance in 2019 [5]. Applying such fertilizer can affect the soil in complex ways, changing the mobility of nutrients and the ability of plants to take them up from the environment [6]. Numerous products, including textiles, paper, rope, cosmetics, and medicines, are also made from leaves, stems, husks, roots, seeds, and oil. All parts of the plant also contain many proteins and bioactive compounds, including polyphenols and flavonoids, widely regarded as a nutritionally complete food source [7, 8]. In addition, fiber hemp extracts can exhibit antimicrobial [9, 10], anti-inflammatory [11, 12], antioxidant [13], antiproliferative [14], and cytotoxic [15] properties. Therefore, the purpose of the present study was to analyze the content of basic composition (water content, protein, and crude fiber) and antioxidant properties in hemp leaves depending on the applied dose of digestate.

2 Materials and Methods

2.1 Plant Material

The studied material consisting of leaves of *Cannabis sativa* L cv. 'Futura75' was purchased by the Lublin University of Life Sciences experimental farm in Czesławice (51.304595 22.252620—DMS: 51°18′16.542″N 22°15′9.432″E) from Polishemp Ltd. It was the single-seeded variety, with a vegetation period of 140 days, height 250–350 cm, oiliness 28–30%, fiber content 30–35%, grain yield per hectare 700–900 kg, biomass per hectare 10–12 t, weight of 1000 seeds 16–18 g according to the manufacturer's label. Seeds were sown in the first decade of May and harvested in the third decade of September 2022. Harvested leaves from plants (7 sticks each) in respective plots were taken in the final four months of hemp growth, using three experiment variants. Cannabis leaves were harvested at four growth dates. The first harvest date was mid-June BBCH 5; the second was mid-July BBCH 6; the third was mid-August BBCH 7; and the fourth was mid-September BBCH 8.

2.2 Field Experiment Plan

The scope of the field study included four variants of the experiment on which the digestate was sprayed:

- variant I—total dose of digestate 25,000 L/ha (I dose—12,500 L/ha, II—12,500 L/ha);
- variant II—total dose of digestate 37,500 L/ha (I dose—18,750 L/ha, II—18,750 L/ha);
- variant III—total dose of digestate 50000 L/ha (I dose—25,000 L/ha, II—25,000 L/ha);
- variant IV—the control object consisted of sown seeds not fertilized with digestate.

The above doses were determined by analysis of soil samples. Fertilization with digestate was carried out in the spring, and two doses were applied:

- the first dose at the BBCH 15 stage (the fifth pair of leaves—compound leaves);
- the second dose at the BBCH 19 stage (the ninth pair of leaves—compound leaves).

The digestate was taken from the Piaski biogas plant (Lublin Province). The following raw materials were used to produce the digestate: corn silage (70%), sugar beet pulp (15%), fruit pomace (5%), dairy waste (5%) and manure (5%). Before application, the digestate was tested for macronutrients and heavy metals (Table 1). The pH of the digestate applied to the winter rapeseed crop is 8.70, similar to that of cattle manure (7.90).

2.3 Basic Chemical Composition of Hemp Leaves

Hemp leaf was determined according to methods described in The Standard Methods of the Association of Official Analytical Chemists [16]. Parameters tested include moisture content, crude protein, crude fat, crude fiber, and ash. The average values from the three determinations were converted to % dry matter content.

2.4 Extraction of Polyphenols

The extraction was carried out according to the procedure described by Calzolari et al. [17] with minor modifications. 15 mL of methanol was added to 100 mg of the test leaves. The resulting mixture was vortexed for 5 min and then sonicated in the dark at 4 °C for 30 min. The mixture was centrifuged at 5000 × g at 4 °C for 10 min using an OHAUS FC5513 Frontier Microcentrifuge (OHAUS Europe GmbH, Switzerland). The resulting supernatants were combined, filtered through 0.2 μm syringe filters (26 mm, RC membrane, Phenomenex, CastelMaggiore, Italy), and poured into a chromatography vial.

2.5 Profile of Polyphenols

The polyphenol profile was determined according to the method described by Wang et al. [18] with minor modifications. Analysis was performed using high-performance liquid chromatography with a photodiode array (HPLC-UV-VIS) consisting of Perkin Elmer HPLC combined with UV-Vis. Analysis of bioactive compounds was performed for phenolic acids and flavonoids. Before injection on the column, each extracted sample was filtered using a 0.45 μm PTFE filter (Merck Millipore, USA). A C6-phenyl column (250 × 4.6 mm i.d., particle size 3 μm, C6-phenyl 3 μ; Phenomenex, Torrance, CA, USA) was used. The dispensing temperature of the 20 μL sample was 25 °C. The mobile phase consisted of solvent A (2.5%, v/v, acetic acid solution in water) and solvent B (2.5%, v/v, acetic acid solution in methanol) in different proportions. The gradient profile was 100% A at 0–5 min, 90% A after 15 min, 50% A after 45 min, and 100% A after 55 min. For flavonoid analysis, the mobile phase consisted of solvent A (1.0%, v/v, formic acid solution in methanol) and solvent B (1.0%, v/v, formic acid solution in a mixture of methanol and water, 30:70, v/v).

Formic acid was added to increase peak resolution in each case. The gradient profile was 5% A at 0 min, 25% A after 10–15 min, 90% A after 40 min, and 5% A after 50 min. The flow rate was set to 1.0 mL/min. The PDA detector was set at 270 nm to monitor UV/Vis absorption. UV/Vis spectra were recorded from 190 nm to 650 nm. Each compound was quantified by comparing its peak area against a standard calibration curve for standard solutions containing that compound. All phenolic acid and flavonoid standards used were purchased from Sigma-Aldrich, USA.

2.6 Determination of Total Phenolic Content

Total phenolic content was determined according to the Folin-Ciocalteu method described by Singleton and Rossi [19] with slight modifications. To 125 μL of extract sample was added 500 μL of deionized water, then 125 μL of Folin-Ciocalteu reagent was added to the mixture. The sample thus obtained was incubated for 6 min at room temperature. Next, 1.25 mL of 7.5% sodium carbonate solution and 1 mL of deionized water were added to the mixture. The next step was a 90-min incubation of the samples in the dark. After this time, the absorbance of the samples was measured at 760 nm using a UV-2600i spectrophotometer from Shimadzu (Shimadzu, Japan). The results are presented as mg gallic acid equivalents (GAE/g) of a sample.

2.7 Statistical Analysis

In order to determine the significance of the impact that the respective factors had on the analysed values, a one-way variance analysis was performed. The significance of differences between the mean values was established with Tukey's test. The statistical analysis was conducted using Statistica 13 software from StatSoft. The adopted significance level was $\alpha = 0.05$.

3 Results and Discussion

3.1 Analysis of Core Values

Raw hemp leaves are a source of nutrients, among which we can distinguish proteins that are particularly easy to digestate and contain a lot of fiber and potent antioxidants. Therefore, the following table (Table 2) shows the basic parameters of hemp leaves, such as crude protein, fiber, and moisture. The crude protein content of hemp leaves had an increasing character for the control samples, and the values obtained were 22.80, 23.88, 24.30, and 24.59%, respectively. The increase in protein amount was characterized by the first date of leaf harvesting after applying 30 and 50 L of biofertilizer (postfertilizer). The recorded increase was 2.01 and 1.39% concerning the control sample (22.80%). A decreasing trend was observed for each dose of applied fertilizer concerning the corresponding control sample for the remaining leaf harvesting periods.

The crude fiber content for the control samples at successive leaf harvest dates was equal to 18.89, 18.92, 18.96, and 19.42%, meaning that at each successive harvest date, an increase in this parameter was recorded by 1.19, 0.38, and 2.82%, respectively. In all variants of the digestate application experiment, an upward trend was noted concerning the control samples. The most remarkable differences were noted in fiber growth at the second leaf harvesting date. The obtained values differed depending on the applied dose of digestate. For applying 20 L of fertilizer, the noticeable increase in fiber content represented a higher value of 1.62%, 30 L—2.97%, and 50 L—3.62%.

As the cannabis plants grew, leaf water content mainly decreased. It was 6.83, 6.76, 6.82, and 6.59% for the control samples for subsequent harvesting dates, representing values for subsequent harvests lower than the first leaf breaking date by 0.97, 0.19, and 3.56%, respectively.

Table 1. Contents of essential constituents in hemp leaves

Harvest date	Dose of fertilizer/Unit	Crude Protein	Crude Fibre	Moisture
		[%]		
1	Control	22.80 ± 0.01^{a}	18.89 ± 0.04^{a}	6.83 ± 0.02^{a}
	20 L	22.01 ± 0.08^{a}	19.26 ± 0.06^{a}	6.71 ± 0.02^{a}
	30 L	23.28 ± 0.06^{a}	18.91 ± 0.03^{a}	6.79 ± 0.05^{a}
	50 L	23.12 ± 0.02^{a}	19.79 ± 0.03^{a}	6.67 ± 0.04^{a}
2	Control	23.88 ± 0.03^{ab}	18.92 ± 0.03^{aa}	6.76 ± 0.03^{aa}
	20 L	23.56 ± 0.02^{ab}	19.23 ± 0.03^{aa}	6.91 ± 0.03^{aa}
	30 L	23.33 ± 0.03^{aa}	19.49 ± 0.02^{aa}	$6,88 \pm 0.02^{aa}$
	50 L	23.24 ± 0.04^{aa}	19.61 ± 0.02^{aa}	6.75 ± 0.05^{aa}
3	Control	24.30 ± 0.04^{ab}	18.96 ± 0.04^{aa}	6.82 ± 0.02^{aa}
	20 L	$24,11 \pm 0.02^{ab}$	$19,24 \pm 0.02^{aa}$	$6,71 \pm 0.04^{aa}$
	30 L	23.88 ± 0.03^{aa}	19.40 ± 0.04^{aa}	6.64 ± 0.01^{aa}
	50 L	23.78 ± 0.03^{aa}	19.50 ± 0.02^{aa}	6.70 ± 0.01^{aa}
4	Control	24.59 ± 0.05^{ab}	19.42 ± 0.01^{aa}	6.59 ± 0.07^{aa}
	20 L	24.40 ± 0.02^{ab}	19.61 ± 0.03^{aa}	6.43 ± 0.03^{aa}
	30 L	24.21 ± 0.01^{aa}	19.68 ± 0.02^{aa}	6.53 ± 0.03^{aa}
	50 L	24.12 ± 0.02^{ab}	19.70 ± 0.03^{aa}	6.48 ± 0.03^{aa}

Values designated by different small letters in the columns of the table are significantly different ($\alpha = 0.05$).

Polyphenols and flavonoids in hemp leaves are characterized by a wide range of beneficial properties both for human and plant health. Among them, antioxidant, anti-inflammatory, or anti-allergic properties can be distinguished. The total content of flavonoids in hemp leaves and flowers can reach 2.5% of their dry weight, while they are almost absent in seeds and roots [20]. Based on the analyses, it was found that the polyphenol content of hemp leaves (Table 2) decreased as they grew in the control sample. The obtained values for the control samples were 33.51, 28.46, 30.23, and 26.85 mg GAE/g, respectively, which resulted in a decrease of 15.06; 9.76 and 19.87%, respectively, as the plants grew, compared to the sample from the first harvest. A decreasing trend was also observed when a 50 L digestate dose was applied for the first, second, and third leaf harvest dates. The content of polyphenols for this dose of fertilizer was lower than that of the control sample by 10.11%, 1.78%, and 4.54%. The most significant losses in their content were observed in leaves at the second harvest date in all variants of the fertilizer application. These values for subsequent doses of applied fertilizer were lower by 14.06, 11.77, and 18.87% than the control sample.

The flavonoid content of cannabis leaves (Table 2) in the control samples as the plants grew also showed a decreasing trend. The content of the analyzed parameter concerning the first harvest date in the control sample was a value lower for the second date by 1.72%, for the third date by 0.68%, and for the fourth date by 3.68%. Declining trends were also observed when applying the fertilizer at 50 L on the second, third, and fourth leaf harvesting dates. The obtained values were lower than the control sample by 0.71, 0.82, and 0.13%.

The situation was analogous in the case of antioxidant properties measured by the DPPH method (Table 2). A significant reduction in antioxidant properties was observed as the plants grew in all control trials. Subsequent harvesting dates were characterized by lower antioxidant potential than the first date. The values were lower by 11.09, 8.13, and 13.87%, respectively. Furthermore, the 50 L application of post-ferment at each leaf harvest was characterized by lower antioxidant properties. These values were lower than the control sample by 1.55, 1.28, and 0.22%, respectively.

Table 2. Content of polyphenols, flavonoids, and DPPH in hemp leaves

Harvest data	Dose of fertilizer	Polyphenols	Flavonoids	DPPH
		mg (GAE)/g	mg QE/g	mmol trolox/kg
1	Control	33.51 ± 0.15a	55.22 ± 0.04a	68.99 ± 0.24a
	20 L	34.59 ± 0.04b	55.33 ± 0.04a	70.56 ± 0.10b
	30 L	35.84 ± 0.04c	59.30 ± 0.07b	78.36 ± 0.06b
	50 L	30.12 ± 0.03d	56.14 ± 0.03a	73.44 ± 0.10b
2	Control	28.46 ± 0.08^{ab}	54.27 ± 0.05^{ab}	61.34 ± 0.10^{ab}
	20 L	28.80 ± 0.03^{ac}	54.40 ± 0.03^{ab}	63.35 ± 0.07^{bc}
	30 L	29.56 ± 0.04^{ac}	55.18 ± 0.02^{ab}	65.57 ± 0.05^{cc}
	50 L	27.18 ± 0.06^{bc}	53.89 ± 0.04^{ab}	60.38 ± 0.05^{ac}
3	Control	30.23 ± 0.08^{ab}	54.84 ± 0.08^{aa}	63.38 ± 0.08^{ab}
	20 L	31.24 ± 0.05^{bc}	54.92 ± 0.07^{aa}	64.18 ± 0.04^{ac}
	30 L	33.49 ± 0.07^{bc}	55.09 ± 0.03^{ac}	64.85 ± 0.09^{bc}
	50 L	29.70 ± 0.16^{bd}	54.39 ± 0.02^{ab}	62.56 ± 0.06^{ac}
4	Control	26.85 ± 0.15^{ab}	53.18 ± 0.04^{ab}	59.41 ± 0.06^{ab}
	20 L	27.47 ± 0.20^{ab}	53.32 ± 0.05^{ac}	59.78 ± 0.04^{ac}
	30 L	29.22 ± 0.10^{bb}	53.55 ± 0.05^{ac}	59.89 ± 0.05^{ac}
	50 L	25.63 ± 0.08^{cb}	53.12 ± 0.02^{ac}	59.28 ± 0.07^{ac}

Values designated by different small letters in the columns of the table are significantly different ($\alpha = 0.05$).

4 Conclusion

Depending on the parameter considered, the use of digestate as fertilizer under hemp cultivation looked different. In the case of leaf protein content, an increase was noted as the plants grew. However, applying our fertilizer contributed to a decrease in protein content at the second, third, and fourth leaf harvesting dates regardless of the applied rate of digestate. In the case of crude fiber analysis, an increase was observed in all variants of the experiment. The analysis of polyphenols in the leaves showed the most significant impact on the plants with the application of a dose of 50 L of biofertilizer, the application of which contributed in all variants to a reduction in the content in the tested material (10.11, 18.87, 1.78, and 4.54%) concerning the control. A similar observation also occurred regarding flavonoid content; the dose of 50 L of biofertilizer contributed in the second, third, and fourth terms to reducing their content by 0.71, 0.82, and 0.13% compared to the corresponding control sample.

References

1. Nath, M.K.: Benefits of cultivating industrial hemp (Cannabis sativa ssp. sativa)—A versatile plant for a sustainable future. Chem. Proc. **10**(1), 14 (2022)
2. European Commission. Agriculture and Rural Development. Hemp. https://agriculture.ec.europa.eu/farming/crop-productions-and-plant-based-products/hemp_en?ref=nangmantujapateuneoseu. Accessed on 24 April 2024
3. EIHA.: Konopie i zielony ład. https://eiha.org/wp-content/uploads/2020/09/Koopie-i-Zielony-Lad_PL.pdf. Accessed on 19 April 2024
4. Insam, H., Gómez-Brandón, M., Ascher, J.: Manure-based biogas fermentation residues—Friend or foe of soil fertility? Soil Biol. Biochem. **84**, 1–14 (2015)
5. Verdi, L., Kuikman, P.J., Orlandini, S., Mancini, M., Napoli, M., Dalla Marta, A.: Does the use of digestate to replace mineral fertilizers have less emissions of N_2O and NH_3? Agric. For. Meteorol. **269–270**, 112–118 (2019)
6. Berthrong, S.T., Buckley, D.H., Drinkwater, L.E.: Agricultural management and labile carbon additions affect soil microbial community structure and interact with carbon and nitrogen cycling. Microb. Ecol. **66**(1), 158–170 (2013)
7. Tsaliki, E., et al.: Fibre and seed productivity of industrial hemp (Cannabis sativa L.) varieties under Mediterranean conditions. Agronomy **11**(1), 171 (2021)
8. Višković, J., et al.: Industrial hemp (Cannabis sativa L.) agronomy and utilization: a review. Agronomy **13**(3), 931 (2023)
9. Novak, J., Zitterl-Eglseer, K., Deans, S.G., Franz, C.M.: Essential oils of different cultivars of Cannabis sativa L. and their antimicrobial activity. Flavour Fragrrance J. **16**, 259–262 (2001)
10. Zheljazkov, V.D., Sikora, V., Semerdjieva, I.B., Kačániová, M., Astatkie, T., Dincheva, I.: Grinding and fractionation during distillation alter hemp essential oil profile and its antimicrobial activity. Molecules **25**(17), 3943 (2020)
11. Huang, S., et al.: Chemical constituents of industrial hemp roots and their anti-inflammatory activities. J. Cannabis Res. **5**(1), 2–8 (2023)
12. Ferrante, C., et al.: Multiple pharmacognostic characterization on hemp commercial cultivars: focus on inflorescence water extract activity. Food Chem. Toxicol. **125**, 452–461 (2019)
13. Zhu, J., et al.: Anti-fatigue activity of hemp leaves water extract and the related biochemical changes in mice. Food Chem. Toxicol. **150**, 112054 (2021)

14. Moccia, S., et al.: Antiproliferative and antioxidant effect of polar hemp extracts (Cannabis sativa L., Fedora cv.) in human colorectal cell lines. Int. J. Food Sci. Nutr. **71**(4), 410–423 (2019)
15. Zengin, G., et al.: Chromatographic analyses, in vitro biological activities, and cytotoxicity of Cannabis sativa L. essential oil: a multidisciplinary study. Molecules **23**(12), 3266 (2018)
16. AOAC (Association of Official Analytical Chemists). Official Methods of Analysis of the Association of Official Analytical Chemistry. In: Horwitz, W (ed.), 13th Edition Published by AOAC, p. 1141 (1980)
17. Calzolari, D., Magagnini, G., Lucini, L., Grassi, G., Appendino, G., Amaducci, S.: High added-value compounds from Cannabis threshing residues. Ind. Crops Prod. **108**, 558–563 (2017)
18. Wang, L., et al.: The biological activity mechanism of chlorogenic acid and its applications in food industry: a review. Front. Nutr. **9**, 943911 (2022)
19. Singleton, V.L., Rossi, J.A.: Colorimetry of total phenolics with phosphomolybdic-phosphotungstic acid reagents. Am. J. Enol. Vitic. **16**, 144–158 (1965)
20. Cui, Q., Li, J., Yu, C.: Extraction and purification of total flavonoids from Cannabis sativa fibers and their evaluation for antioxidant and antibacterial activities. Text. Res. J. **93**(1–2), 56–69 (2023)

Air Dust Monitoring on a Farm Keeping Cold-Blooded Horses

Katarzyna Karpińska(✉), Joanna Barłowska, Mateusz Ossowski, Łukasz Wlazło, Bożena Nowakowicz-Dębek, and Piotr Maksym

University of Life Sciences in Lublin, Lublin, Poland
katarzyna.karpinska@up.lublin.pl

Abstract. The demands for space where horses are kept should be adapted to optimal conditions that meet the requirements of their welfare. Employees handling these animals spend a significant part of the day there. In order to assess air dustiness, research was carried out on a farm keeping mares for milk production. The research was performed on a farm keeping 30 cold-blooded horses in autumn. The monitoring covered dust PM1, PM2.5, PM10, respirable dust and total dust. Measurements were made in 3 different kinds of space: a stable with foals, a stable with mares, and a room to milk the mares. The highest concentration of total dust (5.37 mg/m^3) and respirable dust (1.17 mg/m^3) was obtained in the room with foals, and the lowest in the mares stable (0.55 mg/m^3 and 0.31 mg/m^3, respectively).

Keywords: dust · pollutants · cold-blooded horses · environmental

1 Introduction

The microclimate in livestock housing is shaped not only by the thermal and humidity system, but also by the concentration of dust and gaseous pollutants. Maintaining appropriate microclimatic conditions in premises with animals determines their level of welfare [Journal of Laws 2019, item 1966; Lühe et al., 2022]. The stable air includes, among other things: gaseous and microbiological pollutants and organic dust of plant or animal origin. Organic dust carries numerous bacteria and fungi, including potentially pathogenic ones, their metabolic products (bacterial toxins), cell wall components (β(1–3)-glucans) and plant particles, such as pollen and fibers. All these components are considered etiological agents of respiratory diseases both in animals themselves and workers who handle animals [Elfman et al., 2011; Rodzyń et al. 2024; Nowakowicz-Dębek et al., 2024]. The consequences of exposure of the respiratory system to dust containing microorganisms are often changes of an inflammatory or allergic nature, which can ultimately be considered an occupational or occupational disease [Szewczyk, 2012]. High concentrations of gases in the air contribute to the formation of the so-called secondary particulate matter (PM), while exposure to them has a significant impact on increasing diagnosed respiratory diseases and even leads to the initiation of tumorigenesis [Ossowski et al., 2022; Wiegand et al., 2022]. Dust, as a carrier of numerous gaseous

E. Lorencowicz et al. (Eds.): FMPMSA 2024, LNCE 609, pp. 211–218, 2024.
https://doi.org/10.1007/978-3-031-70955-5_23

or microbiological pollutants, may additionally intensify their adverse effects [Cambra-López et al., 2010; Hristov 2011], and persistent exposure to such an aerosol leads to the development of, among others, atrophic rhinitis or enzootic bronchitis in animals [Ji et al., 2017]. Moreover, gaseous pollutants such as NH_3 and NO_2 released during animal husbandry negatively affect air quality, aquatic ecosystems and biodiversity. As the ambient temperature decreases, especially in autumn and winter, an increase in CO_2 concentration and relative humidity is recorded in the air of stables. His forces the horse's body to increase gas exchange and, consequently, may lead to their hypoxia [Rodzyń et al., 2024; Nowakowicz-Dębek et al., 2024; Janczarek et al., 2020]. Processes occurring in residual feces and feed with the involvement of bacterial enzymes increase the concentration of harmful gases in the air, the inhalation of which is associated with negative effects on the respiratory system, lowering the body's immune response, making it more susceptible to the action of microorganisms present in the environment [Wålinder et al., 2011]. The most common equine diseases caused by inadequate air parameters include recurrent airway obstruction (RAO), inflammatory airway disease (IAD) [Pirie, 2014] and Chronic Obstructive Pulmonary Disease (COPD) [Kwiatkowska-Stenzel et al., 2017]. Respiratory diseases in horses may have similar symptoms to other diseases, which can make accurate diagnosis difficult. The associated health problems may significantly limit their performance and participation in sports competitions [Ivester et al., 2014]. Therefore, when maintaining horses, you should pay attention to the air quality index AQI (Air Quality Index). If the AQI is between 100 and 150, we can engage horses only in light exercise. However, with AQI above 150, exercise is not recommended. In addition, as AQI increases, horses' behavior changes. Low AQI also reduces the rider's predispositions [Tainio et al., 2020]. The level of particulate contamination indoors depends on many factors, including the housing system, the feed provided, and the bedding used [Lühe et al., 2022]. Straw is considered the most frequently used bedding material in Poland. However, it is characterized by low moisture and ammonia absorption. Moreover, its use is associated with an increase in dust in the stable air. Kwiatkowska-Stenzel et al. [2017], report that the use of straw as bedding may be associated with an up to threefold increase in the concentration of dust and endotoxins in the air compared to other bedding materials. For these reasons, other straw substitutes are constantly being sought or it is technologically processed to produce pellets. This form of bedding material is characterized by lower dustiness and lower concentration of microbiological pollutants compared to traditional straw [Kwiatkowska-Stenzel et al., 2017]. The concentration of dust in the stable air also depends on the type of feed provided and the method of feeding the horses. Silage feeds are characterized by lower dustiness. According to Wyse et al. [2005], feeding horses with silage significantly reduced the dust concentration, resulting in lower markers of respiratory tract inflammation. Stabling facilities, due to the numerous densities of animals in a relatively small area, require efficient ventilation systems that ensure air quality as close to natural as possible. Keeping horses and the employees handling them in the best possible condition requires monitoring the conditions in this environment. Therefore, the aim of the study was to monitor air dustiness on a farm keeping cold-blooded horses.

Silage feeds are characterized by lower dustiness. According to Wyse et al. [2005], feeding silage to horses significantly reduced dust concentrations, resulting in lower markers of respiratory tract inflammation. Stabling facilities, due to the numerous densities of animals in a relatively small area, require efficient ventilation systems to ensure air quality as close to natural as possible. Keeping horses, as well as the workers who handling them, in the best possible condition requires monitoring the conditions in this environment. Therefore, the purpose of this study was to monitor air dustiness in a farm keeping cold-blooded horses.

2 Material and Methods

Air dustiness tests were carried out on a farm breeding Polish Cold-Blooded horse used for milk production. There were 30 horses on the farm during the research. The animals were kept in single stalls on wheat-rye straw, which was added every day, while the stalls were cleaned by hand twice a week. During the research period, the horses were fed in the mornings with hay, oat straw and haylage from permanent grasslands. Additionally, all animals received two doses of crushed oat grain and a mineral and vitamin preparation and had free access to drinking water. All housekeeping were carried out in the morning immediately after feeding the animals. The mare was milked in a separate room in a milking stall without bedding material. Mares and foals spend most of the day on the pasture, but all animals were indoors during air sampling. In the stable buildings, natural (gravity-automatic) ventilation was used through windows, doors and ventilation ducts.

A portable laser photometer, DustTrak DRX, was used to study air dustiness taking real-time measurements of the concentration of the various fractions: PM1, PM2.5, PM10, respirable dust (Resp) and total dust (Total).

Dust concentration tests were carried out twice in the morning and afternoon during a normal working day in each selected room, i.e.:

1 – stable with mares.
2 – stable with foals;
3 – mare milking room.

At the same time, microclimatic indicators were measured using a thermohygrometer. The obtained results were subjected to statistical analysis with Statistica ver.13.3. Average values for individual rooms were determined and given as arithmetic mean (M) and standard deviation (SD) as well as min and max values. Additionally, dust concentration modeling was performed in relation to the permissible value in MS Excel.

3 Results and Discussion

In each of the analyzed horse rooms, the highest concentrations were recorded for PM10 dust (Tables 1, 2 and 3). The fraction of dust pollution referred to as suspended dust PM1 and PM2.5 reached the highest values in the stable with foals and amounted respectively, 1.08 mg/m^3 and 1.09 mg/m^3, respectively (Table 2). Dustiness in this room was also the highest in the case of the respirable fraction and PM10 dust. The highest value of this parameter was recorded in the stable with foals – 2.40 mg/m^3. The lowest concentrations of this dust fraction were obtained in the mare's stable. The lowest concentrations of other measured dust parameters were obtained in the air of this room (Table 1). Milk collection activities in milking rooms resulted in a slight increase in the measured dust fractions, but it still did not reach half of the values recorded in the foal rooms (Table 3).

Table 1. Dust concentration in the stable with mares mg/m^3

Dust fraction	Average	SD	Minimum	Maximum
PM1	0.32	0.01	0.25	0.60
PM2.5	0.31	0.01	0.25	0.61
RESP	0.31	0.01	0.25	0.61
PM10	0.37	0.01	0.27	0.94
TOTAL	0.55	0.03	0.29	2.05

Table 2. Dust concentration in the stable with foals mg/m^3

Dust fraction	Average	SD	Minimum	Maximum
PM1	1.08	0.69	0.46	3.58
PM2.5	1.09	0.69	0.47	3.60
RESP	1.17	0.74	0.51	3.79
PM10	2.40	1.55	1.02	6.98
TOTAL	5.37	4.23	1.60	17.86

Table 3. Dust concentration in the mare milking room mg/m^3

Dust fraction	Average	SD	Minimum	Maximum
PM1	0.46	0.01	0.33	0.64
PM2.5	0.46	0.01	0.33	0.64
RESP	0.49	0.01	0.35	0.66
PM10	0.79	0.01	0.51	1.09
TOTAL	1.54	0.04	0.84	2.50

Table 4. Dust concentrations in research spaces modeled in MS Excel in relation to acceptable limits (3 mg/m^3)

Room	Total dust		Respirable dust	
	Average dust concentration	Minutes	Average dust concentration	Minutes
1	0.55	92	0.31	164
2	5.37	9	1.17	43
3	1.54	33	0.49	104

For comparison with acceptable limits for horses. Modeling in MS Excel was used to determine the length of time that was safe for horses to remain in the research spaces. Horses can remain safely in space 1 for 92 min (about 1.5 h). But for only a few minutes in space 2 and some more time in space 3 (Table 4). Comparison of the results reveals a significant problem arising from the high dust concentration in the stables in relation to the time the horses spend in them.

In Poland there is an act specifying the permissible levels of dust pollution in the air. Which includes PM2.5 and PM10 dust fractions. Among others. Due to the determination of the permissible level of PM2.5 for the entire calendar year. it is only possible to compare the results to the PM10 fraction. Which was set at 50 μg/m^3. In the own study, it can be seen that this value was exceeded at all measurement points. There are also reports specifying the maximum permissible level of PM in stable air. Which is 3 mg/m^3. Only in the case of stables with foals this value was exceeded [Grzyb et al., 2022]. The research by Clements and Pirie [2007] focused on demonstrating the relationship between the concentration of the respirable fraction in the air. The bedding material used and the type of feed. When wood chips were used as bedding and silage as feed. The lowest concentration of this dust fraction was recorded (average 26 μg/m^3). Higher concentrations (average 87 μg/m^3) were obtained when straw was used as bedding and hay as feed. Taking into account the results of our own research. Where the concentration of the respirable fraction was 0.31 to 1.17 mg/m^3. The results of the mentioned authors were much lower.

Siegers et al. [2018], conducted research on total dust concentration in stables. Dust concentration was found to be 0.14 mg/m^3 in horses kept on wood chips and fed with granulated complete feed and 0.97 mg/m^3 in horses with straw bedding and hay. Woods et al. [1993], using the same types of litter and feed. Recorded higher values of total dust. Its concentration was 0.70 mg/m^3 and 2.55 mg/m^3 in the first and second systems. Respectively. Ossowski et al. [2022], in their study fed pigs feed with the addition of bentonite and zeolite. Assessing. Among others. The degree of air pollution with NH_3. CH_4 and CO_2 and the concentration of total and respirable dust. Reducing properties of the tested aluminosilicates against gaseous pollutants and a reduction in dust concentration were noted in the group receiving the bentonite addition. The authors showed an increase in the indoor dust in the group of animals where zeolite was administered. The research presented by the authors requires verification of the possibility of using this type of additives in animal husbandry in the prevention of diseases related to exposure to these agents and as an element of respiratory prevention. In the study by Urbain et al. [1999], analyzed the impact of artificial exposure of animals to dust pollution from feed at a level of 1–15 mg/m^3. Such exposure has been found to be associated with an increased incidence of bronchitis in animals. The results obtained by the authors correspond to their own research. According to Wathes et al. [2004], exposure of pigs to dust with a concentration greater than or equal to 1.5 mg/m^3 is associated with deterioration of their productivity. However, Done et al. [2005], exposed piglets to much higher dust concentrations and did not observe an increased incidence of respiratory diseases. According to Banhazii et al. [2011], subjecting the bedding substrate to the greasing process can significantly reduce dust and harmful gas emissions in farm premises. The use of this method involves the sticking of more fragmented bedding particles to those with smaller fragmentation. Which results in a reduction of this fraction in the air. According to the authors, bedding greasing the litter can reduce dust concentration from 50% [Banhazi et al., 2011] to even 90% [Ji et al., 2017]. At the same time. This process inhibits the microorganisms present in the litter, this reducing NH_3 emissions [Banhazi et al., 2011], which was also confirmed by Mostafa et al. [2017].

4 Conclusion

According to the conducted research. The highest dust levels occur in premises with young animals. Which may result from the increased number of animals per unit area of these rooms and the mobility of young horses. At the same time. The concentration of dust together with other air pollutants predisposes to the occurrence of respiratory diseases. Especially in young organisms with an undeveloped immune system. Organic dust seems to be an inherent element of animal farms. Which is why it requires constant monitoring. Especially where there are young. Highly sensitive horses and where longer workers' presence is required. This will effectively limit exposure and eliminate the occurrence of respiratory diseases.

References

Journal of Laws 2019 item 1966, Announcement of the Minister of Agriculture and Rural Development of September 19, 2019 on the announcement of the consolidated text of the regulation of the Minister of Agriculture and Rural Development on minimum conditions for keeping farm animal species other than those for which protection standards are specified in European Union regulations

Lühe, T., Volkmann, N., Probst, J., Dreyer-Rendelsmann, C., Schulz, J., Kemper, N.: Bacterial burden in the air of indoor riding arenas. Agriculture **12**(12), 2111 (2022)

Elfman, L., Walinder, R., Riihimäki, M., Pringle, J.: Air quality in horse stables. in: chemistry, emission control, radioactive pollution and indoor air quality. IntechOpen. (2011)

Rodzyń, I., et al.: Exposure to chemical pollutants and biological aerosol in indoor facilities for recreational and sport horses. BMC Vet. Res. **20**(1), 78 (2024)

Nowakowicz-Dębek, B., Karpińska, K., Ospałek, W., Kachnič, J., Ossowski, M., Wlazło, Ł: Koncentracja gazów śladowych w pomieszczeniach dla koni rekreacyjno-sportowych (Concentration of trace gases in the rooms for recreational and sports horses). Przemysł Chemiczny **103**(3), 395–398 (2024). https://doi.org/10.15199/62.2024.3.8

Szewczyk, H.: Agricultural occupational and para-occupational diseases. Wieś i Rolnictwo **156**(3), 97–112 (2012)

Ossowski, M., Wlazło, Ł, Bis-Wencel, H., Krzaczek, P., Nowakowicz-Dębek, B.: Zastosowanie naturalnych sorbentów w żywieniu świń jako metoda zmniejszania zanieczyszczeń z pomieszczeń hodowlanych (The use of natural sorbents in the diet of pigs as a method for reducing gaseous pollutants and manure nutrients from live-stock housing). Przemysł Chemiczny **101**(5), 297–303 (2022)

Wiegand, R., Battye, W.H., Myers, C.B., Aneja, V.P.: Particulate matter and ammonia pollution in the animal agricultural-producing regions of north carolina: integrated ground-based measurements and satellite analysis. Atmosphere **13**, 821 (2022). https://doi.org/10.3390/atmos13050821

Cambra-López, M., Aarnink, A.J., Zhao, Y., Calvet, S., Torres, A.G.: Airborne particulate matter from livestock production systems: a review of an air pollution problem. Environ. Pollut. **158**(1), 1–17 (2010)

Hristov, A.N.: Technical note: Contribution of ammonia emitted from livestock to atmospheric fine particulate matter (PM2.5) in the United States. J. Dairy Sci. **94**(6), 3130–3136 (2011). https://doi.org/10.3168/jds.2010-3681

Ji, B., Banhazi, T., Wang, C., Li, B.: A review: the influence and mitigation of ammonia and dust in modern animal buildings. Animal Env Welfare 1–6 (2017)

Janczarek, I., et al.: Influence of air temperature and humidity in the stable on the physiological parameters in horses: preliminary study. Med. Weter. **76**(12), 718–722 (2020)

Wålinder, R., et al.: Installation of mechanical ventilation in a horse stable: effects on air quality and human and equine airways. Environ. Health Prev. Med. **16**(4), 264–272 (2011)

Pirie, R.S.: Recurrent airway obstruction: a review. Equine Vet. J. **46**, 276–288 (2014)

Kwiatkowska-Stenzel, A., Witkowska, D., Sowińska, J., Stopyra, A.: The effect of stable bedding materials on dust levels, microbial air contamination and equine respiratory health. Res. Vet. Sci. **115**, 523–529 (2017)

Ivester, K.M., Couëtil, L.L., Zimmerman, N.J.: Investigating the link between particulate exposure and airway inflammation in the horse. J. Vet. Intern. Med. **28**(6), 1653–1665 (2014)

Tainio, M., et al.: Air pollution, physical activity and health: a mapping review of the evidence. Environ. Int. **147**, 105954 (2021). https://doi.org/10.1016/j.envint.2020.105954

Wyse, C.A., et al.: Effects of changes to the stable environment on the exhalation of ethane, carbon monoxide and hydrogen peroxide by horses with respiratory inflammation. Vet. Rec. **157**(14), 408–412 (2005)

Grzyb, J., Podstawski, Z., Bulski, K.: Bacterial aerosol, particulate matter, and microclimatic parameters in the horse stables in Poland. Environ. Sci. Pollut. Res. **29**(18), 26992–27006 (2022)

Clements, J.M., Pirie, R.S.: Respirable dust concentrations in equine stables. Part 1: validation of equipment and effect of various management systems. Res. Veterinary Sci. **83**, 256–262 (2007)

Siegers, E.W., Anthonisse, M., van Eerdenburg, F.J.C.M., van den Broek, J., Wouters, I.M., Westermann, C.M.: Effect of ionization, bedding, and feeding on air quality in a horse stable. J. Vet. Intern. Med. **32**, 1234–1240 (2018)

Woods, P.S., Robinson, S.E., Swanson, M.C., et al.: Airborne dust and aeroallergen concentration in a horse stable under two different management systems. Equine Vet. J. **25**, 208–213 (1993)

Urbain, B., et al.: Effects of inhalation of dust and endotoxin on respiratory tracts of pigs. Am. J. Vet. Res. **60**(9), 1055–1060 (1999)

Wathes, C.M., Demmers, T.G.M., Teer, N., White, R.P., Taylor, L.L., Bland, V., et al.: Production responses of weaned pigs after chronic exposure to airborne dust and ammonia. Anim. Sci. **78**(1), 87–97 (2004)

Done, S.H., Chennells, D.J., Gresham, A.C.J., Williamson, S., Hunt, B., Taylor, L.L., et al.: Clinical and pathological responses of weaned pigs to atmospheric ammonia and dust. Vet. Record **157**(3), 71–80 (2005)

Banhazi, T.M., Saunders, C., Nieuwe, N., Lu, V., Banhazi, A.: Oil spraying as an air quality improvement technique in livestock buildings: Development and utilisation of a testing device. Aust. J. Multi-Disc. Eng. **8**(2), 169–180 (2011)

Mostafa, E., Hoelscher, R., Diekmann, B., Ghaly, A.E., Buescher, W.: Evaluation of two indoor air pollution abatement techniques in forced-ventilation fattening pig barns. Atmos. Pollut. Res. **8**(3), 428–438 (2017)

The Use of Drones in Agriculture: Perspectives and Limitations

Paweł Karpiński(✉)

Department of Machine Operation and Production Processes Management, University of Life Sciences in Lublin, Akademicka 13, 20-950 Lublin, Poland
pawel.karpinski@up.lublin.pl

Abstract. The use of computer-aided techniques in farm management, including for performing agricultural practices, known as precision agriculture, has become increasingly popular in recent years. This new and rapidly developing field allows for increased crop yields, improved crop quality, and reduced negative impact on the environment compared to conventional agriculture. A special category within precision agriculture is work performed using drones. The most common activities include monitoring crop health using multispectral cameras, spraying, and seeding. This paper presents a review analysis of the use of drones in agriculture, considering both the perspectives and limitations associated with it. Drones for monitoring and mapping crops and for performing agrotechnical treatments were analyzed separately. The most important perspectives include the widespread automation associated with agricultural work and an increase in the efficiency of the work performed. Significant limitations are legal issues, the cost of purchasing equipment and the need to obtain a drone operator's license.

Keywords: agriculture 4.0 · drone · unmanned aerial vehicle · UAV · precision agriculture

1 Introduction

Precise agriculture and Agriculture 4.0 represent cutting-edge advancements in agricultural practices worldwide. These approaches harness the power of digital techniques and tools to enhance crop cultivation processes [1]. Automation, in its various forms, plays a pivotal role in this domain, enabling the execution of tasks with enhanced precision. Additionally, telematics tools facilitate the collection and analysis of data pertaining to the current state of crops. Beyond robots, unmanned aerial vehicles (UAVs), commonly known as drones, have carved a niche in this field. As digital and electronic technologies mature and the production of multicopters becomes more widespread, these devices are increasingly recognized as potential catalysts for improving agricultural efficiency and enabling the substitution of human labor by autonomous machines.

Multicopters powered by electric motors are the most prevalent type of unmanned aerial vehicles (UAVs) employed in agricultural applications [2]. These UAVs offer the capability of vertical takeoff and landing (VTOL) and can maintain a hover state

E. Lorencowicz et al. (Eds.): FMPMSA 2024, LNCE 609, pp. 219–228, 2024.
https://doi.org/10.1007/978-3-031-70955-5_24

with relatively high energy efficiency. Additionally, their design facilitates convenient operation and potential maintenance. Quadcopters and hexacopters with single or dual (coaxial) rotor arrangements are most commonly used, although helicopter (Yamaha RMAX, SAT M20S) [3, 4] and VTOL fixed-wing (DeltaQuad Pro, BZB UAS Koliber VTOL) [5, 6] configurations also exist. A comprehensive review of UAVs employed in agriculture is presented in [7]. According to People's Daily Online, 120,000 drones were utilized in China in 2021 to apply pesticides over 70 million hectares of farmland [8]. This underscores the effectiveness of currently employed UAVs and highlights the substantial potential for global drone utilization in agriculture.

The global market for UAVs in precision agriculture is dominated by Chinese manufacturers. DJI, a Chinese company, holds a commanding position with an estimated 70% share of the global drone market [9]. Their product portfolio includes drones for terrain mapping (Multispectral series) and a diverse range of UAVs for agricultural applications (Agras series). XAG, another prominent Chinese manufacturer, has developed two families of agricultural spray drones (P and V series) [10]. Additionally, they offer drones for capturing images of agricultural fields. To gain a competitive edge, other manufacturers are vying for market share through price competition and enhanced customer support. Examples of companies operating in local markets include ABZ Innovation (Hungary), Drone Volt (France), Hylio (USA), and WohnderDrone (Hungary). Smaller firms, often startups, seek to differentiate themselves from the competition through unique functionalities or designs. One such example is the Pyka Pelican Spray (USA), an electrically powered agricultural fixed-wing aircraft with a maximum takeoff weight of nearly 600 kg [11].

The global adoption of drones in agriculture has witnessed a remarkable surge over the past decade and is projected to maintain its upward trajectory in the foreseeable future. According to a report by Fortune Business Insights, the global agricultural drone market is poised to expand from USD 4.98 billion in 2023 to USD 18.22 billion by 2030, reflecting a CAGR of over 20% during the forecast period [12].

A 2022 survey conducted by DroneDeploy, encompassing 766 respondents, revealed that enhanced productivity (67% of respondents) and improved operational efficiency (64% of respondents) emerged as the primary benefits driving the adoption of agricultural drones. Additionally, respondents emphasized the significance of automation and machine learning in the context of agricultural drone operations [13].

2 Drones for Mapping and Monitoring

Unmanned aerial vehicles (UAVs) equipped with multispectral [14], thermal [15], or, less commonly, hyperspectral [16] cameras are employed for terrain mapping of agricultural fields. For these devices, both spectral resolution, determined by the camera technology, and spatial resolution, determined by the sensor array and drone positioning system, are crucial factors [17]. Drone-based image acquisition serves as an alternative to satellite-based mapping [18]. Drone mapping enables the generation of high-resolution spatial maps for limited areas [19]. Data from geographic information systems (GIS) in the form of orthophotomaps can subsequently be utilized to assess the economic efficiency of agricultural holdings [20].

Multispectral cameras can be integrated with gimbals, which ensure image stabilization during the acquisition of a series of photographs [21]. An example of such a solution is presented in Fig. 1.

Fig. 1. The Sentera 6X multispectral camera mounted on a gimbal attached to the multicopter airframe [22].

Mapping agricultural fields using a camera that captures multiple bands of electromagnetic radiation reflected from the crop surface enables the derivation of a range of indices that inform crop health assessment [23, 24]. Multispectral cameras commonly record the red, green, red edge (RE), and near-infrared (NIR) bands. Blue and medium infrared (MID) bands are less frequently employed. One such parameter is the Normalized Difference Vegetation Index (NDVI), which incorporates the red and near-infrared bands [25]. Spectral analysis involves calculating the index value for each pixel of the captured image and generating a corresponding 2D color map. NDVI values close to 1 indicate healthy vegetation, while a value of 0 represents bare soil [26]. In practice, at least a dozen different indices are used, providing information on various aspects of plant health (e.g., water stress or leaf chlorophyll content) [27, 28] or soil conditions. The generated index maps can be used to assess crop quality [29], forecast yields [30], or estimate damage caused by adverse weather events such as drought or severe storms [31]. Additionally, the derived indices can be employed for land use analysis or as a support method for examining crop phenological development. It is noteworthy that drones equipped with multispectral cameras exhibit high efficiency in fulfilling their mission. The DJI Mavic 3 Multispectral drone can map an area of up to 200 ha in a single flight lasting several tens of minutes [32].

As previously mentioned, drones can be employed to gather data on the current health status and growth progress of crops. This necessitates the acquisition of images with a specific temporal resolution and, potentially, the execution of multi-temporal analysis [33]. The recorded and processed data can be utilized to identify potential crop problems, such as the presence of pests [34], diseases [35], or nutrient deficiencies [36]. This enables the implementation of remedial solutions for further crop protection.

The integration of cameras on drones with neural network-based software facilitates real-time monitoring of livestock on pastures and the localization of wild animals in meadows. In recent years, Convolutional Neural Networks (CNNs) have gained popularity in this domain, enabling object recognition based on algorithms trained on a set of training data [37, 38].

Utilizing drones for crop inventory offers significant time and cost savings for farmers. Information on crop health is obtained rapidly, enabling immediate actions to protect or improve plant health. This contributes to enhanced crop quality and quantity, leading to improved economic efficiency of the cultivated crop.

A terrain scan conducted using a multispectral camera on a drone provides farmers with crucial insights into their crops. This information can be employed to plan agronomic practices and inform decision-making regarding crop cultivation methodologies in subsequent growing seasons.

Employing information acquired from drones equipped with spectral cameras indirectly reduces carbon dioxide emissions. Farmers gain insights into areas of the crop requiring agronomic interventions and those that can be omitted. This minimizes the usage time of conventional vehicles powered by internal combustion engines. The impact is amplified when electrically powered drones are increasingly utilized for agricultural tasks.

An additional advantage arises from the reduced application of pesticides and fertilizers. Precise application of plant protection products is enabled by prior crop analysis using a multispectral camera. The chemical substance is delivered to the required location in the quantity necessary to elicit a positive effect on plant health. This is beneficial for both the environment and the economics of the enterprise due to lower spraying costs.

Drones equipped with cameras generally do not require specialized permits or pilot licenses. This stems from the fact that their maximum takeoff weight does not exceed a few kilograms, and the operation itself does not differ significantly from recreational flying. Nevertheless, it is necessary to hold the appropriate operator's license for unmanned aerial vehicles specific to the country where the observation or mapping mission is planned.

For drones used in mapping and terrain monitoring, the cost of a camera with a gimbal can be significant, comparable to the cost of the drone itself. This is a major limitation for the increased popularity of unmanned aerial vehicles among farmers.

The operation of unmanned aerial vehicles (UAVs) is particularly dependent on weather conditions. When strong winds or heavy rainfall are present, safe flight is not possible. For drones equipped with multispectral cameras, lighting conditions introduce an additional limitation. The scanned area needs uniform illumination, meaning the brightness and color of the light shouldn't change during image capture. This necessitates operation during clear weather or with minimal cloud cover. In some instances, additional radiometric correction may be necessary to eliminate the influence of variable lighting on the obtained results [39].

3 Drones for Agricultural Operations

Unmanned aerial vehicles (UAVs) employed for agricultural tasks differ significantly in design from those used for terrain mapping. They are characterized by considerably larger dimensions due to the requirement for a more powerful propulsion system capable of lifting payloads weighing up to tens of kilograms. These multicopters are equipped with a liquid spraying system comprising nozzles or atomizer disks installed at the ends of the arms. A rotating disk located centrally under the fuselage serves to apply granular materials.

The most frequent applications of agricultural drones are spraying [40], sowing [41], and fertilization [42, 43]. Feeding tasks (e.g., for farmed fish) are employed significantly less often. It is important to emphasize the versatility of these drones. A single aerial vehicle can be adapted to perform several types of work, which reduces the costs associated with cultivating crops. These mentioned tasks are often preceded by mapping conducted using a drone with a multispectral camera. This allows for the identification of areas within the crop that require specific attention due to the health of the plants. Such an approach enables the precise application of pesticides or fertilizers. The amount of the chemical substance can be controlled so that plants in a poorer health state receive an increased dose.

Drones used in agriculture are equipped with RTK (real-time kinematic) technology. This method, also used in geodesy, allows for determining the precise location of the aerial vehicle relative to the ground in real time. Thanks to this, the drone is located much more accurately compared to a conventional satellite navigation system [44]. The RTK system ensures accuracy on the level of single centimeters (usually 2–5 cm). As a result, a flight using a multispectral camera combined with the high spatial resolution of the camera guarantees high-accuracy results.

Agricultural drones can be utilized for applying pesticides (insecticides, herbicides, fungicides) and fertilizers with much greater precision compared to traditional methods [45]. According to Emergen Research, the global market for drone spraying reached a value of $170 million in 2022. The projected compound annual growth rate (CAGR) for this sector is expected to exceed 24% by 2032 [46]. Drone spraying can be performed using continuous, targeted (spot), or variable-rate application methods. This approach helps reduce the amount of chemicals used and minimizes the risk of environmental contamination. Additionally, the autonomous operation of drones ensures the safety of the operator (farmer) by eliminating exposure to pesticide aerosols [47].

Drones can also be employed for the precise application of fertilizers [48]. Similar to spraying, the location for fertilizer application is determined by prior mapping of the agricultural field. This reduces the amount of fertilizer used, leading to lower overall operational costs and indirectly mitigating environmental impact.

Similar to terrain mapping drones, agricultural UAVs are equipped with an RTK system. This system ensures high-precision localization of the drone in space, enabling the precise application of chemicals along the planned route.

The software on agricultural UAVs facilitates route planning to circumvent natural terrain obstacles, construction infrastructure, and power lines. During the automated path determination, obstacles are bypassed, and the algorithm independently establishes a trajectory that offers the best energy efficiency for the planned mission. The issue of optimal path planning for agricultural UAV flight is addressed in [49]. To detect and avoid potential collisions with the surroundings, agricultural UAVs are equipped with radars and surround cameras.

The DJI Agras T50, depicted in Fig. 2, is the largest model in the series designed for spreading bulk fertilizers and applying liquid plant protection products. The drone features a coaxial rotor system in a quadcopter configuration. This means that two pairs of propellers and motors are installed on each arm of the aircraft. The high-performance propulsion system enables the transport of bulk material weighing up to 50 kg or liquid weighing 40 kg (maximum tank capacity is 75 dm^3). When using four spray nozzles, the liquid atomization system provides a rate of up to 24 dm^3/min. This translates into an efficiency of 21 ha/h for field operation and 4 ha/h for orchard work. The granular material spreading system has a capacity of up to 1500 kg/h.

Spraying drones can be equipped with either hydraulic nozzles or an atomization system with a rotating disk that allows for droplet size control [50]. This system is increasingly being adopted by drone manufacturers due to its advantages. The solution enables the control of droplet size in the range of tens to hundreds of μm. Regulation is achieved by changing the rotational speed of the disk through which the liquid flows. The liquid is expelled due to the centrifugal force acting on it. Unlike the hydraulic nozzle system, this solution does not require the use of higher pressure values reaching several hundred Pa. Research is underway to optimize existing designs [51] and develop new solutions in this area [52].

Fig. 2. DJI Agras T50 agricultural drone [53].

The widespread adoption of UAVs for agricultural applications is hindered by varying national regulations across different countries. These regulations restrict the use of

UAVs for spraying activities due to environmental and human health concerns. In the European Union, Directive 2009/128/EC of the European Parliament and of the Council stipulates that member states shall ensure that aerial spraying is prohibited. This directive aims to minimize the potential risks associated with aerial pesticide application, such as environmental contamination and human exposure to harmful chemicals. In the United States, Part 137 regulations issued by the Federal Aviation Administration (FAA) govern UAV operations in agriculture. While these regulations are generally less restrictive than those in the EU, they still impose limitations on UAV use, such as requiring pilots to maintain visual line of sight with the aircraft and restricting operations in certain airspace classes. Apart from spraying, the use of UAVs for tasks like seeding or fertilization is generally subject to less stringent regulations. However, it is crucial to adhere to local aviation and agricultural regulations that may apply in specific areas.

A significant limitation of agricultural UAVs lies in their inherent design and power capabilities. For tasks like spraying, seeding, or fertilization, multirotor drones with electric propulsion are commonly employed. While these drones can carry up to tens of kilograms of working material (pesticide mixtures, fertilizers, or seeds), their payload capacity is still limited compared to traditional agricultural machinery. Furthermore, the maximum flight time of these drones, typically around 20 min using lithium-polymer batteries, poses another constraint. This limited flight time necessitates frequent landings for refilling empty tanks and replacing or charging batteries. This requirement for frequent landings and repetitive maintenance tasks increases the operational complexity of UAVs compared to conventional agricultural machinery.

A further limitation of agricultural UAVs lies in the fragmented development and lack of standardization of their software. Each manufacturer independently develops its own software based on its own experiences and customer feedback. This lack of public and regular exchange of technical knowledge hinders the pace of innovation in mission planning and data analysis software for agricultural UAVs.

The absence of international guidelines for multicopter spraying methodologies limits the effectiveness and widespread adoption of agricultural UAVs. The efficiency of the spraying process depends on many parameters such as: number and location of nozzles on the drone, type of nozzles, liquid pressure, physicochemical properties of the liquid (viscosity, surface tension, volatility), flight speed and altitude, atmospheric conditions (temperature, wind), and crop type. There is a lack of comprehensive research and theoretical models to determine the optimal values of these parameters for effective application of chemical substances to specific crop varieties.

The operation of agricultural UAVs, particularly those with a payload capacity of tens of kilograms and the capability to apply chemical agents, is subject to stringent regulatory requirements. These requirements necessitate specialized training, encompassing both theoretical and practical aspects of UAV operation and pesticide application. Additionally, depending on the specific jurisdiction, pilots may need to obtain additional flight permits from local authorities to conduct aerial spraying operations.

The cost associated with the acquisition and operation of agricultural UAVs presents a significant barrier to their wider adoption. While the initial purchase price may be a substantial investment, ongoing maintenance and repair expenses also play a crucial role in the overall cost of ownership. Once the warranty period expires, the financial burden of repairs falls solely upon the equipment owner. However, it is important to note that self-repairing agricultural UAVs by users is often challenging, if not impossible, due to their high level of technological complexity. This is particularly true for electronic components and software.

4 Conclusions

UAVs have found numerous applications in agriculture and possess the potential for even broader utilization. The first key application is observation and mapping of agricultural land. The second major purpose lies in performing plant protection treatments. In both cases, proper integration of the flight control system with mission planning software and data analysis software from on-board equipment (cameras and sensors) is essential.

For farmers who are also UAV operators, the simplicity of the user interface, service support, and the performance of the aircraft are of primary importance. Therefore, regardless of the application, the development of UAV technology to increase operational range, ensure longer flight time, and equip them with more precise cameras and sensors remains crucial. The future of agricultural UAVs presents two promising directions: integration with Geographic Information Systems (GIS) and other digital farm management tools, and automation of tasks such as object recognition, data analysis, and autonomous execution of plant protection treatments.

References

1. Monteiro, A., Santos, S., Gonçalves, P.: Precision agriculture for crop and livestock farming—Brief review. Animals **11**(8), 2345 (2021)
2. Hafeez, A., et al.: Implementation of drone technology for farm monitoring & pesticide spraying: a review. Inform. Process. Agric. **10**(2), 192–203 (2023)
3. Yamaha Precision Agriculture Website: https://www.yamahamotorsports.com/Precision-Agriculture.php. Last accessed 8 May 2024
4. Supersonic Aviation Technology (SAT) Website: https://www.satuav.com/helicopter-drone/helicopter-sprayer-drone.html. Last accessed 8 May 2024
5. DeltaQuad Website, https://www.deltaquad.com/vtol-drones/map/, last accessed 2024/05/08
6. BZB UAS Website, https://bzbuas.com/sklep/samoloty-bezzalogowe/rtf/koliber-vtol-linia-agro/, last accessed 2024/05/08
7. Del Cerro, J., Cruz Ulloa, C., Barrientos, A., de León Rivas, J.: Unmanned aerial vehicles in agriculture: a survey. Agronomy **11**(2), 203 (2021)
8. People's Daily Online, China's agri-tech industry abuzz as agricultural drones facilitate domestic proliferation of smart farming. http://en.people.cn/n3/2022/0701/c90000-10117605.html. Last accessed 8 May 2024
9. CNBC Website, N. Anwar, World's largest drone maker is unfazed — even if it's blacklisted by the U.S. https://www.cnbc.com/2023/02/08/worlds-largest-drone-maker-dji-is-unfazed-by-challenges-like-us-blacklist.html. Last accessed 8 May 2024
10. XAG Website: https://www.xa.com/en/plant-protection-uas. Last accessed 8 May 2024
11. Pyka Website. https://www.flypyka.com/pelican-spray. Last accessed 8 May 2024
12. Fortune Business Insights Website. Agriculture Drones Market Report, https://www.fortunebusinessinsights.com/agriculture-drones-market-102589. Last accessed 8 May 2024

13. Drone Deploy Website, DroneDeploy's State of the Drone Industry Report 2022. https://www.dronedeploy.com/resources/ebooks/state-of-the-drone-industry-report-2022/. Last accessed 8 May 2024
14. Choubey, A., Reddy, B.C.: Drones in agriculture: multispectral analysis. In: Computational Intelligence in Robotics and Automation, pp. 217–241. CRC Press (2023)
15. Awais, M., et al.: UAV-based remote sensing in plant stress imagine using high-resolution thermal sensor for digital agriculture practices: a meta-review. Int. J. Env. Sci. Technol. **20**, 1135–1152 (2022)
16. Yee Kit, C.H.A.N., Voon Chet, K.O.O., Edmund Hou Kheat Choong, C.S.: The drone based hyperspectral imaging system for precision agriculture. NVEO-Natural Volatiles & Essential Oils J.| NVEO **8**(5), 5561–5573 (2021)
17. Panday, U.S., Pratihast, A.K., Aryal, J., Kayastha, R.B.: A review on drone-based data solutions for cereal crops. Drones **4**(3), 41 (2020)
18. Bansod, B., Singh, R., Thakur, R., Singhal, G.: A comparision between satellite based and drone based remote sensing technology to achieve sustainable development: a review. J. Agric. Env. Int. Dev. **111**(2), 383–407 (2017)
19. Budiharto, W., Irwansyah, E., Suroso, J.S., Chowanda, A., Ngarianto, H., Gunawan, A.A.S.: Mapping and 3D modelling using quadrotor drone and GIS software. J. Big Data **8**, 1–12 (2021)
20. Sikora, J., Szeląg-Sikora, A., Gródek-Szostak, Z., Niemiec, M., Stuglik, J.: Support of internal transport optimization in farms with the use of spatial information systems. Agric. Eng. **24**(4), 87–94 (2020)
21. Kurbanov, R., Litvinov, M.: Development of a gimbal for the Parrot Sequoia multispectral camera for the UAV DJI Phantom 4 Pro. IOP Conf. Ser.: Mater. Sci. Eng. **1001**(1), 012062 (2020). https://doi.org/10.1088/1757-899X/1001/1/012062
22. Sentera Website. https://sentera.com/hardware/drones/. Last accessed 8 May 2024
23. Candiago, S., Remondino, F., De Giglio, M., Dubbini, M., Gattelli, M.: Evaluating multispectral images and vegetation indices for precision farming applications from UAV images. Remote Sens. **7**(4), 4026–4047 (2015)
24. Voitik, A., Kravchenko, V., Pushka, O., Kutkovetska, T., Shchur, T., Kocira, S.: Comparison of NDVI, NDRE, MSAVI and NDSI indices for early diagnosis of crop problems. Agric. Eng. **27**(1), 47–57 (2023)
25. Radočaj, D., Šiljeg, A., Marinović, R., Jurišić, M.: State of major vegetation indices in precision agriculture studies indexed in web of science: a review. Agriculture **13**(3), 707 (2023)
26. Kharuf-Gutierrez, S., Orozco-Morales, R., Díaz, O.D.L.C.A., Ruiz, E.P.: Multispectral aerial image processing system for precision agriculture. Sistemas y Telemática **16**(47), 45–58 (2018)
27. Zhang, L., Zhang, H., Niu, Y., Han, W.: Mapping maize water stress based on UAV multispectral remote sensing. Remote Sens. **11**(6), 605 (2019)
28. Qi, H., et al.: Monitoring of peanut leaves chlorophyll content based on drone-based multispectral image feature extraction. Comput. Electron. Agric. **187**, 106292 (2021)
29. Ren, D.D., Tripathi, S., Li, L.K.: Low-cost multispectral imaging for remote sensing of lettuce health. J. Appl. Remote Sens. **11**(1), 016006 (2017)
30. Barzin, R., Pathak, R., Lotfi, H., Varco, J., Bora, G.C.: Use of UAS multispectral imagery at different physiological stages for yield prediction and input resource optimization in corn. Remote Sens. **12**(15), 2392 (2020)
31. Zhou, J., Pavek, M.J., Shelton, S.C., Holden, Z.J., Sankaran, S.: Aerial multispectral imaging for crop hail damage assessment in potato. Comput. Electron. Agric. **127**, 406–412 (2016)
32. DJI Agriculture Website. https://ag.dji.com/mavic-3-m. Last accessed 8 May 2024

33. Su, J., Liu, C., Hu, X., Xu, X., Guo, L., Chen, W.H.: Spatio-temporal monitoring of wheat yellow rust using UAV multispectral imagery. Comput. Electron. Agric. **167**, 105035 (2019)
34. Vanegas, F., Bratanov, D., Weiss, J., Powell, K., Gonzalez, F.: Multi and hyperspectral UAV remote sensing: Grapevine phylloxera detection in vineyards. In: 2018 IEEE Aerospace Conference, pp. 1–9. IEEE (2018)
35. Gao, C., et al.: Monitoring of wheat fusarium head blight on spectral and textural analysis of UAV multispectral imagery. Agriculture **13**(2), 293 (2023)
36. Lu, N., et al.: Estimation of nitrogen nutrition status in winter wheat from unmanned aerial vehicle based multi-angular multispectral imagery. Front. Plant Sci. **10**, 1601 (2019)
37. Rivas, A., Chamoso, P., González-Briones, A., Corchado, J.M.: Detection of cattle using drones and convolutional neural networks. Sensors **18**(7), 2048 (2018)
38. Xu, B., et al.: Livestock classification and counting in quadcopter aerial images using Mask R-CNN. Int. J. Remote Sens. **41**(21), 8121–8142 (2020)
39. Assmann, J.J., Kerby, J.T., Cunliffe, A.M., Myers-Smith, I.H.: Vegetation monitoring using multispectral sensors—best practices and lessons learned from high latitudes. J. Unmanned Veh. Syst. **7**(1), 54–75 (2018)
40. Hanif, A.S., Han, X., Yu, S.H.: Independent control spraying system for UAV-based precise variable sprayer: a review. Drones **6**(12), 383 (2022)
41. Worakuldumrongdej, P., Maneewam, T., Ruangwiset, A.: Rice seed sowing drone for agriculture. In: 2019 19th International Conference on Control, Automation and Systems (ICCAS), pp. 980–985. IEEE (2019)
42. Song, C., et al.: Variable-rate control system for UAV-based granular fertilizer spreader. Comput. Electron. Agric. **180**, 105832 (2021)
43. Song, C., et al.: Test and comprehensive evaluation for the performance of UAV-based fertilizer spreaders. IEEE Access **8**, 202153–202163 (2020)
44. González-García, J., Swenson, R.L., Gómez-Espinosa, A.: Real-time kinematics applied at unmanned aerial vehicles positioning for orthophotography in precision agriculture. Comput. Electron. Agric. **177**, 105695 (2020)
45. Carvalho, F.K., Chechetto, R.G., Mota, A.A., Antuniassi, U.R.: Challenges of aircraft and drone spray applications. Outlooks Pest Manag. **31**(2), 83–88 (2020)
46. Emergen Research Website, Drone Spraying Services Market Report. https://www.emergenresearch.com/industry-report/drone-spraying-services-market. Last accessed 8 May 2024
47. Shahrooz, M., Talaeizadeh, A., Alasty, A.:Agricultural spraying drones: advantages and disadvantages. In: 2020 Virtual symposium in plant omics sciences (OMICAS), pp. 1–5. IEEE (2020).
48. Zhou, H., et al.: Application of a centrifugal disc fertilizer spreading system for UAVs in rice fields. Heliyon **10**(8), e29837 (2024)
49. Tevyashov, G.K., et al.: Algorithm for multi-drone path planning and coverage of agricultural fields. In: Ronzhin, A., Berns, K., Kostyaev, A. (eds.) Agriculture Digitalization and Organic Production. SIST, vol. 245, pp. 299–310. Springer, Singapore (2022). https://doi.org/10.1007/978-981-16-3349-2_25
50. Chen, P., et al.: Characteristics of unmanned aerial spraying systems and related spray drift: a review. Front. Plant Sci. **13**, 870956 (2022)
51. Yang, Z., Yu, J., Duan, J., Xu, X., Huang, G.: Optimization-design and atomization-performance study of aerial dual-atomization centrifugal atomizer. Agriculture **13**(2), 430 (2023)
52. Hu, H., et al.: Design and performance test of a novel UAV air-assisted electrostatic centrifugal spraying system. Int. J. Agric. Biol. Eng. **15**(5), 34–40 (2022)
53. DJI Agriculture Website. https://ag.dji.com/t50. Last accessed 8 May 2024

Opportunity Cost (Cost of Lost Profits) as an Element of Decision-Making Analysis in Prosumer Energy

Artur Kraszkiewicz(✉)

Department of Machine Operation and Production Processes Management, University of Life Sciences in Lublin, Głęboka Street, 20-612 Lublin, Poland
artur.kraszkiewicz@up.lublin.pl

Abstract. This work undertook research on the operating costs of energy devices in prosumer energy (simultaneous producer and consumer of energy), which was extended to include the identification of the opportunity cost as an element of decision-making analysis in prosumer energy. During the research in a model single-family building, energy in the prosumer system was produced by a coal-fired central heating (CH) boiler, a heat pump, a photovoltaic installation, and was collected from the power grid. For the technique mentioned, the operating costs were calculated in four variants. The analysis of the obtained results showed that the use of a heat pump supported by a photovoltaic installation was associated with the highest investment costs. However, these costs are much lower than when using a central heating boiler to produce thermal energy at coal price 2022 year. In this analysis be careful the costs of failure and downtime of the used technology were important in this respect, where not only the costs of repairs should be taken into account, but also the costs of lost profits during the downtime of the installation e.g. resulting from weather factors, cloud cover. Failure to take into account opportunity costs in decision-making calculations regarding the use of individual energy generation variants may be costly for the prosumer, because the omission of important elements when choosing an appropriate energy generation technique/system may lead to the selection of an action option that will generate losses.

Keywords: Prosumer energy · Operation costs · Maintenance costs

1 Introduction

Renewable Energy Sources (RES) continue to generate enormous interest in society. This is caused not only by the need to replace fossil fuels in terms of reducing CO_2 emissions into the atmosphere. The popularity of renewable energy technologies is also due to the significant increase in the prices of hard coal and gas or even the lack of availability of these fuels. The prosumer energy industry is particularly sensitive to this situation [1, 2].

E. Lorencowicz et al. (Eds.): FMPMSA 2024, LNCE 609, pp. 229–239, 2024.
https://doi.org/10.1007/978-3-031-70955-5_25

Prosumers, as producers and consumers of energy at the same time, representing mainly the sector of users of single-family buildings, in such dynamically changing conditions, are constantly exposed to the need to make important, even strategic decisions regarding energy management in the building. Additionally, the unstable situation on the energy market directly disrupts their energy security. The above-mentioned factors mean that these entities have recently been exposed to making a number of decisions related to the selection of the appropriate technology for supplying energy to buildings, paying attention not only to operating costs but also to additional factors, work time inputs, which are often not recorded as cash flows. The operating costs of heat and electricity generation systems available on the market include the costs of maintenance and the costs of their use (operation). Maintenance costs include depreciation costs, reimbursement of expenses incurred for the purchase of the equipment used, inspection costs to keep the equipment in the best possible technical condition, and insurance costs. However, operating costs are the costs related to the operation of these devices in the production of energy, and especially in older technologies, they include fuel costs or the costs of energy used to produce energy in another form, e.g. generating thermal energy using electricity [3, 4].

The arguments that determine the use of a given energy source involve taking into account a number of factors that concern the selection of appropriate technology, the type and power of devices in its scope, and the setting options in order to achieve the greatest possible efficiency of their operation. The issue is also complicated by the fact that energy systems composed of several integrated energy sources are used. Recently, the prosumer energy industry has been moving away from the technology of archaic charging coal furnaces and drawing only electricity from a centralized power grid. For some time now, automatic retort boilers powered by eco-pea coal have become very popular, and recently more and more photovoltaic installations and heat pumps have appeared. These devices often work in households, providing energy in various configurations [5, 6].

In management, as one of the methods, optimal decisions are made based on opportunity costs, which are often called costs of lost benefits or opportunities. By defining these costs in terms of the goods and services market, we formulate that the opportunity cost of a given good is the amount of another good that must be given up in order to produce (consume) another unit of the first good. In this approach, we give up part of the production of one good in favor of the production of another. In the case of consumption, we talk about an opportunity cost when, with a limited budget, we can afford to buy a given amount of one good, thereby giving up a certain amount of another good. When discussing the concept of opportunity cost, attention should also be paid to the problem of scarcity. Zalega [7] states that it is the limited availability of a given good or service. It occurs whenever people's demand for goods or services is greater than the ability to produce or provide them. Scarcity is the result of limited resources. Due to the scarcity of resources and unlimited needs, people must constantly make choices about what and how to produce. Recently, such limited resources have occurred on the energy and renewable energy markets. So far, the alternative cost has been considered marginally in the literature on renewable energy sources and distributed, prosumer energy. For this reason, research was undertaken on the costs of operating energy devices in the prosumer energy industry, which was extended to include the identification of the opportunity cost as an element of decision analysis in the prosumer energy industry.

2 Material and Research Methods

The research was carried out in four variants in the aspect of economic realities in September 2021, 2022, and 2023. A single-family building with a usable area of 130 m^2 and constructed according to WT 2014 standards was adopted for the analysis. Partial values of the building's primary energy index (EP) for the purposes of heating and hot water preparation usable capacity for this standard is 120 kWh/m^2/year. The subject of the research was located in the Lublin Voivodeship, which corresponds to climate zone III, which is characterized by a design external temperature of -20 °C and an average annual temperature of 7.6 °C [8]. Individual variants of the analysis taking into account the technical parameters of the devices used to generate electricity and heat are presented in Table 1.

The calculations were made for two annual periods in the economic conditions of September 2021, 2022, and 2023. The prices at that time were respectively: gross electricity PLN 0.64, PLN 0.77 and PLN 0.92/kWh [PGE energy prices] and PLN 1,100 for eco-pea coal, 3,200 and 1,769 PLN/t [4, 9–11].

Table 1. Technical parameters of devices in individual analysis variants.

Analysis variant	Source of heat	Source of electricity
I	Automatic boiler for eco-pea coal with a power of 14 kW, class 5, purchase price including installation PLN 16,000; service life (T) 15 years	Energy network, PGE group
II		Photovoltaics with an installed capacity of 3.8 kWp, connected to the power grid and operating on the basis of discounts, purchase price including installation PLN 17,000; useful life (T) 25 years
III	Air-water heat pump with a capacity of 9 kW in a split system, purchase price including installation PLN 40,000; service life (T) 25 years	Energy network, PGE group
IV		Photovoltaics as above

Annual operating costs were calculated as the sum of the costs of maintaining and using the technique in individual variants, taking into account the standard duration of its use. Maintenance costs included depreciation costs calculated using the straight-line method and settled on an annual basis. Additionally, standard costs of inspections of central heating installations before the heating season were taken into account. at PLN 150, photovoltaic installation PLN 300 and heat pump PLN 300 [12, 13]. However, operating costs included primarily the costs of fuel or electricity purchased from external entities. Unit energy costs were calculated as the quotient of annual operating costs and the sum of energy used in the tested facility.

3 Results

3.1 Operating Costs

The values of energy generated in the prosumer system and consumed from the network are presented in Table 2.

Table 2. Energy generated and consumed in individual variants adopted during the research.

Analysis variant	Energy produced	Energy taken from the power grid
I	thermal energy generated using coal 21,666 kWh	1612 kWh
II	thermal energy generated from coal 18,055 kWh, plus energy production from photovoltaics 3,500 kWh	energy surplus 1798 kWh
III	energy generated using a heat pump 15,600 kWh with electricity demand of 4,500 kWh	5752 kWh
IV	energy generated using a heat pump 15,600 kWh with electricity demand 4,500 kWh, energy production by photovoltaics 3,500 kWh	2252 kWh

Variant I was characterized by the production of thermal energy for central heating (central heating) and domestic hot water (DHW) using coal fuel in a retort boiler. However, electricity was consumed 100% from the low-voltage power grid. This solution was characterized by the lowest investment costs. The cost of the boiler is only PLN 16,000 (Table 1). In this variant, the thermal energy generated from coal is almost 22,000 kWh and the electricity taken from the network is 1,612 kWh (Table 2).

Variant II was a modification of variant I with a photovoltaic installation, which was intended to cover the demand for electricity and additionally enable the production of domestic hot water. Outside the heating season. The cost of the two types of installations in this case amounted to a total of PLN 33,000 (Table 1). In this variant, the thermal energy generated from coal is smaller and amounts to 18,055 kWh, and the electricity taken from the network was settled in a discount system under which there is a photovoltaic installation generating 3,500 kWh of electricity annually. In this arrangement, there was an overproduction of energy worth 1,798 kWh, which was not used by the examined farm (Table 2).

Variant III of the operation of a single-family building in the aspect of prosumer energy management was based on replacing the central heating boiler. Heat pump without taking into account the possibility of generating energy from a photovoltaic installation. The costs of installing a heat pump are PLN 40,000 (Table 1). In such a system, energy production was to cover the building's central heating energy demand. And to heat domestic hot water with ET 2014 requirements of 15,600 kWh. The electricity consumed in such a system to power the heat pump and operate the building is 5,752 kWh, which is a 3.5-fold increase compared to variant I (Table 2).

Variant IV supplemented variant III of the building's operation with the production capacity of the photovoltaic installation at the level of 3,500 kWh per year. In this case, the energy consumed from the grid is only 2252 kWh (Table 2). However, this variant is characterized by the highest investment costs of PLN 57,000 (Table 1).

The costs of use and operation of energy-generating installations in the prosumer energy system in individual analysis variants are presented in Fig. 1 taking into account the prices from September 2021, in Fig. 2 taking into account the prices from September 2022 and in Fig. 3 taking into account the prices from September 2023.

Comparing the individual years of operation of the considered installations, a significant increase in the costs of using a central heating boiler is noticeable in 2022. Due to a significant (3-fold) increase in coal prices and after reducing this price, a decrease in these costs in 2023. Use of a photovoltaic installation (variant II), which provides energy for the production of domestic hot water, partially reduces the costs of using the central heating boiler. Nevertheless, only the use of a heat pump (variant III) significantly (especially recently) reduces the costs of using the installation generating thermal energy (Figs. 1, 2 and 3). Considering the full operating costs of these installations, taking into account the annual depreciation costs of the technology used, we observe quite equal values between the individual variants in the price realities of 2021. However, variants II and IV supported by photovoltaics have the lowest costs. However, the situation changes dramatically in the realities of prices in 2022, where, in the context of the individual variants adopted during the research, a significant reduction in the building's operating costs was observed, which, compared to the base variant I, was 2.6 times smaller for variant IV.

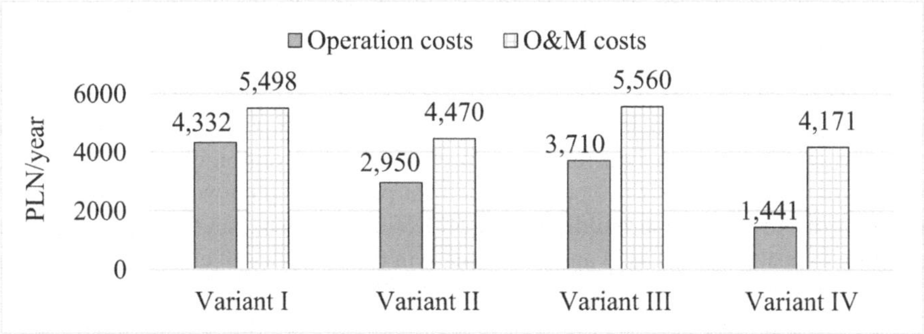

Fig. 1. Costs of energy production according to energy prices at the beginning of the heating season in 2021.

The unit costs of providing heat and electricity for the building in the individual variants of the technique used are shown in Fig. 4. In the realities of prices in 2021, the unit costs in the individual variants were quite equal and ranged from PLN 0.24 (variant 1) to 0.27 PLN (variant 3). However, a significant increase in coal prices in 2022 means that for variants using central heating boiler unit costs are almost twice as high as for

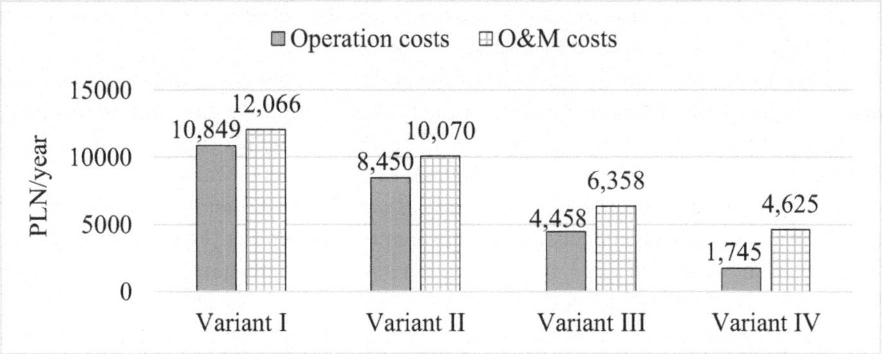

Fig. 2. Costs of energy production according to energy prices at the beginning of the heating season in 2022.

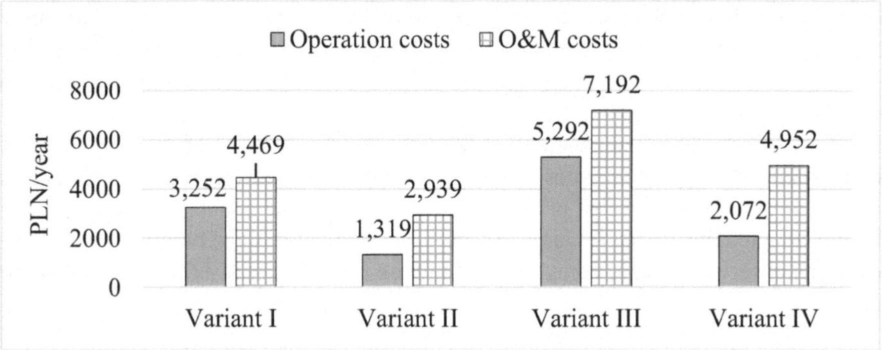

Fig. 3. Costs of energy production according to energy prices at the beginning of the heating season in 2023.

variants using a heat pump. It is also important that the use of photovoltaic installations in variant II did not result in a reduction in unit costs, this only took place in variant IV, for which the unit costs of operating the heat pump and photovoltaics enabled the provision of energy for the building at PLN 0.26 per 1 kWh. Nevertheless, after the reduction in coal prices in 2023 and the increase in the unit price of electricity, the costs of using the technique in question in variants III and IV have increased.

3.2 Opportunity Costs

The identified opportunity costs of operating the considered installations are presented in Table 3.

Producing heat energy in a building using a prosumer system certainly generates significant investment costs – the most technically and ecologically advanced variant IV in this respect was PLN 41,000 more expensive than variant I based on fossil fuels. Nevertheless, in terms of prices from September 2022, this solution had PLN 7,641 lower operating costs (Table 3). Additionally, variant IV ensured the avoidance of activities

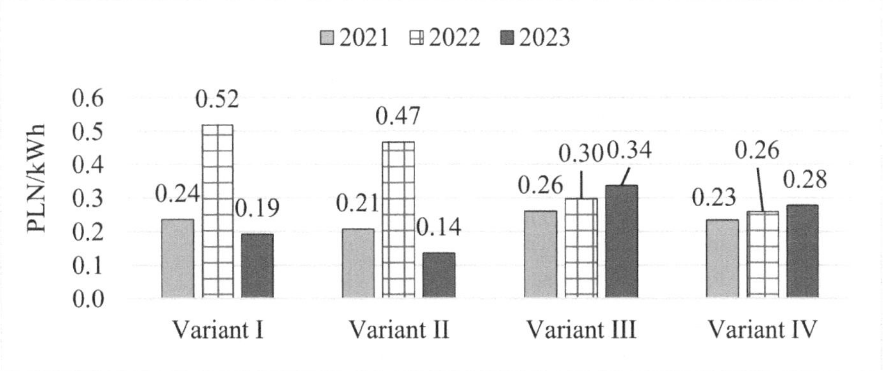

Fig. 4. Unit costs of operating the analyzed energy generating devices.

related to the operation of the technology in variant I, which are difficult to estimate, including: disproportionately less time spent on supervising the photovoltaic installation and the heat pump than on the central heating boiler, no fossil fuel used for heating, and therefore lower CO_2 emissions. Nevertheless, a significant advantage of using a heat pump was its integration with a photovoltaic installation, which resulted in a much lower demand for electricity from the power grid. Variant III was characterized by operating conditions in which electricity came 100% from the grid, which meant that the demand for it was much higher than in the base variant I. Nevertheless, the investment costs were PLN 24,000 higher than in variant I but thus the operating costs were PLN 5,908 lower (Table 3). Nevertheless, the main advantage of this solution was that it avoided daily maintenance of the boiler and weekly cleaning and removal of ash. In terms of the conducted analyses, it was important that variant II, in relation to variant I, was not a variant that significantly improved the efficiency of the building's operation. Under test conditions, the photovoltaic installation, or more precisely, the energy generated from it, is not fully used. Some of the energy remains in the network and will not be used by the prosumer. Of course, in global terms, this is a benefit that persists and leads to less energy production from fossil fuels. However, this variant is characterized by time-consuming boiler maintenance activities during the heating season, and additionally promotes CO_2 emissions. By investing an additional PLN 7,000 in the implementation of variant III, you can further reduce operating costs and avoid time-consuming maintenance of the central heating boiler.

4 Discussion

The analyzes performed on the functioning of a single-family building as part of prosumer energy, in which attention was paid to the possibilities of providing heat and electricity using own RES installations, indicate differences in both the operating costs of these installations as well as the unit costs of consumed energy. Comparing such values required taking into account the original operating parameters of the building that influenced energy demand. Therefore, the interpretation of the results should refer to the

Table 3. Opportunity costs identified in the price realities of September 2023.

Analysis variant	Non-accounting, additional aspects of the functioning of the technique in a given variant	Opportunity costs
IV vs III	In variant IV, shorter time spent on supervising the photovoltaic installation, lower CO_2 emissions due to lower coal involvement	In variant IV, PLN 17,000 higher investment outlays – PLN 680 per year and PLN 2,240 lower annual operating costs
IV vs II	In variant IV, the time spent on supervising the photovoltaic installation is much shorter than when servicing the central heating installation, no fossil fuel is used to heat domestic hot water due to the use of energy from photovoltaics, there is less demand for coal, resulting in lower CO_2 emissions	In variant IV, PLN 17,000 higher investment outlays – PLN 680 per year and PLN 2,013 higher annual operating costs
IV vs I	In variant IV, disproportionately shorter time spent on supervising the photovoltaic installation and the heat pump than on the central heating boiler, the lack of use of fossil fuel for heating leads to lower CO_2 emissions	In variant IV, PLN 41,000 higher investment outlays – PLN 1,640 per year and PLN 483 higher annual operating costs
III vs II	In variant III, no time-consuming cleaning of the boiler, time spent on supervising the photovoltaic installation unclear CO_2 values in the atmosphere due to the involvement of fossil fuels in the energy system, increasing demand for electricity	In variant III, PLN 7,000 higher investment outlays – PLN 960 per year and PLN 4,253 higher annual operating costs
III vs I	In variant III, no time-consuming cleaning of the boiler, increased demand for electricity from the network	In variant III, PLN 24,000 higher investment outlays – PLN 280 per year and PLN 2,723 higher annual operating costs
II vs I	Variant II is characterized by the lack of involvement of the central heating boiler. For heating domestic hot water due to the use of energy from photovoltaics, this results in lower demand for coal, lower CO_2 emissions, shorter supervision time of the central heating installation, no boiler cleaning in the summer	In variant II, PLN 17,000 higher investment outlays – PLN 680 per year and PLN 1,530 lower annual operating costs

system of billing energy from photovoltaic installations, which was based on a system of discounts. The net-billing system would certainly slightly affect the obtained values. Additionally, investment costs were 100% borne from own funds, without taking into account possible subsidies. Therefore, referring to the presented relationships in terms of cost intensity of renewable energy production, the observed trends are consistent with the data available in the literature [4–6, 10, 14–17].

It is worth noting here that the situation of changes in energy prices has a significant impact on the result of the economic calculation. An additional criterion is the work input and environmental effects that are identified with a given technical solution. When assessing individual variants of the technique used, an additional important argument was the possibility of using a certain solution, for example due to the lack of fuel. The presented research conditions indicate a number of criteria that become important under certain conditions. In management, when making decisions in the area of non-accounting costs, the opportunity cost account (lost opportunity benefits) is used. We almost always deal with opportunity costs – even when we initially see no choice. Although, according to Wolska [18], the concept of cost causes many problems because the definition of this term is different among economists and accountants. For people who are accountants, costs called accounting costs are important. These costs concern expenses that are incurred during the production process and the company has appropriate financial documents for them. Accounting costs are key here, but there is no information on alternative costs. However, opportunity costs, apart from accounting costs, are of interest to economists. Therefore, it was important to introduce a new category of costs, called economic costs.

In the course of own research, additional activities were identified that were related to the use of a given energy generation variant in the building in question. It was important to commit financial resources to the selected type of technical installation and compare them with benefits not only expressed in monetary terms, but also those that often relate to social or ecological aspects.

Additionally, the costs of failure and downtime of the used technology were important in this respect, where not only the costs of repairs should be taken into account, but also the costs of lost profits during the downtime of the installation. Failure to take into account opportunity costs in decision-making calculations regarding the use of individual energy generation variants may be costly for the prosumer, because the omission of important elements when choosing an appropriate energy generation technique/system may lead to the selection of an action option that will generate losses.

5 Conclusion

As a result of the research conducted, the following conclusions were formulated:

1. In the conditions of testing a single-family building, the increase in prices for coal and electricity in 2022 resulted in the costs of use and, therefore, operation of devices based on such fuel being much higher than those of a heat pump supported or not by a photovoltaic installation. In other years, the costs varied much less.
2. The use of a photovoltaic installation should be correlated with higher self-consumption of electricity to be able to fully use the amount of energy produced by it.
3. The use of a heat pump supported by a photovoltaic installation involves the highest investment costs, which, however, translates into the lowest operating costs. Resignation from photovoltaics contributes to higher energy consumption from the power grid and, therefore, to higher operating costs.

4. The identification of alternative costs complements the accounting and operating costs. However, their analysis also supports decision-making in the field of prosumer energy.
5. It would be advisable to conduct further research taking into account other energy carriers and valuable settlements of the surplus energy produced by the prosumer.

References

1. Energy Policy of the European Union: https://www.europarl.europa.eu/factsheets/pl/sheet/68/polityka-energetyczna-zasady-ogolne. Access 01 Sep 2022. (in Polish)
2. Poland's Energy Policy until 2040. https://www.gov.pl/web/klimat/polityka-energetyczna-polski. Access 5 Sep 2022. (in Polish)
3. Methodological tips for calculating machine operating costs. https://poznan.cdr.gov.pl/normatywy/public/pdf/11_A_1.pdf. Access 06 Sep 2022. in Polish
4. Biernaciak, E.: Market price of electricity and net-billing. Sale price – September 2022. https://enerad.pl/aktualnosci/rynkowa-cena-energii-elektrycznej-a-net-billing-cena-sprzedazy/ (2022). Access 05 Sep 2022. (in Polish)
5. Kwaśniewski, D., Akdeniz, C., Durmaz, F., Kömekçi, F.: Economic analysis of the photovoltaic installation use possibilities in farms. Agric. Eng. **24**(4), 47–60 (2020). https://doi.org/10.1515/agriceng-2020-0035
6. Topolańska, J.; Teleszewski, T.J.: Heat pump compared to other heat sources in multi-family buildings. Construct. Env. Eng.. **3** (2012). (in Polish)
7. Zalega, T.: Contemporary microeconomics, Collection of tasks, Scientific Publishing House of the Faculty of Management of the University of Warsaw, Warsaw (2015). (in Polish)
8. Climatic zones of Poland. https://strefaklimatyzacji.pl/baza-wiedzy/artykuly/strefy-klimatyczne-polski/. Access 01.09.2022. (in Polish)
9. Coal prices in Poland are up to 4 times higher! There are no queues in coal depots, no queues in mines. https://www.muratorplus.pl/biznes/wiesci-z-rynku/ceny-wegla-2022-w-polsce-w-skladach-na-swiecie-cena-tony-wegla-pgg-aa-2Wgb-NFp1-kFbJ.html. Access 08 Sep 2022. (in Polish)
10. Renewable energy costs have dropped again in 2021, but this year will be different. https://www.gramwzielone.pl/energia-sloneczna/108496/koszty-oze-w-2021-roku-znowu-spadly-ale-w-tym-roku-bedzie-inaczej. (in Polish)
11. Prices of coal, eco-pea coal and firewood are falling at the end of 2023. https://www.farmer.pl/farmer-po-godzinach/ceny-wegla-ekogroszku-drewna-opalowego-spadaja-pod-koniec-2023,137817.html. Access 30 Jan 2024 (in Polish).
12. How much does photovoltaics cost? Cost of installation and equipment. https://lepiej.tauron.pl/zielona-energia/ile-kosztuje-fotowoltaika-koszt-instalacji-i-urzadzen/ access 12 Sep 2022. (in Polish)
13. Heat pump service – what does it involve and how often does it need to be performed? How much does heat pump maintenance cost? https://muratordom.pl/instalacje/instalacja-grzewcza/serwis-pompy-ciepla-kiedy-i-kto-moze-go-wykonac-ile-kosztuje-aa-qCSA-nbgZ-gvro.html (access 12.08.2022 r.) (in Polish)
14. Brodziński, Z., Brodzińska, K., Szadziun, M.: Photovoltaic Farms – Economic Efficiency Of Investments In North-East Poland. Energies **14**, 2087 (2021). https://doi.org/10.3390/en14082087
15. Ogonowska-Rejer, A.: Is photovoltaics under the new rules still profitable in 2022. https://enerad.pl/aktualnosci/czy-fotowoltaika-sie-oplaca-w-2022-roku-liczymy/ (in Polish). Access 30 Jan 2024

16. Prosumers will sell energy for over PLN 1,000/MWh. https://gramwzielone.pl/energia-sloneczna/108878/prosumenci-sprzedadza-energie-za-ponad-1000-zlmwh. access 13 Sep 2022. (in Polish)
17. Raczkiewicz, M.: Economic profitability of a photovoltaic farm up to 1 MW in Polish conditions. https://www.cire.pl/pliki/2/2019/fotowoltaika_producent.pdf (2019). (in Polish). access 23.09.2022
18. Wolska, G. (ed.), Microeconomics, concepts – subject – evolution, Polish Economic Publishing House" Warsaw (2014). (in Polish)

Analysis of Water Intended for Drinking in Terms of the Content of Nitrogen Compounds (Ammonium Nitrogen, Nitrates (III) and (V))

Luiza Kubisiak-Banaszkiewicz[1], Wioletta Żukiewicz-Sobczak[1,2](✉), Agnieszka Starek-Wójcicka[2], and Paweł Sobczak[3]

[1] Department of Nutrition and Food, University of Kalisz, Kalisz, Poland
wiola.zukiewiczsobczak@gamil.com

[2] Department of Biological Bases of Food and Feed Technologies, University of Life Science in Lublin, Lublin, Poland

[3] Department of Food Engineering and Machines, University of Life Science in Lublin, Lublin, Poland
wiola.zukiewiczsobczak@gamil.com

Abstract. Nitrogen compounds dissolved in water are serious and still incessant problem for environmental protection. Therefore, because of the runoff of nitrogen compounds into surface water and their leaching into deep groundwater, the European Union Directive of 1991 concerning the protection of waters against pollution caused by nitrates from agricultural sources limits their use. The aim of the study was to analyse the content of nitrogen compounds (ammonium nitrogen, nitrates (III and V) in water samples from well as drilled deep wells and dug shallow wells. The research revealed that in 90% of the tested dug (household) wells and tap water, the nitrates (V) concentration limit was exceeded. In the case of nitrates (III), the permissible values for this parameter were not surpassed in either of the tested water samples. Ammonium nitrogen concentration exceeded the threshold level in a single water sample from a deep well. Water from drilled and dug wells is not regularly tested. It is therefore necessary to educate the public about the quality of drinking water, especially in the case of domestic wells. Owners of such water intakes should be aware of the importance of regular testing in this regard, especially when the water is consumed.

Keywords: Nitrogen compounds · Water · Sustainability · Environmental protection · Health

1 Introduction

The quality of drinking water should meet the requirements specified in the Regulation of the Minister of Health of 7 December 2017 on the quality of water for human consumption (Journal of Laws of 2017, item 2294). Water is safe for human consumption when it is free from pathogenic microorganisms and parasites, as well as from other substances, e.g. uranium [1, 5] and nitrates, in amounts that pose a potential health hazard

E. Lorencowicz et al. (Eds.): FMPMSA 2024, LNCE 609, pp. 240–245, 2024.
https://doi.org/10.1007/978-3-031-70955-5_26

[2]. The main anthropogenic source of nitrogen in the environment is nitrogen fertilisers used in agriculture. About half of runoff from fields pollutes surface and groundwater, increasing the concentration of nitrates (V) in our aquatic resources [3]. The maximum nitrate (V) concentration level in municipal water sources in the United States is 10 mg/L of nitrogen (NO_3-N), that is 44.3 mg/L of nitrate (V) (NO_3), which is roughly equivalent to the recommendations of the World Health Organization of 50 mg/L [4]. The aim of the study was to analyse the content of nitrogen compounds (ammonium nitrogen, nitrates (III and V)) in water samples from well as drilled deep wells and dug shallow wells.

2 Material and Methods

The methods used to determine the water composition are consistent with the guidelines of RMZ and have been validated and verified under laboratory conditions at the PSSE in Kalisz. The https://www.facebook.com/groups/167444088549084/programme for monitoring the validity of the results included yearly checking of: the standard curves over the entire range of the curve, correctness and repeatability at each point of the standard curve, limits of detection and quantification, intermediate precision, repeatability of the actual sample of potable water, recovery of 10%, 50% 90% of the standard curve and at the highest permissible concentration for a given parameter. Each result was presented with uncertainty for the given value and was expressed as an expanded uncertainty for a confidence interval of approximately 95% and coverage factor $k = 2$. The uncertainty quoted takes into account sampling only in the case of analyses of water from drilled and dug wells requested by external customers.

The results presented here are from the laboratory of the PSSE in Kalisz and were performed under an external client's order (Figs. 1 and 2). The latter included companies conducting service, commercial or agricultural activities as well as private individuals. As part of the commissioned work, tests were performed for drilled deep wells and old household wells located in the city of Kalisz and the Kalisz district.

The results

Drilled deep wells are supplied with deep groundwater which is underneath impermeable layers. These layers protect the water body against unfavourable external factors and ensure its bacteriological and chemical purity. Drilled deep wells have a depth of 15–30 m. For such a well, one must obtain a water permit issued by the State Water Holding *Polish Waters*. Only household dug wells with a depth of less than 30 m do not require such a permit [14]. The examined drilled deep wells are located in the city of Kalisz and the Kalisz district (Figs. 1 and 2). They supply water to crop fields, horticultural farms and, in some cases, households (when there is no collective drinking water supply or when its output is too low). From the results obtained, it can be concluded that the water quality in drilled wells located in the city of Kalisz and the Kalisz district is satisfactory. Only in one tested well was the threshold concentration of nitrates (V) exceeded, amounting to 92 mg/L (Figs. 1 and 2). Dug wells are built of concrete rings, and their depth ranges from 5 to 15 m. These wells are supplied with groundwater, less exposed to external environment contact than near-surface water. Nevertheless, these groundwaters are so shallow that their composition is often questionable. The investigated dug wells (Figs. 1 and 2) are localised in the city of Kalisz and the Kalisz district.

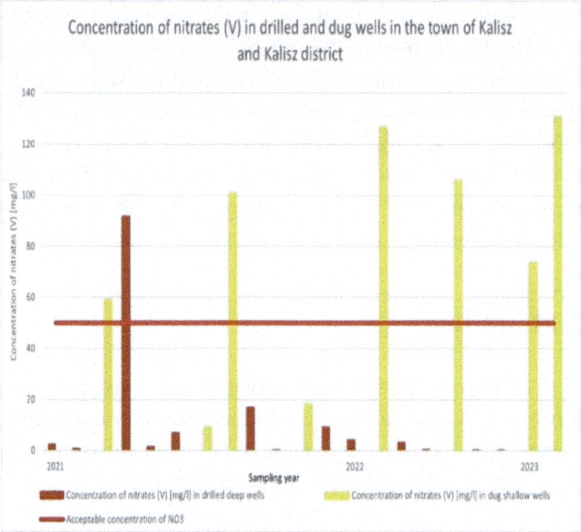

Fig. 1. Concentration of nitrates (V) in drilled and dug wells in the town of Kalisz and Kalisz district

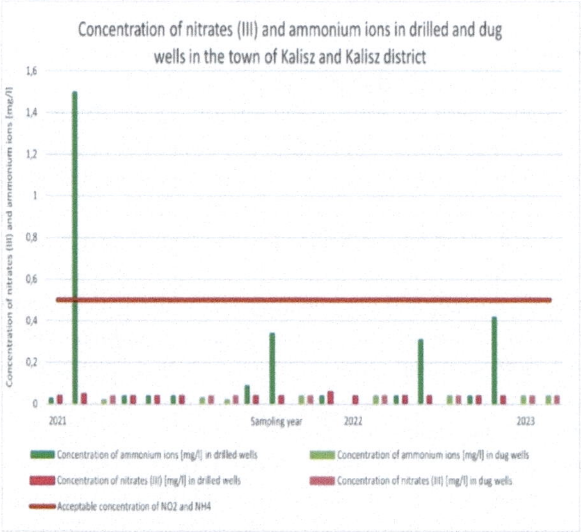

Fig. 2. Concentration of nitrates (III) and ammonium ions in drilled and dug wells in the town of Kalisz and Kalisz district

The analysed water was of poor quality; in 9 out of 11 wells, the permissible values of nitrates (V) were exceeded. The water from almost 80% of these wells was unsuitable for consumption. It contained persistent nitrogen compounds. Only in two dug wells from the Kalisz district did the concentration of nitrates (V) not exceed the recommended parametric value [6]. In one of the investigated drilled deep wells from the Kalisz district, the value of ammonium ions exceeding the recommended value (1.5 mg/L) was observed. Seven wells had ammonium ions concentration below the detection limit. In four of them, the concentration ranged from 0.09 to 0.42 mg/L, and in just one examined well, the value of this parameter was exceeded. As for nitrates (III), no values above this parameter threshold were observed. In most cases, the concentration was below the limit of quantification. In dug wells, concentrations of nitrogen compounds (ammonium ions and nitrates (III)) were at or below the quantification limit. Examination of the wells (four from the Kalisz district and five from the city of Kalisz) showed that regardless of their location, be it in rural or urban area, water from dug wells is not potable. The water from these wells is only suitable for watering crops and flowers.

People, especially from rural areas, also use private intakes (household wells), which provide water of an unknown physicochemical and bacteriological composition [9]. According to the literature [8, 9], in Poland, the concentration of nitrates (V) exceeds acceptable levels, especially in the water from dug wells. Drinking water can, therefore, become a threat. This usually happens when users of private water intakes rely solely on the organoleptic assessment of its quality [9]. This results from the fact that private drinking water intakes in our country are not subject to mandatory sanitary supervision [10]. This increases the risk of drinking water that is hazardous to health, of low quality, and contaminated, which may cause many diseases [7]. Typically, contaminated water is unsuitable for consumption and satisfying other human needs, while its treatment remains too costly and complicated to be performed in a single household [11, 12]. This poses a threat mainly to the poorer part of the population and to rural areas where it is not possible to connect to the collective water supply network [13]. In our research on dug wells, it was observed that the acceptable concentration of nitrates (V) in drinking water was exceeded over twice (131 mg/L in a well located in Kalisz). The commonly observed increase in levels of nitrogen compounds in these wells is caused by their shallow depth and runoff of water polluted with fertiliser from agricultural fields. The present study revealed that water samples from 78% of the dug wells, 8% of the drilled wells tested as well as 25% of tap water intakes contained nitrates (V) and did not meet the requirements of the Regulation of the Minister of Health [6]. It is necessary to create public awareness about the quality of the water we consume, especially from our private water intakes. Water from old, dug household wells is usually not tested. The quality of well water is the owner's responsibility, and it is their responsibility when they connect to the domestic water supply and consume water of questionable quality. If it is used for consumption, it should be tested at least once a year, especially regarding health parameters (nitrogen compounds). Otherwise, it should be used only for watering crops. As shown by our research, well water contains very large amounts of nitrates (V) (ranging from 9.3 mg/L to 131 mg/L) and ammonium nitrogen (up to 1.5 mg/L of NH^{4+}). It should be noted that water from domestic and drilled wells used for personal needs is usually not subjected to purification. Drilled wells in agricultural fields are becoming more and more frequent. These wells are intended for agricultural and horticultural production and not for consumption. Fees are charged for using groundwater and reducing natural land retention [14]. Climatic factors affecting the presence of nitrates in groundwater include amount, intensity and timing of rainfall. Increasing rainfall intensity, especially in winter and spring, may lead to increased nitrate (V) runoff from crop fields and penetration into groundwater.

3 Conclusions

1. Nitrate (V) limit values were exceeded in 90% of the dug wells located in the district of Kalisz, in the south-eastern part of Wielkopolska (Poland).
2. It was found that water from dug shallow wells located in Kalisz and the Kalisz district is unsuitable for human consumption due to large amounts of nitrates (V).

The research revealed that in 90% of the tested dug (household) wells and tap water, the nitrates (V) concentration limit was exceeded. In the case of nitrates (III), the permissible values for this parameter were not surpassed in either of the tested water samples. Ammonium nitrogen concentration exceeded the threshold level in a single water sample from a deep well. Water from drilled and dug wells is not regularly tested. It is therefore necessary to educate the public about the quality of drinking water, especially in the case of domestic wells. Owners of such water intakes should be aware of the importance of regular testing in this regard, especially when the water is consumed.

References

1. Davidson, E.A., et al.: Excess nitrogen in the U.S. environment: Trends, risks, and solutions. In: Issues in Ecology. Ecological Society of America, Washington, DC, USA (2012)
2. LNCS Homepage USEPA. Regulated Drinking Water Contaminants: Inorganic Chemicals. Available online: https://www.epa.gov/ground-water-and-drinking-water/table-regulated-drinking-water-ontaminants. Last accessed 23 September 2017
3. Bilek, M., Małek, K., Sosnowski, S.: The physicochemical parameters of drinking water from dug wells in the area of Podkarpacie. Bromatol. Chem. Toksykol. **48**(4), 40–646 (2005). (in Polish)
4. Raczuk, J.: Nitrogen compounds in well water as a factor of a health risk to the Maciejowice commune inhabitants (Mazowieckie Voivodeship). Archives of Environmental Protection **36**(4), 31–39 (2010)
5. Ustawa z dnia 20 lipca 2017 r. Prawo wodne (Dz. U. 2017, poz. 1566 z późn. zm.) (Dział III, rozdział 4)
6. Rozporządzenie Ministra Zdrowia z 7 grudnia 2017 roku „W sprawie jakości wody przeznaczonej do spożycia przez ludzi" Dz.U. 2017 poz. 2294
7. Królak, E., Raczuk, J.: Nitrate concentration-related safety of drinking water from various sources intended for consumption by neonates and infants. Archives of Environm. Protect. **44**(1), 3–9 (2018)
8. Wójcik-Jackowski, S., Bilek, M.: Woda z "prywatnych" ujęć wody pitnej, jako czynnik ryzyka zdrowia człowieka, w świetle badań nad jej jakością na tle obowiązujących uregulowań prawnych. Bromat. Chem. Toksykol. **48**(2), 216–222 (2015)
9. Raczuk, J., Dziuban, E., Biardzka, E.: Azotany w wodzie do picia jako czynnik ryzyka zdrowotnego mieszkańców gminy Platerów (województwo mazowieckie). Ochr. Środ. i Zasob. Natur. **24**(1), 5–9 (2013)
10. Raczuk, J., Biardzka, E., Michalczyk, M.: Związki azotu w wodzie studziennej w świetle ryzyka zdrowotnego mieszkańców gminy Wodynie (woj. Mazowieckie). Woda Środ. Obsz. Wiej. **9**(1), 87–97 (2009)

11. Raczuk, J., Sarnowska, K.: Jakość wód studni wiejskich w wybranych gminach województwa lubelskiego. Archiwum Ochrony Środowiska **28**(3), 63–75 (2002)
12. Nowak, A., Libiudzisz, Z.: Karcynogeny w przewodzie pokarmowym człowieka, Zywność. Nauka Technologia. Jakość **4**(59), 9–25 (2008)
13. Zeilmaker, M.J., Bakker, M.I., Schothorst, R., Slob, W.: Risk assessment of N-nitrosodimethylamine formed endogenously after fish-with-vegetable meals. Toxicol. Sci. **116**, 323–335 (2010)
14. PN-EN 26777:1999 -Jakość wody. Oznaczanie azotynów - Metoda absorpcyjnej spektrometrii cząsteczkowej

Comparison of Three Spot-Spraying Machines to Control Dock (*Rumex obtusofolius L.*) on Grassland

Quentin Limbourg[1(✉)], Jérôme Gennen[2], and Fabienne Rabier[1]

[1] Walloon Agricultural Research Centre, Gembloux, Belgium
q.limbourg@cra.wallonie.be
[2] Agra-Ost, Sankt Vith, Belgium

Abstract. In Europe, there is a commitment to reduce pesticide use by 50% by 2030 (The European Green Deal, 2018) [1], which is implemented in Wallonia by the Walloon Pesticide Reduction Programme. For herbicides, one solution is the use of mechanical tools for complete weed control [2] or in combination with chemicals [3]. However, there are several obstacles to the use of this technique in conventional agriculture, such as the need for specific tools, the availability of working time for mechanical operations, weather conditions, etc. [4]. In grassland, the problem is even greater, as there are no mechanical tools to control dock. The best option for herbicide-free dock control is to remove the roots manually [5]. Another possibility for reducing herbicide use is artificial intelligence associated with spot spraying [6]. Indeed, some machines are capable of recognizing only the dock plant and opening the spray nozzles located just above it [7].

The aim of this trial is to compare three spot spraying tools for destroying dock in grassland. Results focus on dock destruction efficiency (%) and the amount of product reduction (%).

In conclusion, it has been demonstrated that it was possible to achieve efficiencies of between 88 and 93% with herbicide's reduction of up to 95%.

Keywords: weeding · spot spraying · dock · reduction · efficiency

1 Introduction

Dock (*Rumex obtusofolius L.*) is described as one of the main weeds in intensively cultivated meadows in temperate regions, following thistles and buttercups [8]. Its presence in meadows can affect dry matter yield and the nutritional quality of the grass [9]. There are chemical control options available with approved products. However, these are almost always selective products with moderate and generally short-term effectiveness. Therefore, multiple applications are often necessary [10]. Non-selective herbicides can be used more effectively against dock, but they also affect the grass and other species [11]. Several studies highlight the difficulty of controlling dock without herbicides. Manual removal is labor-intensive and time-consuming. Frequent mowing is also insufficient to prevent dock from spreading [12]. Mechanical alternatives do not seem adequate either, as they are costly and not always successful [13].

E. Lorencowicz et al. (Eds.): FMPMSA 2024, LNCE 609, pp. 246–253, 2024.
https://doi.org/10.1007/978-3-031-70955-5_27

One possibility for using non-selective herbicides is to combine targeted spraying with artificial intelligence recognition to treat only the dock [7]. The advantage of these solutions is to reduce the amount of herbicide used while specifically targeting the intended weed.

2 Materials and Methods

The experiment took place in September 2022 in the municipality of Lontzen in the Walloon Region (Belgium) as part of the Innoveau project funded by the Public Water Management Company (SPGE). The trial was conducted on a 2.38 ha meadow.

2.1 Description of the 3 Spot Spraying Machines

1° The first machine is a 6-m trailed sprayer. It has a 500 L water tank and a 200 L tank for the herbicide mixture, both transported at the front of the tractor. The spraying system, located at the rear of the tractor, consists of 3 units, each measuring 2 m. Each unit comprises a large box containing a 52-nozzle boom spaced 4 cm apart, along with two identical weed detection modules. Each detection module is equipped with an RGB camera that transmits real-time images to an onboard NVIDIA Jetson computer. Using artificial intelligence, the system processes the images, detects the targeted weeds, and determines which nozzle should open and when. The detection modules also feature a flashing LED lamp, ensuring uniform illumination of the box for optimal weed detection. The images are also used to adjust the ramp height to maintain a 26 cm nozzle height relative to the target. The boxes are equipped with a lower cover to increase light tightness and limit the effect of wind during spraying. The sprayer does not have a proportional spraying system based on speed; instead, the flow rate remains constant and is determined by the tractor's velocity. For a spray rate of 400 L/ha, the forward speed is 3.2 km/h at three bars of pressure.

2° The second machine is a trailed sprayer measuring 8.80 m. It consists of 3 fully covered units: the two outer units are 3 m long, and the central unit is 2.80 m long. Each unit features a precision nozzle boom spaced 10 cm apart and two detection modules. These modules include Omnivision 5 MP 5647 near-infrared (NIR) cameras that capture images just below the surface, along with an onboard computer. Using artificial intelligence, the system processes the images, detects the targeted weeds, and determines which nozzle should open and when. The detection modules also have LED lamps for uniform illumination of the box, ensuring optimal weed detection. The 600 L spraying tank is located on the machine just after the drawbar. The unit covers limit drift due to wind. The ramp height is fixed at 45 cm above the ground. This height remains constant thanks to gauge wheels on the outer units and the floating position of the cylinders that support them. The forward speed can vary between 5 km/h and 10 km/h, and the application rate per hectare can be adjusted by modifying the pressure in the spraying boom using the pump located under the spraying tank.

3° The third machine is a standard trailed agricultural sprayer measuring 25 m, to which a detection system was later added. The spraying boom consists of 50 pulse-width modulation (PWM) nozzles spaced 50 cm apart. PWM technology controls the opening and closing of the nozzles at a frequency of 20 Hz using an electromagnetic valve on each nozzle. Since each nozzle operates independently, it allows for localized spraying. The detection system comprises 9 boxes distributed along the entire boom, except for the central box, which is mounted on a support attached to the front lift system of the tractor. These boxes contain RGB cameras and IR and hyperspectral sensors (300 - 1000 nm), all oriented forward at a 45° angle to the ground. This arrangement prevents mounting the central box directly on the central part of the sprayer. The images are sent to the central unit located on the sprayer for processing by artificial intelligence to detect weeds and open the corresponding nozzles. Currently, only RGB images are used by the system. The central unit communicates with the sprayer to control nozzle opening and closing via the ISOBUS protocol, which manages communication between the tractor, machine, and console. Relevant messages, such as nozzle status, real-time flow rate, and working pressure, are recorded via a Data Logger (Proemion CanLink GSM 5102). The boxes are mounted on masts to position the sensors at a height of 1.80 m relative to the target. The boom height is fixed at 50 cm above the target and is maintained constant using ultrasonic sensors. An in-cabin control screen allows adjustment of parameters, including the detection threshold (minimum plant size for sensor detection). The sprayer has a 4200 L tank. Thanks to PWM technology and a flow meter mounted on the sprayer, the flow rate remains proportional to the tractor's speed. An RTK GNSS antenna (John Deere Starfire 3000) on the tractor axle provides precise positioning for each box and geolocates the weeds. The Data Logger also records real-time position data from this antenna.

Unlike the other two localized spraying solutions presented in this article, this method creates a mapping that captures detection zones and the spraying order. The generated detection map does not represent each individual weed but provides a pixel-based detection. For every 20 cm pixel, the system indicates the presence or absence of individuals without specifying their exact count. The spraying order represents a theoretical spraying surface. The system calculates the number of nozzles that should open based on detections and the position of each nozzle on the boom. The maps can be downloaded after each spraying session.

2.2 Calculating Herbicide Reduction

Each tool sprayed a different surface area based on its width (Table 1).

Table 1. Characteristics of surfaces treated by each tool

	Surface (ha)	Length (m)	Width (m)
Tool 1	0.13	72	18
Tool 2	0.13	50	26.4
Tool 3	0.25	100	25

The product used is Galistop with a concentration of 200 g/L of Fluroxypir, applied at a rate of 1.8 L/ha. Tools 1, 2, and 3 were set for spraying rates of 400, 300, and 250 L/ha of the herbicide, respectively. Galistop has a rapid visual effect (causing growth abnormalities and morphological disturbances in the plant), but it only acts on young dock plants.

Tool 1 calculates the quantity of sprayed herbicide based on the pressure measured on the boom, the number of open nozzles, and their opening times.

For Tool 2, the machine does not provide information on the quantity of herbicide used or remaining in the sprayer. In this trial, the operator measured the initial quantity prepared and the final quantity remaining in the sprayer.

The tank of Tool 3 has an electronic gauge that indicates the remaining herbicide. The difference between the initial and post-treatment quantities determines the actual amount used.

For Tool 3, herbicide reduction can also be calculated from the spraying order map, which represents the theoretical spraying surface of the system and can be compared to the total area to be treated. A third method involves using the ISOBUS messages recorded on the Data Logger to create a nozzle opening map. By using the RTK GNSS antenna position, a theoretical position for each nozzle can be deduced throughout the spraying process and associated with an ON/OFF state.

The actual quantity of product used, associated with the surface treated in localized spraying, allows us to calculate the quantity per hectare of herbicide used and compare it with the prescribed spraying rate for each sprayer.

2.3 Measurement of Treatment Efficacy and Selectivity

Plots measuring $3 \times 3\ m^2$ were delimited and distributed within the areas sprayed by each machine. The number of plots varied depending on the tool, covering different passages based on their width. For tools 1, 2, and 3, there were 10, 11, and 12 plots, respectively. For each plot, a photo was taken using a drone (DJI Mini Pro 3, 48 MP) at a height of 3 m to capture the entire vegetation. The four corners of the plots were surveyed using a GNSS RTK pole (SOKKIA GRX3) with an estimated precision of $\pm$ 2 cm. The measured coordinates were in Lambert 72 projection. The photos were then imported into a mapping program (ArcGIS Pro) and georeferenced using the geolocated corner points of the plots. An operator georeferenced each dock plant present in the plot. A first flight was conducted before treatment, followed by a second flight approximately 10 days after treatment.

Efficacy was calculated as follow:

$$\text{Efficacity} = \frac{\text{Number of rumex at the first flight} - \text{Number of rumex at the second flight)}}{\text{Number of rumex at the first flight}} * 100 \quad (1)$$

The images also allowed identification of other species that were affected during spraying.

3 Results

3.1 Herbicide Reduction

The Table 2 presents the reduction in herbicide usage for each tool.

Table 2. Reduction of volume/ha by spot spraying system

	Initial volume per ha (L/ha)	Sprayed volume per ha (L/ha)	Reduction (%)
Tool 1	400	10.4	97,4
Tool 2	300	8	97.4
Tool 3	200	/	/

The reductions for tools 1 and 2 are identical. However, for machine 3, an issue occurred during spraying. Despite nozzle openings displayed on the screen, it appears that the pressure at the nozzles was insufficient for proper spraying. Consequently, it is impossible to calculate treatment efficacy based on drone images. The herbicide reduction will be based on comparing the surface area from the spraying order or nozzle openings with the theoretical area to be treated (Table 3).

Table 3. Shows the reduction in spraying surface area tool

	Theoretical surface to be treated (ha)	Estimated treated surface using the system (ha)	Reduction (%)
Spraying order	0.25	0.2	20
Nozzle openings	0.25	0.187	25

The reduction calculated for tool 3 based on the estimated treated surface from the spraying order or nozzle openings is significantly lower than that for tools 1 and 2.

Figures 1 and 2 respectively compare the spraying order or nozzle openings with the actual position of dock and the detection by the system.

These figures (Figs. 1 and 2) illustrate that the dock detections (in blue) by the system do not perfectly fit with the actual positioning of the dock (black dots). There are also differences between the spraying order (in red) and the nozzle opening map (in blue) created by the sprayer.

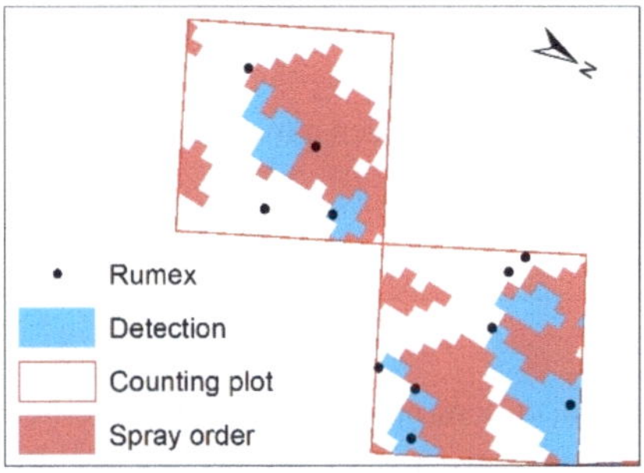

Fig. 1. Detection and spraying order map for two counting plots

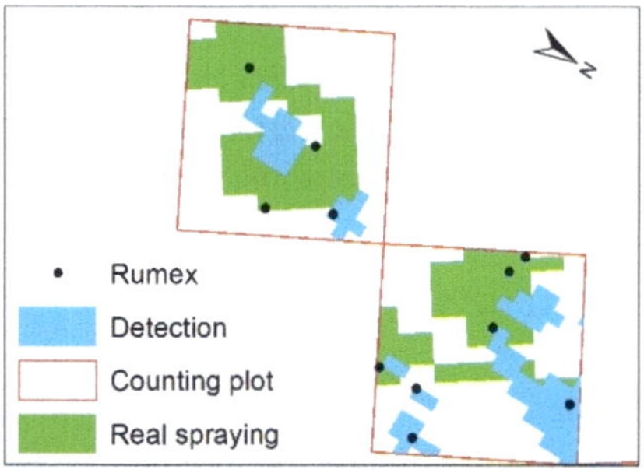

Fig. 2. Nozzle opening and detection map for two counting plots

3.2 Treatment Efficacity and Selectivity

Table 4 presents the average number of weeds per flight and the weeding efficiency for each tool.

Table 4. Average number of weeds per flight and weeding efficiency

	Average number flight 1	Average number flight 2	Weeding efficiency (%)
Tool 1	5,3	0,4	94.14
Tool 2	5,5	1,727	69.20
Tool 3	15,33	15.25	1.39

Tool 3 exhibits very poor weeding efficiency, which is logical given the inadequate spraying during the trial. To determine whether the efficacy of the different tools is statistically significant, a paired test was performed using the Games-Howell method with a confidence level of 95%. This method adapts the student's t-test for unequal variances across multiple groups from the same population. The variance inequality is based on the Levene test (Table 5).

Table 5. Grouping information with the Games-Howell method

	N	Average	Group
Tool 1	9	94.14	A
Tool 2	11	69.20	A
Tool 3	12	1.39	B

Means not sharing the same letter are significantly different. Therefore, the efficacy of tools 1 and 2 is not significantly different.

For Tool 3, a comparison can be made between the detection map provided by the system and the geolocated positioning of the dock. With spraying based on the spraying order and detections, the tool's efficacy would have been a maximum of 62.5%, corresponding to the actual dock detection rate.

4 Discussion

In general, tools 1 and 2 exhibit similar performance, demonstrating very good efficacy and a high percentage of herbicide reduction. However, nuances should be considered regarding treatment efficacy. The product used in this trial was chosen because of its rapid visual effect, which is advantageous for drone-based counting. However, it is effective only on young dock plants. Efficacy could have been improved by using active ingredients such as methsulfuron-methyl or the combination of Fluroxypir and aminopyralid, which act on more developed dock (as observed in our trials).

For Tool 3, the herbicide's reduction is much lower. This is due to the fact that its nozzle spacing is 50 cm compared to 4 cm and 8.8 cm for sprayers 1 and 2, respectively. Consequently, the reduction in spraying surface area will be smaller. However, a theoretical calculation of the spraying surface area with Tool 3 was performed by creating a polygon around each dock plant recorded in the field. The polygon had a width of 50 cm and a length of 30 cm, corresponding to the average nozzle opening distance around a point. By summing up the areas of all polygons and removing overlaps, the sprayed surface area accounted for 25% of the trial area. In other words, a 75% reduction was achieved. The nozzle spacing alone does not explain the very low reduction rate observed in the Tool 3 trial. Other factors, such as over-detection, excessive spraying order relative to detection, or a delay between the order and nozzle openings, may contribute to the poor results.

The efficacy of solution 3 could only be theoretically evaluated based on the dock detection percentage within the counting plots. This detection rate was 62.5%, which is relatively low. However, other tests conducted at CRA-W (Walloon Agricultural Research Center) revealed an average mapping error of 28.21 cm. Considering this error, the theoretical efficacy percentage increases to 94%. It's important to note that this assumes perfect coordination between spraying and detection, which was not the case in this trial.

5 Conclusion

The study demonstrated that solutions exist for identifying dock and targeting them with herbicides. Among the three tested machines, those with a detection system located in a closed compartment appear to have better detection rates. This is likely because lighting conditions are controlled. Another potential improvement in efficacy is the use of a non-selective product for treating dock. A "total herbicide" type of product should enhance efficacy regardless of the dock growth stage.

References

1. European Green Deal Available online: https://ec.europa.eu/info/strategy/priorities-2019-2024/european-green-deal_pl. Accessed on 16 July 2022
2. Cloutier, D., van der Weide, Y., Perruzi, A., Leblanc, M.: Mechanical weed management. Non-chemical Weed Management: Principles Concepts and Technology, pp. 111–134. CAB International, Wallingford, UK
3. Pleasant, J.M., Burt, R.F., Frisch, J.C.: Integrating mechanical and chemical weed management in corn (Zea mays). Weed Technol. **8**(2), 217–223 (1994)
4. Munier-Jolain, N.M., Deytieux, V., Guillemin, J.P., Granger, S., Gaba, S.: Multi-criteria evaluation of cropping systems prototypes based on integrated weed management. In: XIII th Colloque International sur la biologie des mauvaises herbes, Dijon (2010) (2009)
5. Van Evert, F.K., et al.: A robot to detect and control broad-leaved dock (*Rumex obtusifolius L.*) in grassland. Journal of Field Robotics **28**(2), 264–277 (2011)
6. Gerhards, R., Andujar Sanchez, D., Hamouz, P., Peteinatos, G.G., Christensen, S., Fernandez-Quintanilla, C.: Advances in site-specific weed management in agriculture—A review. Weed Res. **62**(2), 123–133 (2022)
7. Anken, T., Latsch, A.: Characteristics of a spot sprayer for the treatment of *Rumex obstusifolius* in meadows. Landtechnik **78**(3), 139–151 (2023)
8. Hopkins, A.: Botanical composition of permanent grassland in England and Wales in relation to soil, environment and management factors. Grass Forage Sci. **41**(3), 237–246 (1986)
9. Hejduk, S., Dolelezal, P.: Nutritive value of broad-leafed dock (*Rumex obtusifolius L.*) and its effect on the quality of grass silages. Czech J. Animal Sci. **49**, 144–150 (2004)
10. Hopkins, A., Johnson, R.: Impact of control of dock (*Rumex obtusifolius*) on potential production from N-fertilised swards. Grass Science in Europe **8**, 200–203 (2003)
11. Mitchell, B.: Herbicide efficacy evaluation. End of project report 44884 (2001)
12. Van Evert, F., et al.: A robot to detect and control broad-leaved dock (*Rumex obtusifolius L.*) in Grassland
13. Stilmant, D., et al.: Rumex à feuilles obtuses dans les systèmes herbagers : importance de la problématique, lutte chimique et méthodes alternatives. Fourrages **192**, 477–793 (2007)

On the Path to Sustainability: Seeking New Competences to Face Challenges of Implementing Sustainable Processes. Polish Chemical Industry Case Study

Aleksandra Lis[1(✉)], Aneta Oleksy-Gębczyk[1,3], Anna Szeląg-Sikora[1,2], and Katarzyna Kowalska-Jarnot[4]

[1] University of Agriculture in Krakow, Mickiewicza 24, 31-00 Krakow, Poland
aleksandra.lis@urk.edu.pl
[2] Cavalry Captain Witold Pilecki State University of Małopolska in Oświęcim, Oświęcim, Poland
[3] University of Applied Sciences in Nowy Sącz, Nowy Sącz, Poland
[4] Department of Management and Accounting, School of Economics and Computer Science in Cracow, Cracow, Poland

Abstract. Sustainable development (SD) is influencing every part of the global economy, there are though industries that are much more exposed to changes and challenges than others, especially industries in which safety plays a crucial role. Those are for instance agriculture and chemistry industries, and even more entities that are operating between those two industries or for them. In the presented study, we have investigated the need of so called "green competences" in the representation of the chemical industry in Poland, including big companies from the chemical industry that relate to the sustainability of farming industry: fertilizer producer, petroleum company and food packaging company. Presented study aimed at exploring the subject, thus we have used a qualitative approach and the focused group interview (FGI) to investigate participants' awareness of SD policies influence on the processes in their companies and thus the need of seeking new competences answering their new needs. Content analysis was enhanced by using the MAXQDA Analytics Pro software. The study shows the emergence of new competences requirements related to implementing SD as well as the need for further investigation of this topic.

Keywords: Chemical Industry in Poland · Sustainable Development · Green Competences · Qualitative Study

1 Introduction

Economy is turning towards sustainability and thus sustainable development (SD). This is not a trend anymore but rather mainstream approach in the global economy. The effect of this approach is not only the protection of the natural environment, but also progressive technological development in many industries, including the agricultural sector [1,

E. Lorencowicz et al. (Eds.): FMPMSA 2024, LNCE 609, pp. 254–261, 2024.
https://doi.org/10.1007/978-3-031-70955-5_28

2]. However, the development of agriculture depends on the development of other economic sectors. Innovations in the fields of technology, bio-technology or chemistry are successively implemented into agriculture [3]. In Poland, agriculture is one of the most dynamically developing sectors of the national economy [4]. The development of this industry is also accompanied by social changes, generational exchange and an increase in the professional qualifications of people working within this industry. Therefore, the aspect of sustainable social development also applies to people employed in agriculture and in industries supplying agriculture, with a significant role of chemical industry. Thus, providing agricultural sector with supplies tailored for the sustainable agriculture, companies from supplying sectors have to implement SD into their practice and seek for competences that support this implementation. The main objective of this paper is to present a chemical industry in Poland' perspective as a case study of companies' competency needs related to SD.

Striving for a sustainable economy has a significant impact on competence needs. Just as SD policies have become horizontal, the competences resulting from the implementation of these policies in the vast majority will be common to many sectors. At the same time, it should be remembered that the implementation of tasks for which the same competences will be needed may be slightly or even significantly different in individual industries. Competences for SD are the so-called green competences. The UN defines green competences as "(..) knowledge, skills, values and attitudes needed to live, develop and support a sustainable and resource-efficient society" [5]. There are also definitions relating to green competences to the competence needs of enterprises. In the context of business management, "green competences of employees are the result of their knowledge and skills in the application of pro-environmental solutions in the enterprise" [6].

In the last decade, the world has faced challenges that affect economic, social and political processes. There has been a decline in economic activity and an unprecedented increase in inflation, social tensions, restrictions related to the COVID -19 pandemic and the war in Ukraine [7]. Changes regarding ways of working and expectations of employees occurred. There has been technological development of artificial intelligence that affect labour. Experience of remote work has changed the working environment and the expectations of employees [8]. The impact of technological development on the labour market applies to all industries, including the chemical and agricultural industries. It is a result of processes such as automation, robotization or the use of artificial intelligence. As technology continues to advance, employees need to acquire new skills and qualifications to adapt to changing professional roles and remain competitive in the job market. The knowledge-based society shows a constant need to improve its competences COVID-19 pandemic and the war in Ukraine are just examples of events that have significantly influenced the market [9]. Thus, there should be an emphasis, in both practice and research, on business development strategies and employers' expectations towards the competences that should be developed by current and potential employees.

Technological changes that are rapidly and radically affect the economy used to be recognised as industrial revolutions. The current one is named a fourth industrial revolution. It refers to the "application of advanced information and communication technologies (ICT) in manufacturing, logistics, marketing and other areas of the economy. In the 4.0 economy, these technologies are used to optimize processes, reduce costs, increase quality and flexibility of production, and to create new business models" [10]. It is also closely related to sustainable development, as new technologies allow companies to reduce their impact on natural environment and implement innovations that reduces their impact on the nature. Implementing sustainable development requires appropriate social and professional competences. This requires employers and employees to redefine their current roles and tasks, including the development of skills in working with technologies and machines, predicting the environmental impact of given processes or their parts, integrating advanced technologies with traditional work models [11]. Therefore, the change in competence needs is a result of implementing sustainability and technological development. The two, though coming from different socio-political sources, enhance emergence of new competences.

IT competences are competences requiring the ability to use modern technologies such as software, robots or artificial intelligence. It is also the ability to program, analyse data and solve technical problems. Social competences, in turn, are related to teamwork requiring emotional intelligence, good communication in teams (also remote), stress management, goal management and managing people. Cognitive competence is intellectual competence such as creativity, logical reasoning and solving complex problems [12]. Resilience [13] and adaptability in a dynamically changing work environment, where technologies and processes are constantly updated, are crucial. As well as continuous development and lifelong learning [14].

There is a common understanding among policymakers, representatives of public institutions on labour market and education as well as employers, that SD requires monitoring and providing new competences to the economy. The need to develop key competences is also noticed by students. Research conducted by Szelag-Sikora, A., Oleksy-Gebczyk, A., and Kowal-ska-Jarnot, K. shows that there are clear expectations of students as to the needs that the university should meet. Universities that offer high quality education, courses that increase the chances of finding a job and orientation to the student (practical teaching, individualization of studies, modern teaching methods, cooperation with the environment, and, friendly approach) have much more recruitment [15]. Thus, while it is crucial to understand the necessity of providing competences for SD in modern agriculture, it is also important to study the needs of employers within the whole economic system, including supplying industries.

2 Methodology

We have used a qualitative approach and a focused group interview method (FGI). This approach has already been successfully applied to the study of competence needs for SD. Due to the purpose of the study, the funnel approach for the FGI questionnaire was used. The interview started with general questions, which then lead to more detailed questions about the competences and requirements of the market in terms of competences. Terms such as sustainability often remain unclear to the general public, even in companies that operates many SD related initiatives. The advantage of the group format is the possibility of finding common understanding of the terms. Moreover, it allows immediate recognition of the needs and possible differences between the participants in terms of competence requirements.

This study presents the results of the first pilot FGI conducted in December 2023 among representatives of the chemical and energy industries. The studied group were entrepreneurs who implement SD policies, projects related to SD and attempt performing circular economy. The interviewees were representatives of two leading companies from the chemical sector, a large company from energy sector (renewable energy), as well as two – industry associations from the chemical sector and the energy sector.

The study was carried out by a moderator using a questionnaire prepared for the purpose of this study. The sound was recorded and transcribed. In accordance with the adopted ethical standards, the study participants were informed about the procedure of the study and asked to agree to the recording as well as further analysis of the collected research material. The collected research material was analysed using MAXQDA Analytics 2022 Pro software.

The analysis of the interview included two stages, both with the use of analytical software. The first step of the analysis is coding. The content of the interview regarding specific issues relevant to the objectives of the study was marked and the issues to which they relate were defined. The questions and comments of the moderator were not coded. The codes used for the interview analysis were not pre-determined and adopted either before the interview or before the coding phase began. Following the guidelines from grounded theory, which was taken as an approach to analysis in these studies, the code tree was created in the course of the analysis. The codes resulted from the content of the interview and recognition of what was crucial for the participants /important in the context of the questions asked. Thus, the code tree is a part of the method and yet also the result.

3 Results

The code set consisted of following codes: flexibility /learning, interdisciplinarity, IT competence, problem solving, industry knowledge and collaboration /communication. The interview encoding including code names and the number of encoded snippets for a given code is presented in Table 1 below.

Table 1. Code Tree

Code	Frequency	Remarks
Flexibility/learning	13	Encoded fragments about the need for competences that are related to flexibility and learning
Industry competences	13	Encoded fragments about the need for industry competences
IT competences	8	Encoded fragments about the need for IT competences
Cooperation and communication	19	Encoded fragments about the need for communication and cooperation related competences
Problem solving	3	Encoded fragments about the need for cognitive competences related to problem solving
Interdisciplinarity	8	Encoded fragments about the need for combining knowledge and skills from different fields
TOTAL		Total number of encoded fragments

The most frequently discussed competence was cooperation /communication, which is currently a highly valued competence, and missing on the market, according to interviewees:

"I see (…) that young people cannot talk to each other. (…) Even if there is a problem, the transfer of information and I do not know, we work in a twenty-four-hour, four-brigade system, [communication doesn't work]. Therefore, even the issue of transferring information from the brigade to the brigade is sometimes a problem, (…) such a simple communication." (FGI 1:7:55.411 -- > 1:8:16.991).

"(…) responsibility for their actions and the ability to cooperate, that is, just saying I,t collapsed. [Saying] 'let's fix it together, yes, I don't know, help me, I would do it like this, [is hard for employees]. But also [saying it] in a [polite] way, well not offending." (FGI 1:14:34.501 -- > 1:15:7.321).

Another two competences (codes) with a comparable number of appearances are industry knowledge and flexibility /learning. As the participants of the interview pointed out, in the chemical industry a foundation from basic knowledge is needed, and even a basis in the scope of industry knowledge is necessary. The possession or lack of this knowledge, according to some of the respondents, determines employment in many departments of chemical companies. It is slightly different in the energy industry, where field studies do not always condition employment. Equally important as industry knowledge is learning skills and flexibility. According to the participants, this competence will be of great importance in the future.

"(..) so that these people have these competences, capabilities and this flexibility for this type [innovative / SD related] of action." (FGI 1:3:41.241 -- > 1:4:6.861).

"(..) the question is essential, are we as people teaching the next generations of them [flexibility, learning new things and forgetting yet not up to date knowledge and skills] - are we able to teach these young people a quick adaptation to rapid changes?" (FGI 0:31:35.871 -- > 0:31:56.151).

Interdisciplinarity and IT competences were discussed slightly less frequently. Although they were not discussed as frequently as communication, industry knowledge, or flexibility and learning, participants agreed, that they are crucial now and will gain in value in the future. High value will be given to people who will combine interdisciplinary skills, for example in computer science or programming and chemistry or energy. That is a requirement for innovations: understanding the needs of SD, combining it with the industry knowledge and usage of hi-tech tools.

Another important measure that was used to present the dominant content in the interview is the portrait of the document, presented in Fig. 1. Portrait was developed for a set of codes. This tool allows to see the codes, not by the frequency of codes but the percentage of coded content. Each colour represents the share of text falling on a given code, in the context of the entire text, which was encoded in the document.

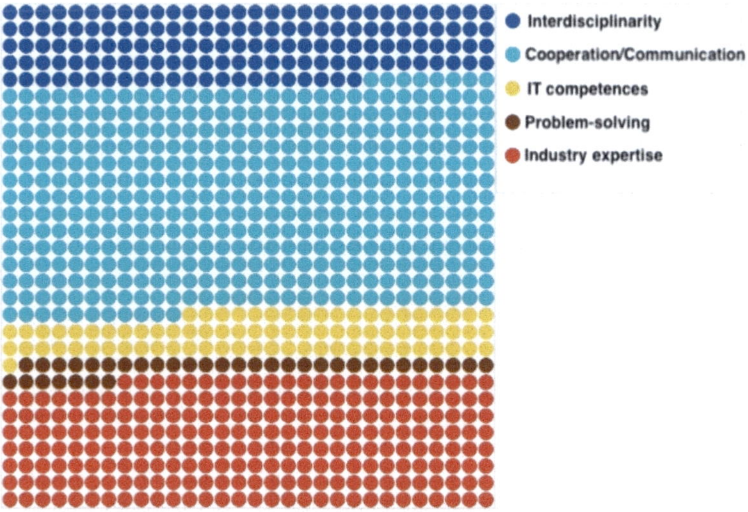

Fig. 1. A figure caption is always placed below the illustration. Short captions are centered, while long ones are justified. The macro button chooses the correct format automatically.

As the portrait of the transcription of the analysed interview shows, not only most often, but also the most talked about topic were communication and cooperation competences. In the second place in terms of the encoded area there are competences related to industry knowledge, and in the third –interdisciplinarity. The least area is coded for solving problems. However, this is the competence, according to the participants, that may gain more meaning in the future.

"soft skills in problem solving, that is, creating groups and leaving them when the project is completed and again flexibly addressing another problem in the company, considering what [should be next step / proper solution] (..) we said, that this transformation [in] direction of industry 5.0 [and it] is already happening, (…). In short, these are essential skills." (FGI 1:4:6.931 -- > 1:5:11.661).

4 Discussion

The perception of SD as a holistic concept, affecting horizontally all areas of life, including all sectors of the economy, implies the need to redefine the requirements for economies and thus the need for competences. The introduction of the European Green Deal, as well as the implementation of the UN Sustainable Development Goals, affect and will also affect the labour market. Some of the current professions will disappear, others will be redefined, and still others will be created. SD is a crucial issue for both the present and future of the industry in national and global context. Especially in conditions of uncertainty related to the current geopolitical situation, increasing the possibilities of recovering raw material from waste and flexible changes in supply chains – green competences may become even more important.

To conclude, green competences are an essential precondition for building a sustainable economy. These competences are very diverse and include both numerous skills and expertise as well as general competences, needed in numerous sectors and being the basis for social and economic development in the chosen direction. Strengthening SD in any industry means also supplying it from companies following SD solutions. New processes and innovative actions require technological and green competences. They might differ between sectors and vary from very transversal skills to particular knowledge and attitudes for the industry. Therefore, further research on competency needs is needed to embrace change on a labour market and enhance industries in their path to sustainability.

References

1. Niemiec, M., Chowaniak, M., Sikora, J., Szeląg-Sikora, A., Gródek-Szostak, Z., Komorowska, M.: Selected properties of soils for long-term use in organic farming. Sustainability **12**, 2509 (2020). https://doi.org/10.3390/su12062509
2. Komorowska, M., et al.: Closed-loop agricultural production and its environmental efficiency: a case study of sheep wool production in northwestern Kyrgyzstan. Energies **15**, 6358 (2022). https://doi.org/10.3390/en15176358
3. Gródek-Szostak, Z., et al.: Promotion of RES in a technology transfer network. Case study of the enterprise europe network. Energies **13**, 3445 (2020). https://doi.org/10.3390/en13133445
4. Zieliński, M.: Znaczenie rolnictwa w polsce na obszarach ONW jako źródła dóbr publicznych w kontekście polityki rolnej UE. Wieś i Rolnictwo. **199**(2), 7–39 (2023)
5. What are green competences? https://www.unido.org/stories/what-are-green-skills dn. 20 March 2024
6. Kozar, Ł.: Kształtowanie zielonych kompetencji pracowników w gospodarce ukierunkowanej na zrównoważony rozwój. Zarządzanie Zasobami Ludzkimi **61**(6), 57–69 (2017)
7. Oleksy-Gębczyk, A.: Preferences and consumer choices: a case of polish markets for goods and services. Asian J. Appl. Econ. **30**(2). BANGKOK (2023)

8. Ćwiertniak, R., Stach, P., Kowalska-Jarnot, K., Worytkiewicz-Raś, K., Wachułka-Kościuszko, B.: Addressing students' perceived value with the virtual university concept. e-mentor **2**(94), 65–76 (2022)
9. Szeląg-Sikora, A., Oleksy-Gębczyk, A., Kowalska-Jarnot, K.: The importance of marketing activities for the functioning of the university. Society. Integration. Education. Proceedings of the International Scientific Conference, vol. 1, pp. 245–253 (2024)
10. Shelest-Szumilas, O.: Praca i kompetencje w gospodarce 4.0, W. In: Andrzejczak, A. (ed.) Rozwój kompetencji dla gospodarki 4.0, pp. 37–57. Wydawnictwo Naukowe NFCE, Poznań (2022)
11. Tomaszewska, R., Pawlicka, A.: Praca w warunkach czwartej rewolucji przemysłowej. Nowe wyzwania dla humanizacji. Human Resource Management/Zarządzanie zasobami ludzkimi **141**(4) (2021)
12. Strategia Rozwoju Kapitału Ludzkiego 2030, Ministerstwo Rodziny, Pracy i Polityki Socjalnej, Warszawa 2019 dostęp: https://www.gov.pl/web/rozwoj-technologia/strategia-rozwoju-kapitalu-ludzkiego-2030
13. Zalecenie Rady w Sprawie Kształcenia i Szkolenia Zawodowego na Rzecz Zrównoważonej Konkurencyjności, Sprawiedliwości Społecznej i Odporności, https://eur-lex.europa.eu/legal-content/PL/TXT/PDF/?uri=CELEX:52020IP0373&from=EN
14. Zintegrowana Strategia Umiejętności 2030 (część szczegółowa). Polityka na rzecz rozwijania umiejętności zgodnie z ideą uczenia się przez całe życie. Załącznik do uchwały nr 195/2020 z dnia 28 grudnia 2020 r. dostęp: https://www.gov.pl/web/edukacja-i-nauka/zintegrowana-strategia-umiejetnosci-2030-czesc-szczegolowa-dokument-przyjety-przez-rade-ministrow
15. Szeląg-Sikora, A., Oleksy-Gębczyk, A., Kowalska-Jarnot, K.: Factors determining the choice of university by candidates. Society. Integration. Education. Proce. Int. Sci. Conf. **1**, 254–263 (2024)

Threats to Road Safety by Agricultural Tractors

Edmund Lorencowicz[1(✉)] and Jacek Uziak[2]

[1] Department of Machine Operation and Production Processes Management, University of Life Sciences, Lublin, Poland
edmund.lorencowicz@up.lublin.pl
[2] College of Management and Enterprise, Wałbrzych, Poland

Abstract. Farmer's safety in workplace as well as during travel related to work is one of the crucial aspects of sustainable agriculture. There are approximately 1.3 million farms in Poland, with almost 1,480 thousand tractors, 165 thousand combine harvesters and thousands on other farm-related machines vehicles and used. Those machines travel on public roads in order to reach outlying fields and to transport products to market or processing facilities, creating danger for all road users. According to Polish National Police Headquarters, accidents involving tractors and self-propelled agricultural machines accounted for approximately 0.5–0.8% of all road accidents in 2016–2022. The year fatalities ranged from 16 to 22, whereas the number of injuries, from 144 to 166. The specific nature of agricultural vehicles and machines, including dimensions and weight, but also vehicle aggressivity, limited visibility, and poor operator skills and lack of experience are all contributing factors to these accidents. Tractors are the major contributor to the accidents involving agriculture machinery. The paper presents an analysis of selected cases of farm-related vehicles and equipment accidents. It shows that the most common causes of accidents are excessive speed; failure to maintain a safe distance between vehicles; and general lack of caution driving around farm equipment.

Keywords: agricultural tractor accidents · road safety · health and safety

1 Introduction

There are approximately 1.3 million farms in Poland, with almost 1,480 thousand agricultural tractors. During the fieldwork season, these tractors must move between the fields and the farm, most often on public roads. Due to their size and weight, they pose a great threat to other road users. There are statistically few road incidents involving agricultural tractors and in 2023, they accounted for only 0.5% of all accidents (Table 1). However, the danger caused by their movement, in most cases, causes serious consequences, also due to their dimensions. They are particularly dangerous for road users driving much smaller vehicles, such as passenger cars or motorcycles. They often cause damage to people's health and even death. Most accidents occur in the summertime due to work in fields and increased tractor traffic on public roads. The problem increases when drivers of these vehicles pose an additional threat through poor lighting or no marking of the machine.

E. Lorencowicz et al. (Eds.): FMPMSA 2024, LNCE 609, pp. 262–269, 2024.
https://doi.org/10.1007/978-3-031-70955-5_29

Table 1. Road accidents by offender's vehicle - agricultural tractors in 2016–2023.

Year	Accidents involving an agricultural tractor (number)	Percentage of accidents involving agricultural tractors	Deaths (numbers)	Injured (numbers)
2016	118	0.4	16	133
2017	152	0.5	16	167
2018	130	0.5	19	143
2019	99	0.4	12	105
2020	113	0.5	10	116
2021	109	0.5	15	126
2022	97	0.5	16	97
2023	97	0.5	13	94

Source:[9–11]

A common cause of road accidents is the psychophysical condition of the driver, including delayed reflexes, poor skills, lack of experience and forecasting skills. The driver's health is also an important aspect when it comes to road safety. Another important condition is the ever-changing road conditions, such as bad weather, blinding sun, rain, snow, poor or no road lighting. Drivers' behavior is another important safety factor - excessive speeding, priority violations or poor judgments make it more difficult for the driver to drive safely. Tractor failures may also cause road accidents. An unexpected stop of the tractor due to an engine failure or even a burned-out bulb may lead to an accident.

Research conducted in Switzerland in 2011–2017 identified five (5) main causes of accidents involving agricultural tractors [1]. The three most common causes of accidents were tractor overturning (24%), behavior of other road users (21%) and limited visibility (15%). The others were the poor condition of the machine (13%) and incorrect driver reactions (11%).

2 Materials and Methods

2.1 Technical Conditions of Tractors in Poland

According to the law in Poland [7], an agricultural tractor is a motor vehicle designed to be used together with equipment for agricultural, forestry or horticultural work. Such a tractor can also be adapted to pull trailers and for earthworks. It is permissible to participate in road traffic with an agricultural tractor with a mounted machine or a special trailer with a width not exceeding 3 m. An agricultural tractor combined with a machine or trailer with a width exceeding 3 m but not more than 3.5 m may obtain permission to travel as non-standard vehicle. Road traffic legislation prohibits the use of agricultural tractors wider than 3.5 m. The permissible total weight of an agricultural tractor with tires (two-axle) is 18 tons, and the permissible length of a vehicle combination consisting of an agricultural tractor and two trailers is 22 m. The maximum speed of the tractor can be 30 km/h.

2.2 Methodology of Data Collection and Analysis

In the first stage, an analysis of the available literature was carried out to formulate the threats resulting from the movement of tractors on roads. In order to show the problem of the threat caused by driving agricultural tractors on roads, the case study method was used. Selected accidents, involving tractors described in information on the Internet, were analyzed. In total, 21 cases in the years 2016 - 2021 were selected and were used to conduct a comparative analysis.

3 Hazards Resulting from the Tractors' Design

3.1 Hazards Resulting from Limited Visibility from the Vehicle Cabin

Due to the size of the agricultural tractor and its specific purpose (e.g. attaching large agricultural machines), the driver has limited visibility around the vehicle [4]. A blind spot is the area around the vehicle that the driver cannot see directly or can only see indirectly. The direct view, where the driver can see an area through one of the windows, is often blocked by the structural components of the attached equipment. Mirrors or cameras provide the driver with an indirect view. The size and location of blind spots depend on the type of vehicle, including the height of the cabin and the presence of mirrors or cameras. Implements or field machines coupled to a tractor may limit direct visibility to such an extent that other road users become largely invisible to the driver of an agricultural tractor.

3.2 Threats Resulting from Limited Visibility of the Tractor by Other Road Users

The problem of tractor visibility by other road users increases at night. According to the law, the tractor should be equipped with 4 lights, including: 2 lights placed at a height of less than 900 mm and with a minimum distance between them of 400 mm, and 2 lights placed at a maximum height of 2300 mm and with a minimum distance between them of 600 mm (it is allowed 400 mm for tractors less than 1.3 m wide). Rear lights, warning signs and lighting for mounted machinery must also be properly visible on the road. In rear, red tail, brake lights and red reflectors are used. On trailers, reflectors must be triangular, with one end of the triangle pointing upwards. Trailed implements have the same rear lighting as trailers, but non-triangular red reflectors may also be used. If attached machines cover the tractor's lights (e.g. fertilizer spreaders or sprayers), the vehicle's rear lighting should be repeated on them. The Road Traffic Law details precise guidelines for the use of special warning signs and machine lighting [7].

3.3 Hazards Resulting from the Weight and Dimensions of the Vehicle

An agricultural tractor is a much larger vehicle than a passenger car, which is a real danger to drivers in the event of an accident. One of the important parameters is the size of tractor wheels, which are many times larger than those used in passenger cars. If a tractor runs into such a vehicle, there may be a serious threat to the driver and passengers. Additional attached trailers, machines or agricultural tools increase the weight and dimensions of the unit [5].

A clear example of problems related to excessive dimensions of agricultural vehicles in comparison to passenger cars is shown in Fig. 1. The pictures shows, that due to the limited road width, the passenger car is in clear danger of frontal collision with the tractor.

Fig. 1. Dimensions comparison of the tractor towing the trailer and the passenger car [8].

Driving speed plays a crucial role in the event of a collision because it affects braking distances. Although the maximum legal speed of agricultural tractors in Poland is 30 km/h, considering additional factors such as the operator's reaction time and the masses moved (up to 40 tons), the risk of collision and the resulting serious consequences increases significantly.

4 Analysis of Selected Cases

4.1 Effects of the Analyzed Accidents

There were 21 cases of road accidents involving agricultural tractors analyzed [2]. In most cases, the effects were very serious as shown in examples in Figs. 2, 3, 4 and 5. The most common cause of accidents in the analyzed cases was excessive speed of other road users and lack of caution by drivers.

Fig. 2. Effects of a collision between a tractor and a truck.

Fig. 3. Effects of a collision between a passenger car and a tractor with a sprayer attached.

Fig. 4. Effects of a collision between a tractor and a passenger car.

As shown in Figs. 2, 3, 4 and 5, the major problem is the size of the agricultural equipment in comparison to the other road users, mainly passenger cars. The suspicion may be that car drivers do not immediately recognize the agricultural vehicles' dimensions and unusual speed which differs from what they would normally expect, i.e. another passenger car. This normally leads to sever consequences.

4.2 Causes of the Analyzed Accidents

According to the analysis of the causes of accidents (Table 2), the most common reason for accidents was excessive speed of other road users - 10 cases.

Fig. 5. Effects of a collision between a passenger car and a tractor at night.

Lack of a safe distance between vehicles (7 cases) was the second most common factor. The next ones were lack of caution and driver distraction. The next factor influencing road accidents was poor visibility and incorrect overtaking (2 cases). There were also incidents caused by the lack of a warning triangle, when the driver of an agricultural tractor did not deploy the triangle during an emergency stop. As a result, the driver of a passenger car hit a plow used for cultivating. Another case that generated an additional threat was the lack of proper protection of moving vehicles. In this case, the driver of an agricultural tractor "lost" a trailer with hay that hits a moving car. The next ones were forcing the right of way, lack of appropriate reaction of the driver and loss of control of the vehicle (one case each).

Table 2. Causes of analyzed accidents (do not add up to 100%, as several causes could have occurred)

Cause of the accident	Number	Percentage
Excessive speed	10	47%
No safe distance between vehicles	7	33%
Lack of caution	4	19%
Driver distraction	4	19%
Incorrect overtaking	2	9%
Poor visibility	2	9%
Loss of vehicle control	1	4%
Poorly secured attached vehicle	1	4%
Lack of appropriate response from the driver	1	4%
Forcing the driver to give priority	1	4%
No warning triangle on the agricultural tractor	1	4%

Note: The percentage does not add up to 100%, as several causes could have occurred

According to the police recommendations and the conducted analysis, road users must observe the following basic rules: reduce the speed of his vehicle, increase the distance to the vehicle in front, adapt his driving skills to the road conditions, strictly follow road traffic regulations, overtake prudently when the opportunity arises, always check the technical condition of his vehicle and possess basic vehicle equipment.

5 Summary

An agricultural tractor is a vehicle that travels on both public roads and agricultural land, hence it generates more hazards than an ordinary road vehicle. The analysis of the research results allows the following conclusions to be formulated.

1. Excessive speed by drivers of passenger vehicles is one of the main causes of road accidents. Passenger vehicles can legally drive at higher speeds than agricultural tractors, which increases the likelihood of causing accidents on the roads. The driving speed of agricultural vehicles is an important factor that directly affects the risk of collision resulting from vehicle speed. In the cases examined, that accounted for 47% of the incidents.
2. The lack of a safe distance between vehicles and the lack of caution on the part of drivers increases the risk of collisions and road accidents involving agricultural vehicles. There is a general problem of the lack of proper estimation of the dimensions of tractors with agricultural machinery. In the cases examined, the reasons for the accidents were insufficient distance and the lack of caution constituted 33% and 19%, respectively.
3. Driver distraction is a major problem on the roads due to the high risk of accidents and collisions. A common problem in this case is the use of a mobile phone while driving and the driver's drowsiness, which may result in a collision or accident. In the analyzed cases, these were the reason for the accidents in 19%.
4. Incorrect overtaking and poor visibility generate additional risk. Overtaking in an unauthorized place is often associated with a lack of proper visibility, e.g. in the case of overtaking vehicles driving slowly uphill, which is a huge danger on the roads. In studies, it was 9% of cases.

Various reports and expert opinions [1–3, 6, 8, 11] indicate additional threats.

- Limited, unsuitable visibility around the vehicle due to the construction of an agricultural tractor.
- Poor visibility of the agricultural tractor at night and in conditions of limited visibility.
- Lack of appropriate reaction from drivers of passenger vehicles to moving tractor and machine sets.
- Lack of warning elements (triangle and boards) and poor protection of machines connected to the agricultural tractor.
- Inadequate road infrastructure, in particular road width, especially in rural areas where the movement of tractors and machines increases the risk to other road users.

References

1. Accident tracteur: Agro-Service Magazine, CEMA, 34–35 (July 2017) https://www.landbouwservice.be/sites/default/files/70/bulletin_1-9_2017-03.pdf. Last accessed 18 June 2024
2. Czerwonka, R.: Ocena zagrożenie spowodowanego przez ciągniki rolnicze podczas przejazdów po drogach publicznych. Final Year Project (Supervised by E. Lorencowicz). University of Life Sciences in Lublin (2022)
3. Hostens, I., Adam, U.: Agricultural vehicle accidents: history, forgotten facts, 5 measures. CEMA (2017), https://circabc.europa.eu/sd/a/7592df7a-b42d-4bd8-93ba-3f4020df3951/4a_CEMA_views_ABS_updated(0).pdf. Last accessed 18 June 2024
4. Jaarsma, C.F., de Vries, J.R.: Agricultural Vehicles and Rural Road Traffic Safety: An Engineering Challenge! CIGR-AgEng2012, Agriculture & Engineering for a Healthier Life, Valencia, Spain (2012)
5. Kompatybilność aut podczas zderzenia. https://motoryzacja.interia.pl/samochody-nowe/testy/news-kompatybilnosc-aut-podczas-zderzenia,nId,624902. Last accessed 20 January 2021
6. Koralewski, G., Wrona, R.: Zagrożenia drogowe spowodowane ruchem ciągników rolniczych i pojazdów wolnobieżnych. Autobusy: technika, eksploatacja, systemy transportowe **18**, 267–270 (2017)
7. Prawo o ruchu drogowym. Dziennik Ustaw. 1997 r. numer 98 poz. 602 http://isap.sejm.gov.pl/isap.nsf/download.xsp/WDU19970980602/U/D19970602Lj.pdf. Last accessed 11 November 2021
8. Road traffic accidents with agricultural and construction vehicles - Thematic study. The Dutch Safety Board, Hague (2010)
9. Symon, E.: Wypadki drogowe. KGP Biuro Ruchu Drogowego, Warszawa, 2016–2021. https://statystyka.policja.pl/st/ruch-drogowy/76562,wypadki-drogowe-raporty-roczne.html. Last accessed 15 April 2024
10. Symon, E., Rzepka, P.: Wypadki drogowe. KGP Biuro Ruchu Drogowego, Warszawa, 2022–2023. https://statystyka.policja.pl/st/ruch-drogowy/76562,wypadki-drogowe-raporty-roczne.html. Last accessed 15 May 2024
11. Vaupel, M.: LU Verkehrs-Tipp: Beleuchtung checken. Lohnunternehmen, https://lu-web.de/redaktion/news/lu-verkehrs-tipp-beleuchtung-checken/?utm_source=Lohnunternehmen&utm_campaign=8cfc39d15b-EMAIL_CAMPAIGN_2021_12_22_08_35&utm_medium=email&utm_term=0_262b1bae07-8cfc39d15b-59905665. Last accessed 15 January 2024

Use of Fruit Pomace in the Production of Selected Cosmetics

Patrycja Łusiak, Paweł Sobczak(✉), Jacek Mazur, Monika Wójcik, and Adrianna Romańska

Department of Food Engineering and Machines, University of Life Sciences, Głęboka 28, 20-612 Lublin, Poland
pawel.sobczak@up.lublin.pl

Abstract. Constantly developing branches of the food industry, increasing environmental pollution as well as the fashion for smart-, novel- and superfoods, which appeared some time ago - force modern scientists to constantly look for new, more creative solutions for effective management of waste as additives to food, or functional food. What is regarded as waste in one factory may be a valuable raw material with high nutritional value in another. The current goal of food producers is completely waste-free production, using post-production waste as an enriching raw material for popular, widely available food products, and being used for new food sources (e.g. edible insects), which can be used both as food - in itself, as well as a valuable feed additive.

Keywords: Sustainability · Production · Utilization · Pomace · Cosmetology · Shampoo

1 Introduction

Shampoo is an essential product for keeping scalp and hair clean. Products dedicated to scalp and hair care have been one of the largest segments of the cosmetics industry for years. The cosmetics market offers shampoos in a variety of forms, usually the liquid form or the solid form, like shampoos in bars that have increasingly gained popularity. Shampoo was developed to replace soap in cleansing the scalp and hair by removing unwanted impurities. Most of the impurities in daily life, including sebum produced by the sebaceous glands of the scalp, are insoluble in water, so it is extremely difficult to remove them from the surface of the skin and the hair shaft. It is, therefore, necessary to use a product that will help remove these impurities to cleanse the skin and hair, give freshness, and improve the visual effect of the hairstyle [1].

The cosmetics market is constantly evolving and adapting to current trends. One of these is the waterless trend, an eco-friendly approach to products and care. It responds to environmental changes caused by the overexploitation of naturally occurring bodies of water. The term "waterless" refers primarily to the reduction in water consumption and applies not only to the production of cosmetics (waterless formulations) but also to packaging and the subsequent use of cosmetics. The increase in popularity of this

E. Lorencowicz et al. (Eds.): FMPMSA 2024, LNCE 609, pp. 270–276, 2024.
https://doi.org/10.1007/978-3-031-70955-5_30

type of products and the trend that is still prevailing was observed in Europe already in 2015. The reduction of water in cosmetics was supposed to benefit the active ingredients by increasing their concentration in the product and thus achieving more effective formulations. Today, waterless cosmetics combine aspects of conscious care, sustainable production, as well as concern for the environment and natural resources. Growing consumer awareness of sustainability means that natural products are steadily gaining in popularity because they are safe for the environment and for human health [1, 2].

Shampoo can be defined as a cosmetic preparation, packed in a convenient form for use, used to cleanse the scalp and hair of dirt, residues of hair styling products, and environmental pollutants. The main objectives of shampoo include: maintaining the physiological balance of the skin and essential components of the hair and making hair shiny for beautification and improved aesthetics. It is worth noting that shampoo is selected according to the needs of the scalp. The shampoo should be easy to remove by rinsing with water and produce sufficient lather to meet the users' needs, and it should also be non-toxic and non-irritating to the hair and scalp to avoid side effects as well as skin and eye irritation [3, 4].

On the cosmetics market, one can find products for daily care of the scalp and hair that differ primarily in their properties and composition. Among the commercial shampoos, in addition to the standard cleaning and conditioning shampoos, there are "specific" shampoos referred to as *specialised*. In their composition, there are additional functional ingredients that address, for example, scalp problems. Nowadays, shampoo is not just a product to keep the skin clean but a product that is able to sustain the effects of a scalp and hair treatment in specific disease states. Specialised shampoos contain antibacterial agents, natural essential oils, or extracts effective in the treatment of dandruff, dermatitis, as well as other hair diseases. Balancing good hair cleansing and beautifying is an art achieved by mixing different ingredients in the right proportions in a formulation. Modern advances in chemistry and technology have made it possible to replace soap bases with complex formulations that contain cleansing agents and conditioning agents along with functional additives, preservatives, aesthetic additives, and sometimes even medically active ingredients [1, 2].

Shampoos in a bar form are relatively new products in the cosmetics industry. Such shampoos fit in with current trends in the cosmetics market. Consumers are looking for products with the shortest composition, if possible natural, consisting of quality constituents, and with desirable application properties. The prevailing climate crisis has made it necessary to take care of the environment. As a result, cosmetics manufacturers are introducing new ideas already in the formulation stage. Currently, the most popular ones in the cosmetics industry include:

- **the zero waste concept**, which aims to minimise waste generation,
- **the waterless concept**, which focuses on conserving water resources
- **the plasticless concept**, which pertains to the reduction of plastic use
- **the eco-friendly concept,** which refers to the use of completely safe and environmentally friendly ingredients, e.g. raw plant-based materials must come from certified cultivation.

Analysing the composition of shampoo bars, it can be seen that they fit in with the above ideas. The main difference between traditional shampoos and bar shampoos is the water content of the composition. Traditional shampoos contain around 90% of water, whereas the content of water in shampoo bars is only around 15%. The surfactant in solid form is the most abundant constituent and accounts for about 40–50% of the formulation. It has cleansing and foaming properties, which contribute to the effective washing of the scalp and hair [5–7].

The aim and methodological scope of the study were to design a solid-state shampoo recipe and modify it by adding crushed hazelnut shells, make the shampoos under laboratory conditions, and assess their quality. The shampoos obtained were compared with shampoo bars available in the drugstore.

The quality of the shampoo was assessed by measuring its cut and penetration resistance, as well as hardness.

2 Material and Methods

The first stage of the work consisted of selecting raw materials appropriate for the production of a solid shampoo. Based on the literature data, the following ingredients were chosen for the formulation.

Ingredients for 60 g shampoo bar with cocoa butter for dry hair:

- 18 g Honeyquat – honey derivative
- 23.4 g Theobroma Cacao Seed Butter
- 18 g Decyl Glucoside
- 180 g Sodium Cocoyl Isethionate (SCI) powder
- 30.6 g Sacha Inchi organic oil
- 10.8 g Cetyl alcohol
- 77.4 g *Kaolin Clay*
- 1.8 g Cymbopogon Martini (Palmarosa) Oil

The recipe was modified by adding 10 g of ground hazelnut shells (particle size of 200 μm) and 5 ml of palmarosa oil.

Method of production:

1. Two glass beakers, of 100 ml and 500 ml, and a laboratory balance were prepared. In the 100 ml beaker, cocoa seed butter, decyl glucoside, Sacha Inchi oil and cetyl alcohol were weighed out using the balance. In the 500 ml beaker, clay and SCI were weighed out.
2. A water bath was prepared. A container was filled with boiling water to ¾ of its height and placed on a burner. The beaker containing cocoa butter, decyl glucoside, and Sacha Inchi oil was put into the water bath and heated until the ingredients dissolved and combined.
3. The beaker was removed from the water bath once the ingredients were dissolved. The entire mixture was poured into the second beaker of larger capacity, which contained clay and SCI.
4. Finally, honeyquat and palmarosa essential oil were added. The entire mixture was blended.
5. The resulting substance was removed from the beaker and placed in a flat dish for thorough mixing. The mixture for the shampoos was kneaded with hands.

6. The finished mass, of consistency similar to wet sand, was divided into 2 equal parts.
7. One part of the mixture was the basic recipe of the shampoo, while the other was enriched by adding 10 g of ground hazelnut shells and 5 ml of palmarosa oil. The mixture was blended again.
8. The resulting masses were placed in separate moulds.
9. Once cooled, they were transferred to a freezer for 3 h and then set aside to dry for 2 days.
10. Assessment of the quality of the obtained shampoos included testing penetrator force and work, cutting force and work, and crushing resistance (hardness) in a Micro Stable Pro TA.XT PLUS texture analyser.
11. Tests were performed on samples at room temperature. Shampoo samples were subjected to penetration (piercing), cutting and hardness measurement, respectively, in a horizontal stacking position. The way in which samples were placed is shown in Fig. 1.

The figures below show a schematic drawing of the test bench used to measure the cutting process. By replacing the texture analyser testing head, further parameters, such as hardness and penetration force, were determined on the same bench.

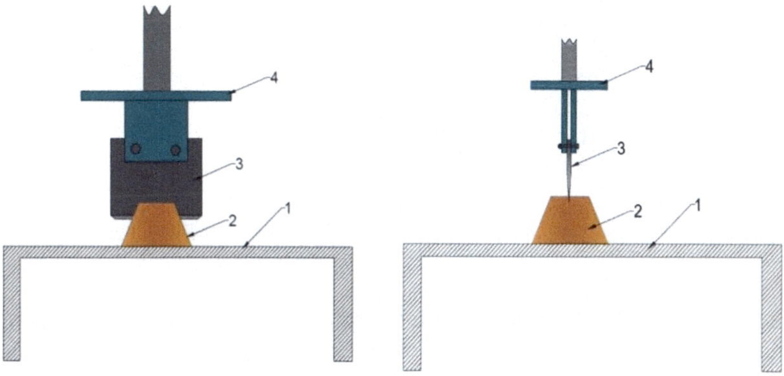

Fig. 1. The process of cutting shampoo horizontally [own elaboration].

1. Texturometer base
2. Shampoo
3. Texturometer blade
4. Texturometer arm

The hardness measurement was carried out on the bench shown above, by selecting the appropriate measurement parameters: distance 6.5 mm, sample diameter 27 mm, and sample height 13 mm. The measurement was carried out at a speed of 1 mm/s, with a cylinder having a 35 mm diameter. The cut resistance test was carried out at a speed of 5 mm/s, with a knife thickness of 2.5 mm and with a distance of 18 mm. The diameter of the samples was 27 mm for the basic and modified versions of the shampoo and 25 mm for the drugstore shampoo sample. The sample height was 13 mm for the basic and modified versions and 10 mm for the drugstore sample. In the case of penetration forces, the measurements were carried out at a speed of 1 mm/s and a needle thickness of 2 mm. The diameter and height of the basic and modified formulation shampoo samples were 27 mm and 13 mm, respectively, while for the drugstore sample, they equalled 25 mm and 10 mm, respectively. All tests were carried out in six repetitions.

3 Research Results and Discussion

The following figures present the results of the quality assessment of the shampoos that were produced.

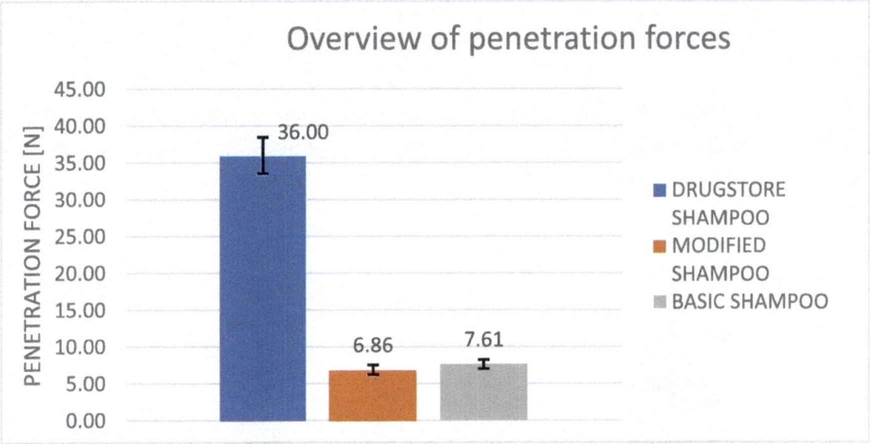

Fig. 2. Overview of penetration forces of shampoo bars [own elaboration].

The obtained results for penetration force measurements for all shampoos indicate that the highest one occurred in the case of the drugstore shampoo and equalled 36 N. This value is more than 5 times higher than the value obtained for the shampoo with the modified formulation. The differences in the composition of the products may cause the disparity in values obtained. The basic and modified shampoo had the same base ingredients, based on a mild surfactant powder derived from coconut oil - Sodium Cocoyl Isethionate (SCI). The composition of the drugstore shampoo was based on a mixture of three surfactants: Sodium Coco Sulfate (SCS), Disodium Lauryl Sulfosuccinate, and the auxiliary surfactant - Cocamidopropyl Betaine. Using more powdered raw materials as well as fatty and auxiliary substances may have influenced the differences in consistency of the products, increasing the hardness and thus resulting in obtaining the measurements in significant deviation, as shown in Fig. 2.

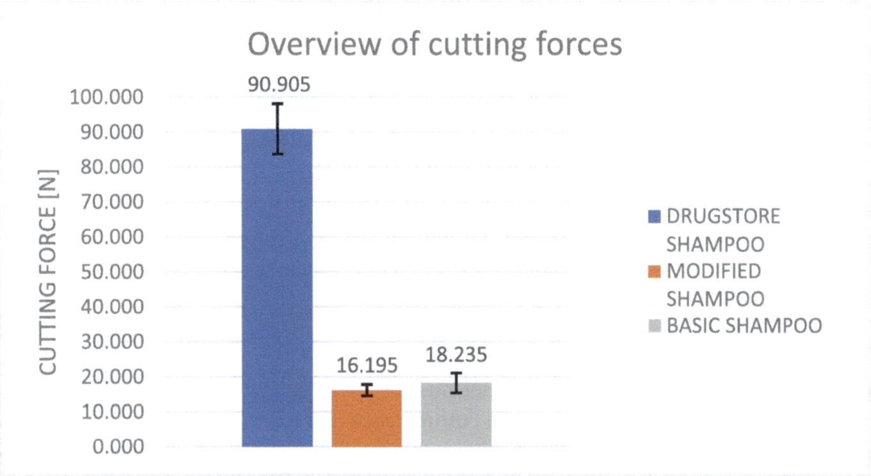

Fig. 3. Overview of the cutting forces of shampoo bars [own elaboration].

The measurements of the cutting force values of the shampoos, recorded at room temperature of 22 °C, are shown in Fig. 3. All three samples subject to the cutting process were placed horizontally relative to the texturometer. The graph shows that the highest cutting force characterised the drugstore shampoo and equalled 90.9 N, while the lowest was recorded for the shampoo with the modified formulation. The result obtained for the drugstore shampoo may be caused by the difference in the formulation of the product, as it contained other fatty components, including coconut oil, which is characterised by hard consistency at room temperature. This might have influenced the hardness of the shampoo bar and explains the need to use more force to cut it.

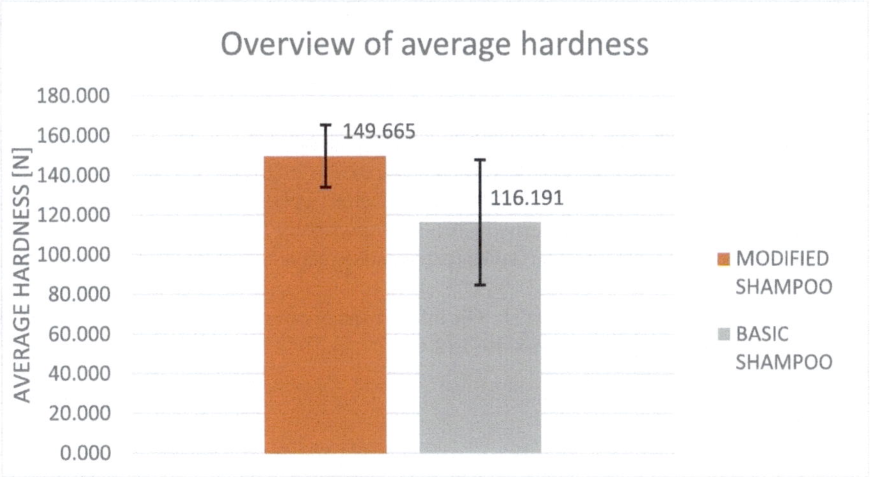

Fig. 4. Overview of average hardness of shampoo bars [own elaboration].

As shown in Fig. 4, the results reveal the difference in hardness between the basic shampoo and the shampoo with the modified formulation. By comparing the values obtained, one can see that the texture analyser had to use a higher force to compress the sample with the modified formulation. The value measured was 149.6 N and was 28.7% higher than the value obtained for the shampoo with the basic formulation. This may be related to the addition of more oil and dry ground hazelnut shells, which affected the consistency of the product by increasing its hardness. The hardness of the drugstore shampoo was not determined due to its high hardness exceeding the measurement range.

4 Conclusions

The shampoos designed and made under this research were subject to evaluation concerning the penetration force and work, cutting force and work, as well as hardness. The results obtained for the shampoo bars were compared to those obtained for the drugstore shampoo in the same form. The tests carried out made it possible to evaluate the final products and to conclude that the most important aspect in the case of shampoo bars is the correct composition of the formulation and selection of raw materials. The composition of raw materials has a major impact on the durability characteristics of shampoo bars. Adding natural raw materials to the recipe to improve the consistency and thus influence the hardness of the bar will increase the performance of the product. This, in turn, will make it more resistant to moisture and temperature when the cosmetic is kept in a bathroom.

References

1. Zhang, Y., Alsop, R.J., Soomro, A., Yang, F.C., Rheinstadter, M.C.: Effect of shampoo, conditioner and permanent waving on the molecular structure of human hair. Peer J. 1296 (2015)
2. Bhushan, B.: Introduction – human hair, skin and hair care products. Biological and Medical Physics. Biomedical Engineering, 1–19 (2010)
3. Robbins, C.: Chemical and Physical Behavior of Human Hair. Chemistry, Materials Science (1994)
4. Koch, S., Tridico, S.R., Bernard, B.A., Shriver, M.D., Jablonski, N.G.: The biology of human hair: A multidisciplinary review. Am. J. Hum. Biol. **32**(2) (2020)
5. Maciejszyk, E.: Trendy na rynku kosmetycznym: duży wpływ technologii i influencerów, Chemia i Biznes. Rynek Kosmetyczny i Chemii Gospodarczej **1**(32), 12–15 (2021)
6. Szampony w kostce i ich charakterystyka (przemyslkosmetyczny.pl). Last accessed 24 April 2024
7. Duragkar, A., Chande, S., Pradhan, J.: Oil infused shampoo: current hair care trend. Int. J. Current Eng. Scient. Res. **6**(1), 1026–1031 (2019)

Innovations in Corn Planting: A Comparison of Strip-Till, No-Till, and Traditional Technologies in the Context of Climate Adaptation and Sustainable Development

Katarzyna Łyp-Wrońska[1](✉), Krystian Nowak[2], and Agnieszka Dudziak[3]

[1] AGH University of Krakow, Krakow, Poland
klyp@agh.edu.pl
[2] AGRO-MASZ Agriculture Sp. z o.o., Łódzkie, Poland
[3] University of Life Sciences, Lublin, Poland

Abstract. The goal of the article is to assess the effectiveness of three leading corn-planting technologies in Poland—traditional cultivation, no-till, and strip-till—in the context of ongoing climate and market changes. The study compared the performance of these methods in terms of yield, fuel consumption, and machinery operating costs. Additionally, the focus was on optimizing the nitrogen dose and its utilization by plants depending on the application method. Experiments were conducted using different models of New Holland tractors, and Agro-Masz brand machinery, including the Salvis Z3800 strip-till planter, the Runner 30 + Ikar 1800 no-till cultivator with a front-mounted fertilizer tank, and the Falcon 6 precision planter, POR plough and Vibro 80 cultivator for traditional cultivation method – in small experimental fields that together covered 7 hectares area. The article presents not only yield results but also potential savings from investing in modern cultivation technologies. These results can serve as valuable guidance for farmers aiming to increase production efficiency while reducing environmental impact.

Keywords: Sustainable agriculture · Energy efficiency · Fertilization optimization · Precision agriculture · Climate adaptation

1 Introduction

In contemporary global crop production, maize is one of the primary cereal crops. Over the last 30–50 years, this crop has ranked first in the world in terms of yields and gross grain harvests. The rapid increase in maize cultivation is due to its high feed, nutritional, and technical values, as well as its extremely positive response to genetic changes and technological development. Maize is characterized by the efficient use of soil moisture for the formation of organic matter [1]. Maize yields are high on clean, loose, breathable soils with a deep humus layer, well supplied with nutrients and moisture [2]. The most favorable soils are black soils, dark chestnut soils, loamy and sandy-loamy soils, and alluvial soils [3]. By producing a large organic mass, maize removes many mineral components from the soil [4].

E. Lorencowicz et al. (Eds.): FMPMSA 2024, LNCE 609, pp. 277–283, 2024.
https://doi.org/10.1007/978-3-031-70955-5_31

For example, one considers that harvesting maize remove per ha about 150–160 kg of nitrogen, 45–50 kg of phosphorus, and 125–130 kg of potassium. The absorption of the main nutrients follows a single-peaked curve and corresponds to the progress of dry matter accumulation. Maize uses most of the nutrients in the second half of the growing season [7].

Maize (Zea mays L.) is one of the most important crops in the world. In Poland, it is mainly cultivated for animal feed (e.g., silage, feed mix components), as food for humans (e.g., corn oil, corn flakes, cornmeal), and for the energy sector (biomass, biogas, bioethanol, briquettes, etc.) [5]. Over the past decade, both the area of maize cultivation (over 350,000 ha) and its yield (2,000,000 tons) have been steadily increasing, significantly driven by the development of the bioenergy sector in Poland [6].

Maize is the most commonly used crop for the production of high-bromatological-value silage, intended for feeding medium- and high-yielding dairy cows.

1.1 Methodology of the Field Experiment

The field studies encompassed 18 plots, with 6 plots allocated to each technology – see Fig. 1. A detailed description was provided on how each plot was managed.

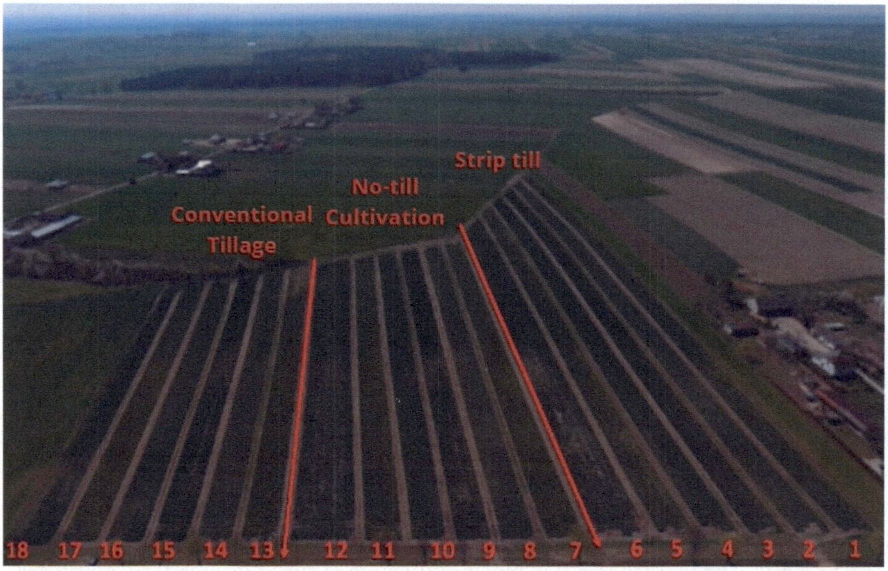

Fig. 1. Photo. Strip-till Cultivation. Source: Own elaboration

Strip-Till Cultivation:
Sowing was conducted using a strip-till cultivation unit. Fertilizer doses and sowing depth:

ABS CORN: 225 kg/ha

SuperN46: 200 kg/ha, 300 kg/ha, 400 kg/ha
Fertilizer was applied at depths ranging from 12.5 cm to 25 cm.

No-till Cultivation:
A no-till runner unit with fertilizer applicators was used. Fertilizer doses:

ABS CORN: 225 kg/ha
SuperN46: 200 kg/ha, 300 kg/ha, 400 kg/ha

Conventional Tillage:
Various fertilization and sowing configurations were applied:
Different combinations and doses of SuperN46 and ABS CORN.

Machine Costs and Efficiency
Strip-Till Cultivation:

Fuel: 36.2 L/ha
Time: 43.6 min/ha
Machine costs: 415,000 PLN

No-Till Cultivation:

Fuel: 43.2 L/ha
Time: 62.4 min/ha
Machine costs: 260,000 PLN

Conventional Tillage:

Fuel: 62.7 L/ha
Time: 125.9 min/ha
Machine costs: 175,000 PLN

2 Results

The results of the yield analysis are presented in Table 1.

Table 1. Results of Yield Analysis

Experimental Plots	Wet (yield) [t/ha]	Dry (yield) [t/ha]	Area [ha]	Yield per plot [t]	Variety	Moisture [%]
1 Strip-till cultivation	16.073	12.376	0.4467	7.180	ES Bond	38.0
2 Strip-till cultivation	15.658	13.027	0.4279	6.700	ES Bond	31.8
3 Strip-till cultivation	17.089	14.372	0.3909	6.680	ES Bond	30.9
4 Strip-till cultivation	16.106	13.835	0.3688	5.940	ES Bond	29.1
5 Strip-till cultivation	14.910	12.822	0.3434	5.120	ES Bond	29.0
6 Strip-till cultivation	16.246	13.972	0.4321	7.020	ES Bond	29.0
7 No-till cultivation	16.711	14.656	0.2633	4.400	ES Bond	27.3
8 No-till cultivation	15.378	13.225	0.2419	3.720	ES Bond	29.0
9 No-till cultivation	15.948	13.731	0.2320	3.700	ES Bond	28.9
10 No-till cultivation	16.871	14.509	0.2205	3.720	ES Bond	29.0
11 No-till cultivation	17.282	14.966	0.2384	4.120	ES Bond	28.4
12 No-till cultivation	16.626	14.531	0.3717	6.180	ES Bond	27.6
13 Conventional Tillage	6.807	5.813	0.1998	1.360	ES Bond	29.6
14 Conventional Tillage	7.356	6.252	0.2012	1.480	ES Bond	30.0
15 Conventional Tillage	16.542	14.425	0.1874	3.100	ES Bond	27.8
16 Conventional Tillage	13.713	12.068	0.1896	2.600	ES Bond	27.0

(*continued*)

Table 1. (*continued*)

Experimental Plots	Wet (yield) [t/ha]	Dry (yield) [t/ha]	Area [ha]	Yield per plot [t]	Variety	Moisture [%]
17 Conventional Tillage	16.493	14.184	0.1819	3.000	ES Bond	29.0
18 Conventional Tillage	14.806	12.674	0.1729	2.560	ES Bond	29.4

Source: Own elaboration

3 Discussion of Results

Based on the experiment, it can be concluded that:

1. The yield results from the experimental plots using Strip-till technology show that the application of SuperN46 fertilizer in various doses has a direct impact on yield. An average increase in yield of about 1 ton of dry grain was observed between the doses of 200, 300, and 400 kg. Increasing the input of nitrogen fertilizers translates to higher profit per hectare, which is confirmed by the higher yield obtained with higher fertilizer doses.
2. These results are consistent with studies conducted by Al-Kaisi and Licht [8], who found that Strip-till technology can lead to higher corn yields, particularly with proper nitrogen fertilizer management. Similarly, studies by Duiker and Lal [9] indicate that strip-till cultivation can improve the efficiency of water and nutrient use, contributing to higher yields.
3. The yield results from the no-till technology indicate that the close placement of nitrogen fertilizers near the seeds results in reduced yield, which may be due to soil salinity and germination issues. In this technology, the yield increase depending on the amount of fertilizer applied was less noticeable.
4. Similar observations can be found in the literature. Studies by Govaerts et al. [10] showed that no-till technology can lead to salt accumulation in the soil, negatively affecting plant germination. Furthermore, Hobbs et al. [11] noted that no-till cultivation requires careful fertilizer management to avoid problems associated with salinity and improper nutrient distribution.
5. The lack of fertilizer and herbicide protection in conventional tillage resulted in very low yields. However, with proper fertilization, conventional tillage has high yield potential, comparable to other technologies. Fuel costs and working time are higher, though, which affects the profitability of this method.
6. Studies conducted by Lal [12] confirm that traditional conventional tillage can be effective in terms of yield but is associated with higher operational costs and fuel consumption. Additionally, it is often less environmentally sustainable compared to no-till and strip-till technologies [13].

Yield values are needed to complete the Table 2.

Table 2. Comparison of Costs and Efficiency of Cultivation Technologies

Technology	Machine costs [PLN]	Fuel [l/ha]	Working time [min/ha]	Average Yield [t/ha]	Savings [PLN/ha]
Strip-till cultivation	415,000	36.2	43.6	14.17	400–500
No-till cultivation	260,000	43.2	62.4	13.72	300–400
Conventional Tillage	175,000	62.7	125.9	13.25	-

4 General Conclusions

Based on the conducted research, it can be concluded that conventional tillage for maize cultivation becomes uneconomical due to high fuel costs and labor time. No-till and strip-till technologies prove to be more cost-effective. Strip-till cultivation allows for higher yields with increased nitrogen fertilizer doses, leading to higher profits per hectare. Therefore, for maize cultivation, it is recommended to use no-till and strip-till technologies. The choice of technology depends on farm capabilities, availability of suitable machinery, and field size.

In summary, no-till and strip-till technologies are more economically viable and efficient, especially in large farms with appropriate machinery infrastructure.

References

1. Allison, F.E.: Soil organic matter and its role in crop production. Elsevier (1973)
2. Scheub, U.: Terra preta: How the world's most fertile soil can help reverse climate change and reduce world hunger. Greystone Books Ltd. (2016)
3. Agrawal, R.P., Ramamoorthy, B.: Morphological and chemical characteristics of alkali and normal soils from black and red soils of India. Geoderma **4**(4), 403–415 (1970)
4. Fageria, N.K.: Role of soil organic matter in maintaining sustainability of cropping systems. Commun. Soil Sci. Plant Anal. **43**(16), 2063–2113 (2012)
5. Piątek, E., Nowacka, T., Podkówka, Z., Waligóra, H.: Kukurydza. Czasopismo wydawane przez Polski Związek Producentów Kukurydzy (2007)
6. Hoffmann, A., Zenelt, W., Krawczyk, K.: Wpływ metody inokulacji roślin kukurydzy patogenami bakteryjnymi na powstawanie i rozwój objawów chorobowych. Progress in Plant Protection **64**(1), 28–35 (2024)
7. Anatoliiovych, K.O., Viktorivna, O.N., Mykhailivna, T.A.: Wpływ technologii uprawy oraz środków antystresowych na parametry biometryczne roślin mieszańcowych kukurydzy. European Research Materials (5) (2024)
8. Al-Kaisi, M.M., Licht, M.A.: Strip-tillage effect on corn nitrogen uptake and yield with nitrogen rate. Agron. J. **106**(1), 179–185 (2014)
9. Duiker, S.W., Lal, R.: Carbon sequestration potential of no-till farming in the tropics. Agron. J. **92**(3), 381–386 (2000)
10. Govaerts, B., Sayre, K.D., Deckers, J.: Conservation agriculture as a sustainable option for the central Mexican highlands. Soil and Tillage Res. **103**(2), 222–230 (2009)

11. Hobbs, P.R., Sayre, K., Gupta, R.: The role of conservation agriculture in sustainable agriculture. Philosophical Transact. Royal Soc. Biolog. Sci. **363**(1491), 543–555 (2008)
12. Lal, R.: Soil carbon sequestration impacts on global climate change and food security. Science **304**(5677), 1623–1627 (2004)
13. Montgomery, D.R.: Soil erosion and agricultural sustainability. Proc. Natl. Acad. Sci. **104**(33), 13268–13272 (2007)

Assessment of Slagging and Fouling Indicators for Ashes from Vine Shoots Regent Cultivar

Grzegorz Maj[1(✉)], Kamila Klimek[2], and Magdalena Kapłan[3]

[1] Department of Power Engineering and Transportation, University of Life Sciences in Lublin, Głęboka 28, 20-612 Lublin, Poland
grzegorz.maj@up.lublin.pl

[2] Department of Applied Mathematics and Computer Sciences, University of Life Sciences in Lublin, Głęboka 28, 20-612 Lublin, Poland

[3] Institute of Horticulture Production, University of Life Sciences in Lublin, Głęboka 28, 20-612 Lublin, Poland

Abstract. The article attempts to assess the degree of fouling and slagging of boilers when burning shoots of the Regent grape variety grown on the 101-14 and 125AA root-stocks and with own roots as a control. The study used an indicators assessment based on a proximate and ultimate analysis of the biofuel. A LECO TGA 701 thermogravimetric was used to incinerate the biomass samples. The content of elemental oxides in the ash was determined using the EDXRF method using the RIGAKU NEX QC+ QuantEZ elemental analyzer. The content of elements in the form of oxides was determined in the ash, i.e.: SiO_2, Fe_2O_3, Al_2O_3, CaO, MgO, P_2O5, Na_2O, K_2O and Ti_2O. Based on the chemical composition of ash expressed in oxides, the slagging and pollution indicators were determined, i.e. Base to Acid index (*Rb/a*), Slagging index (Sulfur ratio) (*Rs*), Fouling ratio (*Fu*), Alkali Index (*AI*), Sintering index (*SI*), Iron to calcium ratio (*IC*), Slag viscosity index (*Sr*). Research has shown that cultivation on rootstock does not reduce the potential for fouling and slagging of the boiler. Nevertheless, all indicators except *SI* and *IC* showed an extremely high probability of boiler fouling and slagging.

Keywords: Slagging · Fouling · Ash Composition · Biomass · Vine Shoots

1 Introduction

Viticulture requires proper management of the plantation in order to obtain a crop of specific quantitative and qualitative values every year. Considering this, various rootstocks are used in cultivation. The use of the appropriate type of rootstock may contribute to achieving specific cultivation parameters, including: impact on grapevine resistance to phylloxera, nematodes, pH, salinity and temperature drops. The use of an appropriate rootstock for a specific grape variety can stimulate growth, yield, fruit quality and, consequently, influence the quality of the wine obtained [1, 2].

One of the care procedures in viticulture is pruning shoots. Bio-waste generated during this procedure should be collected and removed from the cultivation site so that

E. Lorencowicz et al. (Eds.): FMPMSA 2024, LNCE 609, pp. 284–294, 2024.
https://doi.org/10.1007/978-3-031-70955-5_32

the required agrotechnical procedures can be performed [3–5]. Practical management of pruned shoots involves mulching them on site, storing them outside the crop or burning them [4, 6]. Mulching itself, apart from having a positive effect on the soil (enriching it with organic matter, maintaining soil moisture), may contribute to the transmission of diseases in the cultivation area [7]. Removing organic matter by burning, apart from low disposal costs, may lead to air pollution, but it also provides the opportunity to obtain additional energy [8, 9].

Apart from the positive aspect of obtaining energy from plant biomass, the use of this type of energy carrier is associated with combustion problems. Due to the inorganic matter of biomass fuels, technical complications arise when burning this type of fuel in boilers [10]. Due to the ease of formation of compounds with a low melting point from ash components using biomass fuels, there is a need to analyze the degree of possibility of ash sintering, boiler slagging, and potential high-temperature corrosion, affecting the boiler operation efficiency [11]. Taking into account the above, the aim of the work is to analyze the slagging and fouling indicators for Regent grapevine shoots depending on the rootstock used in the cultivation.

2 Material and Methods

2.1 Material

The study was conducted in 2021 at the NOBILIS Vineyard (50°39′ N; 21°34′ E) in the Sandomierska Upland in southeastern Poland. The research material consisted of vines of the Regent variety, which were planted in the spring of 2010 at a spacing of 2.0 × 1.0 m (5000 plants ha^{-1}) on loess soil. Plants were managed as a single Guyot twine with a trunk height of 40 cm, a single arm of about 0.9 m and 1 two-eyed pivot.

The experiment evaluated the effect of the type of rootstock in cultivation of the Regent variety on the ash composition and slagging and fouling indicators. Grapevines of the studied variety grew on 2 types of rootstocks: 125 AA, 101-14 and the control was made up of ungrafted vines growing on their own roots. All annual shoots were cut on the shrubs included in the experiment in winter during winter pruning of vine bushes.

2.2 Ash Chemical Composition

Ash was obtained by incinerating biomass samples in an air stream in a Nabatherm B180 muffle furnace. In the furnace, 2 g of the tested biomass were placed in crucibles and heated from ambient temperature to 550 °C at a heating rate of 10 $°C \cdot min^{-1}$. The incineration period lasted for 4 h, and then after the ash reached a temperature of approximately 100 °C. The crucibles with ashes were then removed from the furnace and placed in a desiccator until ambient temperature was reached. The prepared ash samples were placed in the measuring cups of the device and analyzed. The content of elements expressed in oxides in the ash was determined using the EDXRF method using the RIGAKU NEX QC+ QuantEZ elemental analyzer. The content of the ash was determined: silicon as SiO_2, iron as Fe_2O_3, aluminum as Al_2O_3, calcium as CaO, magnesium as MgO, phosphorus as P_2O_5, sodium as Na_2O, potassium as K_2O and titanium as Ti_2O.

2.3 Slagging and Fouling Indicators

Based on the ash chemical composition, higher heating value, sulfur content and ash content, the ash slagging and fouling indices were determined according to the formulas (Eq. 1–7) [12–16]:

- Base to Acid index ($R_{b/a}$):

$$R_{b/a} = \frac{Fe_2O_3 + CaO + MgO + Na_2O + K_2O}{SiO_2 + Al_2O_3} \tag{1}$$

- Slagging index (Sulfur ratio) (R_s):

$$R_s = R_{b/a} \cdot S^d \tag{2}$$

where:
S^d – is the Percentage of S in Dry Fuel

- Fouling ratio (F_u):

$$Fu = R_{b/a} \cdot (Na_2O + K_2O) \tag{3}$$

- Alkali Index (AI):

$$AI = \frac{(K_2O + Na_2O) \cdot A}{HHV} \tag{4}$$

where:

A – percentage of ash obtained at 550 °C (%),
HHV –Higher Heating Value ($MJ \cdot kg^{-1}$).

- Sintering index (SI):

$$SI = \frac{CaO + MgO}{Na_2O + K_2O} \tag{5}$$

- Iron to calcium ratio (IC):

$$IC = \frac{Fe_2O_3}{CaO} \tag{6}$$

- Slag viscosity index (Sr):

$$Sr = \frac{SiO_2 \cdot 100}{SiO_2 + Fe_2O_3 + CaO + MgO} \tag{7}$$

Determination of the higher heat value (HHV, $MJ \cdot kg^{-1}$) was done using calorimeter LECO AC 600 with EN-ISO 1928:2009 standard [17]. Ash content (A, %) was measured using a thermogravimetric analyzer LECO TGA 701 with EN-ISO 18122 standard [18]. The sulfur content (S, %) was determined using LECO S 628 analyzer according to the EN-ISO 16994:2016–10 standards [19].

Table 1 presents the ranges of indicators determining the level of risk of slagging and ash slagging for the tested biomass, calculated using formulas 1–7.

Table 1. Relationships between the values of indicators and the levels of tendency to slagging and fouling [20].

Index	Range	Slagging and fouling inclinations
Rb/a	<0.5 0.5–1 1–1.75 >1.75	Low Medium High Extremely high
Rs	<0.6 0.6–2 2–2.6 >2.6	Low Medium High Extremely high
AI	<0.17 0.17–0.34 >0.34	Low Medium High
SI	>0.6 0.6–2 <2	Low Medium High
IC	<0.3 0.3–3.0	Low High
Sr	>72 65–72 <65	Low Medium High
Fu	<0.6 0.6–40 >40	Low High Extremely high

Depending on the obtained values of the tendency to slagging and fouling of ashes from the tested biomass, the risk was marked as extremely high (E), high (H), medium (M) and low (L).

2.4 Statistical Analysis

The normality of distribution was checked by the Shapiro–Wilk compliance test. The impact of a tested biomass upon the value of obtained features was assessed by means of the ANOVA test. The significance differences were assessed by Tukey's HSD test. The analysis of the principal components was performed in order to graphically visualize significant differences and similarities between the ash composition and slagging and fouling indicators on the rootstock on the dendrograms. Ward's method was chosen for the grouping of the three rootstocks. The results obtained in the experiment were statistically analyzed in STATISTICA 13.

3 Results

Table 2 presents data characterizing Regent vine shoots in terms of higher heating value, ash content and sulfur content, necessary to estimate slagging and fouling indices.

Table 2. Results of higher heating value (HHV, Mg·kg^{-1}), ash content (A, %) and sulfur (S, %) for the tested Regen vine shoots, necessary to calculate slagging and fouling indices.

	HHV	A	S
101-14	15,69 ± 0,07 C	3,20 ± 0,12 B	0,43 ± 0,01 B
125AA	16,40 ± 0,08 B	4,19 ± 0,07 A	0,44 ± 0,12 A
Control	16,48 ± 0,06 A	3,17 ± 0,16 C	0,41 ± 0,02 C
p-value	0,001	0,001	0,001

± standard deviation, means with the same letter in column show no significant difference at $\alpha = 0.05$

The tests showed the highest higher heating value (HHV) for the Control among the analyzed materials, and it was approximately 5% lower than the lowest value obtained for vine shoots grown on the 101-14 rootstock. The highest ash content was found for shoots from the 125AA rootstock, and the lowest for the Control with a difference of approximately 1%. The sulfur content in individual samples was at a similar level (0.41–0.44%). The cultivation of shoots on the 125AA rootstock showed the highest concentration of this element, and the lowest for the control. Statistical analysis showed significant differences between rootstocks for each of the analyzed parameters. Similar values for the ash content in shoots were obtained in the work of Rico et al. [21] and Ganan et al. [22]. However, for HHV, the conducted research obtained lower values for all analyzed rootstocks compared to the data presented for grapevine shoots in the work [21, 23] and similar to those in papers [22, 24]. When comparing the sulfur content in the analyzed materials in relation to literature data, its level was significantly increased [22, 25, 26].

Table 3 lists the content of elements in the tested ash for vine shoots of the Regent variety, grown on two rootstocks and the control, i.e. grown on its own roots.

Table 3. Ash composition for tested materials (%).

	Al_2O_3	MgO	NaO	SiO_2	P_2O_5	K_2O	CaO	TiO_2	Fe_2O_3
101–14	nd	10,79 ± 0,05 B	nd	2,21 ± 0,01 A	7,10 ± 0,02 A	24,95 ± 0,04 A	49,47 ± 0,09 C	0,01 ± 0,00A	0,34 ± 0,01 A
125AA	nd	8,97 ± 0,05 A	nd	1,41 ± 0.02 B	6,29 ±0,01 B	24,84 ± 0,04 B	51,72 ±0,01 B	0,02 ± 0,00 A	0,25 ± 0.02 B
Control	nd	10,57 ± 0,06 C	nd	2,58 ± 0.01C	8,06 ±0,01 C	25,33 ± 0,02 C	47,19 ± 0,02 A	0,01 ± 0,00 A	0,36 ±0,01 C
p-value	-	0,001	-	0,001	0,001	0,001	0,001	0,5892	0,001

nd – not detected, ± standard deviation, means with the same letter in column show no significant difference at $\alpha = 0.05$.

The analyzed materials showed varied elemental composition of the ash. Al and Na elements were not detected in all analyzed cases. In the case of analysis of the MgO content, similar contents were recorded between the cultivation on the 101-14 rootstock and the Control. The difference in the MgO content in the ash for shoots between the rootstocks used in cultivation was not greater than 2%. When analyzing the SiO_2 content in the obtained ash, its amount was found to be in the range of 1.41–2.58%, hence the difference between the tested samples did not differ. Similar values were demonstrated regardless of the pad used for P_2O_5. Its lowest concentration was recorded in the ash from shoots grown on the 125AA rootstock, and the highest in the Control. However, the difference did not exceed 2% in the tested materials. The K_2O content in the ash of the tested shoots is noteworthy, as the level ranged from 24.84 to 25.33%, hence the difference was a maximum of 0.5%. The highest CaO content was found in the ash from shoots grown on the 125AA rootstock (51.72%), and the lowest for self-root cultivation (47.19%). Titanium dioxide was detected in low concentration (0.01–0.02%), while Fe_2O_3 for the 101-14 rootstock and the Control showed a very similar content with a difference of only 0.02%. The content of iron oxide in the ash from shoots from rootstock 125AA was 0.1% lower than the others. Statistical analysis showed significant differences between rootstocks for each of the analyzed parameters. When comparing the ash composition for the tested materials, it should be noted that the detection of Al_2O_3 and NaO is low compared to other types of biofuels, such as Softwood chips, Amaranth wood, hay and cereal straw [13]. The content of magnesium is comparable to Wood pellets (spruce), silicon to that of Amaranth wood or Rye grain, phosphorus to that of Sunflower pellets, and potassium to that of Safflower grains [13]. Comparing the obtained results to Vine shoots (*Vitis vinifera L.*), a similar CaO content was observed, but the remaining composition differed significantly [27]. The K_2O content in the tested samples was approximately 7% higher, while MgO was approximately 3% lower, SiO_2 was 2.5% lower, P_2O_5 was 6% lower, and Fe_2O_3 was 10 times lower [21, 27]. The observed differences may result from the different nature of cultivation and cultivation on different types of rootstocks.

Figure 1 shows the results of the cluster analysis for the ash composition expressed in oxides.

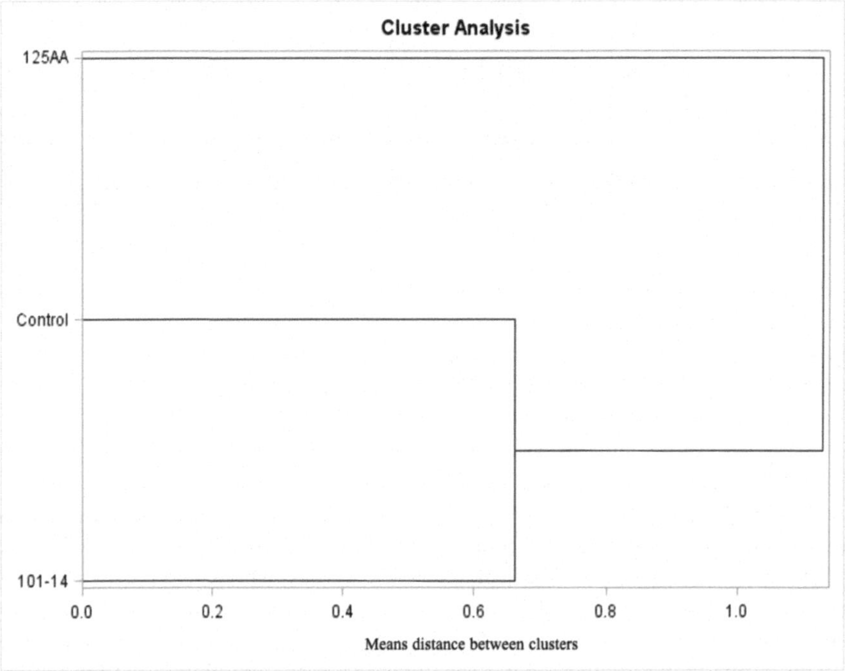

Fig. 1. Comparison of analyzed waste shoots of selected grapevine varieties in relation to ash composition

The dendogram shows a division into two clusters, of which a single cluster consists of rootstock 125AA, and the next cluster consists of rootstock 101-14 and the control. Similarities were observed between rootstock 101-14 and the Control in the composition of ash expressed in oxides. Table 4 lists the estimated slagging and fouling indicators for the analyzed materials.

Table 4. Indicators of contamination of heating devices by ash from the tested biomass.

Root-stock	*Rb/a*	*Rs*	*Fu*	*AI*		*SI*	*IC*	*Sr*
101-14	value	38,79 ± 0,14 B	16,58 ± 0,06 B	967,86 ± 4,74 B	5,09 ± 0,26 B	2,41 ± 0,01 B	0,01 ± 0,00 B	3,51 ± 0,01 B
	risk	H	E	E	H	H	L	H
125AA	value	60,99 ± 01,7 A	26,94 ± 0,07 A	1514,79 ± 2,42 A	6,34 ± 0,12 A	2,44 ± 0,01 A	0,00 ± C	2,26 ± 0,01 C
	risk	H	E	E	H	H	L	H
Control	value	32,32 ± 0,07 C	13,17 ± 0,03 C	818,40 ± 1,17 C	4,87 ± 0,26 B	2,28 ± 0,01 C	0,01 ± 0,00A	4,25 ± 0,01 A
	risk	H	E	E	H	H	L	H
p-value		0,001	0,001	0,001	0,001	0,001	0,001	0,001

± standard deviation, means with the same letter in column show no significant difference at $\alpha = 0.05$

The estimation of slagging and fouling indices for the tested Regent vine shoots showed a very high probability of the indicated phenomena, regardless of the rootstock used in cultivation. Slagging index (*Rs*) for all tested shoots showed an extremely high risk of boiler slagging due to the high concentration of sulfur in the material. The same situation was demonstrated for the Fouling ratio (*Fu*). In this case, such a high risk is related to the high K_2O content in the ashes of the tested materials. A high risk of slagging and slagging was noted for all primers in the case of alkali index (*AI*), sintering index (*SI*) and Slag viscosity index (*Sr*). In the case of *AI*, its value was influenced by the high K_2O content, which determined its high value. A high risk of sintering with CaO and low NaO content was demonstrated for all analyzed materials. However, the high risk of slag viscosity may result from the average content of silicon, iron or magnesium with a high calcium content. A low risk of slagging and fouling was demonstrated only by the *IC* index, which was influenced by a low iron concentration with a high calcium content. Statistical analysis showed significant differences between all estimated indicators, regardless of the rootstock used in viticulture. A high or very high risk of slagging or fouling was shown, similarly to the tested vine shoots, by Hardwood chips, Amaranth wood, Safflower grains, Rye grains, and White mustard seed [13]. In the case of vine shoots, the high risk of fouling is confirmed by the results of Rico et al. [21].

Figure 2 shows the result of the cluster analysis for the estimated slagging and fouling indicators. The dendogram shows a division into two clusters, of which a single cluster consists of rootstock 125AA, and the next cluster consists of rootstock 101-14 and the Control. Similarities were observed between rootstocks 101-14 and the Control in estimated slagging and fouling rates. When comparing the dendrograms from Figs. 1 and 2, indications of connections between individual objects obtained from the analysis of the similarity matrix were observed.

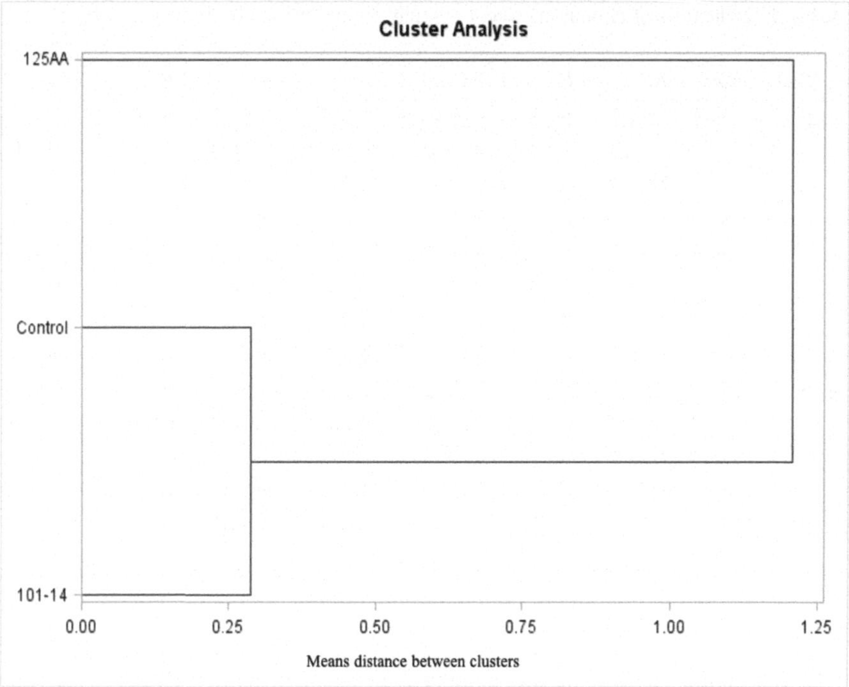

Fig. 2. Comparison of analyzed waste shoots of selected grapevine varieties in relation to slagging and fouling indictors

4 Conclusions

The research showed that Regent vine shoots, regardless of the rootstock used, have a high or extremely high risk of slagging and fouling when using this material as a biofuel. This is due to the high content of CaO and K_2O in the ashes. Research has proven that the use of rootstock in the cultivation of Regent grapevines significantly affects the energy parameters obtained as well as the composition of the ash itself. From the point of view of micro and macro elements, the best option is Control, i.e. growing on your own roots. For this reason, it should be considered whether the use of specific rootstock should not be correlated with the desire to obtain ashes of a specific composition.

References

1. Sabır, A., Doğan, Y., Tangolar, S., Kafkas, S.: Analysis of genetic relatedness among grapevine rootstocks by AFLP (Amplified Fragment Length Polymorphism) markers. J. Food Agric. Environ. **8**, 210–213 (2010)
2. Sivilotti, P., Zulini, L., Peterlunger, E., Petrussi, C.: Sensory properties of "Cabernet sauvignon" wines as affected by rootstock and season. Acta Hortic. 443–448 (2007). https://doi.org/10.17660/ActaHortic.2007.754.58
3. González-García, S., et al.: Comparative environmental and energy profiles of potential bioenergy production chains in Southern Europe. J. Clean. Prod. **76**, 42–54 (2014)

4. Manzone, M., Paravidino, E., Bonifacino, G., Balsari, P.: Biomass availability and quality produced by vineyard management during a period of 15 years. Renew. Energy **99**, 465–471 (2016)
5. Mendívil, M.A., Muñoz, P., Morales, M.P., Juárez, M.C., García-Escudero, E.: Chemical characterization of pruned vine shoots from La Rioja (Spain) for obtaining solid bio-fuels. J. Renew. Sustain. Energy. **5**, 033113 (2013)
6. Spinelli, R., Lombardini, C., Pari, L., Sadauskiene, L.: An alternative to field burning of pruning residues in mountain vineyards. Ecol. Eng. **70**, 212–216 (2014)
7. Scarlat, N., Blujdea, V., Dallemand, J.-F.: Assessment of the availability of agricultural and forest residues for bioenergy production in Romania. Biomass Bioenerg. **35**, 1995–2005 (2011)
8. Magagnotti, N., Nati, C., Spinelli, R., Vieri, M.: Technical protocol for the utilization of pruning residues from vineyards and olive groves. The Forest-Wood-Energy Chain: Results from the International Project Woodland Energy (2009)
9. Keshtkar, H., Ashbaugh, L.L.: Size distribution of polycyclic aromatic hydrocarbon particulate emission factors from agricultural burning. Atmos. Environ. **41**, 2729–2739 (2007)
10. Okot, D.K., Bilsborrow, P.E., Phan, A.N.: Briquetting characteristics of bean straw-maize cob blend. Biomass Bioenerg. **126**, 150–158 (2019). https://doi.org/10.1016/j.biombioe.2019.05.009
11. Wang, L., Skreiberg, Ø., Becidan, M.: Investigation of additives for preventing ash fouling and sintering during barley straw combustion. Appl. Therm. Eng. **70**, 1262–1269 (2014). https://doi.org/10.1016/j.applthermaleng.2014.05.075
12. Ghazidin, H., et al.: A comprehensive evaluation of slagging and fouling indicators for solid fuel combustion. Thermal Sci. Eng. Prog. **40**, 101769 (2023). https://doi.org/10.1016/j.tsep.2023.101769
13. Lachman, J., Baláš, M., Lisý, M., Lisá, H., Milčák, P., Elbl, P.: An overview of slagging and fouling indicators and their applicability to biomass fuels. Fuel Process. Technol. **217**, 106804 (2021). https://doi.org/10.1016/j.fuproc.2021.106804
14. Yang, H., Jin, J., Hou, F., He, X., Hang, Y.: An ANN-based method for predicting Zhundong and other Chinese coal slagging potential. Fuel **293**, 120271 (2021). https://doi.org/10.1016/j.fuel.2021.120271
15. Zhu, C., Tu, H., Bai, Y., Ma, D., Zhao, Y.: Evaluation of slagging and fouling characteristics during Zhundong coal co-firing with a Si/Al dominated low rank coal. Fuel **254**, 115730 (2019). https://doi.org/10.1016/j.fuel.2019.115730
16. Maj, G., Szyszlak-Bargłowicz, J., Zając, G., Słowik, T., Krzaczek, P., Piekarski, W.: Energy and emission characteristics of biowaste from the corn grain drying process. Energies **12**, 4383 (2019). https://doi.org/10.3390/en12224383
17. EN-ISO 1928:2009: Solid mineral fuels—Determination of gross calorific value by the bomb calorimetric method and calculation of net calorific value (2010)
18. EN-ISO 18122:2016-01: Solid Biofuels. Determination of Ash Content (2016)
19. ISO 16994:2016: Solid Biofuels—Determination of Total Content of Sulphur and Chlorine (2016)
20. Pronobis, M.: Evaluation of the influence of biomass co-combustion on boiler furnace slagging by means of fusibility correlations. Biomass Bioenerg. **28**, 375–383 (2005). https://doi.org/10.1016/j.biombioe.2004.11.003
21. Rico, J.J., Pérez-Orozco, R., Cid, N., Larrañaga, A., Míguez Tabarés, J.L.: Viability of agricultural and forestry residues as biomass fuels in the Galicia-North Portugal region: an experimental study. Sustainability. **12**, 8206 (2020). https://doi.org/10.3390/su12198206
22. Gañán, J., Al-Kassir Abdulla, A., Cuerda Correa, E.M., Macías-García, A.: Energetic exploitation of vine shoot by gasification processes: a preliminary study. Fuel Process. Technol. **87**, 891–897 (2006). https://doi.org/10.1016/j.fuproc.2006.06.004

23. Duranay, N.D., Akkuş, G.: Solid fuel production with torrefaction from vineyard pruning waste. Biomass Conv. Bioref. **11**, 2335–2346 (2021). https://doi.org/10.1007/s13399-019-00496-0
24. González, J.F., Encinar, J.M., Canito, J.L., Sabio, E., Chacón, M.: Pyrolysis of cherry stones: energy uses of the different fractions and kinetic study. J. Anal. Appl. Pyrol. **67**, 165–190 (2003). https://doi.org/10.1016/S0165-2370(02)00060-8
25. Maj, G., Klimek, K., Kapłan, M., Wrzesińska-Jędrusiak, E.: Using wood-based waste from grapevine cultivation for energy purposes. Energies **15**, 890 (2022). https://doi.org/10.3390/en15030890
26. Florindo, T., Ferraz, A.I., Rodrigues, A.C., Nunes, L.J.R.: Residual biomass recovery in the wine sector: creation of value chains for vine pruning. Agriculture **12**, 670 (2022). https://doi.org/10.3390/agriculture12050670
27. Manyà, J.J., Alvira, D., Videgain, M., Duman, G., Yanik, J.: Assessing the importance of pyrolysis process conditions and feedstock type on the combustion performance of agricultural-residue-derived chars. Energy Fuels **35**, 3174–3185 (2021). https://doi.org/10.1021/acs.energyfuels.0c04180

ESG: Carbon and Biodiversity a Problem or an Opportunity for Smart Agriculture?

José Rafael Marques da Silva[1,2](✉)

[1] MED—Mediterranean Institute for Agriculture, Environment and Development and CHANGE—Global Change and Sustainability Institute, Institute for Advanced Studies and Research, Universidade de Évora, Pólo da Mitra, Ap. 94, 7006-554 Évora, Portugal
jmsilva@uevora.pt

[2] AgroInsider Lda., 7005-841 Évora, Portugal

Abstract. The concept of ESG and smart agriculture is revolutionizing food production by integrating cutting-edge technologies to optimize agricultural processes. This ESG holistic approach not only aims to enhance efficiency and resilience in food production, but also mitigation of environmental impact, addressing concerns such as carbon emissions and biodiversity loss. By maximizing crop yields while minimizing the utilization of finite natural resources, smart agriculture contributes significantly to environmental and social sustainability.

By prioritizing the adoption of ESG and sustainable agricultural practices, agricultural organizations not only bolster food security, but also foster biodiversity conservation, thereby generating a positive impact on local communities. This leads to the creation of economic opportunities and an overall improvement in quality of life.

Furthermore, the ESG European regulation aligned with the voluntary carbon markets presents new business prospects for farmers, particularly those facing challenges in competitive food production due to factors such as poor soil quality or steep terrain. At the farm level, farmers can be categorized into three main groups: those dedicated solely to food production, those solely focused on environmental conservation, and those capable of balancing both aspects simultaneously. Nevertheless, all farms should align with ESG and smart agriculture principles in order to reduce risk competitive issues.

Keywords: ESG · Smart agriculture · Food · Carbon · Biodiversity

1 ESG and Agriculture

ESG is all about making a difference: to the planet, to your people and to your organization. So, what is it, and how is it important to any economic activity such as agriculture?

ESG stands for **E**nvironmental, **S**ocial, and **G**overnance. It encapsulates several standards that measure how an organisation's activities contribute to sustainable, viable development, and how that organisation minimises its negative impacts on the planet, its people, and itself [1].

E. Lorencowicz et al. (Eds.): FMPMSA 2024, LNCE 609, pp. 295–299, 2024.
https://doi.org/10.1007/978-3-031-70955-5_33

ESG is also known as the non-financial report that the largest European companies will be required to include in their annual reports starting in 2024. This practice has already been adopted by various companies globally for several years. Gradually, by 2026, it is estimated that 54,000 companies in Europe will be mandated to present this type of non-financial report, with well-defined obligations and specific characteristics. One of the requirements of ESG reporting is to monitor the value chain to understand the upstream and downstream ESG footprints of the company's activities.

Companies that are not yet required to publish this type of annual report are already anticipating this new standard within their operations, as they recognize that ESG indicators will influence the cost of bank credit, making it cheaper or more expensive depending on the company's scores in these three areas (Table 1).

Table 1. ESG rating

ESG Score	ESG rating: Calculated by cumulating total ESG performance		
Pilar score	Environment	Social	Governance
Theme score	Climate change Ecological footprint Resource use Pollution	Health & Safety Customer responsibility Community impact Labour standards	Shareholder rights Risk management Tax transparency Anti-corruption

Agricultural companies are not required to submit their ESG reports. However, their value chain, specifically distribution, needs to consult their suppliers, requiring a set of data that agricultural companies will need to start providing if they wish to remain operational in the market.

The question remains: What type of data will be most requested from agricultural companies by their value chain, and what is the relative value of this data for their business?

2 ESG and Agriculture Major Footprints

What type of ESG data from agricultural activities will be valuable for the value chain considering the characteristics of the agricultural activity itself?

If we focus solely on environmental footprints (Table 2), those companies that can respond with objective and certified data to the listed footprints will have a competitive advantage. Examples of such data include:

i) Ton CO_2eq / Ton of food produced;
ii) m^3 of water used / Ton of food produced;
iii) Ton of NPK used / Ton of food produced;
iv) Ton of substances of concern / Ton of food produced;
v) Ton of plastics/Ton of food produced;
vi) Areas of discontinuity in the landscape/Specialized productive area;
vii) Tilled area/Non-tilled area;

Table 2. ESG rating and major potential agriculture footprints

Major potential agriculture footprints		
Environmental	Social	Governance
CO_2 emissions, Water productivity, Water contamination, NPK non-source pollution, Plastics and other residues, Pesticides, Soil degradation, Soil erosion, Biodiversity losses…	Slavery work	Animal welfare, Landscape, Climate resilience

viii) Tilled area/Ton of food produced, etc.

Organizations that achieve an **excellent** rating (>70%) comply with all best ESG practices and present few or no internal and external risks, they will gain the visibility that will drive them to greater commercial success, as well as greater success in accessing low-cost credit. Organizations with a **good** rating comply with the best ESG practices and have a low impact on people and the environment. Organizations with an **average** score do not meet significant ESG objectives. Organizations with a **low** score (<50%) indicate that they do not follow relevant ESG criteria and probably negatively impact the planet and do not treat their employees and suppliers correctly. Consequently, the last, compared to the first, will potentially have the worst commercial performance and the most difficulty in accessing low-cost credit.

3 ESG and Precision Agriculture

Half a dozen years ago, if we had to define Precision Agriculture, it would surely have been described as "…inputs optimization from an economic and **environmental** perspective…". Currently, in light of the ESG aspects mentioned earlier, the definition would be more along the lines of "…**resources** optimization from an economic and **financial risk** perspective…". Yes, the environment and sustainability are no longer solely associated with ecology enthusiasts but increasingly with financial risk, which in agriculture is undoubtedly influenced greatly by the environment and ecology.

Thus, precision agriculture now has new scales and dimensions of analysis. In addition to concerns associated with the plot scale, it must also address the farm scale. At the plot level, inputs differentiation is important for maximum factor efficiency. However, at the farm level, differentiation of land use and crops is also relevant to increase the resilience of the activity and reduce economic, financial, environmental, social, and governance risks (Fig. 1).

In this type of land use diversification, new business opportunities may arise for agricultural companies, not only due to the diversification of products (food, fibres, energy) and reduction of risks, but also because of the possibility of creating new types of environmental services (e.g., by sequestering CO_2 in forests and soil) that can be sold in voluntary carbon and biodiversity markets.

Landowners with land located in difficult-to-farm areas (e.g., very poor soils, steep slopes, etc.) now have new opportunities to capitalize on their assets, not so much on the food production side, but rather on the production of environmental assets that sustainable companies' ESG will require.

If smart agriculture was already a type of agriculture supported by data, now the importance of data collection and information is even greater. Only through these can most of the risks associated with the activity be calculated and support reliable certifications for the activity throughout the value chain. If data recording was never a major concern in agricultural activity, it will now become central, as the financial credit associated with the activity will assess the ESG risks of each agriculture company and differentiate the interest rate associated with each company's ESG performance.

In this context, what concerns should agricultural entrepreneurs have to structure their businesses based on resilience and not solely on efficiency?

4 ESG and the Road Map from Efficiency to Resilience

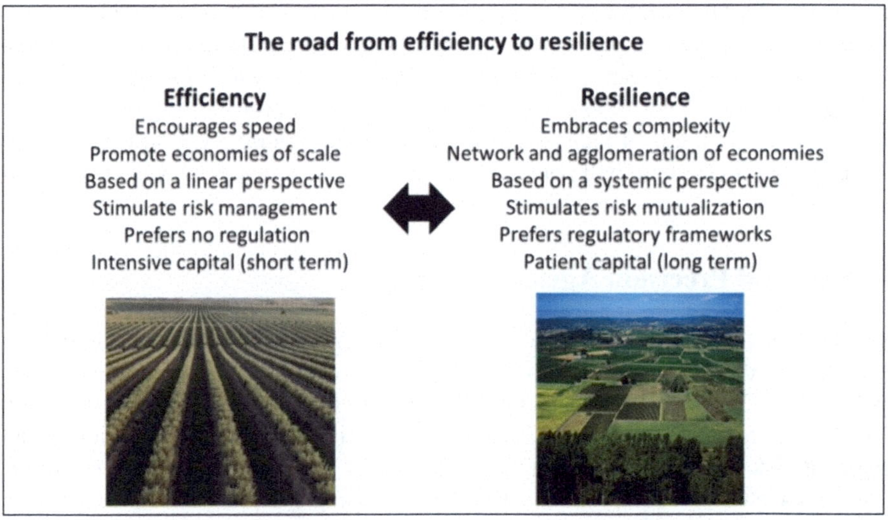

Fig. 1. The road from efficiency to resilience

The road map from efficiency to resilience in agriculture can be clearly illustrated with Fig. 1, on the left side of the figure one can sees an example of a super-intensive olive oil yard in Portugal. It is a highly efficient system in terms of GWP (Global Warming Potential), since from the perspective of CO_2eq emissions per ton of product produced, it is a system that has very low emissions when compared with less intensive systems. However, it presents serious biodiversity (E) and landscape footprints, and potentially a very high financial risk if a multi-resistant pest to approved pesticides appears and could decimate the thousands of hectares installed in the Iberian Peninsula, thus causing almost all accumulated value to be lost up to that date. We can also look for other types of high risks in the present, such as a persistent heatwave during flowering impacting the widespread productivity of these types of systems.

A resilient system, on the right side of Fig. 1, should indeed include super-intensive olive oil yards as part of the solution because it is an efficient system, however, integrated into a crop MOSAIC with other types of crops and activities where reducing financial risk is achieved not through efficiency, but through resilience (Fig. 1) and the mix of associated activities and circularities.

Agricultural companies, according to European regulations, are not required to present their ESG, however, they will be highly influenced by their value chain and especially by distribution, banking, and insurance companies, as financial risk does not forgive. If you are an agricultural entrepreneur, I believe you should already be seriously considering your ESG and the data collection it will require, as well as learning how to do the transition from efficiency to resilience in your company.

5 Conclusion Remarks: Why ESG and Smart Agriculture?

What does an agricultural entrepreneur gain by considering ESG principles and data-driven sustainability in their activity and company?

In the 21st century, consumers are becoming more selective in their purchases, typically valuing companies that are environmentally (E) and socially (S) responsible. According to research by Forbes [2], 62% of Generation Z respondents prefer to buy sustainable brands, and 87% of consumers said they would purchase a product if the organization producing it considers a certain aspect they value.

There is a strong connection between brand value and product value and the sustainability practices of the organization. According to data from Bain & Company [3], brands with higher sustainability scores experience five times the revenue growth compared to those with lower scores.

Caring for the land and others ethically has both local and global impacts on the reputation of companies, brands, and products. Recognition in these areas will attract not only more customers and suppliers but also the best employees.

When organizations care about the world around them, they must ensure that their employees not only feel safe and protected in their tasks but also proud of the organization they work for.

In conclusion, although agriculture activity doesn't require to present an ESG report, it will in some way be dragged into the societal dynamics already being generated by companies that are obligated to do so. Therefore, the discussion is no longer whether ESG will enter or not in agriculture activity, but rather, at what speed it will occur. To care for the land and others ethically, we must monitor the path, and for this, data and principles of smart agriculture are essential in decision-making. What are you waiting for?

References

1. Qualtrix Homepage: https://www.qualtrics.com/uk/experience-management/employee/esg/. Last accessed 27 May 2024
2. Forbes homepage: https://www.forbes.com/sites/gregpetro/2020/01/31/sustainable-retail-how-gen-z-is-leading-the-pack/. Last accessed 28 May 2024
3. Bain & Company home page: https://www.bain.com/insights/how-sustainable-brands-add-value-snap-chart/. Last accessed 28 May 2024

Consumption of Plant Protection Products as an Indicator of Agriculture Intensity – A Case Study of Low Risk Active Substances

Ewa Matyjaszczyk(✉)

Bydgoszcz University of Science and Technology, Al. prof. S. Kaliskiego 7, 85-796 Bydgoszcz, Poland
Ewa.Matyjaszczyk@pbs.edu.pl

Abstract. Reduction of the volume of plant protection products consumed in agriculture is mentioned frequently in the European Union. It is an important aim of the European Green Deal. Consequently, the volume of plant protection products was set as one of Harmonized Risk Indicators of from Farm to Fork Strategy to quantify the progress in reducing the risks linked to pesticides. In 2020 the Commission decided to reduce the overall use and risk of chemical pesticides by 50% before 2030. At the beginning of 2024, the Commission decided that it was necessary to continue discussions on the subject. Therefore, the consumption of plant protection products will very likely remain the topic under discussion.

The volume of plant protection products consumed in agriculture is being used as an indicator of agriculture intensity not only in the European Union, but also around the world. Is however the indicator clear and reliable? The paper focuses on the topic using low-risk active substances approved in the European Union as the example.

Keywords: Pesticides · Plant Protection Products · Volume · Consumption · Measure · Low-Risk · Active Substances

1 Introduction

Throughout the world, there is a strong emphasis on more environmentally friendly agricultural production systems. Varied measures are taken in different regions to make the food systems more sustainable [1, 2].

In the European Union all professional users of pesticides are by law obliged to use integrated pest management in their daily farming routine since 2014 [3]. It raises interest in preventive methods of pest control used in traditional agriculture and forestry [4, 5]. There is also a growing interest in microorganisms and botanical substances that may be effective in plant protection [6–8]. The public is concerned about food safety. The awareness of public regarding food components and food impurities are topics of numerous studies [9–11].

E. Lorencowicz et al. (Eds.): FMPMSA 2024, LNCE 609, pp. 300–307, 2024.
https://doi.org/10.1007/978-3-031-70955-5_34

The European Directive 2009/128 required all the EU Member States to adopt national action plans for the establishment of quantitative objectives, targets and measures to reduce the risks and impact of pesticide use for human health and the environment [12]. Reduction of use of plant protection products, is the target of European Green Deal [13]. Consequently, there has been a lot of emphasis on reducing the use of plant protection products recently.

Use of plant protection products is logically related to their volume. Hence the volume of plant protection products applied in agriculture, expressed in grams or kilograms per hectare, is used as the main or as an important indicator of environmental concerns. From scientific point of view this indicator is not very fortunate. The volume of plant protection products may be a reliable indicator in certain cases only, namely, to compare the use of the same active substance (e.g. in certain periods of time or in various regions), otherwise it may be misleading. This has been proved by Matyjaszczyk [14] who compared doses of three groups of active substances approved for use in the European Union in 2023: candidates for substitution, basic substances and microorganisms. The study shows that difference among recommended doses of products under study is very significant and often reaches several orders of magnitude. Adding together the volumes of substances recommended for use at very low and very high doses (as different as milligrams and kilograms) may give a bias to statistical calculations. In the study mentioned above [14] Matyjaszczyk calculated that replacing active substances considered riskier (candidates for substitution) with active substances considered safer (basic substances) may give the statistical effect of increasing the volume of plant protection products used.

It would be interesting to study if significant differences in recommended doses of plant protection products concern also another group of active substances approved for use in the European Union, namely low-risk active substances.

2 Recommended Doses of Low-Risk Active Substances

2.1 What Are Low-Risk Active Substances?

In the European Union, the active substances in plant protection products are divided into several categories, depending on their properties. An active substance can be approved for use in plant protection as a low-risk active substance if it meets the regular approval criteria and additionally meets the low-risk criteria specified the Regulation 1107/2009.

What are the low-risk criteria? They are related to the properties of the active substance. Following those criteria the low-risk active substance cannot be persistent, bioaccumulative, endocrine disrupter and neither have neurotoxic or immunotoxic effects. In addition, the low-risk active substance cannot be classified as: carcinogenic, mutagenic, toxic to reproduction, sensitising chemicals, very toxic or toxic, explosive and corrosive [15].

Plant protection products that contain only low-risk substances can then be authorized as a low-risk plant protection product. Following recommendations of integrated pest management, due to their favorable properties, the low-risk plant protection products should be preferred by farmers and other users in their approach to manage pests [16].

2.2 What Are the Recommended Doses of Low-Risk Active Substances?

The recommended doses of low-risk active substances are presented in Table 1. The source for calculations is official data of the European Union from October 2023. It's worth mentioning that the doses used in practice are often lower than shown in Table 1. The doses recommended for use in plant protection products depend on their formulation [17] and the farmers can further reduce the registered doses based on the pest infestation. Table 1 gives data on maximal limits of low-risk active substances in kilograms per hectare, that can be registered in plant protection products.

Table 1. Low-risk active substances. Maximum doses in kilograms per hectare.

No.	Active Substance	Maximum Dose (kg of active substance/ha)
1	(E)-11-Tetradecen-1-yl acetate	0.025
2	(E)-5-Decen-1-yl acetate	0.025
3	(E)-8-Dodecen-1-yl acetate	0.025
4	(E,E)-7,9-Dodecadien-1-yl acetate	0.025
5	(E,E)-8,10-Dodecadien-1-yl acetate	0.025
6	(E,Z)-2,13-Octadecadien-1-yl acetate	0.025
7	(E,Z)-3,13-Octadecadien-1-yl acetate	0.025
8	(E,Z)-3,8-Tetradecadien-1-yl acetate	0.025
9	(E,Z)-7,9-Dodecadien-1-yl acetate	0.025
10	(E,Z,Z)-3,8,11-Tetradecatrien-1-yl acetate	0.025
11	(Z)-11-Hexadecen-1-yl acetate	0.025
12	(Z)-11-Tetradecen-1-yl acetate	0.025
13	(Z)-7-dodecen-1-yl acetate	0.025
14	(Z)-8-Dodecen-1-yl acetate	0.025
15	(Z)-8-Tetradecen-1-yl acetate	0.025
16	(Z)-9-Dodecen-1-yl acetate	0.025
17	(Z)-9-Tetradecen-1-yl acetate	0.025
18	(Z,E)-7,11-Hexadecadien-1-yl acetate	0.025
19	(Z,E)-9,11-tetradecadien-1-yl-acetate	0.025
20	(Z,E)-9,12-Tetradecadien-1-yl acetate	0.025
21	(Z,Z)-3,13-Octadecadien-1-yl acetate	0.025
22	(Z,Z)-7,11-Hexadecadien-1-yl acetate	0.025
23	24-Epibrassinolide	0.00004

(*continued*)

Table 1. (*continued*)

No.	Active Substance	Maximum Dose (kg of active substance/ha)
24	ABE-IT 56	0.088
25	Akanthomyces muscarius Ve6 (formerly Lecanicillium muscarium strain Ve6)	0.144
26	Ampelomyces quisqualis strain AQ10	0.04
27	Aqueous extract from the germinated seeds of sweet Lupinus albus	4.016
28	Bacillus amyloliquefaciens AH2	not available
29	Bacillus amyloliquefaciens IT-45	0.2
30	Bacillus amyloliquefaciens strain FZB24	0.37
31	Bacillus subtilis strain IAB/BS03	1.5
32	Blood meal	19.96
33	COS-OGA	0.063
34	Calcium carbonate	1.352
35	Cerevisane	0.75
36	Clonostachys rosea strain J1446 (Gliocladium catenulatum strain J1446)	1.2
37	Coniothyrium minitans Strain CON/M/91-08 (DSM 9660)	0.3
38	Dodecyl acetate	0.025
39	Ferric phosphate	1.5
40	Ferric pyrophosphate	1.5
41	Heptamaloxyloglucan	0.000437
42	Hexadecyl acetate	0.025
43	Isaria fumosorosea Apopka strain 97 (formerly Paecilomyces fumosoroseus)	0.6
44	Laminarin	0.135
45	Lavandulyl senecioate	0.093
46	Metarhizium brunneum strain Ma 43 (formerly Metarhizium anisopliae var anisopliae)	not available
47	Mild Pepino Mosaic Virus isolate VC 1	0.00002
48	Mild Pepino Mosaic Virus isolate VX 1	0.00002
49	Pasteuria nishizawae Pn1	0.00811
50	Pepino mosaic virus (PepMV) Chilean (CH2) strain, mild isolate Abp2 (PEPMVO)	not available

(*continued*)

Table 1. (*continued*)

No.	Active Substance	Maximum Dose (kg of active substance/ha)
51	Pepino mosaic virus (PepMV) European (EU) strain, mild isolate Abp1 (PEPMVO)	not available
52	Pepino mosaic virus strain CH2 isolate 1906	not available
53	Phlebiopsis gigantea strain FOC PG 410.3	0.0072
54	Phlebiopsis gigantea strain VRA 1835	0.0072
55	Phlebiopsis gigantea strain VRA 1984	0.0072
56	Potassium hydrogen carbonate	15
57	Purpureocillium lilacinum PL 11	not available
58	Repellents by smell of animal or plant origin/fish oil	7.2
59	Repellents by smell of animal or plant origin/sheep fat	1.292
60	Saccharomyces cerevisiae strain LAS02	2.47
61	Sodium hydrogen carbonate (low risk active substance)	12
62	Spodoptera exigua multicapsid nucleopolyhedrovirus (SeMNPV), isolate BV-0004	0.105
63	Trichoderma atroviride AGR2	not available
64	Trichoderma atroviride AT10	not available
65	Trichoderma atroviride strain SC1	not available
66	Verticillium albo-atrum (formerly Verticillium dahliae) strain WCS850	not available
	Average	**1.29479**

Source: EU Pesticides Database (2023) https://food.ec.europa.eu/plants/pesticides/eu-pesticides-database_en Accessed in October 2023

The source of data for the analysis was EU Pesticides Database where official data regarding active substances of plant protection products used in the European Union are available. If more than one review report was available in the EU Pesticide Database, the most recent data were considered.

2.3 The Recommended Doses of Low-Risk Active Substances – Discussion

In the autumn 2023 there were 66 low-risk active substances approved for use in European Union. Among them were microorganisms (including viruses), a group of straight chain *Lepidopteran* pheromones, repellents by smell (such as sheep fat or fish oil) and some other substances.

It is worth stressing that the data on maximum recommended dose of active substance per hectare were not available in all cases in spite of the fact that it is the legal obligation. The data were only available for 56 out of 66 (84.8%) low-risk active substances approved for use in the EU in October 2023. This very likely follows the fact that some review reports were prepared earlier, when the current rules were not yet in force. The low-risk active substances for which the data on maximum recommended dose per hectare were not available were all microorganisms (including three strains of Pepino mosaic virus).

On analysing the Table 1 we may observe very significant differences among maximum recommended doses of low-risk active substances. The lowest maximum recommended dose was noted for two mild isolates of Pepino mosaic virus –0.00002 kg/ha. The highest maximum recommended dose was for blood meal (a substance of animal origin used as a repellent), 19.96 kg/ha. The difference between the lowest and the highest maximum recommended dose in the group of low-risk active substances was therefore six orders of magnitude. The average maximum recommended dose was 1.29 kg/ha. The maximum dose recommended most often in the case of low-risk active substances was 0.025 kg/ha. Such maximum dose was recommended for 24 low-risk active substances, many of them belonging to the group straight chain *Lepidopteran* pheromones. It is also worth noting that the maximum dose exceeding 1 kg of active substance per hectare was in fact recommended for minority of low-risk active substances, namely for 12 out of 66, in spite of the fact that average maximum recommended dose for the whole group was 1.29 kg/ha. It may illustrate the fact that simple statistical calculations, may give a bit biased picture in case of such a diverse group.

It's worth stressing that the author decided to draw the average among such different values as the volumes of very different active substances are added for statistical purposes in practice and the results are used for quite serious calculations. The results of those calculations are then used in decision-making processes regarding agriculture.

3 Conclusions

There is a lot of focus on more environmentally friendly methods in different areas of economics [18–20]. The focus is also visible in the studies on the products used in plant production [21, 22].

The results presented in the paper show that there are very significant differences among recommended doses of low-risk active substances, as high as six orders of magnitude. The statistical calculations based on data of such diversity may lead to biased conclusions. In more general terms, the amount of active substance of plant protection applied per hectare may not be an ideal indicator in case of comparisons between varied active substances and should be not treated as indicator of environ-mental impact of plant protection products.

References

1. Veres, A., Wyckhuys, K.A.G., Kiss, J., et al.: An update of the Worldwide Integrated Assessment (WIA) on systemic pesticides. Part 4: alternatives in major cropping systems. Environ. Sci. Pollut. Res. **27**, 29867–29899 (2020). https://doi.org/10.1007/s11356-020-09279-x
2. Wysokińska-Senkus, A., Górna, J., Kaźmierczak, M., Mielcarek, P., Senkus, P.: CARVER+Shock and Business Process Management in Improving Food Safety of Primary Production Agriculture (Switzerland), 2022, **12**(7), 1018. Springer, Heidelberg (2016)

3. Directive 2009/128/EC of the European Parliament and of the Council of 21 October 2009 establishing a framework for Community action to achieve the sustainable use of pesticides OJ L 309, pp. 71–86 24 Nov 2009
4. Matyjaszczyk, E., Skrzecz, I.: How European Union accession and implementation of obligatory integrated pest management influenced forest protection against diseases and weeds: a case study from Poland. Crop Protect. **127** (2020). https://doi.org/10.1016/j.cropro.2019.104986.
5. Matyjaszczyk, E.: Selected aspects of plant protection in Poland, five years on from EU accession. Outlook Agric. **40**(2), 119–123 (2011). https://doi.org/10.5367/oa.2011.0042
6. Feldmann, F., Pandey, A.K., Rajabpour, A., et al.: Botanical active substances: a prospering field of research. J. Plant. Dis. Prot. **130**, 439–441 (2023). https://doi.org/10.1007/s41348-023-00739-3
7. Matyjaszczyk, E.: Legislative situation of botanicals used in plant protection in the European Union. J. Plant Dis. Prot. **130**, 443–447 (2023). https://doi.org/10.1007/s41348-023-00724-w
8. Le Gall, C., Hidrot, D., Girolami, P., Marchand, P.: Seeds and plant protection substances for French organic oilseed crops: what are the challenges for tomorrow? OCL **31**, 8 (2024). https://doi.org/10.1051/ocl/2024002
9. Marakis, G., Katsioulis, A., Kontopoulou, L., et al.: Knowledge, attitude and behaviour of university students regarding salt and iodine: a multicentre cross-sectional study in six countries in Europe and Asia. Arch. Public Health **79**, 68 (2021). https://doi.org/10.1186/s13690-021-00593-5
10. Matyjaszczyk, E., Schumann, R.: Risk assessment of white willow (Salix alba) in food. EFSA J. **16**(S1), e16081, 9 pp (2018). The European Food Risk Assessment Fellowship Programme, Series 1, 2017–2018 (2018). https://doi.org/10.2903/j.efsa.2018.e16081
11. Matyjaszczyk, E., Schumann, R.: Cadmium contamination in food supplements containing white willow (Salix alba) bark. J. Consum. Prot. Food Saf. **14**, 179–182 (2019). https://doi.org/10.1007/s00003-018-1199-0
12. European Parliament 'Chemicals and pesticides'. https://www.europarl.europa.eu/factsheets/en/sheet/78/chemicals-and-pesticides. Accessed 15 May 2024
13. European Commission Communication from the Commission to the European Parliament, the European Council, the Council, the European Economic and Social Committee and the Committee of the Regions The European Green Deal, Brussels, (2019), 11 Dec 2019 COM(2019) 640 final https://eur-lex.europa.eu/resource.html?uri=cellar:b828d165-1c22-11ea-8c1f-01aa75ed71a1.0002.02/DOC_1&format=PDF
14. Matyjaszczyk, E.: Is the volume of pesticides consumed in agriculture a good indicator of environmental concerns? Environ. Sci. Pollut. Res. (2024). https://doi.org/10.1007/s11356-024-33365-z
15. Regulation (EC) No 1107/2009 of the European Parliament and of the Council of 21 October 2009 concerning the placing of plant protection products on the market and repealing Council Directives 79/117/EEC and 91/414/EEC OJ L 309, 24 Nov 2009
16. European Commission 2024 Low Risk Plant Protection Products https://food.ec.europa.eu/system/files/2017-09/pesticides_sup_low-risk-ppps.pdf. Accessed 15 May 2024
17. Matyjaszczyk, E.: Comparison between seed and foliar treatment as a tool in integrated pest management. J. Agric. Food Chem. **65**(30), 6081–6086 (2017). https://doi.org/10.1021/acs.jafc.7b01095
18. Łącka, I.: The role of green energy in the economic growth of the world. In: Bąk, I., Cheba, K., (eds.) Green Energy. Meta-analysis of the Research Results. Springer, pp. 41–57, (2023). https://link.springer.com/book/https://doi.org/10.1007/978-3-031-12531-7
19. Supron, B., Łącka, I.: Research on the Relationship between CO2 Emissions, Road Transport, Economic Growth and Energy Consumption on the Example of the Visegrad Group Countries. Energies **16**(3), 1340, (2023). https://doi.org/10.3390/en16031340

20. Matuszak-Flejszman, A., Hadryjańska, B., Górna, J. Changes in environmental management after Poland's entry to the EU - the example of the dairy sector. Polish J. Environ. Stud. **22**(3), 781–791 (2013)
21. Rys, M., Miastkowska, M., Łętocha, A., Wajs – Bonikowska A., Lorenzo P., Synowiec, A.: The effect of caraway oil-loaded bio-nanoemulsions on the growth and performance of barnyard grass and maize. Sci. Rep. **14**, 4313 (2024). https://doi.org/10.1038/s41598-024-54721-6
22. Klocke, B., Sommerfeldt, N., Wagner, C., et al.: Disease threshold-based fungicide applications: potential of multi-disease resistance in winter wheat cultivars in Germany. Eur. J. Plant Pathol. **165**, 363–383 (2023). https://doi.org/10.1007/s10658-022-02611-w

Field Traffic-Induced Compaction Effects on Physical Soil Properties, Plant Vegetation Index and Crop Yield on a Chernozem Soil

Gerhard Moitzi[1(✉)], Paul Riedl[1], Matthias Konzett[2], Thomas Weninger[2], Gernot Bodner[3], and Helmut Wagentristl[1]

[1] Department of Crop Sciences, University of Natural Resources and Life Sciences (BOKU), Experimental Farm Groß-Enzersdorf, Schloßhofer Straße 31, 2301 Groß-Enzersdorf, Austria
gerhard.moitzi@boku.ac.at

[2] Federal Agency for Water Management, Institute for Land and Water Management Research Petzenkirchen, Pollnbergstraße 1, 3252 Petzenkirchen, Austria

[3] Department of Crop Sciences, Institute of Agronomy, University of Natural Resources and Life Sciences (BOKU), Konrad Lorenz-Straße 24, 3430 Tulln an Der Donau, Austria

Abstract. Soil compaction induced by in-field traffic affects key soil functions and plant growth. In the context of the SoilCompaC project (EJP SOIL), a three factorial field experiment in Lower Austria was established to investigate the effects of field traffic (slurry tanker-tractor combination wheeling with different loads in spring), soil tillage for cover crop (ploughing, cultivating, no-tillage) and cover crop m (deep rooted, shallow rooted, no cover crop) on soil physical and agronomic parameter. NDVI (Normalized Difference Vegetation Index) were calculated using multispectral imagery mounted on an unmanned aerial vehicle (hexacopter). The measured indicators (track depth, soil penetration resistance, bulk/packing density, crop yield) detect soil compaction well. Soil compaction monitoring via multispectral imagery mounted hexacopter is sensitive and could be a non-destructive method monitoring in-field compaction during vegetation period.

This one-year field experiment shows that the testing factor wheel load affects the measured parameters more than the factors tillage for cover crop and cover-crop mixture. The strong multi-passing effect on soil compaction and above-ground biomass yield (–27%), indicated that Controlled-Traffic Farming (CTF) is a suitable technical solution to separate traffic lines from plant cropping areas in the field. Physical soil protection with its technical und agronomic measurements is a keystone in knowledge-based cropping (smart farming).

Keywords: Wheel load · Cover crop · Tillage · Track depth · Soil penetration resistance · Bulk density · NDVI · Crop yield · Pannonian climate

1 Introduction

Field traffic is responsible for soil compaction with severe impacts on the soil functioning and negative consequences on crop production. Soil compaction is a serious global problem and is a major cause of inadequate rooting and poor yield in crops around the

E. Lorencowicz et al. (Eds.): FMPMSA 2024, LNCE 609, pp. 308–322, 2024.
https://doi.org/10.1007/978-3-031-70955-5_35

world [1]. The published report of the mission board for soil health and food, "Caring for soil is caring for life" [2] estimates that the area of land failing soil health due to compaction is 23–33%.

Agricultural soils have a wide diversity of ecological services (e.g., root growth, water movement, aeration) that are directly influenced by the physical properties of soils [3]. Farm vehicles should not impose stresses that exceed soil strength. Heavy agricultural vehicles passing on low load-bearing and soft soil (loosened, wet) can generate tire tracks in the field with the potential risk of subsoil compaction. Studies by Keller et al. [4] stated that future agricultural operations must consider the inherent mechanical limits of soil due to the increasing average weight of farm machinery exceeding loads bearable by the ground. Tracks caused by field traffic are visible indicators of soil deformation/compaction in the field. Technical mitigation options are comprehensive discussed in Moitzi et al. [5].

Soil compaction has a major effect on the rooting environment. Especially the root-restricting soil layers have a direct impact on crop yield. More than half of Germany's agricultural land reveals root-restricting soil layers with a direct impact on crop yield, provoking considerable potential in the melioration of affected sites [6]. Therefore, it is of considerable importance that the further spread of root-restricting, compacted soil layers should be prevented.

1.1 Soil Pore Surface and Pore Size Distribution

Soil pores are responsible for the ecological soil functions. The inner pore-surface is mainly contributed by the fine pores (Table 1), which are mainly responsible for chemical and physical soil processes. The total inner pore-surface in a fertile arable soil (0–40 cm) is more than one million hectare per hectare arable land!

Table 1. Pore volume und inner pore-surface in a Chernozem soil (silty loam) at the experimental site Gross Enzersdorf (East of Austria). Dry bulk density: 1.2 t/m^{-3} in 0–20 cm and 1.4 t m^{-3} in 20–40 cm).

	Pore volume (%)		Inner pore-surface (ha t^{-1} dry soil)	
	0–20 cm	20–40 cm	0–20 cm	20–40 cm
Large pores – wide (>50 μm)	20	12	1.1	0.7
Large pores - small (50–10 μm)	7	6	0.7	0.6
Medium pores (10–0.2 μm)	16	14	9.0	7.8
Fine pores (<0.2 μm)	13	14	185.6	214.2
Total	**56**	**47**	**196.4**	**223.3**

Physical soil degradation changes physical soil functions with consequences for ecological properties (infiltration, aeration, water storage, root growth, water and nutrient uptake, microbial activity, crop yield and soil erosion). Farm machinery induced soil stress results in soil deformation with rearrangement of particles and pores. The generated platy soil results in anisotropy of pore functions. The soil deformation caused by in-field traffic changes mainly the pore size distribution. A large pore is changed to many small pores without a significant change of the total pore volume. Soil compaction changes the pore size dramatically with significant effect on flux rate (Fig. 1) – calculated by the Hagen-Poiseuille-equation.

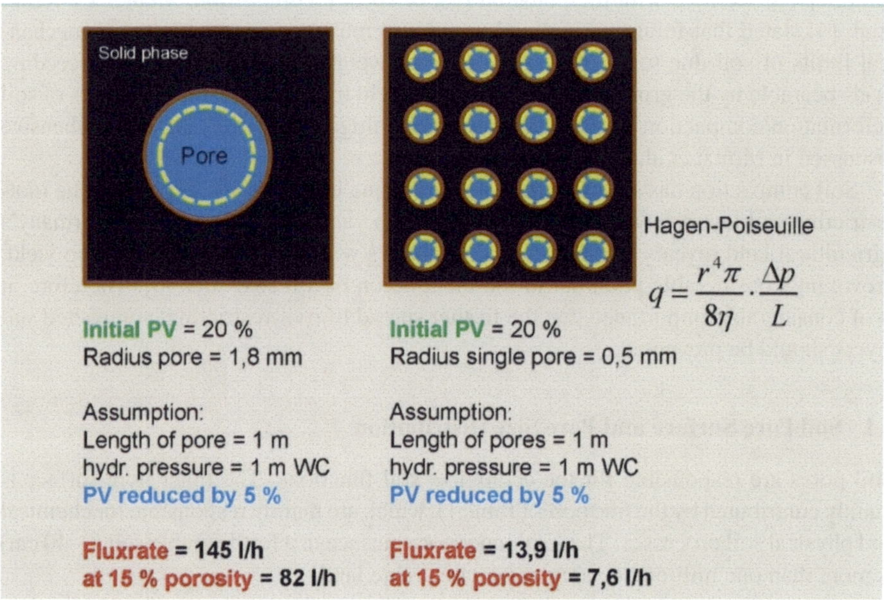

Fig. 1. Effect of pore space dynamics on hydraulic conductivity [7].

1.2 Plough Pan Formation

In conventional ploughing the furrow-wheels are running on the furrow-ground, with pressure effects to the ground soil and formation of a plough pan. Conventional ploughing with a 2 × 3 mouldboard plough (working width: 99 cm) requires 53% of the furrow area for where the furrow wheels, with a maximum wheel load of 1511 daN (=1540 kg) on sandy loam, are running (Fig. 2). The increased working width reduces this area to 20% but with increased load of the rear furrow wheel for the 2 × 5 plough (working width: 250 cm) with 4097 daN (=4176 kg) and 2 x 7 plough (working width: 350 cm) with 4522 daN (=4610 kg). Mouldboard ploughs with traction control are equipped with an additional hydraulic cylinder to increase the axle load of the rear wheel. This is due to the dynamic load transfer from the plough and front axle of the tractor. Mouldboard ploughs equipped with traction control improve traction through lowering slip. The dynamic furrow wheel load also increases. For preventing the risk of subsoil compaction on the 20% of the field through the furrow wheeling, on-land ploughing is a recommended as a technical solution. On-land ploughing also allows the usage of wider tyres or dual tyres, which is not limited by the furrow width as it is given in conventional ploughing within furrows.

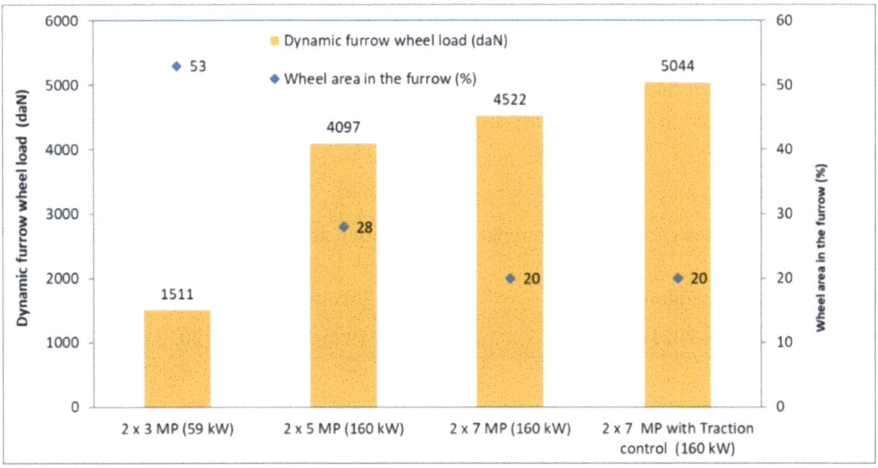

Fig. 2. Rear dynamic furrow wheel load on sandy loam for three mouldboard ploughs with 4-WD tractor [8].

1.3 UAV Measurements

The decrease in soil pore size, as well as the formation of a plough pan, is hypothesized to affect plant physiological parameters such as plant height or chlorophyll content in the leaves. Using multispectral imagery mounted on unmanned aerial vehicles (UAV), these plant physiological parameters can be observed on a field scale. Using the various bands as multispectral image offers, it is possible to calculate vegetation indices, which allow the assessment of plant vitality, stress, chlorophyll content, or overall biomass.

1.4 Objective of the Field Experiment

In the context of the SoilCompaC Project (EJP SOIL), a three-factorial field experiment at the Experimental Farm Groß-Enzersdorf was established in the year 2022 to investigate the effects of field traffic, soil tillage (ploughing, cultivating, no-tillage) for cover crop and cover crop mixture (deep rooted, shallow rooted, no cover crop) on agronomic and soil physical parameters. In this study, the results of the soil physical parameters and the crop results of the year 2023 are presented and the UAV measurements results of the year 2022. Soil physical results and crop results of the year 2022 are published in Moitzi et al. [9].

2 Materials and Methods

2.1 Field Experiment

The field experiment was carried out at the Experimental Farm Groß-Enzersdorf on a Calcaric Chernozem in Eastern Austria. The soil (0–30 cm) is a silty loam with 19.9% clay, 44.1% silt, and 36.1% sand and soil organic carbon of 16.3 g kg^{-1}. The year 2022 was very dry with a mean annual precipitation of 392 mm and temperature of 11.9 °C. The year 2023 was characterized by high rainfall in the months April, May and June (Table 2).

Table 2. Mean temperature and mean precipitation at the experimental site [10]

Month	Temperature (°C)			Precipitation (mm)		
	1996–2021	2022	2023	1996–2021	2022	2023
I	0.5	2.9	3.7	27	25	20
II	2.3	5.9	3.9	21	14	13
III	5.8	5	7.4	31	14	5
IV	11.1	9.6	8.9	39	27	106
V	15.8	17.6	14.9	67	31	80
VI	19.9	21.7	19.6	70	89	103
VII	21.6	22.5	23.3	73	37	22
VIII	20.9	22.2	21.4	71	45	90
IX	15.9	15.2	19.3	62	49	28
X	10.7	12.2	13.9	40	7	52
XI	6.0	6.3	6.8	34	26	77
XII	1.5	2.1	3.6	32	28	89
I – XII	**10.9**	**11.9**	**12.2**	**566**	**392**	**685**

In the three-factorial design (Table 3), cover crop mixture and load were nested in tillage of the cover crop. The plot size for a tillage treatment was 30 m × 27 m and for cover crop within tillage of the cover crop 30 m × 9 m. The total field trial size was 7290 m^2, with three replicates for each factor.

After the harvest of the previous crop winter wheat, soil tillage (ploughing, cultivating, harrowing) was carried out at the end of July. Cover crop seeding was done on the 27th of July 2022 with a seed drill. Wheeling with a tractor-slurry tanker combination (10 m^3, pendulum tandem axle) was done on the 24th of April 2023 with three load levels: empty slurry tanker (low load), total filled slurry tanker (high load), double-wheeled with total filled slurry tanker (high load – double wheeled). Tire inflation pressure was fixed for the tractor wheels with 300 kPa and for the slurry tanker wheels with 400 kPa. The average soil moisture content was 15.9%. After passing the plots with the slurry

Table 3. Testing factors

Cover crop	No Cover crop (**No CC**)	Deep rooted cover crop mixture[1)] (**CCd**)	Shallow rooted cover crop mixture[2)] (**CCs**)
Tillage– Cover crop	Mouldboard plough (25 cm) – (**MP**)	Wing sweep cultivator (25 cm) – (**CT**)	No tillage – (**NT**)
Load	**Low** (TM[3)]: 10030 kg WL_{max}: 22 kN)	**High** (TM[3)]: 19540 kg, WL_{max}[4)]: 30 kN)	**High-double wheeled** (TM[3)]: 19540 kg, WL_{max}[4)]: 30 kN)

[1)] fodder radish – *Raphanus sativus L. oleifromis Pers.*, safflower thistle – *Carthamus tinctorius*), ramtil – *Guizotia abyssinica; 33* kg ha^{-1}; [2)] pea–*Pisum sativum*), flaxseed – *Linum usitatissimus L.*, phacelia – *Phacelia tanacetifolia*; 45 kg ha^{-1}, [3)] total mass, [4)] maximum wheel load

tanker-tractor combination, the track depth as a vertical distance between the middle tire print and the horizontal rigid slat was measured with a ruler. Maize (cv. P9610, FAO 370) was seeded (without tillage before!) directly on the 26th of April 2023 in tracks (generated by the tractor slurry tanker combination) and on the area with no tracks using a precision seeder in a row distance of 75 cm. N-fertilization was carried out with 100 kg N ha^{-1} as calcium ammonium nitrate in end of May. The measured parameters during the vegetation period of maize are shown in Table 4.

Table 4. Measured parameters in the vegetation period 2023

Parameter	Method	Date
Track depth (cm)	Measuring the vertical distance between the middle tire print and the horizontal rigid slat with a ruler (6 values/treatment)	25th April
Soil penetration resistance (MPa)	Cone penetrometer (Eijkelkamp, The Netherlands) (60° cone and 1 cm^2), 5 penetrations per treatment	26th April
Bulk density (g cm^{-3})	Ring cylinder (203.6 cm^3) in soil depths 0–10 cm, 10–20 cm, 20–30 cm, 30–40 cm; 3 replicates/soil depth	28th April 25th May
Grain maize yield (kg ha^{-1})	Cutting the whole maize plants of 2 m rows in each treatment manually and separating and drying the kernels	26th September

The measured bulk density was used for the calculation of packing density (PD) after Renger et al. [11]:

$$PD = \text{Bulk density} + 0.005 * \text{clay} + 0 : 001 * \text{silt}$$

where both PD and Bulk density are given in g cm^{-3}, and clay and silt contents are given in mass-%. PD describes the combined effect of soil texture and bulk density on the degree of soil compaction [11].

2.2 UAV Measurements

UAV surveys were performed using a DJI M600 hexacopter on which a MicaSense RedEdge-M multispectral camera was mounted. The camera acquired data in five spectral bands, including visible bands (red, green, and blue), red-edge band, and near-infrared band. The central wavelength (wavelength widths) of the bands are 475 nm (20 nm), 560 nm (20 nm), 668 nm (10 nm), 717 nm (10 nm), and 840 nm (40 nm), respectively. Before and after each flight, camera calibration images were taken using a calibration panel provided by the camera manufacturer.

Between the 28th of April 2022 (right after planting) and the 14th of September 2022, 7 flights were performed, whereas the flight timing was focused on the earlier stages of plant development before they were fully withered. Each flight was performed at an altitude of 45 m above ground, which resulted in a ground sampling distance of 3 cm. Forward camera overlap was set to 80% and side overlap to 85%. The autonomous flight was performed according to the planned route using the software UgCS (Version: 4.0.187; www.sphengineering.com).

The data was processed using Agisoft Metashape (Version 2.0; www.agisoft.com). The tie point cloud was calculated after image calibration using the images taken of the calibration panel. The tie point cloud was filtered iteratively by deleting points with a reconstruction uncertainty higher than 10 and a projection accuracy greater than 3. Then, markers were placed and georeferenced, followed by the last filtering step where points with a reprojection error higher than 0.3 were deleted. After that, the dense point cloud, digital elevation model (DEM), and orthomosaics (OM) were calculated. DEM and OM were exported with a cell size of 10 cm.

A Python script (Version: 3.9.13) was used to calculate 17 different vegetation indices, including NDVI (normalized difference vegetation index), CG (chlorophyll green), and NDRE (normalized difference red edge). Additionally, a DEM of Difference (DoD) was calculated for each flight. As a reference model for the DoD, the elevation data of the first flight was used.

For the analysis, 5 plots per compaction degree (no compaction, compacted) were defined in the field. Each plot was approximately 1 m^2 in size and was measured using a Topcon GPS. The cell values of each vegetation index and DoD within the boundaries of each plot were then extracted and used for further analysis.

2.3 Statistical Analysis

All analyses were conducted using IBM® SPSS® Statistics 21. The requirements for analysis of variance were tested with the Levene test and assessment of normal distribution of residues. Three-factorial analysis of variance (ANOVA) was carried out for the measured parameters and a multiple comparison test (Student-Newman-Keuls procedure ($p < 0.05$)) to separate means.

3 Results and Discussion

3.1 Track Depth

Overall, the track depth was significantly influenced by the load of the slurry-tanker-tractor combination. The highest mean track depth was generated by the high-double wheeled load with 8.12 cm, followed by the high load treatment with 7.48 and low load treatment with 5.81 cm.

The overall tillage effect showed the significant lowest track depth with 5.95 cm for NT. CT and MP with 7.49 cm and 7.97 cm, respectively, did not differ signignificantly.

The track depth was not influenced significantly by the cover crop: No CC (7.44 cm) = CCd (6.93 cm) = CCs (7.04 cm).

The separated means for the three factors of the experimental design are shown in Table 5.

Table 5. Mean tire track depth (cm) affected by tillage of cover crop (MP = Mouldboard Plough, CT = Cultivator, NT = No Till), cover crop mixture (No CC = No cover crop, CCd = deep rooted cover crop, CCs = shallow rooted cover crops) and wheel load (High-double wheeled, High and Low)

Load	MP				CT				NT			
	No CC	CCd	CCs	Mean	No CC	CCd	CCs	Mean	No CC	CCd	CCs	Mean
High-double wheeled	8.74	9.22	8.83	8.93^{b}	8.99	8.77	8.51	8.76^{c}	7.03	6.51	6.51	6.69^{b}
High	7.71	7.31	7.34	7.46^{a}	8.08	7.77	7.65	7.83^{b}	7.28	6.70	6.70	7.14^{b}
Low	9.28	6.44	6.84	7.52^{a}	6.11	5.44	6.09	5.88^{a}	3.74	4.24	4.24	4.04^{a}
Mean	**8.58**	**7.66**	**7.67**	**7.97^{B}**	**7.73**	**7.33**	**7.42**	**7.49^{B}**	**6.02**	**5.82**	**6.03**	**5.95^{A}**

Statistically significant differences ($p < 0.05$) are shown for the tillage effect with capital letters and for the wheel-load effect with small letters.

3.2 Soil Penetration Resistance

The soil deformation in the tire track is also linked with higher soil penetration resistance (Table 6 and Fig. 3). A high load in the track results is significant higher soil penetration resistance than in un-wheeled area. Till about 16 cm soil depth, there was not significant difference between high-load and high-double wheeled treatment. A significant differentiation between high-load treatment and high-double wheeled treatment was below 16 cm, indicating that multiple traffic compacted more the subsoil than topsoil.

A cover crop effect was not found. A possible explanation is the mitigated soil loosening effect by the lower biomass production (about 4900 kg DM/ha above groundbiomass) of the cover crops due to precipitation shortage till end of November.

A tillage effect was discovered. MP revealed significant highest soil penetration resistance in the soil layer 0–5 cm, which could be explained by a natural re-compaction of the loosened soil after power harrowing. The combination of soil-overloosening in conventional tillage with MP and power harrow with soil texture (silty loam) and random heavy rainfall is responsible for this effect.

MP, showed from 6–40 cm soil depth the lowest soil penetration (Table 4).

Table 6. Mean soil penetration resistance (MPa) affected by tillage, cover crop and load. Date: 26th April 2023, Mean soil moisture content: 16.7% (0–10 cm), 16,3% (10–20 cm), 17.7% (20–30 cm), 18,4% (30–40 cm), cone penetrometer (Eijkelkamp, The Netherlands) (60° cone and 1 cm^2), 5 penetrations per plot.

	0–5 cm	6–10 cm	11–15 cm	16–20 cm	21–25 cm	26–30 cm	31–35 cm	36–40 cm	**0–40 cm**
Tillage n = 135									
MP	1.46^b	2.12^a	2.67^a	3.10^a	3.42^a	3.82^a	4.28^a	4.58^a	3.18^a
CT	1.32^a	2.10^a	2.75^a	3.31^b	3.73^b	4.15^b	4.62^b	4.97^b	3.37^b
NT	1.31^a	2.26^b	2.88^b	3.32^b	3.60^b	3.88^b	4.22^a	4.44^a	3.24^a
Cover crop n = 135									
CCs	1.35	2.15	2.75	3.23	3.56	3.93	4.32	4.57	3.23
CCd	1.41	2.21	2.83	3.30	3.63	3.97	4.39	4.74	3.31
No CC	1.32	2.13	2.72	3.21	3.56	3.97	4.42	4.68	3.25
Load n = 135									
Un-wheeled	0.68^a	1.15^a	1.53^a	1.98^a	2.43^a	2.88^a	3.32^a	3.66^a	2.21^a
High	1.73^b	2.62^b	3.36^b	3.77^b	4.23^b	4.13^b	4.39^b	4.60^b	3.57^b
High-double	1.67^b	2.71^b	3.42^b	3.99^c	4.38^c	4.85^c	5.41^c	5.74^c	4.02^c

Statistically significant differences ($p < 0.05$) are shown for the tillage and load effect with small letters.

Soil physical analysis from the same experimental design at the same field in the year 2022 showed significant effects of the field traffic on infiltration rate and air permeability in the subsoil (30–40 cm): Field-measured saturated hydraulic conductivity in the subsoil (30–40 cm) decreased by 42%. Laboratory-based results measured at −100 hPa revealed that compaction decreased soil gas transport by convection (air permeability) and diffusion (gas diffusivity) by 70% and 51%, respectively [9].

3.3 Soil Bulk Density

Separated means of packing density (PD) and the gravimetric soil moisture content are presented for the sampling date 28th April 2023 in Table 7 and for the sampling date 25th May 2023 in Table 8. The overall PD (n = 64) at the first sampling date was 1.58 g cm^{-3} and was significantly lower than the PD at the second sampling date with 1.64 g cm^{-3}.

At the first sampling date (28th April 2023), the packing density (PD) was significantly influenced by the factor load. Overall (n = 32) the un-wheeled treatment showed

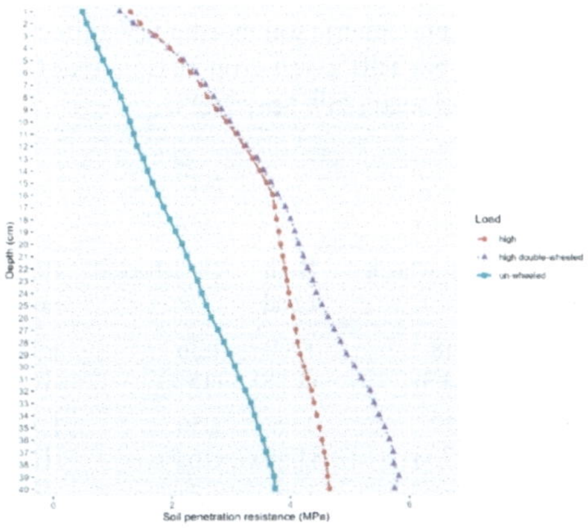

Fig. 3. Overall soil penetration resistance (n = 135) in different load treatments.

for the soil depth (0–40 cm) a mean PD of 1.48 g cm^{-3} and was much lower than the high-load treatment with 1.68 g cm^{-3}. The cover crop has no effect of PD: CCd = 1.57 g cm^{-3}, NoCC = 1.59 g cm^{-3}. NT resulted in higher PD (= 1.60 g cm^{-3}) than MP (= 1.56 g cm^{-3}) but the difference was not significant (p = 0.052).

Table 7. Packing density (g cm^{-3}) and gravimetric soil moisture (%) content affected by tillage of cover crop (MP = Mouldboard Plough, NT = No Till), cover crop mixture (No CC = No cover crop, CCd = deep rooted cover crop) and wheel load (28th April 2023)

Depth	MP				NT			
	CCd		No CC		CCd		No CC	
	Un-wheeled	High Load	Un-wheeled	High Load	Un-wheeled	High Load	Un-wheeled	High Load
0–10 cm	1.48 (13.4)	1.82 (14.4)	1.38 (16.3)	1.62 (15.8)	1.31 (15.3)	1.66 (17.1)	1.38 (15.8)	1.69 (16.9)
10–20 cm	1.34 (15.8)	1.72 (15.7)	1.49 (16.6)	1.60 (16.6)	1.49 (15.7)	1.70 (16.4)	1.52 (16.0)	1.76 (17.0)
20–30 cm	1.36 (15.9)	1.60 (17.1)	1.53 (15.6)	1.72 (15.9)	1.54 (15.4)	1.74 (16.4)	1.62 (15.7)	1.69 (16.8)
30–40 cm	1.45 (16.04)	1.63 (15.5)	1.54 (15.7)	1.63 (16.4)	1.62 (15.4)	1.65 (15.6)	1.63 (15.8)	1.64 (16.2)
0–40 cm	**1.41^a**	**1.69^b**	**1.49^a**	**1.64^b**	**1.49^a**	**1.69^b**	**1.54^a**	**1.69^b**

Also, at the second sampling date (25th May 2023), the packing density (PD) was significantly influenced by the factor load. Overall (n = 32) the un-wheeled treatment showed for the soil depth (0–40 cm) a mean PD of 1.54 g cm^{-3} and was much lower than the high-load treatment with 1.73 g cm^{-3}. The cover crop has no effect on PD: CCd = 1.62 g cm^{-3}, NoCC = 1.65 g cm^{-3}. Also, tillage has no effect on PD: NT = 1.65 g cm^{-3}), MP = 1.63 g cm^{-3}).

Table 8. Packing density (g cm^{-3}) gravimetric soil moisture (%) affected by tillage of cover crop (MP = Mouldboard Plough, NT = No Till), cover crop mixture (No CC = No cover crop, CCd = deep rooted cover crop) and wheel load (25th May 2023)

Depth	MP				NT			
	CCd		No CC		CCd		No CC	
	Un-wheeled	High Load	Un-wheeled	High Load	Un-wheeled	High Load	Un-wheeled	High Load
0–10 cm	1.46 (12.9)	1.73 (12.4)	1.48 (13.3)	1.87 (12.8)	1.36 (15.4)	1.71 (16.4)	1.40 (17.2)	1.76 (14.6)
10–20 cm	1.42 (14.6)	1.76 (15.1)	1.50 (17.1)	1.65 (17.9)	1.56 (16.2)	1.78 (15.8)	1.63 (16.4)	1.79 (16.4)
20–30 cm	1.51 (15.3)	1.71 (16.3)	1.55 (15.9)	1.75 (16.5)	1.64 (15.6)	1.79 (15.9)	1.54 (15.8)	1.78 (15.8)
30–40 cm	1.63 (15.2)	1.66 (15.3)	1.79 (15.4)	1.58 (16.5)	1.58 (16.0)	1.70 (16.2)	1.69 (16.0)	1.73 (16.4)
0–40 cm	**1.50^a^**	**1.71^b^**	**1.58^a^**	**1.71^b^**	**1.53^a^**	**1.74^b^**	**1.56^a^**	**1.76^b^**

3.4 Crop Yield

Separated means for grain of maize yield affected by tillage of cover crop, cover crop mixture and wheel load are shown in Table 9.

Table 9. Grain of maize yield[1] (kg ha^{-1}) affected by tillage of cover crop (MP = Mouldboard Plough, CT = Cultivator, NT = No Till), cover crop mix-ture (No CC = No cover crop, CCd = deep rooted cover crop, CCs = shallow rooted cover crops) and wheel load

Load	MP				CT				NT			
	No CC	CCd	CCs	Mean	No CC	CCd	CCs	Mean	No CC	CCd	CCs	**Mean**
Un-wheeled	4469	6184	5033	**5229^b^**	4827	5907	6730	**5821**	4720	5944	5110	**5258**
Low (empty)	4628	5858	5583	**5356^b^**	5979	7031	7008	**6673**	5353	5197	6007	**5519**
High load	4333	4579	4635	**4516^ab^**	5677	6159	5093	**5643**	4452	5323	5803	**5193**
High load (double)	3569	3674	3871	**3705^a^**	4385	6076	5729	**5397**	3044	4664	4536	**4081**
Mean	**4250**	**5074**	**4781**		**5217**	**6293**	**6140**		**4392**	**5282**	**5364**	
	4701^A^				**5883^B^**				**5013^A^**			

[1]Dry maize yield; Statistically significant differences ($p < 0.05$) are shown for the tillage effect with capital letters and for the wheel-load effect with small letters

Overall, the tillage effect (Table 9), the cover crop effect, and the load effect were significant. CCd with 5550 kg ha^{-1} and CCs with 5428 kg ha^{-1} were significantly higher than No CC with 4620 kg ha^{-1}. Also, the above-ground biomass was in the same order: CCd with 11023 kg ha^{-1} = CCs with 10867 kg ha^{-1} > No CC with 9604 kg ha^{-1}.

The load effect differentiated significantly for the grain of maize yield as followed: un-wheeled with 5436 kg ha^{-1} = low load with 5849 kg ha^{-1} > high load with 5117 kg ha^{-1} > high load double wheeled with 4394 kg ha^{-1}. The order for the above-ground biomass was: un-wheeled with 10844 kg ha^{-1} = low load with 11939 kg ha^{-1} = high load with 10572 kg ha^{-1} > high load double wheeled with 8637 kg ha^{-1}.

The results indicated that a low load treatment (=empty slurry tanker-tractor-combination with max. Wheel load of 22 kN) increased the crop yield by trend. This positive yield response is mainly explained by a plant-root suitable packing density (Table 7 and Table 8).

Crop yield depression due soil compaction begin in the high load treatment (=full slurry tanker-tractor-combination with max. Wheel load of 30 kN) and was highest in the high load double wheeled treatment with 24% for grain of maize yield and with 27% for the above biomass yield in comparison to the low load treatment.

3.5 NDVI – Mapping with Multispectral Camera

The year 2022, where the UAV-measurements were carried out, was an extremely dry year with 278 mm rainfall between April and September 2022 and was about 27% lower than in the average rainfall in the long term (1996–2021). This should be considered when interpreting and discussing the results.

The maximum NDVI values of 0.7 and 0.64 for no and high load levels, respectively, were recorded at the end of June (Fig. 4). Afterwards, NDVI values decreased and reached their lowest point on August 24th when withering was almost complete. On the 14th of September, the NDVI values slightly increased despite the plants being completely withered by then.

Plant growth was observed until August 9th, after then the UAV-measured plant height decreased continuously due to withering. The maximum median height of 1.89 m was recorded for plants growing on plots with no load level, while it was 1.68 m for plants on high load level plots.

Although the high load level negatively affected the NDVI and plant height measurements at almost all time points, the effects on both parameters were small and not statistically significant. However, this small differentiation of NDVI and plant height is corresponding with the load effect of the grain yield of maize [9].

We hypothesize that maize grown in compacted soil needs more energy and time for root prolongation, resulting in lower growth and vitality of the above-ground-biomass, which is indicated by lower NDVI values compared to uncompacted soil. This effect was observed in maize during the blooming stage around 28th July. Results from comparable experiments under dry conditions show, nevertheless, that rooting in compacted soil leads to better water uptake from smaller pores and improved penetration through plough pan.

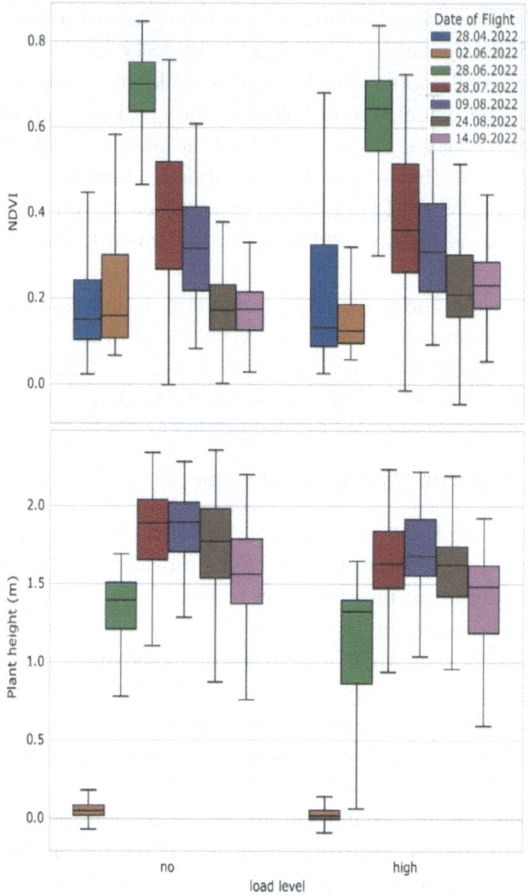

Fig. 4. NDVI and Plant height extracted from UAV data per load and date of flight. Date 28.04.2022 represent only the cover crops and bare. Other dates represent maize plants and bare.

4 Conclusions

Technical induced soil compaction impacts on key soil functions and is a result of tyre/soil interaction. Consequently, there are machinery variables (wheel load, wheel dimension, wheel type, inflation pressure) and soil management variables (crop rotation, tillage, time of operation) which are responsible for soil compaction.

The change of pore size distribution with its effect on hydraulic conductivity affects the soil functions significantly. There are many indicators which can detect soil compaction in the field. A visible indicator is the depth of track in the field as result of soil deformation. A higher wheel load generates deeper wheel tracks, which affects soil packing density and soil penetration resistance.

Layer-specific soil compaction indicators are the soil penetration resistance and bulk/packing density. The above-ground biomass production in different soil compaction sites can be also detected by monitoring plant physiological parameters (e.g. NDVI) and using multispectral imagery mounted on unmanned aerial vehicles (UAV) and crop yield detection at harvest.

Whereas the state of art indicators (track depth, soil penetration resistance, bulk/packing density, crop yield) detect the effects of technical soil compaction well, the novel UAV-measurements with multispectral cameras are less sensitive to detecting soil compaction. More research is needed for identifying the right spectral bands, vegetation indices and solutions for UAV-measurements.

The one-year field experiment shows that the testing factor wheel load affects the measured parameters more than the factors tillage for cover crop and cover-crop mixture.

The strong multi-passing effect on soil compaction and above-ground biomass yield (–27%) indicated that controlled-traffic farming (CTF) is a suitable solution to separate traffic lines from plant cropping areas in the field.

Another detected result was that one pass with an empty slurry-tanker tractor combination (max. Wheel load 22 kN) increased the crop yield moderately. This indicates that re-compacting after seeding for reaching an optimum bulk density is growth stimulation because of better root/soil interaction.

Physical soil protection with its technical und agronomic measurements is a keystone in knowledge-based cropping (smart farming).

Long-term field experiments are necessary to show the effectiveness of recovery techniques (mechanical subsoiling, biosubsoiling, natural recovery) of compacted (sub)-soil (e.g. long-term NT-fields).

Funding. EJP SOIL has received funding from the European Union's Horizon 2020 research and innovation program: Grant agreement No 862695.

References

1. Correa, J., Postma, J.A., Watt, M., Wojciechowski, T.: Soil compaction and the architectural plasticity of root systems. J. Exp. Botany **70**(21), S. 6019–6034 (2019)
2. European Commission: Caring for soil is caring for life – Ensure 75% of soils are healthy by 2030 for food, people, nature and climate, Brussels (2020)
3. Lamandé, M., Greve, M.H., Schjønning, P.: Risk assessment of soil compaction in Europe – Rubber tracks or wheels on machinery. CATENA **167**, 353–362 (2018)
4. Keller, T., Sandin, M., Colombi, T., Horn, R., Or, D.: Historical increase in agricultural machinery weights enhanced soil stress levels and adversely affected soil functioning. Soil Tillage Res. **194**, 104293 (2019)
5. Moitzi, G., Košutić, S., Kumhála, F., Nozdrovicky, L., Martinov, M., Gronauer, A.: Machinery induced compaction of agricultural soil and mitigation strategies in the Danube Region. [44th International Symposium "Actual Tasks on Agricultural Engineering", Opatija; CROATIA, 23–26 Feb 2016] In: Sveučilište u Zagrebu, Agronomski fakultet, Zavod za mehanizaciju poljoprivrede, Svetošimunska 25, 10000 Zagreb, Proceedings of the 44th International Symposium on Agricultural Engineering – Actual Tasks on Agricultural Engineering, http://atae.agr.hr; ISBN: 1848-4425 (2016)
6. Schneider, F., Don, A.: Root-restricting layers in German agricultural soils. Part I: extent and cause. Plant Soil **442**, 433–451 (2019)
7. Horn, R.: Soil deformation in agriculture and forestry- what are the consequences for hydraulic processes and how far can we forecast them in various landscapes. In: ERC-Symposium, 21–22 Oct 2014, Vienna (2014)

8. Moitzi, G., Wagentristl, H., Boxberger, J., Gronauer, A.: Energy consumption in cultiving and ploughing with traction improvement system and consideration of the rear furrow wheel-load in ploughing. Soil Tillage Res. **134**(2013), 56–60 (2013)
9. Moitzi, G., Arthur, E., Fouladidorhani, M., Lamandé, M., Wagentristl, H.: Effect of wheel load, soil tillage and cover crop on track depth, soil structural quality parameters and maize yield in Pannonian basin in 2022. ALVA-Jahrestagung 2023, Linz, AUSTRIA, 22.05.2023–23.05.2023. In: Arbeitsgemeinschaft für Lebensmittel, Veterinär- und Agrarwesen, „Innovativer Pflanzenschutz – neue Technologien zur Versorgungssicherung" Tagungsbericht 2023; ISBN: 2791-51, www.alfa.at, Linz, Austria (2023)
10. GeoSphere Austria: www.geosphere.at. Vienna, Austria (2024)
11. Renger, M., Bohne, K., Wessolek, G.: Bestimmung und Aussagemöglichkeiten der effektiven Lagerungsdichte. In: Wessolek, G., Kaupenjohann, M., Renger, M. (eds.) Bodenphysikalische Kennwerte und Berechnungsverfahren für die Praxis, Teil II. TU Berlin, Berlin (2014)

Influence of Worldview Factors on Food Consumers' Purchasing Decisions

Aneta Oleksy-Gębczyk[1](✉), Anna Szeląg-Sikora[2,3], Katarzyna Kowalska-Jarnot[4], Aleksandra Lis[3], Jakub Sikora[2,3], and Michał Cupiał[3]

[1] University of Applied Sciences, Nowy Sącz, Poland
aoleksy-gebczyk@ans-ns.edu.pl
[2] Cavalry Captain Witold Pilecki State University of Małopolska, Oświęcim, Poland
[3] University of Agriculture in Krakow, al. Adama Mickiewicza 21, 31-120 Krakow, Poland
[4] Department of Management and Accounting, School of Economics and Computer Science, Cracow, Poland

Abstract. In times of hyper-competition in the food market, monitoring and understanding the motives behind consumer decisions is becoming an important factor in building a company's competitive advantage. Today's poultry meat market is an interesting example of an area in which – in addition to classic factors influencing the purchase of food products such as price, quality or brand – ideological factors play a special role. This article and the research conducted for it aimed to investigate what factors, and in particular what beliefs and values, influence consumers' decisions related to the purchase of poultry meat. A questionnaire survey was conducted among 1180 poultry consumers nationwide. The results obtained can be applied and provide food producers with important information related to real consumer needs and concerns.

Keywords: consumer decision-making process · safe consumption · competitive advantages · sustainability

1 Introduction

The pursuit of sustainable agricultural production leads to the implementation of innovative technological solutions. The development of agricultural production is a process stimulated by the growing needs of consumers and access to modern technologies [1], but it should not be forgotten, this development should take into account the necessity of maintaining the principles of sustainable development in both plant and animal production [2, 3]. The food market is determined by consumer preferences, whose awareness of the environmental impact of agricultural production is constantly increasing [4].

Consumer behaviour in the food market is an ever-changing category, dependent on many internal (such as habits, values, income level, education, health, among others) and external factors (such as trends, availability, quality, brand, price and others) [5]. It

E. Lorencowicz et al. (Eds.): FMPMSA 2024, LNCE 609, pp. 323–332, 2024.
https://doi.org/10.1007/978-3-031-70955-5_36

requires continuous monitoring to meet the changing needs of buyers. The Covid-19 pandemic, widespread access to information and the increasing persuasive power of social media means that buyers are heavily influenced by consumer trend factors, including worldview factors. The arena where these influences are particularly visible is the meat market, including poultry meat. Meat consumption in Poland has a tradition. Polish cuisine has been based on this product for years and its consumption continues to increase. On the other hand, there are opposing trends related to an environmentally friendly approach, concerns about the impact of meat farming on the environment, health concerns and concerns about the quality of meat (the content of harmful chemicals in it, such as antibiotics or hormones. The market for meat substitutes and entomophagy is growing, although at the moment it does not yet threaten the position of meat in the shopping basket of the Polish consumer. A. Orkusz et al. [6] conducted an entomophagy study, unusual for the Polish consumer market. Data obtained from several survey sessions and tastings indicate a low willingness to use edible insects as a meat substitute among Polish schoolchildren due to psychological barriers such as neophobia and aversion.

In the context of the ideological decisions of purchasers, the study by A. Cegiełka [7] on the growing demand for environmentally friendly products. The authors proved the need for legislative approval of the "clean label" concept using the example of the meat processing industry. The dependence of Polish customers' behaviour on worldview factors was also studied by N. Drejerska et al. [8]. The authors concluded that the organic food market is undergoing evolutionary changes as the supply and demand sides are slowly tuning into each other. Food choices have also been influenced by the global concept of sustainability, as noted by A. Barska and J. Wojciechowska-Solis [9]. According to them, there is a growing belief that local food is the best and healthiest. The choice of where to buy food has also been influenced by the Covid-19 pandemic. Poles already shop a little differently and pay more attention to the safety of their purchases.

An important voice in these considerations belongs to A. Zielińska-Chmielewska et al. [10]. In her analysis of the future of the Polish food market in the coming years, she points to the likelihood of an inevitable decline in the price advantage of Polish producers, both with regard to raw materials and ready-to-eat products.

The above-mentioned studies point to changes looming on the horizon and the need to monitor changes in consumer behaviour. At the same time, there is a lack of up-to-date, post-covid research reviewing the influence of various factors, including worldview factors, on consumer behaviour in Poland. This article should fill this gap.

2 Research Methodology

The study is based on primary data. Using the author's survey questionnaire, 1180 consumers from different regions of Poland were interviewed using the CAWI (Computer Assisted Web Interviewing) method. The demographic composition of the respondents is shown in Fig. 1.[1]

The survey provided information on key parameters of ideological consumer behaviour. Respondents answered the following questions:

[1] For a detailed analysis of the respondent group, see the article 'Development of poultry meat production and consumption levels in the light of environmental sustainability goals'.

- What factors influence choice when buying meat?
- What factors could change your eating habits?
- Do your beliefs influence your poultry consumption?
- Do traditions or customs influence your decision to buy meat?
- Does the environmental impact of animal husbandry affect your decision to buy meat?
- Do the feelings of animals (their welfare) influence your decision to buy meat?
- Does the way animals are raised influence your decision to buy meat?
- Does the long-term health impact of meat influence the decision to buy it?
- Have you ever heard the statement that poultry meat is unhealthy? If so, where exactly?
- How do you personally assess the risk of poultry consumers being adversely affected by factors such as hormones, antibiotics, salmonella, various chemicals and GMOs (genetically modified organisms)?
- What sources of nutrition information do you consider reliable?

The data obtained were analysed, systematised and compared, allowing the identification of dominant patterns of poultry consumer behaviour.

3 Results

As a first step, respondents were asked whether worldview factors influence their poultry meat purchasing decisions. As many as 74% of respondents answered that they do not; only 26% of respondents answered in the affirmative. The distribution of responses to this question is presented in Fig. 1.

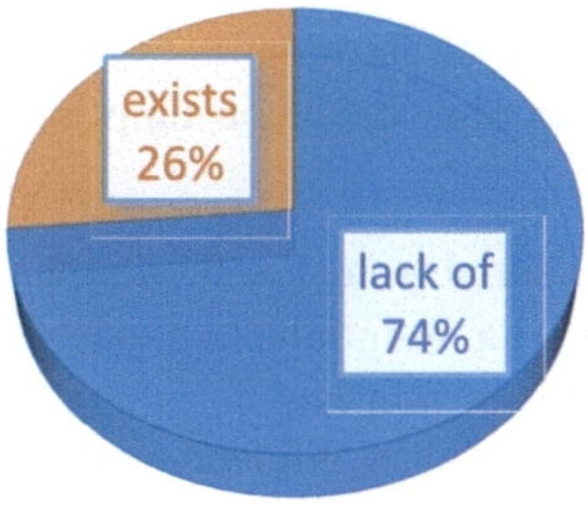

Fig. 1. Declared influence of worldview factors on decision to choose poultry meat

It is worth noting that both in the group declaring the influence of worldview factors on the decision to choose poultry meat and in the other group declaring that such factors do not influence the purchase decision, the vast majority of respondents eat poultry meat frequently, i.e. at least once a week (92% of 'ideologues' and 80% of 'non-ideologues').

Comparing the data on the influence of ideology on choice and frequency of consumption, it was found that representatives of both categories of consumers, those declaring that their worldview is related to their choice of poultry meat and those denying such influence, overwhelmingly consume this meat species at least once a week (Fig. 2).

The findings are supplemented by further responses. Consumers who believe that they are guided by ideological factors when choosing poultry meat were able to assess in detail the strength of the influence of individual factors on their decision.

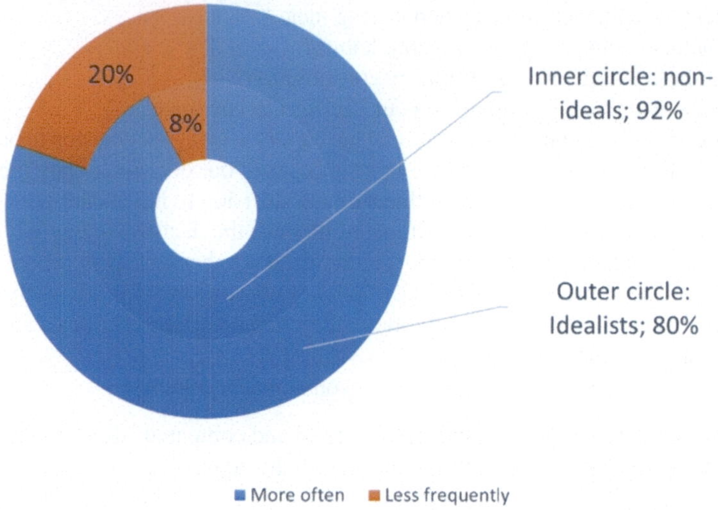

Fig. 2. Number of consumers in the 'ideologues' and 'non-ideologues' categories who eat poultry meat at least once a week

They were asked to assess the importance of the following subjective determinants of demand:

- Animal feelings
- Method of animal husbandry
- Environmental impact of livestock farming
- Long-term health effects of poultry meat consumption
- Tradition and customs

Each response option was scored from 1 to 5 (where 1 has no impact, 2 has a weak impact, 3 has a medium impact, 4 has a significant impact and 5 has a very strong impact).

The results of the study are shown in Fig. 3

The ideological factors most strongly influencing decisions to buy meat were found to be 'long-term health effects' (almost two-thirds of respondents rated the influence as at least significant) and 'farming method' (more than half of respondents rated the influence as at least significant). The lowest impact was attributed to the factor 'animal feelings'. One can venture to say that respondents' rate as more important those factors that have a significant direct impact on them.

In order to predict consumer behaviour in the future with greater accuracy, it is also necessary to know what factors consumers believe might change their attitudes. To this end, respondents were asked to rate the likelihood of abandoning poultry meat if certain factors were to change. A scale of 1–5 was used (where 1 means that a factor would not influence behaviour change and 5 means that it would definitely influence behaviour change). The results are shown in Fig. 4.

The most relevant factors likely to influence consumer behaviour change are 'health factors' and 'changes in the quality of poultry meat'. Respondents' concerns can also be

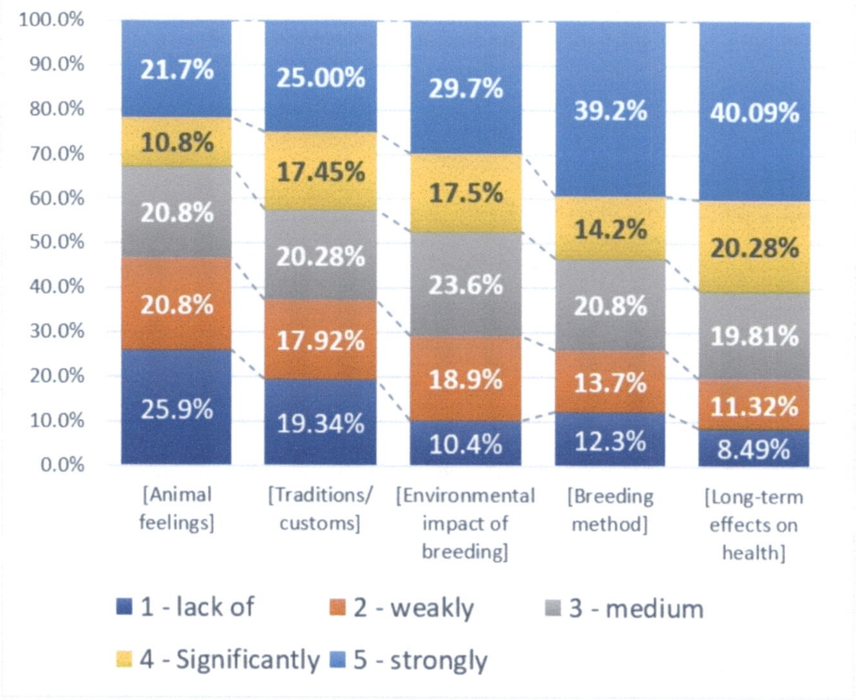

Fig. 3. Distribution of the impact of the factors Distribution of the impact of the determinants of demand for poultry of a worldview nature among respondents for whom it matters

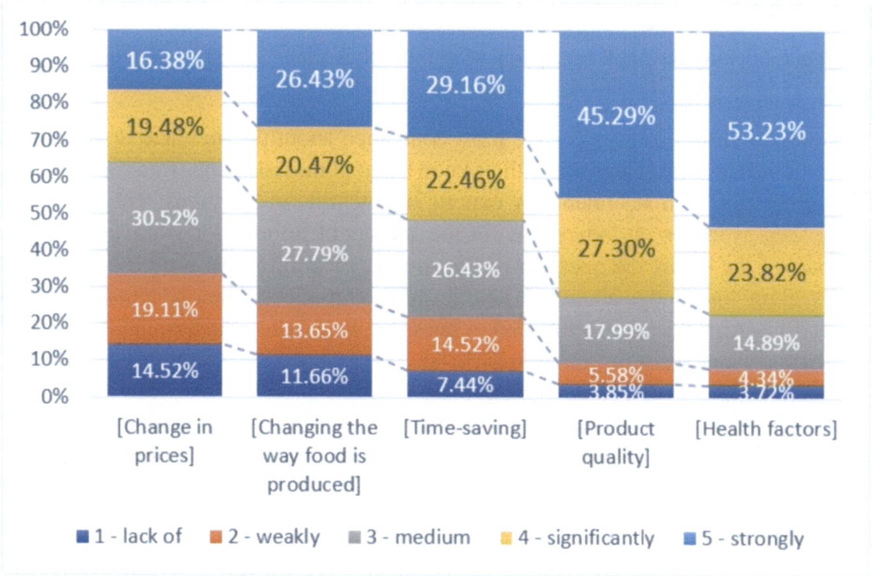

Fig. 4. Factors for changing eating habits according to respondents

considered as change factors. Respondents were asked to rate the risks associated with the consumption of poultry meat. The data are presented in Fig. 5.

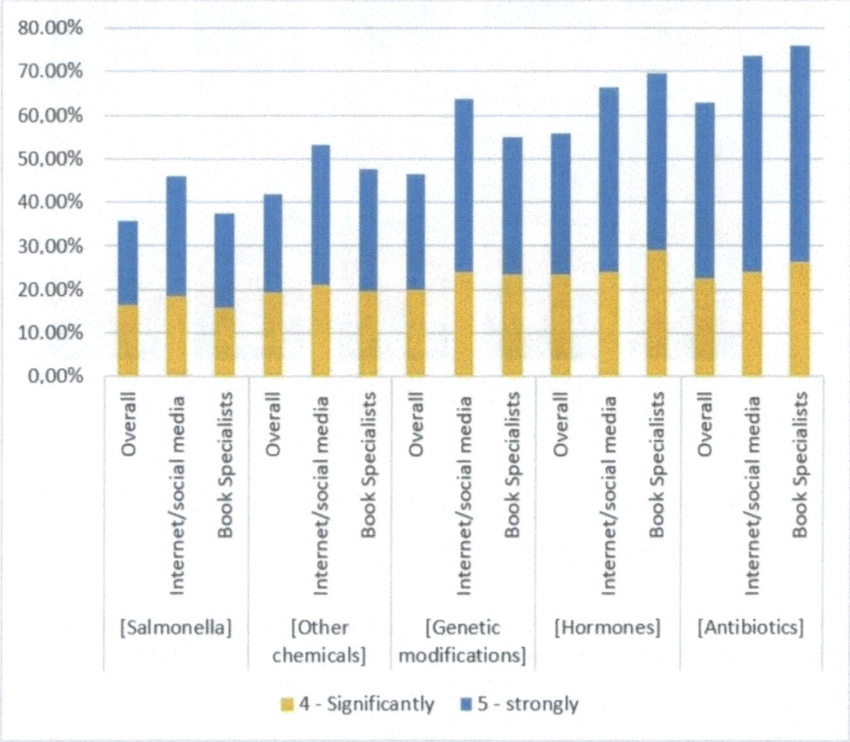

Fig. 5. Poultry meat risk perception. Most influential sources of information on risks

As Fig. 5 shows, consumers are most concerned about the presence of antibiotics and hormones in meat. They derive their knowledge of the risks both from social media, the internet and from statements by experts or scientific publications. These findings can be seen as important indications for the authors of information campaigns. In the next part of the study, the question of the sources of information used by consumers when seeking knowledge about the qualities of food products (their benefits/risks) was explored in greater depth. To this end, a question was asked to determine the level of trust for such sources of information as:

- Social media
- Friends and acquaintances
- Nutritionists
- Internet resources
- Literature
- Press.

The results of the study are shown in Fig. 6.

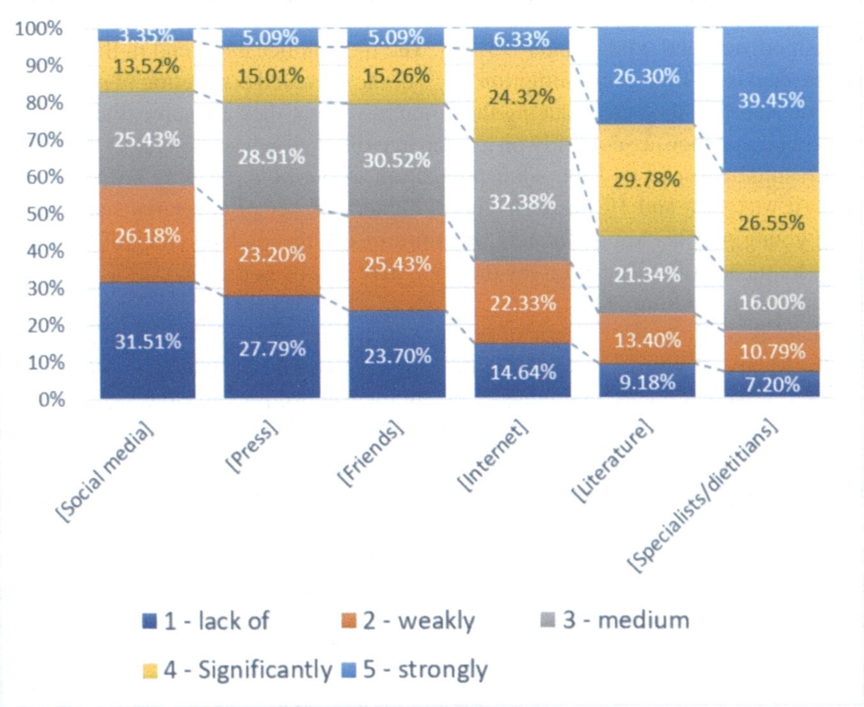

Fig. 6. Assessment of the reliability of different sources of information in % by source

According to the data obtained, respondents consider experts and literature to be reliable sources of information on the benefits or harms of food products, while social and traditional media are considered less reliable. However, as the answers to the previous question show, they derive their knowledge both from sources they judge as reliable and from sources they judge as unreliable.

4 Discussion

Constantly updating knowledge of consumer preferences is crucial for producers [4], because in the modern world, attention to the quality of the offer and an appropriate campaign informing about specific product characteristics can significantly influence consumer behaviour. Meanwhile, consumer behaviour is constantly influenced by both environmental and internal factors. Worldview factors such as environmental concerns, farming methods and animal welfare, among others, are beginning to play a particular role. As the study shows at this point, ideological factors concern a smaller group of consumers. Nevertheless, it can be assumed that their pressure will increase, possibly as a result of the increasingly loud debate on environmental protection, global warming, ecology and food awareness. A new generation is also bringing in new patterns of behaviour. The survey on the attitudes of young Polish women and men towards meat consumption was conducted using the CAWI (online survey) method in October 2021 on a sample of 1,300 people (reflecting the gender and age structure of the population of Polish citizens and women aged 15–29). The most noticeable trend in their attitude towards meat is to limit its consumption. 44% of the surveyed population of people aged 15–29 declare that they are reducing the amount of meat they consume, including 8% who declare that they are currently on a meat-free diet – vegetarian or vegan. Compared to the results of surveys of the adult population as a whole, such as last year's IQS survey (it was 43% and 1% respectively), a higher number of vegetarians are striking among young people. Approximately one third of our respondents, regardless of their current frequency of meat eating, say they will give up (4%) or reduce (27%) their meat consumption in the future [11].

Poland is the largest producer of poultry meat in the EU (53.8% of national production) Meat consumption is increasing. Cultural patterns (including Polish culinary tradition), as well as habits, undoubtedly favour meat consumption. However, it is worth bearing in mind that vegan and vegetarian alternatives to meat are rapidly gaining popularity. According to research, an area such as humane and less environmentally damaging ways of raising animals matters to consumer 'ideologues'. In the aforementioned Atlas Mięsa study, almost 50 per cent of young Poles are in favour of improving animal husbandry conditions, animal welfare and even reducing it if it were to harm the environment. So, although meat sales do not currently appear to be under threat, it is worthwhile for producers to modify their approach to take into account the growing ideological trends.

In a study of the impact of different poultry production systems on consumers' purchase perceptions, P. Pinto da Rosa et al. [12] conducted a blind tasting and questionnaire that identified differences in taste and texture between samples from organic and conventional systems. The qualitative data obtained through the questionnaire showed that consumers are interested in buying the products, having been assured of good animal husbandry conditions. Poultry companies have the opportunity to make their products more attractive by taking care of the animals, the conditions in which they are kept and informing consumers about this. These theses are directly in line with the results of the study. Presented in this study. The literature also highlights the role of eco-labelling. Among consumers of the millennial generation, according to S. Siddiqui et al. [13], the so-called "clean label" significantly influences the decision to purchase products.

5 Conclusions

When considering the sustainability of agriculture, it is important to bear in mind that it is not a question of production technology, but also the indirect influence of consumer purchasing decisions. It is consumer preferences that cost the assortment development of agricultural production. Food choices are one of the key areas for observing and analysing consumer behaviour. This is because they are made basically every day, while being subject to numerous influences. The aim of this study was to verify to what extent various worldview factors exert pressure on poultry meat purchasing decisions. Of those surveyed, 74% of respondents answered that their decision to purchase meat is not influenced by worldview factors. It does, however, matter to 26% of respondents. Among this latter group of 'ideologues', the most important worldview purchasing factors were 'long-term health effects' (almost two-thirds of respondents rated the influence as at least significant) and 'farming method' (more than half of respondents rated the influence as at least significant). The lowest impact was attributed to the factor 'animal feelings'. One can venture to say that respondents' rate as more important those factors that have a significant direct impact on them. At the same time, respondents are aware of the risks associated with eating meat, and they get their information both from experts, books and unverified internet sources. Juxtaposing this study with the aforementioned survey of young Poles, it is impossible to resist the impression that a certain stabilisation of habits applies to older and less willing to change generations. Younger generations declare to be more conscious of purchases and take worldview factors more into account. It seems, therefore, that change is coming, although it will probably not mean abandoning meat purchases. There is a growing group of consumers for whom the quality of meat, how it is reared and the environmental impact of rearing will be crucial. For producers, this is at the same time a recommendation as to the direction of their offer, as well as their information activities.

References

1. Gródek-Szostak, Z., Luc, M., Szeląg-Sikora, A., Sikora, J., Niemiec, M., Ochoa Siguencia, L., Velinov, E.: Promotion of RES in a technology transfer network. a case study of the enterprise Europe network. Energies **13**, 3445 (2020). https://doi.org/10.3390/en13133445
2. Niemiec, M., Chowaniak, M., Sikora, J., Szeląg-Sikora, A., Gródek-Szostak, Z., Komorowska, M.: Selected properties of soils for long-term use in organic farming. Sustainability **12**, 2509 (2020). https://doi.org/10.3390/su12062509
3. Komorowska, M., et al.: Closed-loop agricultural production and its environmental efficiency: a case study of sheep wool production in Northwestern Kyrgyzstan. Energies **15**, 6358 (2022). https://doi.org/10.3390/en15176358
4. Oleksy-Gebczyk, A.: Preferences and consumer choices: a case of Polish Markets for Goods and Services, Asian J. Appl. Econ./Appl. Econ. J. **30**(2) (2023)
5. Kovalenko, N., Kovalenko, V., Hutsol, T., Ievstafiieva, Y., Polishchuk, A.: Economic Efficiency and Internal Competitive Advantages of Grain Production in the Central Region of Ukraine. Agric. Eng. (2021). https://intapi.sciendo.com/pdf/https://doi.org/10.2478/agriceng-2021-0004
6. Orkusz, A., Wolańska, W., Harasym, J., Piwowar, A., Kapelko, M.: Consumers' attitudes facing entomophagy: Polish case perspectives. Int. J. Environ. Res. Public Health **17**(7), 2427 (2020). https://doi.org/10.3390/ijerph17072427
7. Cegiełka, A.: "Clean label" as one of the leading trends in the meat industry in the world and in Poland. Rocznik Państwowego Zakładu Higieny **71**(1), 43–55 (2020). https://doi.org/10.32394/rpzh.2020.0098
8. Drejerska, N., Sobczak, W., Gołębiewski, J., Gierula, W.A.: Does organic means health for consumers? Selected issues of organic food market. British Food J. **123**(8), 2622–2640 (2021). https://doi.org/10.1108/BFJ-12-2020-1175
9. Barska, A., Wojciechowska-Solis, J.: E-consumers and local food products: a perspective for developing online shopping for local goods in Poland. Sustainability (2020). https://doi.org/10.3390/su12124958
10. Zielińska-Chmielewska, A., Olszańska, A., Kaźmierczyk, J., Andrianova, E.V.: Advantages and constraints of eco-efficiency measures: the case of the Polish Food Industry. Agronomy **11**(2), 299 (2021). https://doi.org/10.3390/agronomy11020299

11. Meat Atlas 2022 Report: Meat Atlas 2022 | Heinrich Böll Stiftung | Regional Office Central Europe; Poland, Warsaw (boell.org) [Read date: 10.08.2023]
12. Pinto da Rosa, P., et al.: Impact of different chicken meat production systems on consumers' purchase perception. Br. Poult. Sci. **62**(3), 387–395 (2021). https://doi.org/10.1080/00071668.2020.1857335
13. Siddiqui, S.A., Pahmeyer, M.J., Mehdizadeh, M., Nagdalian, A.A., Oboturova, N.P., Taha, A.: Consumer Behavior and Industry Implications. In: Galanakis, C.M. (ed.) The Age of Clean Label Foods, pp. 209–247. Springer International Publishing, Cham (2022). https://doi.org/10.1007/978-3-030-96698-0_7

Development of a Battery Swapping and Charging Unit in Servicing Station for Farming Robot: A Review

Jüri Olt(✉), Yevhen Ihnatiev, Tormi Lillerand, and Indrek Virro

Institute of Forestry and Engineering, Estonian University of Life Sciences, 56 F.R. Kreutzwaldi Street, 51006 Tartu, Estonia
jyri.olt@emu.ee

Abstract. The aim of this study was to increase the productivity of a farming robot in the field by increasing the effective working time of the machine during the shift required for recharging the depleted battery, optimizing the idle time of the farming robot during movement from the servicing station to the work site and back, as well as reducing the overall specific cost of agricultural work operations. For this purpose, an automatic Battery Swapping Unit (BSU) and charging unit (BCU) for the robot was developed, which replaces the depleted battery with a charged one during the robot's operation, allowing the robot's work in the field and battery recharging to occur independently of each other. The farming robot's battery is placed in a container with rollers attached underneath, and the movement of the battery container within the unit between different positions, charging slots, and the fixed location on the robot device frame is along linear guides on the same plane. The battery container is moved within the device by a cart equipped with an electric drive and a linear actuator attached to the cart, to which an electrically locked gripper is also attached. The duration of one cycle of the original battery swapping unit for the farming robot, as presented in the article, is $t_s = 225$–238 s, including the movement times of the linear actuator, the cart, and the activation times of relays and locks. The developed battery swapping unit for the farming robot meets the requirements set for it.

Keywords: Actuator · Battery Swapping Unit (BSU) · Battery Charging Unit (BCU) · Cart · Electric Drive · Farming Robot · Servicing Station

1 Introduction

Modern agricultural development is moving towards optimizing production efficiency and quality [1, 2], as well as reducing environmental impacts and production risks [3, 4]. To achieve this, specific production costs are optimized by reducing labor costs and improving the utilization of technological machine operating time [5]. This goal can be achieved by the wider use of autonomous agricultural machines or agricultural robots [6–12].

E. Lorencowicz et al. (Eds.): FMPMSA 2024, LNCE 609, pp. 333–345, 2024.
https://doi.org/10.1007/978-3-031-70955-5_37

Agricultural machines predominantly use internal combustion engines, the exhaust gases of which pollute the environment during operation, which poses a significant problem in modern times. To reduce environmental pollution, agricultural machines must be equipped with environmentally friendly drives. One such drive, which does not produce exhaust gases containing components hazardous to humans and the environment, is the electric drive [13, 14].

In the technical solution described in this article, the agricultural robot is equipped with an electric drive [15]. The use of electric vehicles and electric work machines is also expanding [16–18]. For the movement of the agricultural robot in the field and the operation of its technological devices during the working process, an electric mobile machine, including the electric drive of the agricultural robot, requires a continuous supply of electricity. Concentrated sources, specifically batteries, are used as energy sources in electric mobile machines. During operation, the batteries of electric machines discharge and depleted batteries need to be recharged.

The battery is recharged using the energy produced by the mobile servicing station's energy supply system. According to patent research and professional literature [19–24], known methods and technological solutions for replacing batteries in mobile machines vary depending on the location of the battery in the mobile machine as follows: 1) beneath the mobile machine; 2) on the side of the mobile machine; 3) on top of the mobile machine, including the roof; 4) at the rear of the mobile machine, including from the trunk of a vehicle.

Some agricultural machines have working organs located at the front, such as harvesting machines (combines etc.), loaders etc. In such cases, the suitable location for the replaceable battery on these machines is behind them. However, in most cases, the working organs are at the rear of the machine (soil cultivation tools, plant protection and fertilization equipment, etc.), or between the two axles (row crop cultivators, etc.), hence the suitable location for the replaceable battery on these machines could be at the front of the machine.

Known BSU's are mostly built-in systems where the entire electric swapping cart mechanism is placed underground [22, 23]. However, they have their drawbacks: such an automatic battery swapping station cannot be relocated, it has high construction costs, it is inconvenient to maintain, usually used indoors, does not work in outdoor conditions in bad weather, and therefore its adoption is limited.

From the patent US20170259675 [24], the known system is simple and safe, and the solution would be an autonomous BSU for electric vehicles, consisting of various small, lightweight, and easily installable parts that are convenient to transport. In this known solution, the electric swapping cart is placed on a guide rail and moves back and forth along the guide rail to complete the battery transfer between the mobile machine and the BSU The electric swapping cart mechanism also includes a mechanism that can be vertically raised and lowered to raise it to a height suitable for swapping the vehicle's battery. However, the main drawbacks of battery swapping systems here are: storing and moving batteries with large masses are complicated, and the quantity of batteries to be replaced is large. The system itself is bulky because to swap the battery, the vehicle must be raised to working height to remove the battery from under the mobile machine.

In an article on mobile machine BSU's [25], an automatic battery swapping device is described where the electric vehicle is lifted with a lifting device, the battery is removed from the vehicle with a special lifting and transport device, and moved to a cabinet equipped with multiple charging slots, from where the same device takes a fully charged battery and brings it under the electric vehicle and installs it in place. The speed of battery swapping in this device is limited by the need to lift the vehicle, and the lifting device makes this system bulky and complex. It should be added that this solution described is a stationary BSU, so it is not suitable for servicing agricultural robots.

The placement of the battery on top of the mobile machine and swapping it there, as described in literature [26], is not feasible in agriculture. Indeed, it would raise the center of mass of the agricultural mobile machine high, and considering that agricultural mobile machines often operate on uneven and soft ground, its stability of movement is not always adequately ensured, and various risks may arise during operation.

Justification: Since the specific cost of agricultural production c_A consists of fixed and variable production costs, and depends on the machine, including the agricultural robot productivity $W_{A.d}$, being inversely proportional to it, and represents the costs of agricultural work operations per unit of land area (€ ha^{-1}) k and can be calculated as follows [4, 27]:

$$c_A = \frac{1}{W_{A.d}}(C_F + C_V), \tag{1}$$

where, $W_{A.d}$ – productivity of the agricultural robot's ha h^{-1},
C_F – fixed cost € h^{-1},
C_V – variable cost € h^{-1}.

During the operation of the electric agricultural robot in the field, its battery becomes discharged. Recharging the battery requires time, which negatively affects the utilization of working time and, consequently, the productivity of the agricultural robot. The productivity of the agricultural machine or robot is better characterized by hourly productivity $W_{A,d}$, in hectares, which is expressed as follows [28]:

$$W_{A,d} = 0.1 \cdot v_m \cdot B_h \cdot \tau, \tag{2}$$

where v_m – movement speed km h^{-1},
B_h – operation with m,
τ – utilization factor of working time, that is calculated as:

$$\tau = \frac{t_w}{t_d} = \frac{t_w}{t_w + t_{0.1} + t_{0.2} + t_p + t_r}, \tag{3}$$

where, t_w – effective working time of the agricultural robot h,
$t_{o.1}$ – travel time from the maintenance station to the working row h,
$t_{o.2}$ – travel time from working row to the maintenance station h,
t_p – time spent on turning on the pivot and moving from one working row to another h,
t_r – time required for technological auxiliary operations and recharging the battery of the agricultural robot h.

From Eq. (1), it can be seen that an increase in the area productivity of the agricultural robot reduces the specific cost of the operation, resulting in an increase in the efficiency of the agricultural robot. The movement speed of the agricultural robot during operation is limited because an increase in speed compromises the quality of work [28]. Travel movement needs to be limited, optimized, and the time required for technological auxiliary operations and charging the depleted battery should be minimized. The most efficient way to save working time is to replace the depleted battery with a charged one during the work process, rather than spending working time on recharging the battery. Since manually swapping the battery is time and labor-intensive, an automated BSU should be used to save labor costs and working time.

The need, concept, and purpose of an automatic BSU was to increase the productivity per shift (which can last up to 14 h per day, for example, from dewpoint to dewpoint) $W_{A.d}$, by improving the utilization factor τ, thus saving auxiliary time t_r.

2 Material and Methods

The development activities of the agricultural robot's BSU include compiling the requirements for the device, followed by patent research and determining the state of the art. Then, the basic and detailed functions were defined, solution principles were selected, and from among them, a specific technical solution was chosen. A 3D model, a sketch project, and an initial prototype were then developed [29].

The following requirements [29] were set for the developed robot battery swapping device: 1) the battery swapping must occur automatically, without operator intervention; 2) the battery swapping duration must not exceed 4 min; 3) the battery swapping must occur smoothly and without faults; 4) the battery swapping must occur without mechanically damaging the battery.

Idea Generation: The principle diagram of the developed BSU and BCU is shown in Fig. 1, which illustrates all the technological components of the device.

According to the principle depicted in Fig. 1, the BSU is located on a platform in the mobile servicing station, with the floor of the platform equipped with guides 1 for the agricultural robot's wheels (Fig. 2), which are intended for the initial alignment and positioning of the agricultural robot's wheels relative to the battery swapping device when the agricultural robot drives from the field to the servicing station for docking with the BSU. For further precise alignment, conical guides 9 and locking elements 10 (similar to pos. 3) are mounted on the frame of the BSU, intended for securing the robot in a specific position during servicing.

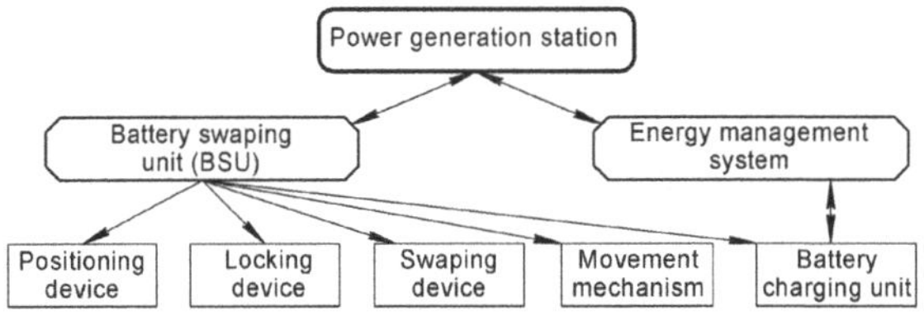

Fig. 1. Schematic of the rapid BSU for the agricultural robot.

During the preliminary patent search and analysis of existing scientific and technical literature, it became apparent that there are currently no ready-made technical solutions for automatic battery replacement in agricultural robots worldwide. Existing options are mainly tailored for drones or electric vehicles, which are not suitable for autonomous agricultural robots with their servicing stations due to specific technical and technological characteristics.

The goal of this article was to present the developed prototype BSU (Fig. 2), which was created during scientific research and design work and can be easily integrated into the system of an autonomous agricultural robot and its servicing station. Importantly, the requirement was to ensure that the battery replacement time does not exceed the time needed for technological servicing (e.g., filling the bunker with fertilizers, seeds, herbicides, etc.).

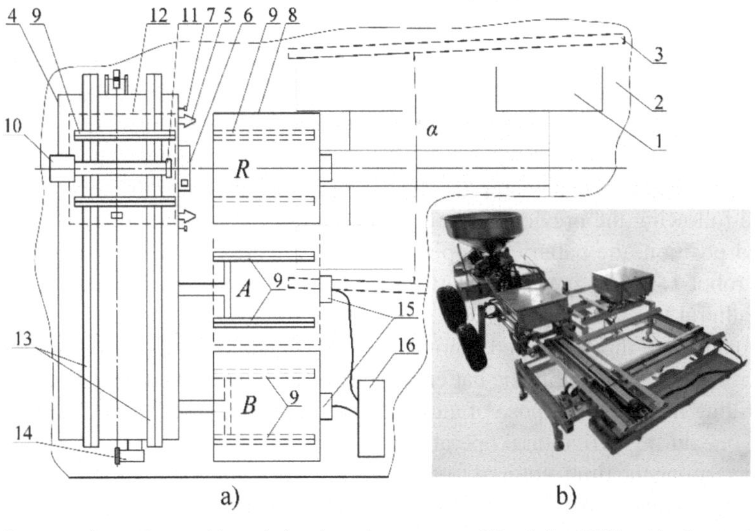

Fig. 2. Constructive scheme (a) and developed prototype (b) of the BSU with the agricultural robot docked to it: 1 – agricultural robot; 2 – mobile servicing station; 3 – guides for the robot's wheels; 4 – BSU; 5 – conical pin; 6 – electromagnetic gripper; 7 – mechanical sensors; 8 – battery container; 9 – guides; 10 – linear actuator; 11 – gripper; 12 – cart; 13 – cart guides; 14 – stepper motor with gearbox; 15 – connectors; 16 – automatic battery charging unit (BCU).

To determine the time required for battery replacement, it as the mobile servicing station is placed on a wheeled trailer for transportation to the required field, the platform 2 on which the Battery Swapping Unit (BSU) 4 is located is positioned at a certain height above ground level. The agricultural robot 1 with front steerable wheels is guided onto it via inclined ramps. To facilitate its positioning relative to the battery swapping device 4, converging guide rails 3 are attached to the platform 2. Further precision alignment is ensured by two conical guides 5 mounted on the frame of the battery swapping device 4 and corresponding parts installed at the front of the agricultural robot 1. Their joint immobilization is achieved by an electromagnetic grip 6 on the BSU side and electrically controlled locks on the agricultural robot side 3. Once successfully parked, confirmed by signals from two mechanical sensors 7 and a sensor indicating the presence of a closed magnetic field, which is combined with the electromagnetic grip 6, the cycle of replacing the discharged battery begins.

The discharged battery, housed in a protective box 8 equipped with special rollers for movement along rhombic guide rails 9, is first gripped by a linear actuator 10 using a special clamp 11. This clamp 11 has a battery holder, a position sensor relative to the robot frame 1, and a sensor for detecting the presence of the battery in the grip. Subsequently, the actuator 10 moves the battery 8 onto the longitudinal cart 12, which also has similar guides 9 and rollers for transverse movement along the frame of the battery swapping device 4.

The movement towards the empty charging slot A is facilitated by a stepper motor with a reducer 14, and precise positioning is ensured by optical sensors. Once in the correct position, the actuator 10 pushes the battery 8 into the vacant charging slot, where it is detected by the battery swapping device 4 using an optical sensor and secured with an electrically controlled lock. The gripper 11 of the actuator 10 is released and returns to the cart 12. Thus, the discharged battery is connected via connector 15 to the automatic charging device 16, which immediately initiates diagnostics and recharge.

The cart 12 moves to charging slot B, where a fully charged battery identical to battery 8 is located. The actuator 10 grips it and transports it to the robot, where it is installed following the previously described algorithm. After the gripper 11 returns to its initial position, the battery swapping device 4 releases the electromagnetic grip 6, and the robot 1, reversing off the platform 2 of the mobile servicing station, continues its agricultural tasks.

In conclusion, the developed innovative BSU, integrated into the robot's servicing station, is designed for automatic battery replacement in unmanned agricultural machinery, fulfilling its crucial purpose of automatically replacing a discharged battery with a charged one during agricultural operations.

To determine the time required for battery replacement, it is necessary to understand the movements performed by the BSU mechanisms. For this purpose, we will create a diagram (Fig. 3) illustrating the movement of the gripper 11 (numbering corresponds to Fig. 2), actuator 10, and cart 12 throughout the entire working cycle.

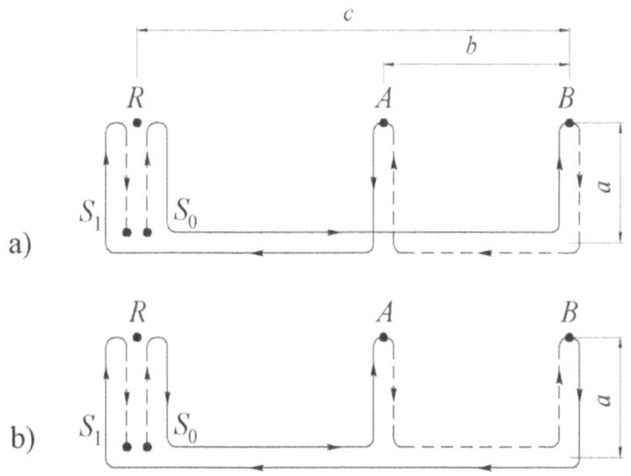

Fig. 3. Battery replacement scheme of the robot by the developed BSU: S_0 – start of the route; S_1 – end of the route; a – distance covered by the linear actuator rod; b – distance between charging cells of BSU; c – distance between the parking position of the robot and charging position B; --- – idle movement of the gripper without a battery.

At the beginning of the battery replacement cycle, the cart 12 is always positioned opposite the robot's parking spot R, while the gripper 11 actuator is in position S_0. Then, the actuator's rod undergoes idle extension to a distance a to grip the battery, followed by retraction over the same distance a with the battery extracted. Subsequent movements occur according to the previously described principle and depending on which charging bay is available at the moment, either Fig. 3a or Fig. 3b. The end of the gripper's 11 actuator's 10 movements in the battery replacement cycle is marked as point S_1 on the diagram, which essentially coincides with point S_0. The diagram illustrates the gripper's 11 movements with and without the battery with solid and dashed lines, respectively.

Principles of Selecting the Drive for the Battery Swapping Device. According to the requirements set for the unit, the total cycle time for battery swapping t_s must not exceed the time required for other technological operations, such as filling the fertilizer hopper of the robot or filling the tank of the robot sprayer, which is equal $t_t = 240$ s, i.e.,

$$t_s \leq t_t, \tag{4}$$

The total cycle time for battery swap t_s consists of three components

$$t_s = t_a + t_c + t_l, \tag{5}$$

where, t_a – time for longitudinal movement of the battery cart s;
t_c – time for transverse movement of the battery cart s;
t_l – time for the operation of electromechanical devices (locks and relays) s.

The analysis of the diagrams (Fig. 3) shows that for both cases, the actuator's grip, along with the battery, undergoes four forward and four reverse movements in the longitudinal direction over a distance a. To increase positioning accuracy at points R, A, and B during battery gripping or releasing, the actuator slows down near these points. Additionally, due to the design features of the applied actuator model (модель актуатора), the reverse movement of the actuator occurs slightly faster. Therefore, the extension

V_{out} and retraction V_m speeds of the actuator were determined experimentally based on the average time taken for the actuator to cover the distance during gripping. The time was automatically recorded by the built-in measuring tools of the BSU with millisecond accuracy and is equal to $V_{out} = 0.024$ m s^{-1} and $V_m = 0.028$ m s^{-1}.

For Eq. (5), the following logic was used: since a linear actuator was used as a drive for the longitudinal movement of the battery box, whose speed is practically constant. The movement time t_a can be found as follows:

$$t_a = 4\frac{a}{V_{out}} + 4\frac{a}{V_{in}} \text{ or } t_a = 4a\left(\frac{1}{V_{out}} + \frac{1}{V_{in}}\right) \tag{6}$$

where, a – length of the battery box movement path longitudinally m (see Fig. 3);
V_{out} – speed of movement of the actuator's gripper at reverse travel m min^{-1};
V_m – speed of movement of the actuator's gripper at forward travel m min^{-1}.

The distance travelled by the battery cart during transverse movements of the battery during the replacement cycle according to Fig. 3 for both presented variants will be equal to $c + b\,(c - b)$, or after simplification, $2c$. Essentially, the cart travels the distance between the extreme positions of the route points R and B twice in the optimal movement method. Since a stepper motor is used in the cart drive, it is possible to adjust the movement speed within wide limits while maintaining high precision of movement. Battery cart transverse movement time t_c can be found as follows:

$$t_c = \frac{2c}{V_c}, \tag{7}$$

where c – distance between points R and B (Fig. 3),
V_c – linear speed of the cart m s^{-1}.

The time t_c for the transverse movement of the cart based on the characteristics' of the electric motor speed. Due to the rotational frequency of the stepper motor, the linear speed V_c of the battery cart can be expressed as:

$$V_c = \frac{n \cdot \pi \cdot d}{i}, \tag{8}$$

where, n – rotational frequency of the stepper motor shaft s^{-1};
d – diameter of the belt pulley, m;
i – gear ratio of the gearbox.

Substituting Eq. (8) into Eq. (7), the following expression for calculating the time t_c for the transverse movement of the cart is found:

$$t_c = \frac{i \cdot 2c}{n \cdot \pi \cdot d} \tag{9}$$

The actual sufficient time t_c for the transverse movement of the battery cart 8 may differ from the calculated one, as it does not consider the additional time required for accelerating and decelerating the battery cart (with a total mass of 75 kg) carrying the container 4 with the battery when starting and stopping the movement. This time is planned to be controlled during initial tests. The time t_c depends on the rotational frequency n_g of the motor shaft, the significance of which in the total cycle time t_s of the battery swap can be characterized by the indicator value λ as follows:

$$\lambda = \frac{t_c}{t_s} \cdot 100. \tag{10}$$

3 Results and Discussion

Figure 2b shows the developed prototype of the BSU. The battery swap cycle begins with the successful parking of the agricultural robot to the BSU and ends then fully charged battery was installed in the robot, the locking mechanisms fixing the robot's position are open, and the robot is ready to return to the field to continue its unfinished work.

The operational cycle should be considered as the movement back and forth of the battery cart, so if the length of one operational cycle is $a = 0.65$ m, then the longitudinal movement time of the battery cart $t_a = 202$ s.

Considering $d = 0.062$ m; $i = 15.3$; $c = 1.15$ m and the suitable rotational frequency of the stepper motor shaft was searched within the limits $n = 5\ldots20\ \text{s}^{-1}$, then the transverse movement time t_c of the battery cart according to Eq. (9) remains in the range of $t_c = 36.1\ldots9.0$ s.

Through the conducted verification test, it is evident that the stable operation of the battery cart drive motor and belt transmission is ensured up to stepper motor shaft rotation frequency of 17.5 s^{-1}. Higher rotation frequencies than this value generate sudden accelerations that overload the motor controller and lead to step losses in the stepper motor (at a voltage of 30V DC and a current of 2A), which in turn affects the positioning accuracy of the battery cart. Analyzing the trend, it can be seen in Fig. 4 that higher rotation frequency n does not result in a significant reduction in the battery swap cycle time t_s. When the stepper motor shaft rotation frequency of 7.5 s^{-1}, the duration of the battery swap cycle t_s is 236.3 s, what corresponds to condition eq. (4).

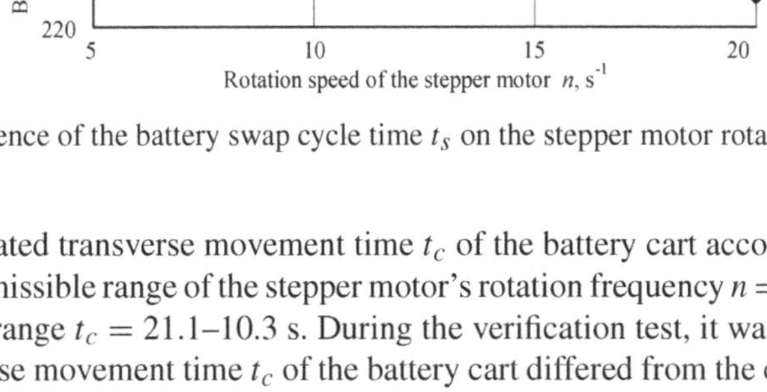

Fig. 4. Dependence of the battery swap cycle time t_s on the stepper motor rotation frequency n.

The calculated transverse movement time t_c of the battery cart according to Eq. (9) within the permissible range of the stepper motor's rotation frequency $n = 7.5\ldots17.5\ \text{s}^{-1}$, it will vary at range $t_c = 21.1$–10.3 s. During the verification test, it was found that the actual transverse movement time t_c of the battery cart differed from the calculated time.

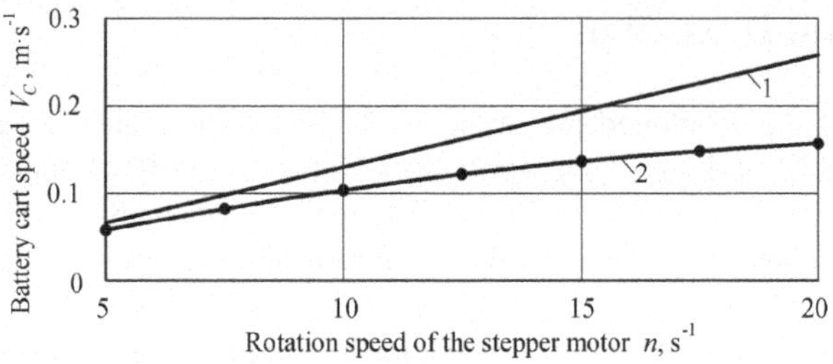

Fig. 5. Comparison of theoretical (1) and experimental (2) speeds of battery cart depending on the stepper motor rotation frequency n.

Figure 5 shows the comparison between the calculated and measured linear speed of the battery cart depending on the stepper motor rotation frequency n. The total time spent on acceleration and deceleration during the entire battery swap cycle is constant, with a duration of 9 s (1.5 s for each acceleration and deceleration of the cart) and only the resulting movement speed of the battery cart changes. For the calculated speed, the time spent on acceleration and deceleration is not considered. The greater the stepper motor rotation frequency, the greater the difference between the calculated and actual time, which is related to the increased proportion of time spent on acceleration and deceleration of the battery cart in the total cycle time.

Figure 6 shows the time taken for the cart's movement depending on the rotation frequency of the stepper motor n. It can be seen that when the rotation frequency exceeds 12.5 s^{-1}, the cart's movement time decreases insignificantly. Even with subsequent increases in the rotation speed n of the stepper motor to 20 s^{-1} (1.6 times), the cycle time t_s for battery replacement decreases by only 4.9 s or 2.4%. Therefore, we can conclude that increasing the rotation speed n of the stepper motor beyond 12.5 s^{-1} is not rational and does not significantly affect the cycle time t_s for battery replacement.

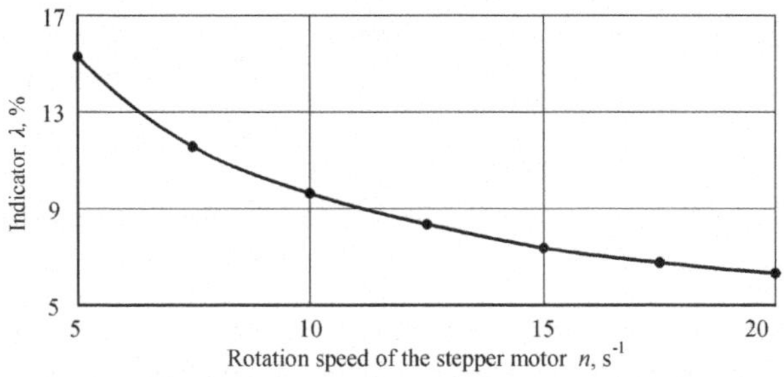

Fig. 6. The indicator value λ depending on the stepper motor rotation frequency n.

The operations comprising the battery swap cycle and their numerical values are presented in Table 1.

Table 1. Durations t_s of battery swap cycle operations at $n = 7.5–17.5\ s^{-1}$.

Operation Name	Time, s
Linear actuator movement time *ta*	202.0 ± 1.6
Transverse movement time of battery cart *tc*	(16.1–29.0) ± 0.05
Summed time of activation of electromechanical devices (locks, relay, etc.) *ted*	7.0
Total cycle duration *ts*	(225.1–238.0) ± 1.65

4 Conclusions

As a result of the analysis of literary sources, a structural scheme and algorithm for the operation of an autonomous system for automatic battery replacement in agricultural robots have been developed.

The article presents the design of the developed prototype Battery Swapping Unit (BSU), which can be easily integrated into the system of an autonomous agricultural robot and its servicing station. This BSU always keeps one battery in reserve, fully charged and serviced until replacement.

Through the analysis of the developed battery replacement scheme, various factors influencing the total duration of the battery replacement cycle, t_s, have been examined. Optimal ranges of the rotation frequency n of the stepper motor shaft have been selected based on the condition of limiting the total duration t_s of the battery replacement cycle (less than 240 s) and avoiding stepper motor skipping due to increased inertial loads during the movement of a 75 kg battery.

Calculations have determined that the above conditions correspond to a rotation frequency range n of the stepper motor shaft of $7.5–17.5\ s^{-1}$. In this case, the transverse movement time of the battery cart t_c is approximately (16.1–29.0) ± 0.05 s, and the total cycle duration t_s is approximately (225.1–238.0) ± 1.65 s.

Thus, the fast BSU for agricultural robots enables increased productivity of the agricultural robot by saving time on recharging its battery.

Funding. This publication was supported by development fund project PM210001TIBT from the Estonian University of Life Sciences.

References

1. Kutzbach, H.K.: Trends in power and machinery. J. Agric. Eng. Mach. **76**(3), 237–247 (2000). https://doi.org/10.1006/jaer.2000.0574
2. Moise, E., Biris, S., Grecu, B., Ungureanu, N., David, L., Paraschiv, G.: Determination of the minimum surface of an agricultural farm from witch a certain power range of tractors become profitable. In: Proceedings of the 38. International Symposium on Agricultural Engineering, Actual Tasks on Agricultural Engineering, Opatia, Croatia, 22–26 Feb 2010, pp. 99–110 (2016)

3. Mandal, M., et al.: Food waste-based bio-fertilizers production by bio-based fermenters and their potential impact on the environment. Chemosphere **353**, 141539 (2024). https://doi.org/10.1016/j.chenosphere.2024.141539
4. De Laporte, A., Schuurman, D., Skolrud, T., Slade, P., Weersink, A.: Business risk management programs and the adoption of beneficial management practices in Canadian crop agriculture. Canadian J. Agric. Econ./Revue canadienne d'agroeconomie (2024). https://doi.org/10.1111/cjag.12349
5. Olt, J., Küüt, K., Ilves, R., Küüt, A.: Assessment of the harvesting costs of different combine harvester fleets. Res. Agr. Eng. **65**, 25–32 (2019)
6. Sharma, P., Sashte, N., Phadke, S.: Review of Various Steering Systems for Unmanned Ground Vehicle. Int. J. Innov. Res. Sci. Eng. Technol. **5**(4), 4836–4840 (2016). http://www.ijirset.com/upload/2016/april/33_Review_new.pdf
7. King, A.: Technology: the future of agriculture. Nature **544**, 521–523 (2017). https://doi.org/10.1038/544S21a
8. Duckett, T., Pearson, S., Blackmore, S., Grive, B.: Agricultural robotics: the future of robotic agriculture. UK-RAS Netw. White Papers (2018). https://doi.org/10.31256/WP2018.2
9. Fountas, S., Mylonas, N., Malounas, I., Rodias, E., Santosh, C.H., Pekkeriet, E.: Agricultural robotics for field operations. Sensors **20**(9), 2672 (2020). https://doi.org/10.3390/s20092672
10. Khan, R., Mumtaz Malik, F., Raza, A., Mazhar, N.: Comprehensive study of skid-steer wheeled mobile robots: development and challenges. Ind. Robot.: The Int. J. Robot. Res. Appl. **48**(1), 142–156 (2021). https://doi.org/10.1108/IR-04-2020-0082
11. Zhao, J., Yang, Y., Zheng, H., Dong, Y.: Global agricultural robotics research and development: trend forecasts. J. Phys.: Conf. Ser. **1693**(1), 012227 (2020). https://doi.org/10.1088/1742-6596/1693/1/012227
12. Kägo, R., Vellak, P., Karofeld, E., Noorma, M., Olt, J.: Assessment of using state of the art unmanned ground vehicles for operations on peat fields. Mires and Peat **27**, 20 (2021). https://doi.org/10.19189/MaP.2020.OMB.StA.2128
13. Bi, Z., Luo, C., Miao, Z., Zhang, B., Zhang, C.W.J.: Automatic robotic recharging system – development and challenges. Ind. Robot. **48**(1), 95–109 (2021). https://doi.org/10.1108/IR-05-2020-0109
14. Harik, E.H.C.: Design and implementation of an autonomous charging station for agricultural electrical vehicles. Appl. Sci. **11**, 6168 (2021). https://doi.org/10.3390/app11136168
15. Virro, I., Arak, M., Maksrov, V., Olt, J.: Precision fertilization technologies for berry plantation. Agron. Res. **18**(S4), 2797–2810 (2020). https://doi.org/10.15159/AR.20.207
16. Redpath, D.A.G., McIlveen-Wright, D., Kattakayam, T., Hewitt, N.J., Karlowski, J., Bardi, U.: Battery powered electric vehicles charged via solar photovoltaic arrays developed for light agricultural duties in remote hilly areas in the Southern Mediterranean region. J. Clean. Prod. **19**, 2034–2048 (2011). https://doi.org/10.1016/j.jclepro.2011.07.015
17. Bechar, A., Vigneault, C.: Agricultural robots for field operations: concepts and components. Biosyst. Eng. **149**, 94–111 (2016). https://doi.org/10.1016/j.biosystemseng.2016.06.014
18. Bechar, A., Vigneault, C.: Agricultural robots for field operations: part 2: operation and systems. Biosys. Eng. **153**, 110–128 (2017). https://doi.org/10.1016/j.biosystemseng.2026.11.004
19. Zhang, J., Song, G., Li, Y., Qiao, G., Li, Z.: Battery swapping and wireless charging for a home robot system with remote human assistance. IEEE Trans. Consum. Electron. **59**(4), 747–755 (2013). https://doi.org/10.1109/TCE.2013.6689685
20. Yang, S., Yao, J., Kang, T., Zhu, X.: Dynamic operation model of the battery swapping station for EV (electric vehicle) in electricity market. Energy **65**, 544–549 (2014). https://doi.org/10.1016/j.energy.2013.11.010

21. Amiri, S.S., Jadid, S., Saboori, H.: Multi-objective optimum charging Management of electric vehicles through battery swapping station. Energy **165**, 549–562 (2018). https://doi.org/10.1016/j.energy.2018.09.167
22. Patent US2021188113. Method and system for rapid power exchange. IPC: B60L53/80; B60S5/06 (2021)
23. Patent DE4229687. Battery exchange station for electrically driven road vehicle IPC: B60K1/04; B60L11/18; B60S5/06 (1994)
24. Patent US2017259675. Battery swapping system and techniques. IPC: B60L11/16; B60S5/06 (2017)
25. Adegbohun, F., von Jouanne, A., Lee, K.Y.: Autonomous battery swapping system and methodologies of electric vehicles. Energies **12**(4), 667 (2019). https://doi.org/10.3390/en12040667
26. Kim, J., Song, I., Choi, W.: An electric bus with a battery exchange system. Energies **8**(7), 6806–6819 (2015). https://doi.org/10.3390/en8076806
27. Ammann, H.: FAT-Berichte Nr. 539: Maschinenkosten 2000. Kostenansätze Gebäudeteile und mechanische Ein-richtungen. Tänikon, Eidgenössische Forschungsanstalt für Agrarwirtschaft und Landtechnik (1999)
28. Boson, E.S., Verniaev, O.V., Smirnov, I.I., Sultan-Shach, E.G.: Theory, Construction and Calculation of Agricultural Machines, 2nd edn., 810 pp. Scientific Publisher (2019). ISBN: 9789388399838
29. Ulrich, K., Eppinger, S.: Product Design and Development, 7th edn., 448 p. McGraw-Hill Education (2015)

Performance Evaluation of Hybrid-Electric Architectures for Agricultural Tractors

Francesco Paciolla[1,2](✉), Pietro Patimisco[1], Alessia Farella[3], Tommaso Quartarella[3], and Simone Pascuzzi[3]

[1] PolySense Lab-Dipartimento Interateneo di Fisica, Polytechnic and University of Bari, Via Amendola 173, 70126 Bari, Italy
francesco.paciolla@poliba.it

[2] Dipartimento di Ingegneria Elettrica e dell'Informazione, Polytechnic University of Bari, Via Edoardo Orabona 4, 70126 Bari, Italy

[3] Department of Soil, Plant and Food Science, University of Bari, Via Amendola 165/A, 70126 Bari, Italy

Abstract. The European parliament points out that agriculture sector can play a primary role in the green transition, thus it is necessary to exploit new approaches technologies to decrease the environmental impact of agricultural machinery. One of the most interesting approaches is the development of hybrid-electric tractors. These vehicles combine the traditional diesel internal combustion engine with an electric powertrain. This paper reports the modelling and simulation, performed modifying a simulation software typically employed for on-road vehicles, of 70 kW 4-wheel-drive hybrid-electric an orchard tractor, in two different architectures: the parallel and the power-split. The simulated driving cycle is the EPA REP05. The evaluation and the comparison of the performance and the environmental impact, in terms of fuel consumption and CO_2 emissions, between the conventional and the hybrid-electric architectures have been carried out. Moreover, two levels of hybridization have been tested to evaluate the possible downsizing of the internal combustion engine. The presented work points out that the parallel and the power-split hybrid-electric architectures reduce the fuel consumption of 10% and 21%, respectively. The CO_2 emissions lowered of 9% and 22%, respectively. Compared to those of a traditional orchard tractor.

Keywords: Hybrid-electric agricultural tractors · Hybrid-electric architectures · Simulation software · Environmental impact · Performance evaluation

1 Introduction

In 2020, the European Commission has approved the European Green Deal, which is a set of strategic actions to ensure climate neutrality by 2050 [1]. An interim target of 55% reduction in greenhouse gas (GHG) emissions by 2030 has been set [2]. In 2023, agriculture and forestry sectors account for almost the 10% of the annual global emissions of GHGs emissions, with a contribution of 20% in the annual CO_2 emissions [3]. Most of these pollutant emissions is related to intensive animal farming and ground

E. Lorencowicz et al. (Eds.): FMPMSA 2024, LNCE 609, pp. 346–356, 2024.
https://doi.org/10.1007/978-3-031-70955-5_38

working, but a considerable amount is generated by exhaust gases from diesel Internal Combustion Engines (ICEs), which represent the most adopted powertrain in agricultural machines that fulfill several agricultural operations [4–14].

For the aforesaid, several very strict regulations have been imposed to limit the pollutant emissions of Non-Road Mobile Machinery (NRMM), category to which agricultural machineries belong. In this context, to meet the new European (Stage V) and US (Tier 4) regulatory standards, agricultural machines manufacturers have equipped the engines with specific exhaust gas treatment devices [15–18]. The employment of these devices heavies the vehicle and is no longer able to satisfy these standards, thus it is necessary to exploit new approaches and technologies to drastically reduce pollutant emissions and fuel consumption, prioritizing sustainability, and efficiency.

Nowadays, following the automotive industry trend, there is an always increasing interest in powertrain hybridization of agricultural tractors. This process is in its early stages, but academic researchers and industrial manufacturers have recently started to analyze and design new concepts of agricultural tractors, with hybrid-electric powertrain's architectures and equipped with several sensors [19–25]. Hybrid-electric tractors combine the conventional diesel ICE with one or more electric machines and are equipped with an energy storage system (ESS). The presence of electric machines can ensure a power boost to the ICE, or they can independently operate in low-load conditions, also exploiting the regenerative braking, storing the obtained energy in the ESS, optimizing fuel consumption, and reducing pollutant emissions [26–28].

To evaluate the performance, fuel consumption and pollutant emissions of an agricultural tractor, several expensive and time-consuming field tests are required. Therefore, a greater emphasis should be placed on modeling and simulation phases. Nowadays, the use of simulation tools for the design of vehicles' powertrains is rapidly spreading because this approach offers several advantages [29–31]:

- Cost-Effectiveness and Rapid Prototyping: It allows to virtually test and evaluate several desigun configurations and working scenarios, accelerating the design process;
- Performance Analysis: It provides detailed insights about the performance, energy consumption and efficiency of components.
- Environmental Impact Evaluation: It enables to assess the environmental impact of hybrid vehicles, including pollutant emissions, fuel efficiency and energy consumption;
- Customization and Optimization: It is possible easily to modify several vehicles' parameters, analyzing the impact of these modifications on performances.

In literature, few papers have investigated the technical feasibility, the performances, and the environmental impact of hybrid-electric tractors [32, 33]. This paper presents the modeling and the simulation, in a software typically employed for on-road vehicles' simulation, of a hybrid-electric orchard agricultural tractor (70 kW) during the EPA REP05 driving cycle, which accurately represents a driving pattern in an urban area. The two modeled hybrid-electric architectures are the parallel and the power-split configurations.

Moreover, the comparison between the performance of a conventional tractor, powered only by a diesel ICE, and its respective hybrid-electric configurations, in terms of speed profile tracking, and environmental impact, i.e. fuel consumption and CO_2 emissions, has been performed. To investigate the possible downsizing of the ICE, two different levels of hybridization have been analyzed. The employment of a simulation software to model and simulate hybrid-electric agricultural tractors represents the novelty of this work.

2 Materials and Methods

2.1 "Autonomie" Simulation Software

In this study, the modelling and simulations of two architectures of hybrid-electric tractors have been performed using the "Autonomie" software (https://www.anl.gov/taps/autonomie-vehicle-system-simulation-tool, accessed on 20 March 2024), developed by the Argonne National Laboratory Vehicle & Systems Mobility Group (VMS) [34]. The software has been originally designed for of on-road vehicles' simulation but in this study several modifications have been performed in the vehicle propulsion architecture, in the configuration's parameters of the powertrain and in the vehicle dynamics, making the software suitable also for simulation of NRMM, e.g., agricultural tractors. The software's outputs include the performances of hybrid-electric vehicles' powertrains and provide information about fuel, energy consumption and CO_2 emissions. A vehicle's model is composed of some fundamental blocks: i) the driver; ii) the environment; iii) the vehicle powertrain controller, which is a supervisory controller; iv) the vehicle propulsion architecture, which includes the different vehicle's components and manages the low-level controllers.

The software includes pre-built vehicles' models that can be modified and customized by the user through the Graphical User Interface (GUI) of the software. This allows to configure custom powertrains by modifying sizing, power, parameters, and efficiency of pre-defined components. "Autonomie" automatically builds in Simulink® (MathWorks) each model's subsystem and interconnects them to create the vehicle's model [24]. The simulations have been run at 100 Hz, using the fixed-step ode4 Runge–Kutta solver, with a time step of 0.01 s.

The Landini REX 4–100 GT orchard tractor has been modelled in "Autonomie" software. Table 1 points out the main technical specifications of the defined vehicle. It has four-wheel-drive and a diesel ICE with a 6-speed automatic transmission. The engine efficiency has been set to 42%, which is a standard value for a modern diesel engine. The gearbox and the final drive efficiency has been set to 97%. The main parameters of the modeled orchard tractor are reported in Table 1. The connected implement is represented by a trailer which is trailed by the tractor.

Table 1. Main parameters of the modeled conventional orchard tractor

Parameter	Value
ICE Maximum Power	70 kW
Maximum Torque	450 Nm @1400 rpm
Mass	2800 kg
Wheel Radius	0.49 m
Connected implement	2500 kg

2.2 Driving Cycle Definition

International standards such as the U.S. federal test procedure (FTP) and the European test cycle (ECE) are widely employed by automotive companies to define driving cycles for on-road vehicles in different driving conditions. However, for NRMM in the agricultural sector, standardized working cycles have not been yet defined, thus is not possible to define some standard procedures for the evaluation of the task's execution. The lack of standardized working cycles is strictly related to the fact that agricultural tractors perform a great variety of tasks, including both transportation and field work operations with light and heavy duties, indeed their power covers a wide range, from few tens of kW up to hundreds of kW. Imposing a driving cycle is mandatory for the evaluation of CO_2 emissions and fuel consumption. In this work, the simulated driving cycle is the EPA REP05 (Fig. 1). The cycle last 1400 s and its parameters have been modified to adapt it to a driving cycle for an agricultural tractor. In particular, the average speed has been reduced to 25.5 km/h, with a maximum speed of 40 km/h.

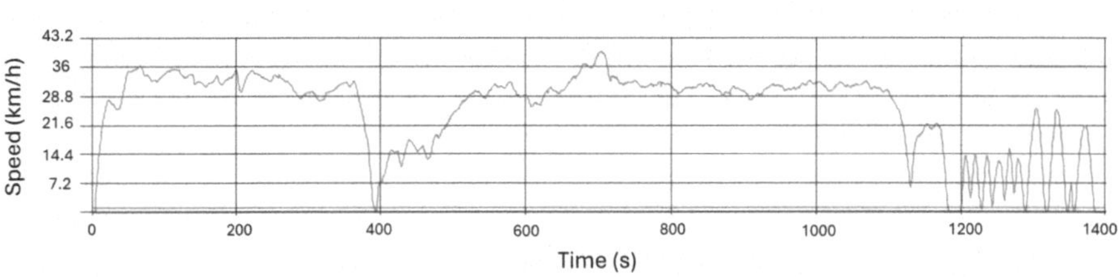

Fig. 1. EPA REP05 driving cycle. Source: Screenshot from the "Autonomie" software.

2.3 Hybrid-Electric Architectures

Hybridization of agricultural machinery powertrains is going to represent the future approach for the design of these machines because it allows to reduce fuel consumption and GHGs emissions, contributing to mitigate the environmental footprint, going towards a greener and more sustainable agriculture. Hybrid-electric tractors ensure improved performances, indeed they can easily switch between the two available power sources, maximizing efficiency and ensuring versatility. Furthermore, hybrid-electric tractors offer lower operating costs, improved working conditions, smoother acceleration and reduced vibration compared to conventional tractors. The two most employed hybrid-electric architectures are the parallel and the power-split configuration [28, 29]. To evaluate the potentiality of hybridization to downsize the ICE, maintaining the same performances and decreasing fuel consumption and environmental impact, two different levels of hybridization (Hybrid 1 and Hybrid 2) of the modeled Landini REX 4–100 GT orchard tractor have been simulated. Hybrid 1 has a 5% downsized ICE, instead Hybrid 2 has a 10% downsized ICE with respect to the conventional orchard tractor powertrain, in line with [34]. Regarding the sizing of the electric machines, several models have been tested and the one which satisfies the imposed driving cycle has been chosen, but at the same time without increasing the vehicle's weight.

Parallel Hybrid-Electric Configuration

The parallel hybrid-electric configuration is widely employed, thanks to its low-cost and simple implementation, allowing to hybridize an existing vehicle only adding one electric machine, which ensures torque boosting. In this way, the ICE can be significantly downsized. In the parallel hybrid-electric architecture, an electric machine is mechanically connected to the ICE shaft, which is linked to the final drive usually through a gear transmission. The mechanical coupling between the ICE and the electric machine can be made with a direct connection between the ICE shaft and the electric ma-chine shaft, or by means of a speed reducer. In this configuration, the electric machine works mainly as a motor: it can provide the starting torque, and it can boost the ICE during high power-demanding duty cycles. Operating as a generator, it can recharge the ESS by kinetic energy recovery, but the process is inefficient. In the parallel architecture, one large ESS, that can store a large amount of energy to guarantee a sufficient autonomy, is mandatory. Li-ion batteries pack are usually chosen [33]. The major drawback of parallel architecture is that the ICE is mechanically connected to the load, thus it cannot work always in its lowest Brake-Specific Fuel Consumption (BSFC) point. Moreover, the final drive speed depends on the ICE speed, thus the introduction of electric auxiliaries is strongly limited. The parallel hybrid-electric architecture, developed in "Autonomie" software is presented in Fig. 2.

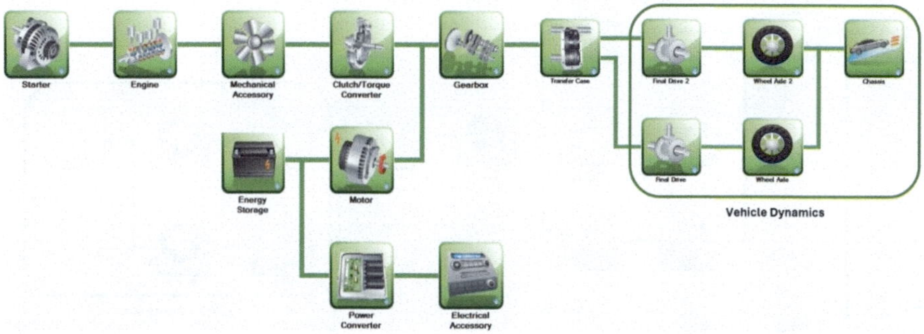

Fig. 2. Model of the simulated parallel hybrid-electric architecture. Each block that has subsystems has a small down blue arrow in the right-bottom. Source: Screenshot from the "Autonomie" software.

The parameters of the two simulated parallel hybrid-electric tractors (Hybrid 1 and Hybrid 2) are reported in Table 2.

Table 2. Simulated parallel hybrid-electric configurations.

Architecture	Parallel Hybrid-Electric	
Component	Hybrid 1	Hybrid 2
ICE [kW]	67	63
Electric Motor [kW]	10	15

Power-Split Hybrid-Electric Configuration

The power-split transmission configuration tries to mitigate the drawbacks of series and parallel hybrid architectures. This configuration avoids double energy conversions and keeps the ICE speed decoupled from the load. It can be considered a good trade-off between reducing sizing of components (mechanical parts, electrical machines, ICE) and good performances. The power split transmission comprises a planetary gear and two electrical machines coupled each other, with the engine and with the final driveline. The power coming from the ICE is split up by the planetary gear into two paths, the main part is sent "directly" to the wheels through mechanical gearing and the "converted" fraction which flows in the electric path through the two electric machines. The main advantages of a planetary gear are higher volumetric power density and higher efficiency compared to conventional gear [34]. The most important drawback of this architecture is its implementation cost and its complexity. The power split hybrid-electric architecture, developed in "Autonomie" software is presented in Fig. 3.

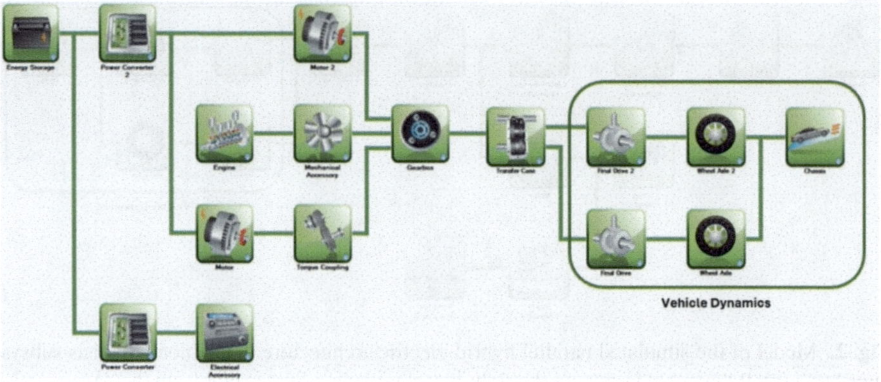

Fig. 3. Model of the simulated power-split hybrid-electric architecture. Each block that has subsystems has a small down blue arrow in the right-bottom. Source: Screenshot from the "Autonomie" software.

The parameters of the two simulated power-split hybrid-electric tractors (Hybrid 1 and Hybrid 2) are reported in Table 3.

Table 3. Simulated power-split hybrid-electric configurations.

Architecture	Power-split Hybrid-Electric	
Component	Hybrid 1	Hybrid 2
ICE [kW]	67	63
Electric Motor 1 [kW]	15	20
Electric Motor 2 [kW]	15	20

3 Results and Discussions

To evaluate the performances of the modeled hybrid-electric architectures of the modeled Landini REX 4-100 GT orchard tractor, the speed profile tracking has been considered. Regarding the environmental impact, the CO_2 emissions and the fuel consumption have been taken into account. Figure 4 reports the speed profile followed by the conventional, the parallel (Hybrid 1) and the power-split (Hybrid 1) architectures during the simulated EPA REP05 driving cycle.

Figure 4 shows that the simulated conventional tractor is not always able to precisely follow the speed profile imposed by the driving cycle. On the contrary, both hybrid-electric architectures follow very good the reference speed profile thanks to electric motors that allow a more flexible speed control.

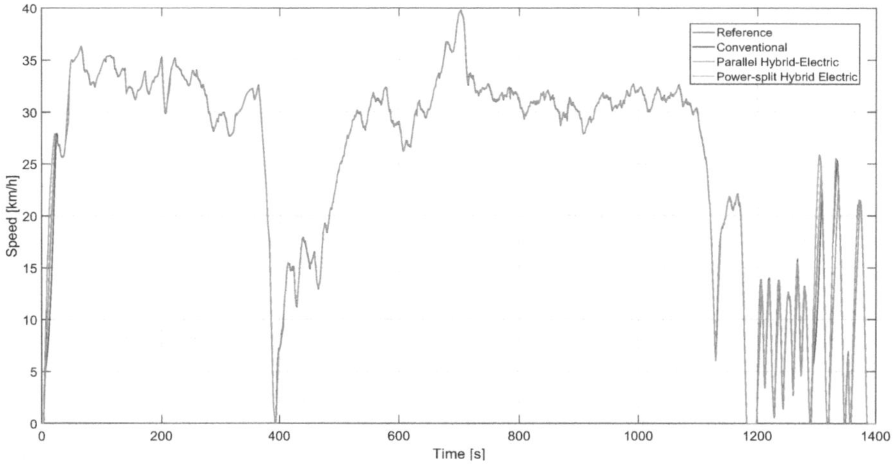

Fig. 4. Speed profile followed by the conventional, the parallel (Hybrid 1) and the power-split (Hybrid 1) architectures of the modeled orchard tractor during the defined driving cycle.

Table 4 reports the CO_2 emissions and fuel consumption comparison among the architectures.

Table 4. CO_2 emissions and fuel consumption comparison among the architectures.

Architecture	Conventional	Parallel Hybrid-Electric		Power-split Hybrid-Electric	
Parameter		Hybrid 1	Hybrid 2	Hybrid 1	Hybrid 2
CO_2 emission [kg/h]	9.7	8.9	8.7	7.8	7.4
Fuel Consumption [L/h]	4.2	3.8	3.7	3.3	3.2

Table 4 points out that CO_2 emissions and the fuel consumption of both the hybrid-electric architectures are relevantly lower than the ones generated by the conventional tractor. CO_2 emissions of the parallel and power-split configurations are on average 9% and 22% lower compared to those of the conventional tractor, respectively.

Moreover, Table 4 highlights that fuel consumption of the parallel and power-split configurations are on average 10% and 21% lower compared to those of the conventional tractor, respectively. Finally, it is possible to notice that the environmental impact of Hybrid 1 and Hybrid 2, which differs for the level of hybridization, are very similar for both the hybrid-electric architectures.

4 Conclusions

This paper presents the modeling and the simulation of the Landini REX 4-100 GT orchard tractor during the execution of the EPA REP 05 driving cycle, for the evaluation and assessment of the performances and the environmental impact of this agricultural tractor. The modelling and the simulations have been performed using the "Autonomie" simulation software, modifying it for the simulation of agricultural tractor. The analyzed configurations have been: i) the conventional one, characterized only by the ICE; ii) the parallel hybrid-electric, which includes an electric motor; iii) the power-split hybrid-electric, which includes two electric motors and a planetary gearbox.

The analysis and the evaluation of the performances of the task execution, the CO_2 emissions and the fuel consumption of the different configurations has been carried out. A detailed study, regarding the considered hybrid-electric configurations, has been carried out to evaluate the potentiality of hybridization to downsize the ICE, maintaining the same performances. Two different levels of hybridization (Hybrid 1 and Hybrid 2) have been simulated; Hybrid 1 has a 5% downsized ICE, Hybrid 2 has a 10% downsized ICE.

The simulation results of the performances show clearly that hybrid-electric architectures follow the reference speed profile in a more precise way compared to the conventional architecture. The results about the environmental impact point out that the parallel hybrid-electric architecture cuts down of 9% and 10% the CO_2 emissions and the fuel consumption. Regarding the power-split architecture, the results highlight that the decrease of CO_2 emissions and fuel consumption are equal to 22% and 21%, respectively, compared to the conventional tractor. This study highlights that hybridization of agricultural tractors improve the performances and reduce the environmental impact of these machines. Hybridization represents one of the most impacting technologies for the development of greener and more sustainable farming machines.

Aknowledgements. *This study was carried out within the Agritech National Research Center and received funding from the European Union Next-GenerationEU (PIANO NAZIONALE DI RIPRESA E RESILIENZA (PNRR)—MISSIONE 4 COMPONENTE 2, INVESTIMENTO 1.4—D.D. 1032 17 June 2022, CN00000022). This manuscript reflects only the authors' views and opinions, neither the European Union nor the European Commission can be considered responsible for them.*

References

1. Hafner, M., Raimondi, P.P.: Priorities and challenges of the EU energy transition: from the European Green Package to the new Green Deal. Russ. J. Econ. **6**(4), 374–389 (2020)
2. Akashi, O., Hanaoka, T.: Technological feasibility and costs of achieving a 50% reduction of global GHG emissions by 2050: mid-and long-term perspectives. Sustain. Sci. **7**, 139–156 (2012)
3. Harsányi, E., et al.: GHGs emission from the agricultural sector within EU-28: a multivariate analysis approach. Energies **14**(20), 6495 (2021)
4. Leach, F., Kalghatgi, G., Stone, R., Miles, P.: The scope for improving the efficiency and environmental impact of internal combustion engines. Transport. Eng. **1**, 100005 (2020)
5. Bulgakov, V., Pascuzzi, S., Nadykto, V., Ivanovs, S., Adamchuk, V.: Experimental study of the implement-and-tractor aggregate used for laying tracks of permanent traffic lanes inside controlled traffic farming systems. Soil Tillage Res. **208**, 104895 (2021)
6. Bulgakov, V., Pascuzzi, S., Ivanovs, S., Nadykto, V., Nowak, J.: Kinematic discrepancy between driving wheels evaluated for a modular traction device. Biosys. Eng. **196**, 88–96 (2020)

7. Pascuzzi, S., Anifantis, A.S., Santoro, F.: The concept of a compact profile agricultural tractor suitable for use on specialised tree crops. Agriculture **10**, 123 (2020)
8. Bulgakov, V., Pascuzzi, S., Beloev, H., Ivanovs, S.: Theoretical investigations of the headland turning agility of a trailed asymmetric implement-and-tractor aggregate. Agriculture **9**, 224 (2019)
9. Bulgakov, V., Pascuzzi, S., Anifantis, A.S., Santoro, F.: Oscillations analysis of front-mounted beet topper machine for biomass harvesting. Energies **12**, 2774 (2019)
10. Bulgakov, V., Pascuzzi, S., Adamchuk, V., Kuvachov, V., Nozdrovicky, L.: Theoretical study of transverse offsets of wide span tractor working implements and their influence on damage to row crops. Agriculture **9**, 144 (2019)
11. Pascuzzi, S., Santoro, F., Manetto, G., Cerruto, E.: Study of the correlation between foliar and patternator deposits in a "Tendone" vineyard. Agric. Eng. Int. CIGR J. **20**(3), 97–107 (2018)
12. Quartarella, T., Farella, A., Paciolla, F., Pascuzzi, S.: UVC Rays devices mounted on autonomous terrestrial rovers: A brief overview. In: Farm Machinery and Processes Management in Sustainable Agriculture XI International Scientific Symposium (2024)
13. Pascuzzi, S.: Outcomes on the spray profiles produced by the feasible adjustments of commonly used sprayers in "Tendone" Vineyards of Apulia (Southern Italy). Sustainability **8**(12), 1307 (2016). https://doi.org/10.3390/su8121307
14. Pascuzzi, S., Cerruto, E.: An innovative pneumatic electrostatic sprayer useful for tendone vineyards. J. Agricult. Engineer. **46**(3), 123 (2015). https://doi.org/10.4081/jae.2015.458
15. Pascuzzi, S., et al.: Theoretical study of the motion of a cut sugar beet tops particle along the inner surface of the conveying and unloading system of a top-ping machine. AgriEngineering **6**, 409–422 (2024)
16. Lovarelli, D., Bacenetti, J.: Exhaust gases emissions from agricultural tractors: State of the art and future perspectives for machinery operators. Biosys. Eng. **186**, 204–213 (2019)
17. Pacura, W., Szramowiat-Sala, K., Gołaś, J.: Emissions from light-duty vehicles—from statistics to emission regulations and vehicle testing in the European union. Energies **17**(1), 209 (2023)
18. Azzara, A., Rutherford, D., Wang, H.: Feasibility of IMO Annex VI Tier III implementation using selective catalytic reduction. Int. Counc. Clean Transp **4**, 1–9 (2014)
19. Scolaro, E., Beligoj, M., Estevez, M.P., Alberti, L., Renzi, M., Mattetti, M.: Electrification of agricultural machinery: A review. IEEE Access **9**, 164520–164541 (2021)
20. Annunziato, A., Godfrey, M., Anelli, F., Dulieu-Barton, J., Holmes, C., Prudenzano, F.: Flexible photonic sensors: investigation of an approach based on ratiometric power in few-mode waveguides for bending measurement. IEEE Access **12**, 29166–29174 (2024). https://doi.org/10.1109/ACCESS.2024.3368885
21. Pascuzzi, S., Łyp-Wrońska, K., Gdowska, K., Paciolla, F.: Sustainability evaluation of hybrid agriculture-tractor powertrains. Sustainability **16**(3), 1184 (2024)
22. Farella, A., Paciolla, F., Quartarella, T., Pascuzzi, S.: Agricultural Unmanned Ground Vehicle (UGV): A brief overview. In: Farm Machinery and Processes Management in Sustainable Agriculture XI International Scientific Symposium (2024)
23. Anelli, F., et al.: Mid-infrared interferometry with non-adiabatic tapered ZBLAN optical fiber. Opt. Express **32**(11), 18944 (2024). https://doi.org/10.1364/OE.521239
24. Moreda, G.P., Muñoz-García, M.A., Barreiro, P.J.E.C.: High voltage electrification of tractor and agricultural machinery–A review. Energy Convers. Manage. **115**, 117–131 (2016)
25. Anelli, F., et al.: Low-loss fluoride optical fiber coupler for mid-infrared applications. J. Lightwave Technol. **42**(7), 2457–2463 (2024). https://doi.org/10.1109/JLT.2023.3337603
26. Alberti, L., Mattetti, M.: Challenges and State of the Art in the Agricultural Machinery Electrification. In: Transportation Electrification: Breakthroughs in Electrified Vehicles, Aircraft, Rolling Stock, and Watercraft, pp. 417–423 (2022)

27. Mocera, F., Martini, V., Somà, A.: Comparative analysis of hybrid electric architectures for specialized agricultural tractors. Energies **15**(5), 1944 (2022)
28. McFadzean, B.; Butters, L.: An investigation into the feasibility of hybrid and all-electric agricultural machines. Scientific Papers. Series A. Agronomy 60 (2017)
29. Briggs, I., Murtagh, M., Kee, R., McCulloug, G., Douglas, R.: Sustainable non-automotive vehicles: the simulation challenges. Renew. Sustain. Energy Rev. **68**, 840–851 (2017)
30. Vogt, H.H., et al.: Electric tractor system for family farming: increased autonomy and economic feasibility for an energy transition. J. Energy Storage **40**, 102744 (2021)
31. Mocera, F., Somà, A.: Analysis of a parallel hybrid electric tractor for agricultural applications. Energies **13**(12), 3055 (2020)
32. Rossi, C., Pontara, D., Falcomer, C., Bertoldi, M., Mandrioli, R.: A hybrid–electric driveline for agricultural tractors based on an e-CVT power-split transmission. Energies **14**(21), 6912 (2021)
33. Beligoj, M., Scolaro, E., Alberti, L., Renzi, M., Mattetti, M.: Feasibility evaluation of hybrid electric agricultural tractors based on life cycle cost analysis. IEEE Access **10**, 28853–28867 (2022)
34. Karbowski, D., Pagerit, S. Autonomie, a plug-and-play software architecture. In: Proceedings of the Vehicle Power and Propulsion Conference, Lille, France (2010)

Quality Evaluation of the Operation of the XAG R 150 Autonomous Sprayer in Currant Cultivation

Stanisław Parafiniuk[1](✉), Łukasz Kopiński[2], Jacek Ogrodniczek[1], Paweł Karpiński[1], Fabienne Rabier[3], and Alaa Subr[4]

[1] Department of Machine Operation and Production Processes Management, University of Life Sciences in Lublin, Głęboka 28, 20-612 Lublin, Poland
stanislaw.parafiniuk@up.lublin.pl

[2] Department of Management and Marketing, University of Life Sciences in Lublin, Dobrzańskiego 37 b, 20-262 Lublin, Poland

[3] Walloon Agricultural Research Centre, Gembloux, Belgium

[4] Department of Agricultural Machines and Equipment, College of Agricultural Engineering Sciences, University of Baghdad, Baghdad 10071, Iraq

Abstract. The aim of the study was to evaluate the use of the XAG R150 autonomous platform for spraying black currant crops. The research was conducted in two stages. In the first part, laboratory tests were performed to determine the uniformity of spray application by the spray module of the platform which consists of a nozzle with adjustable size of droplets generated and a variable-speed fan. The tests were conducted on a vertical spray patternator for evaluating the uniformity of liquid application by orchard sprayers. In the second stage, a trial spray was carried out on a black currant plantation. The quality of liquid application in the sprayed area of black currant bushes was checked using water-sensitive paper.

Laboratory tests showed that the spray stream of liquid carried by the air stream is narrow and is 0.4 m wide, but uniform. Therefore, it is necessary to switch on the variable position of the spray module for spraying. The spray module has three ranges of droplet size. Field tests using water-sensitive papers have shown that the degree of coverage of plants by the spray is equal to at least 80%.

Keywords: Autonomous Vehicle · Berry Spraying · Liquid Application · Spray Uniformity

1 Introduction

In recent years, the use of robots for various agricultural tasks has gained increasing popularity. This is due to the desire to reduce the hard work done by humans, improve worker safety, and implement practices in line with sustainable development policies. In addition, it fits closely into the concepts of precision agriculture and Agriculture 4.0.

E. Lorencowicz et al. (Eds.): FMPMSA 2024, LNCE 609, pp. 357–366, 2024.
https://doi.org/10.1007/978-3-031-70955-5_39

In general, robots can be divided into three main categories: stationary, human-collaborative (so-called cobots), and mobile. The last of the mentioned types is divided into Automated Guided Vehicles (AGVs) and Autonomous Mobile Robots (AMRs). Autonomous robots can move along a programmed route, which is developed based on a prior mapping of the terrain. In order to avoid collisions with the surroundings, the machines are equipped with a number of sensors such as lidars, radars, ultrasonic sensors or vision systems. The dynamic development of robotic technologies and specialized solutions has led to the creation of a number of scientific papers presenting a review of the state of knowledge on robots for agriculture [1–4].

Autonomous robots are widely used for field work or feeding livestock. The devices can be used for tasks such as monitoring crops, spraying pesticides, or transporting crops. The Danish company Agrointelli has proposed the Robotti LR solution, which is capable of performing precise operations on crops such as seeding, weeding, harrowing, ridging, and spraying [5]. Another example is the Danish solution FarmDroid FD20. It is a fully automated robot powered by solar energy. The machine can perform seeding or plant protection treatments [6]. The concept of a solar-powered agricultural robot was also presented in [7]. On the other hand, the French company Naio Technologies offers a range of robotic solutions for performing agricultural tasks [8]. The use of robots for agricultural work allows for the replacement of large and heavy machines (e.g., harvesters) with smaller vehicles that have a lesser impact on the soil [9].

Unmanned aerial vehicles, commonly known as drones, represent a separate category of devices that enable automation in agriculture. These machines allow for mapping agricultural land [10] and performing agricultural tasks such as seeding, spraying, and fertilization [11].

Cobots are a subgroup of robots that require interaction with humans to function properly. The human-robot interaction (HRI) approach was analyzed in the context of agriculture in [12, 13]. A potential benefit of its application is to improve worker safety. One of the key factors influencing the usability of the human-robot interface is the number and placement of camera images showing the situation around the robot [14, 15]. User interface research has been conducted to identify factors that are important for their functionality from the point of view of teleoperation of robots in agriculture [16]. Collaborative robots can be used to perform repetitive tasks that require hard physical labor or pose an increased risk to humans [17]. An example of the use of cobots is presented in [18], where an economic assessment of collaborative robots used in viticulture is presented. A cobot can be equipped with a robotic arm with a spraying head for applying chemicals to crops [19].

The use of robots for spraying offers several advantages. Robots can precisely apply pesticides and fertilizers, which reduces the risk of environmental pollution and increases the efficiency of chemical application. In addition, robots can be used to perform tasks that are unsafe for humans, such as spraying pesticides in areas with high pest pressure. An indirect benefit is the reduction of the costs of performing spraying operations due to the increased efficiency of the process and its automation, as well as the implementation of a sustainable development policy aimed at reducing the impact of agriculture on the environment and reducing dependence on human labor.

An overview of the latest achievements in the development of autonomous robots for spraying pesticides is presented in [20]. A spraying system integrated with a small robot can be used to spray plants in vineyards. The solution can be additionally equipped with additional algorithms that recognize the leaf density in selected regions of the plant [21]. Attempts have also been made to convert a general-purpose mobile robotic platform into a semi-autonomous agricultural sprayer [22]. An electric robot can also be used to spray plants in greenhouses. The intelligent spraying system is able to regulate the amount of products sprayed in terms of quantity and selective activation of nozzles based on planned spraying treatments [23].

The control algorithm for the sprayer installed on the robot can perform spraying in two main ways. The first is point spraying, which involves applying the chemical locally based on prior identification of the areas requiring treatment. The second method is zonal spraying, which involves applying the agent to a larger area where pathogens are present. Regardless of the method chosen, an electronic controller is required to implement the control algorithm in agricultural robots. This can be a microcontroller (e.g., PIC) working with a stepper motor in conjunction with a transmitter, receiver, and wireless camera [24]. It is also possible to use brushless DC motors (BLDC) to drive the vehicle [25]. Vision systems are used to detect areas requiring spot treatment. An example is the use of a color sensor signal in an algorithm for detecting plants requiring spraying [26]. Spraying robots typically have a four-wheel chassis design, however, an original solution in the form of a two-wheel platform was presented in [27].

An example of a robotic solution designed for spraying in agriculture is the XAG R150 unmanned vehicle. The device is equipped with the Jet Sprayer system, which consists of two spray heads with variable liquid flow rates [28]. The brushless direct drive nozzles with built-in gyroscope automatically maintain the spray angle. The vehicle is controlled using a controller or fully autonomously using an integrated route planning algorithm. Centimeter-accurate navigation of the vehicle in field conditions is provided by the RTK (Real Time Kinematic) system.

2 Material and Methods

The research utilized the XAG R150 autonomous platform, equipped with a dual-stream sprayer assisted by an air stream. The spray nozzles were mounted on the left and right sides of the sprayer platform. The range of working movements is possible in two planes. In the horizontal plane, the range of movements is from 0 to 300° when viewed from the longitudinal axis of the vehicle, while in the vertical plane, the working range is 200°. In the vertical plane, the nozzles can each work independently at a set individual tilt angle, or they can work in the so-called swing mode. In this mode, the nozzles perform an up-down movement while maintaining the set tilt angle. When driving using this working mode, the nozzles spray the surface in a sinusoidal motion. The amplitude of the sinusoid can be adjusted by changing the driving speed of the sprayer platform. The unit flow rate of the liquid can be adjusted in each nozzle and there is a possibility of changing the size of the generated droplets.

The spray platform has four modes of operation:

Waypoint mode: adjust parameters in the mobile app, enable autonomous operation along a predetermined route.
Shuttle mode: recording the operational trajectory. A few clicks are required for automatic travel between designated locations.
Follow-me mode: with one hand holding the control stick. The device follows operator.
Remote control mode: flexible hand control with a handheld device.

In the conducted field experiment, the fourth working mode was used with the maximum driving speed. The technical parameters of the spraying platform are shown in Table 1.

Table 1. Technical parameters of the XAG R150 spraying platform.

Basic parameters	Value
Spray range	60–200 μm
Maximum spray width	12 m
Peristaltic pump with high pulse frequency	-
Maximum system discharge rate	4.8 l/min
Maximum discharge rate (single pump)	2.4 l/min
Discharge air velocity	16 m/s
Tank capacity	100 l
Booster fan maximum speed	9000 rpm

3 Research Results

The research was conducted in two stages. The first stage was laboratory testing. The width of the sprayed liquid jet was investigated using a vertical spray patternator dedicated to measuring the sprayed liquid jet from orchard sprayers. The measurement was made for the jet at different angles of inclination of the nozzle. For a fixed working angle, the measurement of the width of the sprayed liquid jet was made from a distance of 1.5 m from the spray patternator at three angles of the nozzle setting. Angle 0° was the setting in the horizontal position and in the remaining two cases for angles of 15 and 45° from the horizontal plane. The results of these measurements are presented in Figs. 1 and 2. The research showed that the width of the sprayed liquid jet at the setting of the droplets to the lowest possible declared value was 40 cm for the setting positions of 0 and 15 degrees, for the setting position of 45° the width of the jet was 50 cm, and the range of the jet was possible from 170 to 210 cm above the ground level. From the conducted research on the width of the jet, it can be seen that it is not uniform throughout its entire range, but it is highly concentrated and most of the liquid in the air jet over 50% is delivered only in the middle of the jet on a width of 10 cm.

With the assumed maximum liquid flow rate and the setting for the largest possible droplet thickness generated by the atomizer, the liquid distribution in the air stream was similar to that with small droplets. At a nozzle tilt angle of 45°, the jet width was larger and the liquid distribution in the jet was more uniform. This phenomenon results from the interaction of the round jet generated by the nozzle set at an angle to the sprayed surface, which in turn results in an elliptical jet shape that reaches the spray patternator or the vertical sprayed object.

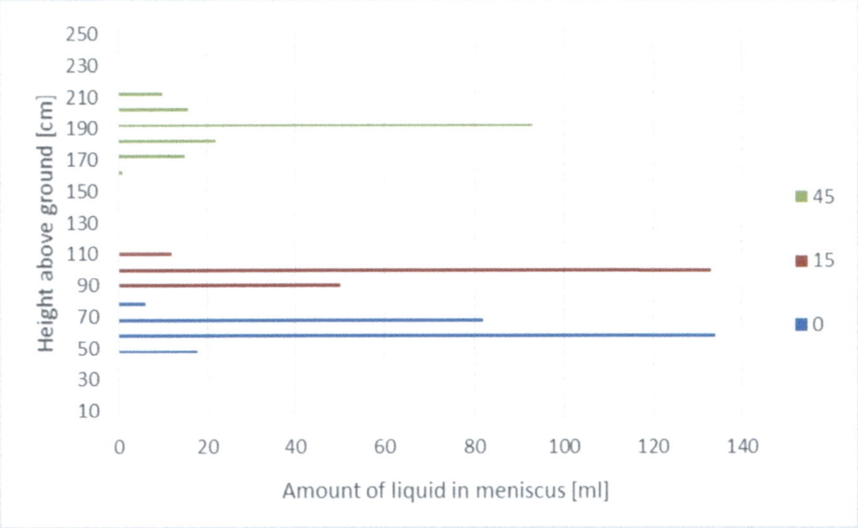

Fig. 1. Amount of liquid depending on the angle of deflection of the sprayer (expressed in degrees) for a fine droplet

For swing mode operation, the nozzle worked in a range from −15° down from the horizontal plane to 45° up. The nozzle worked for one minute performing an up-down movement. The measurement was made for two droplet size settings, for fine droplets and for coarse droplets. The results of these measurements are presented in Figs. 3 and 4. The results obtained indicate that in swing mode operation, the most liquid is applied at the set minimum and maximum nozzle deflections. This phenomenon is similar for the setting of the minimum and maximum possible droplet sizes. This is probably due to the swinging motion of the nozzle and at maximum deflections there is a change in the direction of motion and there are so-called dead spots, in which liquid is applied in the air stream and the nozzle stops and changes direction.

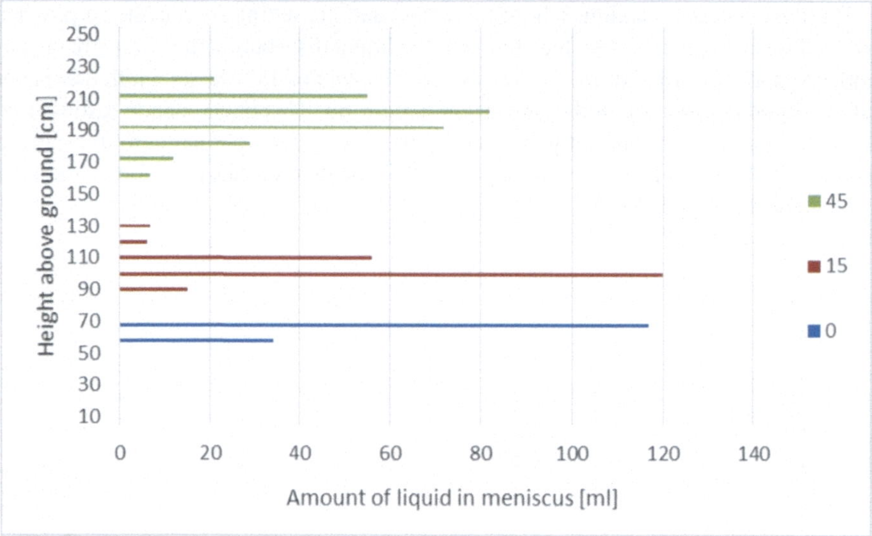

Fig. 2. Amount of liquid depending on the angle of deflection of the sprayer (expressed in degrees) for a coarse droplet

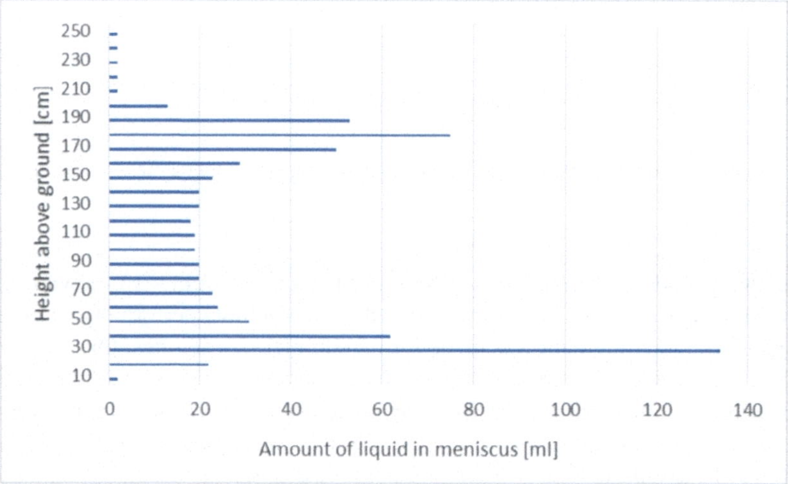

Fig. 3. Amount of liquid for swing mode and fine droplet

The second stage of the research involved field studies conducted on a black currant plantation. The research was conducted in the second half of October. The black currant bushes were foliated at 50%. Water-sensitive papers were placed on bamboo sticks in the central part of the rows of shrubs. The sticks were placed every 50 cm in a row. The water-sensitive papers were placed at three heights: 40, 80, and 120 cm. These settings represented positions above the ground, the middle part of the shrubs, and

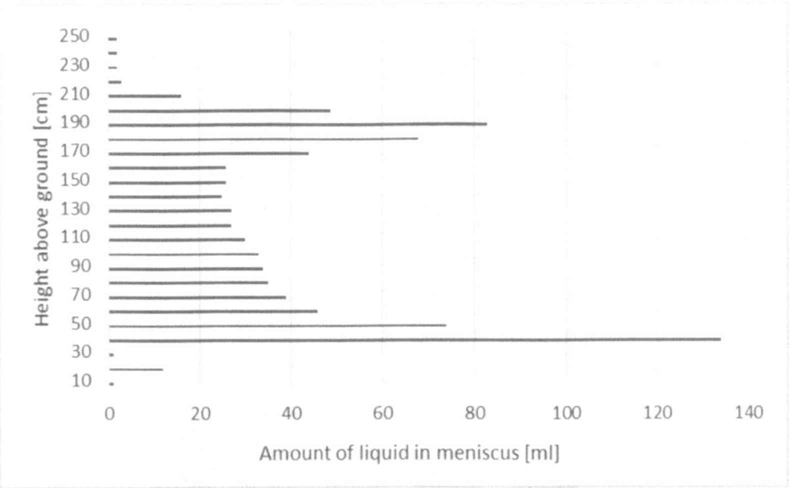

Fig. 4. Amount of liquid for swing mode and coarse droplet

positions in the upper part of the shrubs. For the measurements, the maximum driving speed, maximum droplet size, and working positions in swing mode in the range of nozzle deflection from -15 to $+15°$ were set. The results of the conducted study are presented in Table 2 and Fig. 5. The degree of coverage of the water-sensitive papers was determined visually using a template. The field study indicates that in the assumed swing-working mode, in order to cover the height of the sprayed shrubs with its range, the spray jet was characterized by an uneven application of liquid. Excessive liquid application occurred at the minimum and maximum positions, and in subsequent locations the amount of liquid delivered was over half less. This is due to the vertical movement of the nozzle and the movement of the sprayer platform, which together result in a sinusoidal movement of the sprayed liquid jet.

Table 2. Degree of coverage of water-sensitive papers in one row of black currant [%].

Height [cm]	Arrangement of poles				
	1	2	3	4	5
A 120	90	95	35	85	75
B 80	35	65	55	65	70
C 40	75	65	60	55	80

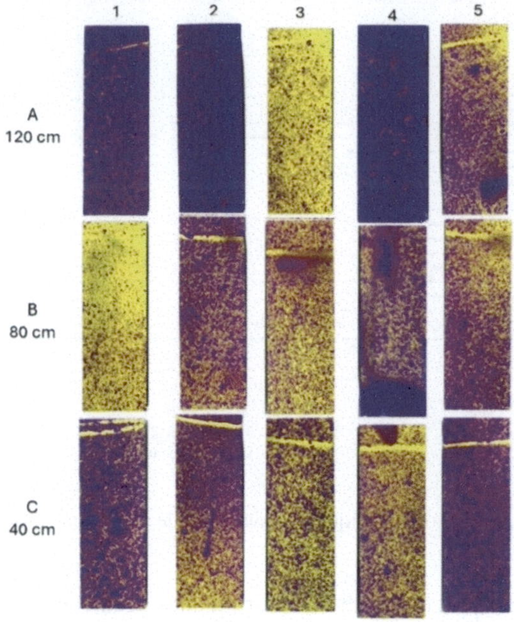

Fig. 5. Image of the degree of coverage of water-sensitive papers in the sprayed row of black currant.

4 Conclusion

Elements of precision agriculture will be increasingly used in crop production. Crop protection treatments are such activities that are easy to automate and target areas where they are needed at a given time without preventive spraying of the entire plantation. Autonomous robots or vehicles with automatic control can be successfully used in agriculture. The autonomous platform with a mounted two-stream sprayer presented in the work can be successfully used for point application of plant sprays. The stream sprayer mounted on the platform with adjustable liquid flow rate has a rather compact stream of sprayed liquid. The use of oscillating spray nozzle in conjunction with the movement of the vehicle gives a sinusoidal pattern of the spray jet. Such a jet does not allow for accurate coverage of the entire high row of sprayed plants. Reducing the speed of the platform movement may cause excessive blowing of the bushes and, consequently, liquid loss. At higher speeds, there may be insufficient application of liquid to the sprayed plants. This platform with a stream sprayer can be perfect for spraying low-growing crops grown in rows.

References

1. Lytridis, C., et al.: An overview of cooperative robotics in agriculture. Agronomy **11**(9), 1818 (2021)
2. Mahmud, M.S.A., Abidin, M.S.Z., Emmanuel, A.A., Hasan, H.S.: Robotics and automation in agriculture: present and future applications. Appl. Model. Simul. **4**, 130–140 (2020)

3. Duckett, T., et al.: Agricultural robotics: the future of robotic agriculture. arXiv preprint arXiv: 1806.06762 (2018)
4. Roldán, J.J., et al.: Robots in agriculture: State of art and practical experiences. Serv. Robots **12**(2), 67–90 (2018)
5. Agro Inteligence Website: https://agrointelli.com/robotti/lr/. Last accessed 1 May 2024
6. FarmDroid Website, https://farmdroid.com/products/farmdroid-fd20/. Last accessed 1 May 2024
7. Kumar, A., Deepak, R.S., Kusuma, D.S., Sreekanth, D.V.: Review on multipurpose agriculture robot. Int. J. Res. Appl. Sci. Eng. Technol. **8**(5), 1314–1318 (2020). https://doi.org/10.22214/ijraset.2020.5209
8. Naïo Technologies Website: https://www.naio-technologies.com/en/home/. Last accessed 1 May 2024
9. King, A.: Technology: the future of agriculture. Nature **544**(7651), S21–S23 (2017)
10. Hafeez, A., et al.: Implementation of drone technology for farm monitoring & pesticide spraying: a review. Inform. Process. Agric. **10**(2), 192–203 (2023)
11. Ghazali, M.H.M., Azmin, A., Rahiman, W.: Drone implementation in precision agriculture—A survey. Int. J. Emerg. Technol. Adv. Eng. **12**(4), 67–77 (2022)
12. Vasconez, J.P., Kantor, G.A., Cheein, F.A.A.: Human–robot interaction in agriculture: a survey and current challenges. Biosys. Eng. **179**, 35–48 (2019)
13. Benos, L., et al.: Human–robot interaction in agriculture: a systematic review. Sensors **23**(15), 6776 (2023)
14. Adamides, G., et al.: HRI usability evaluation of interaction modes for a teleoperated agricultural robotic sprayer. Appl. Ergon. **62**, 237–246 (2017)
15. Adamides, G., et al: A reality-based interaction interface for an agricultural teleoperated robot sprayer. In: International Conference on Robotics and Associated High-Technologies and Equipment for Agriculture and Forestry, vol. 2 (2014)
16. Adamides, G., Katsanos, C., Christou, G., Xenos, M., Papadavid, G., Hadzilacos, T.: User interface considerations for telerobotics: the case of an agricultural robot sprayer. In: Second international conference on remote sensing and geoinformation of the environment (RSCy2014), vol. 9229, pp. 541–548. SPIE (2014)
17. Bissadu, K.D., Sonko, S., Hossain, G.: Society 5.0 enabled agriculture: drivers, enabling technologies, architectures, opportunities, and challenges. Inform. Process. Agric. (2024)
18. Tziolas, E., et al.: Assessing the economic performance of multipurpose collaborative robots toward skillful and sustainable viticultural practices. Sustainability **15**(4), 3866 (2023)
19. Tomazzoli, C., Ponza, A., Cristani, M., Olivieri, F., Scannapieco, S.: A cobot in the vineyard: computer vision for smart chemicals spraying. Appl. Sci. **14**(9), 3777 (2024)
20. Meshram, A.T., Vanalkar, A.V., Kalambe, K.B., Badar, A.M.: Pesticide spraying robot for precision agriculture: a categorical literature review and future trends. J. Field Robot. **39**(2), 153–171 (2022)
21. Baltazar, A.R., Santos, F.N.D., Moreira, A.P., Valente, A., Cunha, J.B.: Smarter robotic sprayer system for precision agriculture. Electronics **10**(17), 2061 (2021)
22. Adamides, G., et al.: Design and development of a semi-autonomous agricultural vineyard sprayer: human–robot interaction aspects. J. Field Robot. **34**(8), 1407–1426 (2017)
23. Cantelli, L., Bonaccorso, F., Longo, D., Melita, C.D., Schillaci, G., Muscato, G.: A small versatile electrical robot for autonomous spraying in agriculture. AgriEng. **1**(3), 391–402 (2019)
24. Aishwarya, B.V., Archana, G., Umayal, C.: Agriculture robotic vehicle based pesticide sprayer with efficiency optimization. In: 2015 IEEE Technological Innovation in ICT for Agriculture and Rural Development (TIAR), pp. 59–65. IEEE (2015)

25. Kassim, A.M., et al.: Design and development of autonomous pesticide sprayer robot for fertigation farm. Int. J. Adv. Comput. Sci. Appl. **11**(2), 545–551 (2020). https://doi.org/10.14569/IJACSA.2020.0110269
26. Tahmasebi, M., Gohari, M., Emami, A.: An autonomous pesticide sprayer robot with a color-based vision system. Int. J. Robot. Control Syst. **2**(1), 115–123 (2022)
27. Ghafar, A.S.A., Hajjaj, S.S.H., Gsangaya, K.R., Sultan, M.T.H., Mail, M.F., Hua, L.S.: Design and development of a robot for spraying fertilizers and pesticides for agriculture. Mater. Today: Proc. **81**, 242–248 (2023)
28. XAG R150 Website: https://www.xa.com/en/xauv_r150. Last accessed 1 May 2024

Improved Design of the Working Body of a Flexible Sectional Screw Conveyor

Simone Pascuzzi[1(✉)], Volodymyr Bulgakov[2], Valerii Adamchuk[3], Oleksandra Trokhaniak[2], Janusz Nowak[4], Adolfs Rucins[5], Zinoviy Ruzhilo[2], and Francesco Paciolla[6,7]

[1] Department of Soil, Plant and Food Science, University of Bari Aldo Moro, Via Amendola 165/A, 70126 Bari, Italy
simone.pascuzzi@uniba.it

[2] Department of Mechanics, Faculty of Construction and Design, National University of Life and Environmental Sciences of Ukraine, 15 Heroiv Oborony Str., Kyiv 03041, Ukraine

[3] Institute of Mechanics and Automation of Agricultural Production of NAAS, 11 Vokzalna Street, Glevakh Village, Fastiv District, Kyiv 08631, Ukraine

[4] Department of Operation of Machines and Management of Production Processes, University of Life Sciences in Lublin, 13, Akademicka Str., 20-950 Lublin, Poland

[5] Institute of Engineering and Energetics, Latvia University of Life Sciences and Technologies, Cakstes Blvd. 5, Jelgava 3001, Latvia

[6] PolySense Lab, Dipartimento Interateneo di Fisica, Polytechnic and University of Bari, Via Amendola 173, 70126 Bari, Italy

[7] Dipartimento di Ingegneria Elettrica e dell'Informazione, Polytechnic University of Bari, Via Edoardo Orabona 4, 70126 Bari, Italy

Abstract. The article outlines the design of a sectional screw working body which can improve the performance of screw conveyors, reduce the material consumption of the working body and energy costs for the transportation process. An analytical dependence between the design parameters of the flexible shaft of the screw working body and the radius of its curvature, which is determined by the angular displacement of adjacent sections of the screw working body, is derived. The graphical relationship between the parameters of the connection of sections and the radius of curvature of the technological line is drawn. To ensure proper functioning, an investigation has been carried out of the design parameters of the hinge assembly, which ensures their mutual rotation and angular displacement of the axes without jamming. An analytical relationship is derived by means of which it is possible to determine the displacements along the arc of the spherical bushing of the ball on both friction surfaces, the speed of the movement of each ball, as well as the efficiency coefficiency of the hinge and the amount of torque on the drive shaft $T = f(\varphi)$ of individual sections.

Keywords: Screw Conveyor · Flexible Shaft · Hinge Joint · Angular Displacement · Radius of Curvature

E. Lorencowicz et al. (Eds.): FMPMSA 2024, LNCE 609, pp. 367–377, 2024.
https://doi.org/10.1007/978-3-031-70955-5_40

1 Introduction

In various branches of industry is arising the need to move bulk and lump agricultural materials to different distances, both horizontally and in vertical and inclined directions [1–5]. Such a need is usually associated with the technological cycle and has the goal to mechanize the process. High mobility of performing technological processes of moving bulk and lump agricultural materials [6–8] is effectuated out by using flexible screw conveyors. From the analysis of literary sources, we can conclude that the least resource-intensive technology of moving bulk and lumpy agricultural materials along curved routes is the use of flexible screw conveyors consisting of separate screw sections that have hinged each other [9–12]. Some papers [13–18] substantiate the design, kinematic and technological parameters of screw working bodies from the point of view of their functional purpose. The results of research of the movement of the material in a screw conveyor with a rotating casing are presented in [19, 20]. The development and research of screw conveyors, as well as their influence on the degree of damage to the transported materials, are provided in few papers [21–24].

However, the existing flexible screw working bodies cannot fully satisfy the operational requirements, set for such types of conveyors. Therefore, the purpose of this investigation is to determine rational design parameters and their justification, which will improve the performance of flexible screw conveyors with low material consumption of the working body, ensuring a reduction in energy costs of the transportation process.

2 Materials and Methods

To improve the performance of flexible screw conveyors, the design of a screw sectional working body was developed, shown in Fig. 1. It contains sections (Fig. 1a–1) of identical diameters and lengths, which are made in the form of welded structures of tubular workpieces (Fig. 1a–2), to one conditional inner end of which a cylindrical sleeve (Fig. 1a–3) is rigidly attached. The latter is made with a system of axial parallel grooves (Fig. 1a–4), evenly spaced in a circle, for example, four, and are parallel to the axis of the tubular workpiece and interact with the rolling elements (Fig. 1a–5). On the other side of the rolling body (Fig. 1a–5) there are interacting with the spherical recess (Fig. 1a–6) of the spherical body (Fig. 1a–7), having a possibility of axial and angular rotation of the spherical drag bar (Fig. 1a–8). In addition, at the top the cylindrical sleeve (Fig. 1a–3) is in rigid interaction with the spherical body (Fig. 1a–7) in which the inner sphere (Fig. 1a–6) interacts with the outer surfaces of the rolling elements (Fig. 1a–5), but at the free inner end of the spherical body (Fig. 1a–7) there is a cylindrical ring with an outer triangular profile (Fig. 1a–9), which is placed in a rigid interaction with a similar recess (Fig. 1a–10) of the cylindrical sleeve (Fig. 1a–3), made at its free end from the outside. From the other end of the welded cylindrical section, there is a connecting sleeve (Fig. 1a–11) with an internal square or shaped hole (Fig. 1a–12) installed rigidly into the internal hole, perpendicular to the axis, which interacts with the corresponding end of the spherical drag bar of the adjacent section and is rigidly fixed in a known manner. To the outer diameter of the connecting sleeve (Fig. 1a–11) there are welded, for example, four jumpers (Fig. 1a–13) evenly around the circumference at one end. The operation of the screw working body takes place as follows. During the rotation of the spiral section (Fig. 1a–14) the rotational motion is transmitted through the rolling elements (Fig. 1a–5) to the spherical body (Fig. 1a–7) and adjacent sections of the screw working body.

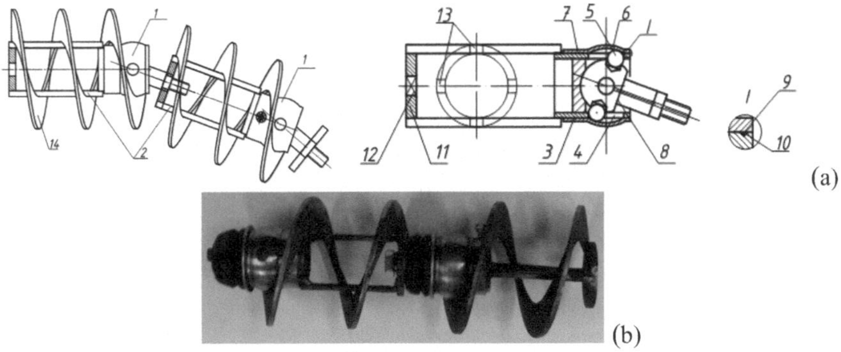

Fig. 1. Structural diagram (a) and general view (b) of the sectional screw working body.

To determine the relationship between the design parameters of the flexible shaft and the radius of its curvature, determined through the angular displacement of adjacent sections, in Fig. 2 there is presented a design diagram. The diagram shows the following symbols: 1 – spherical pin; 2 – cylindrical sleeve; 3 – ball; 4 – sleeve with a spherical surface; α – turning angle of the spherical pin; γ –angle of the generatrix of the conical surface of the hole; R – turning radius of the centres of the balls; R_1 – outer radius of the cylindrical sleeve; R_2 – inner radius of the cylindrical sleeve; R_3 – inner radius of a sleeve with a spherical surface; R_n – radius of the sphere of the pin; r – radius of the ball; h –thickness of the wall of the cylindrical sleeve; H – length of a half of the groove of the cylindrical sleeve; Δ – a necessary guaranteed overlap of the ball and groove of the cylindrical sleeve; λ – distance from the spherical surface of the pin to the centre of the ball; τ – maximum distance from the inner surface of the spherical sleeve to the outer surface of the cylindrical sleeve; δ – gap between the spherical pin and the cylindrical sleeve; s – preset setting depth of the ball.

In the process of angular displacement of the sections the spherical pin turns relative to the cylindrical sleeve at the maximum possible angle φ, at which its contact with the cylindrical sleeve occurs through the balls. In order to ensure the correct functioning of the hinged unit, it is necessary to conduct a research of its structural parameters that ensure mutual turning, the necessary angular displacement of the axes and the absence of jamming.

Let the following characteristics of the construction be known, thus the radius of the ball r, the turning radius R of the centres of the ball and the maximum allowable angle α of deviation of the axis. We determine the dimensions of the cylindrical sleeve, its outer R_1 and inner R_2 radii, the radius of the spherical pin R_n and the dimensions of the hole, the angle of the conical surface γ and the diameter on the outer radius of the spherical pin d (R_4). Appropriate designations are shown in Fig. 2.

When turning, the centre of the ball describes an elliptical line, projected onto a plane, perpendicular to the axis of the pipe. The semimajor axis of the ellipse is $a = R$, and the minor axis is $b = R \cdot \cos\,\alpha$, respectively. Consequently, the inner and the outer radii of the tubular part must be within appropriate limits, considering the required guaranteed overlap Δ:

$$\left.\begin{array}{r} R_1 = R + \Delta; \\ R_2 = R \cdot \cos\alpha - \Delta. \end{array}\right\} \tag{1}$$

Accordingly, the radius of the spherical pin, holding the balls in the holes, must be less than the internal radius of the cylindrical sleeve by a certain guaranteed clearance δ:

$$R_{\Pi} = R_2 - \delta = R \cdot \cos\alpha - \Delta - \delta. \tag{2}$$

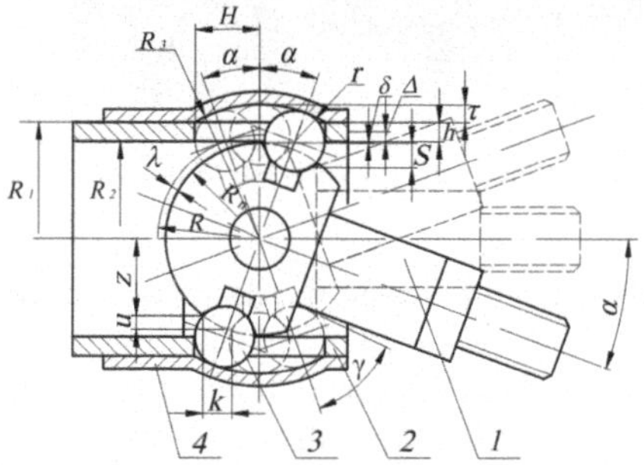

Fig. 2. Calculation diagram of the hinged connection of the screw working body.

To calculate the angle γ of the generatrix of the conical surface of the hole, we use the condition of contact of the ball with the conical surface below the edge at depth s (Fig. 3). From geometric relationships we obtain the relationship:

$$R - R_{\Pi} = r \cdot \sin\gamma - s. \tag{3}$$

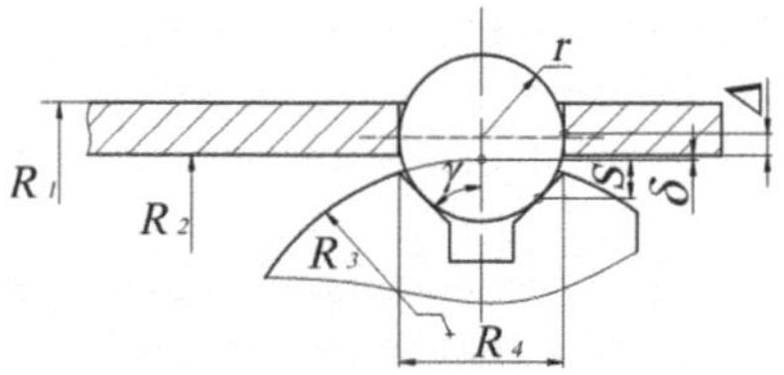

Fig. 3. Calculation scheme for determination of the angle of the generatrix of the conical surface of the hole.

From where we find the required angle of the generatrix of the conical surface of the hole:

$$\sin\gamma = \frac{(R \cdot (1 - \cos\alpha) + \Delta + \delta + s)}{r}. \tag{4}$$

At a smaller value of the angle there will not be ensured correct contact of the ball with the groove surface of the tubular part of the hinge. A larger value of angle γ is possible but this will significantly increase the load upon the ball and hole due to a significant radial component of the reaction in the conical hole. However, for the design and technological reasons of availability of a certain cutting tool, angle γ may be slightly increased. At the selected angle of the generatrix of the conical surface of the hole γ and the preset depth s of the seat of the ball, we determine the diameter d of the hole on the surface of the spherical pin for technological control of manufacture:

$$d = 2 \cdot (r \cdot \cos\gamma + s \cdot \tan\gamma). \tag{5}$$

Writing the inner radius of the spherical sleeve in the form of equalities:

$$R_3 = R_2 + h + \tau = R \cdot \cos\alpha - \Delta + h + \tau, \tag{6}$$

$$R_3 = R + r. \tag{7}$$

Equalizing dependencies eqs. (6) and (7), we obtain:

$$R \cdot \cos\alpha - \Delta + h + \tau = R + r, \tag{8}$$

From where it is possible to determine the maximum distance τ from the inner surface of the spherical sleeve to the outer surface of the cylindrical sleeve:

$$\tau = R \cdot (1 - \cos\alpha) + r - \Delta + h. \tag{9}$$

Value H of a half of the length of the groove of a cylindrical sleeve is determined from the relationship:

$$H = R \cdot \sin\alpha + r = (R_n + \lambda) \cdot \sin\alpha + r. \tag{10}$$

Since $R = R_n + \lambda$, the value λ of the gap between the spherical surface of the pin and the centre of the ball is determined equalizing it with (2), obtaining:

$$\lambda = \frac{R_n + \Delta + \delta}{\cos\alpha} - R_n. \tag{11}$$

The setting depth s of the ball in the spherical pin is determined from the condition:

$$\lambda = R - R_n = r \cdot \sin\gamma - s, \tag{12}$$

and equate it to dependence (12), it is possible to obtain:

$$\frac{R_n + \Delta + \delta}{\cos\alpha} - R_n = r \cdot \sin\gamma - s, \tag{13}$$

From where:

$$s = r \cdot \sin\gamma + R_n - \frac{R_n + \Delta + \delta}{\cos\alpha}. \tag{14}$$

The obtained equalities (10), (12) and (14) make it possible to establish the dependence of the length of half of the groove of the cylindrical sleeve H, the gap between the spherical surface of the pin and the centre of the ball λ, and the seating depth s of the balls in the spherical pin on the running angle α. In order to establish the adequacy of the theoretical study, static experimental research of the sectional working body was done, which made it possible to determine the torque, depending on the relative angular movement of the end surfaces.

The research was conducted using the following methodology. To make the experiments, a section of the working articulated sectional organ was installed in the grips of the KM-50-1 testing machine. Loading onto the section was done by turning the lower gripper with an electric motor through a gear system. The torque value was determined on a circular scale, while the relative displacement of the end surfaces of the working body was recorded using an angular scale. The machine is also equipped with a recording device through which the curve of the torque versus the angle of turnin of the lower grip is crossed out. Three versions of experimental prototypes were made: two single-screw sections, corresponding to two and four longitudinal rods, as well as part of the working body, made of two sections. Photographs of the manufactured versions of the experimental prototypes are shown in Fig. 4.

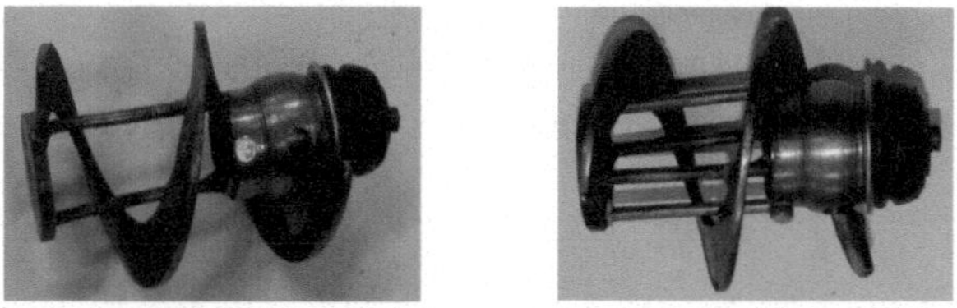

(a) (b)

Fig. 4. General view of the sample sections of helical working tools: (a) Screw section with two longitudinal rods; (b) Screw section with four longitudinal rods.

The produced sections of the helical working body had the following design parameters: outer diameter of the helix $D = 96$ mm; width of the helix $B = 25$ mm; thickness of the helix $t = 4$ mm; length of the section $L = 130$ mm; diameter of the longitudinal bar $d = 5$ mm. The general view of torsion machine KM-50-1 with prototypes of sections of working tools is shown in Fig. 5.

During the investigation a discrete load of the screw section occurred, where the magnitude of the torque and the corresponding angle of twist of the end surfaces of the sections were simultaneously recorded. The research of various options for making sections and their connections were carried out in triplicate repetition.

3 Results

Based on the results of the analysis of the derived analytical dependencies eqs. (10), (11) and (14), graphs of changes in the length of half the groove of the cylindrical sleeve H, the gap between the spherical surface of the pin and the centre λ of the ball, and the

Fig. 5. General view of the KM-50–1 testing machine with experimental prototypes of sections of the working.

seating depth s of the balls s in the spherical pin from the running angle α, shown in Fig. 6. Graphic dependencies are plotted at the following constant values: $R_n = 15$ mm, $r = 4.75$ mm, $\delta = 0.2$ mm, $\Delta = 1.5$ mm, $\gamma = 45°$.

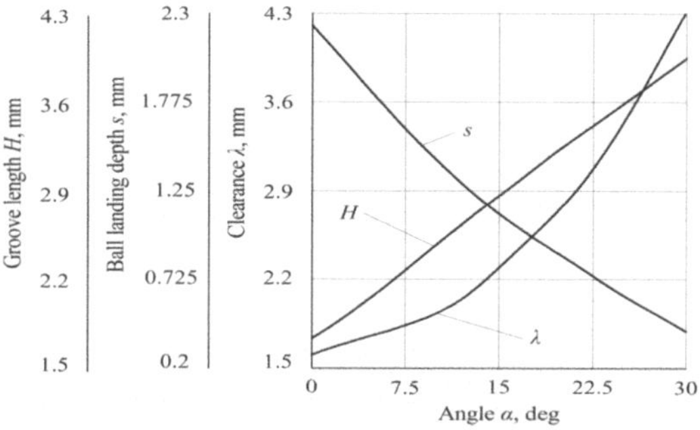

Fig. 6. Dependences of the change in the half of the length of the groove of the cylindrical sleeve H, the gap between the spherical surface of the pin and the centre λ of the ball, and the seating depth s of the balls in the spherical pin upon the running angle α.

From the analysis of graphical dependencies in Fig. 6 it can be established that at an angle α, of more than 25°, the gap between the spherical surface of the pin and the centre λ of the ball becomes large, but the seating depth of the ball s in the spherical pin obtains small values. Therefore, it is advisable to use articulated sectional working bodies with a maximum turning angle of the spherical pin $\alpha_{\max} = 25°$ and a groove length of the cylindrical sleeve of $2H$.

When analysing the influence of the design parameters of the screw sections of the flexible shaft on the change in the value of R_K, we will set the real limits of the values of R_n, r, h, δ, Δ and L. So, for the working body of a flexible conveyor, which is located in a standard elastic casing with an internal diameter of 100 mm, the outer diameter of the flexible shaft should not exceed 50 mm to ensure the installation of a helical spiral. Therefore, the variable parameters are accepted within the limits: $R_n = 10 \div 20$ mm; $r = 4.55 \div 4.95$ mm; $h = 2 \div 4$ mm; $\delta = 0.1 \div 0.3$ mm; $\Delta = 1.3 \div 1.7$ mm; $L = 110 \div 150$ mm. Using the obtained results, it is possible to determine the minimum permissible radius of curvature of the technological line R_n. For the preset length of the section of the working body L, defined as the distance between the hinge holes, the permissible radius of curvature of the technological line R_K is calculated according to the dependence:

$$R_K = \frac{L(\cos\frac{\alpha}{2} + k)}{4\sin\frac{\alpha}{2} \cdot (k+1)}. \tag{15}$$

Figure 7 presents graphical dependences of the minimum permissible radius of curvature of the technological line R_K upon the amount (degree) of angular turning of the adjacent sections at their different lengths L.

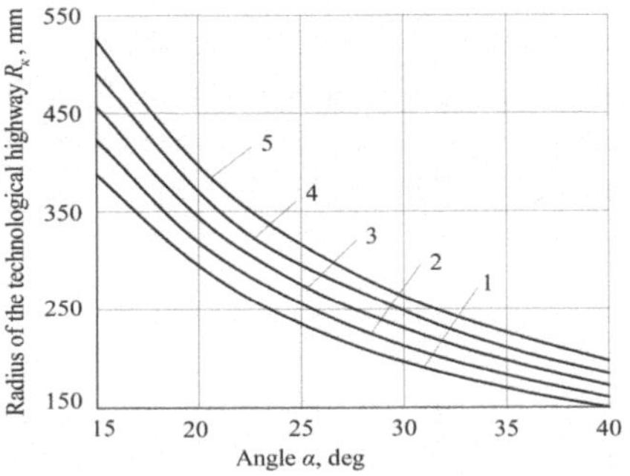

Fig. 7. Graphic dependences of the minimum permissible radius of the technological line R_K on angle α for different lengths: 1 – L = 110 mm; 2 – L = 120 mm; 3 – L = 130 mm; 4 – L = 140 mm; 5 – L = 150 mm.

It should be noted that for efficient operation of the flexible screw conveyor the certain minimum permissible radius of curvature of the technological line must be increased by 20 ÷ 30% to ensure guaranteed contactless turning of the hinged sections, which will eliminate the possibility of emergency situations and increase the service life of the proposed working element. Based on the analysis of graphical dependencies (Figs. 6 and 7), functional dependencies of the radius of curvature of the working body on its design parameters were established. Thus, to ensure the minimum permissible radius of curvature of the working body up to 0.5 m, the following limits for the design parameters have been imposed: $R_n = 10 \div 20$ mm; $h = 2 \div 4$ mm; $r = 4.55 \div 4.95$ mm; $L = 110 \div 150$ mm.

On the basis of the results of experimental investigations there are constructed graphical dependences of the torque on the section and connected sections upon the angle of twist of the end elements, shown in Fig. 8.

From their analysis it can be established that for individual sections without a hinge connection (curves 1 and 2) the functional dependence is of clearly linear nature. In addition, for the above design parameters of the screw sections a further increase in the torque value led to the appearance of residual deformations, which is unacceptable.

The curvilinear nature $T = f(\varphi)$ in the range of up to 40 ÷ 60 Nm for two sections, connected to each other, is explained by the fact that the appearance of additional hinge links leads to a more intense growth of angle φ with increasing T as a result of manufacturing inaccuracy and the occurrence of local gaps. With a further increase in the value of $T > 60$ Nm, the curve becomes a clearly expressed linear dependence.

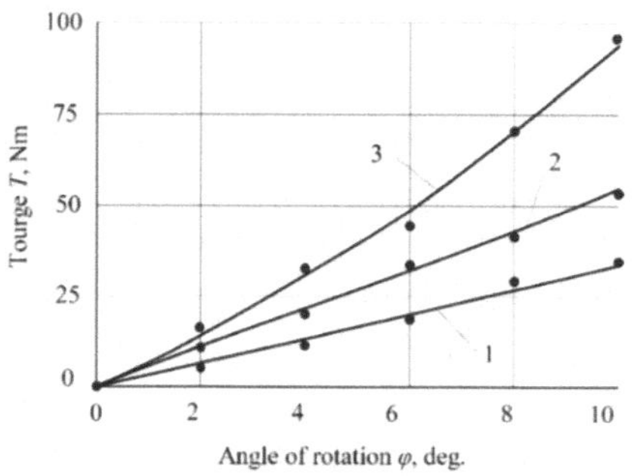

Fig. 8. Graphic dependences of the change in torque T on a section upon the angle of twist φ of the end surfaces: 1 – a screw section with four longitudinal rods; 2 – a screw section with two longitudinal rods; 3 – two sections with two longitudinal rods, connected to each other by an elastic element.

It was also found out that for folded sections an increase in torque of more than 100 Nm leads to the appearance of residual deformations on the sectional elements.

4 Conclusions

Based on the analysis literature and patens, a design of a flexible sectional screw working body which can improve the performance of flexible screw conveyors with low material consumption of the working body and reduced energy costs has been developed and manufactured. A design (calculation) scheme has been developed to determine the relationship between the design parameters of the flexible shaft and its radius of curvature, determined through the angular displacement of adjacent sections of the screw working body. An analytical dependence of the relationship between the design parameters of a flexible shaft and the radius of its curvature has been deduced, recognized through the angular displacement of adjacent sections, and a graphical dependence is presented of the influence of the design parameters of the connection upon the value of the curvature angle of the technological line. There have been constructed graphic dependences of the minimum permissible radius of the technological line upon the amount of angular turning. It has been established that it is advisable to use hinged sectional working bodies with a maximum turning angle of the spherical pin and a groove length of the cylindrical sleeve of $2H$.

In order to ensure correct functioning of the hinge unit, an investigation of its design parameters was made, ensuring mutual rotation, the necessary angular displacement of the axes and the absence of jamming. An analytical relationship has been derived to determine the arc movement of the ball along both friction surfaces, the speed of movement of each ball along the arc of the spherical sleeve, as well as the efficiency of the hinge and the magnitude of the torque on the driving shaft. Thus, to ensure the minimum permissible radius of curvature of the working body up to 0.5 m, the following limits of the design parameters are optimal: the radius of the pin sphere $= 10 \div 20$ mm; the wall thickness of the cylindrical sleeve $h = 2 \div 4$ mm; the ball radius $r = 4.55 \div 4.95$ mm; the length of the working body section $L = 110 \div 150$ mm. Based on the results of static experimental research, it has been established that the dependence of torque on the twist angle ($T = f(\varphi)$) of individual sections is clearly linear. The curvilinear nature within the range of up to $40 \div 60$ Nm for two folded sections is explained by the appearance of local gaps. With a further increase in the value of $T > 60$ Nm, the curve has a linear dependence. It has been established that an increase in torque of more than 100 Nm leads to the appearance of residual deformations on the sectional elements.

References

1. Bulgakov, V., Pascuzzi, S., Nadykto, V., Ivanovs, S., Adamchuk, V.: Experimental study of the implement-and-tractor aggregate used for laying tracks of permanent traffic lanes inside controlled traffic farming systems. Soil Tillage Res. **208**, 104895 (2021)
2. Bulgakov, V., Pascuzzi, S., Ivanovs, S., Nadykto, V., Nowak, J.: Kinematic discrepancy between driving wheels evaluated for a modular traction device. Biosys. Eng. **196**, 88–96 (2020)
3. Pascuzzi, S., Anifantis, A.S., Santoro, F.: The concept of a compact profile agricultural tractor suitable for use on specialised tree crops. Agriculture **10**, 123 (2020)
4. Bulgakov, V., Pascuzzi, S., Beloev, H., Ivanovs, S.: Theoretical investigations of the headland turning agility of a trailed asymmetric implement-and-tractor aggregate. Agriculture **9**, 224 (2019)
5. Bulgakov, V., Pascuzzi, S., Anifantis, A.S., Santoro, F.: Oscillations analysis of front-mounted beet topper machine for biomass harvesting. Energies **12**, 2774 (2019)
6. Bulgakov, V., Pascuzzi, S., Adamchuk, V., Kuvachov, V., Nozdrovicky, L.: Theoretical study of transverse offsets of wide span tractor working implements and their influence on damage to row crops. Agriculture **9**, 144 (2019)
7. Trokhaniak, O.M., Hevko, R.B., Lyashuk, O.L., Dovbush, T.A., Pohrishchuk, B.V., Dobizha, N.V.: Research of the of bulk material movement process in the inactive zone between screw sections. INMATEH Agric. Eng. **60**(1), 261–268 (2020). https://doi.org/10.35633/inmateh-60-29
8. Pascuzzi, S., Łyp-Wrońska, K., Gdowska, K., Paciolla, F.: Sustainability evaluation of hybrid agriculture-tractor powertrains. Sustainability **16**(3), 1184 (2024)
9. Bulgakov, V., Trokhaniak, O., Klendii, M., Ivanovs, S., Dukulis, I.: Research on the impact of the operating modes and main design parameters on t he efficiency of the machine for preparing and packing slaked lime. INMATEH Agric. Eng. **67**, 323–330 (2022). https://doi.org/10.35633/inmateh-67-33
10. Hevko, R.B., Klendiy, M.B., Klendiy, O.M.: Investigation of a transfer branch of a flexible screw conveyer, INMATEH. Agric. Eng. **48**(1), 29–34 (2016)
11. Lech, M.: Mass flow rate measurement in vertical pneumatic conveying of solid. Powder Technol. **114**(1–3), 55–58 (2001)
12. Hevko, R.B., Rohatynskyi, R.M., Rozum, R.I., Klendii, M.B., et al.: Improving the technological level of loading and reloading processes of materials in screw conveyors: a monograph. Ternopil, p.180 (2018). (In Ukrainian)

13. Bulgakov, V., et al.: Research into power and load parameters of flexible screw conveyors for transportation of agricultural materials. In: Pascuzzi, S., Santoro, F. (eds.) Farm Machinery and Processes Management in Sustainable Agriculture: XI International Scientific Symposium 2022, pp. 61–75. Springer International Publishing, Cham (2023). https://doi.org/10.1007/978-3-031-13090-8_6
14. Hevko, R.B., Hladio, Y., Shunkaryk, M.I, Klendii, OM: Dynamic calculation of a screw conveyor safety device. Bull. Eng. Acad. Ukraine **3–4**, 46–52 (2014). (In Ukrainian)
15. Anelli, F., Annunziato, A., Erario, A., Holmes, C., Ciminelli, C., Prudenzano, F.: Design of microstructured flat optical fiber for multiaxial strain monitoring in composite materials. J. Lightwave Technol. **40**(17), 5986–5994 (2022)
16. Vitrovyi, A.O., Hevko, R.B.: Force analysis of the working body of a flexible screw conveyor. Coll. Sci. Art. Lutsk State Tech. Univ. Agric. Mach. **4**, 34–36 (1998). (In Ukrainian)
17. Hevko, R.B., Rozum, R.I., Klendiy, O.M.: Development of design and investigation of operation processes of loading pipes of screw conveyors. INMATEH:d Agric. Eng. **50**(3), 89–94 (2016)
18. Annunziato, A., Godfrey, M., Anelli, F., Dulieu-Barton, J., Holmes, C., Prudenzano, F.: Flexible Photonic Sensors: Investigation of an Approach Based on Ratiometric Power in Few-Mode Waveguides for Bending Measurement. IEEE Access **12**, 29166–29174 (2024). https://doi.org/10.1109/ACCESS.2024.3368885
19. Hevko, R., Rohatynskyi, R., Hevko, M., Lyashuk, O., Trokhaniak, O.: Investigation of sectional operating elements for conveying agricultural materials. Res. Agric. Eng. **66**(1), 18–26 (2020)
20. Pascuzzi, S., et al.: Theoretical study of the motion of a cut sugar beet tops particle along the inner surface of the conveying and unloading system of a topping machine. AgriEngineering **6**, 409–422 (2024)
21. Rohatynskyi, R.M, Diachun, A.I., Varian, A.R.: Investigation of Kinematics of Grain Material in a Screw Conveyor with a Rotating Casing. Bull. Kharkiv Petro Vasylenko Nat. Tech. Univ. Agric. **168**, 24–31 (2016)
22. Lyashuk, O.L., Rogatynska, O.R., Serilko, D.L.: Modeling of the vertical screw conveyer loading. INMATEH: Agric. Eng. **45**(1), 87–94 (2015)
23. Hevko, R., Vitrovyi, A., Klendii, O., Liubezna, I.: Design engineering and substantiation of the parameters of sectional tools of flexible screw conveyers. Bull. Transylvania Univ. Brasov **10**(59), 39–46 (2017)
24. Lyashuk, O., et al.: Investigation of bulk material transportation by screw conveyer with hinge-pan operating device. Key Eng. Mater. **948**, 169–182 (2023)

UVC Rays Devices Mounted on Autonomous Terrestrial Rovers: A Brief Overview

Tommaso Quartarella[1(✉)], Alessia Farella[1], Francesco Paciolla[2,3], and Simone Pascuzzi[1]

[1] Department of Soil, Plant and Food Science, University of Bari Aldo Moro, Via Amendola 165/A, 70126 Bari, Italy
tommaso.quartarella@uniba.com

[2] PolySense Lab-Dipartimento Interateneo di Fisica, Polytechnic and University of Bari, Via Amendola 173, 70126 Bari, Italy

[3] Dipartimento di Ingegneria Elettrica e dell'Informazione, Polytechnic University of Bari, Via Edoardo Orabona 4, 70126 Bari, Italy

Abstract. In recent decades, the focus on sustainable farming practices has revolutionised agriculture. Population growth requires efficient use of resources and higher productivity associated with higher quality production. Technologies must be developed to meet these challenges. In this review, the effect of UVC was studied and its use on unmanned ground vehicles (UGV) was evaluated. The UVC (ultraviolet C-type) effect emerges as a powerful strategy in the fight against pathogens ensuring the disinfection of contaminated environments and surfaces. UVC radiation with wavelengths between 200 and 280 nm damages the DNA and RNA of microorganisms, including pathogenic fungi, hindering their replication, and causing cell death. The use of robotics can improve agricultural management reducing the use of chemicals and ensuring product quality. The possibility of mounting UVC lamps and portable devices on robots offers a promising prospect for a more efficient, sustainable, and future-oriented agriculture. The results of this research highlight the effectiveness of UVC technology combined with robotics in monitoring and treating crops.

Keywords: UV-C light · Agricultural robotics · Unmanned Ground Vehicle (UGV) · Plant defenses · effect of treatments

1 Introduction

At a global level, the agricultural sector faces several challenges such as the pressure of intensive production systems on ecosystems and their services, the impact of climate change on crops, and the respect of safety conditions for consumers, operators, and users of various ecosystems. Major changes in agriculture have taken place through mechanization/automation of different crop operations such as water management, tillage, application of plant protection products and harvesting. The improvement of IPM (Integrated Pest Management) and its integration with integrated agricultural management requires the development of knowledge and techniques that enable the incorporation of

E. Lorencowicz et al. (Eds.): FMPMSA 2024, LNCE 609, pp. 378–386, 2024.
https://doi.org/10.1007/978-3-031-70955-5_41

sensing and precision farming tools to make cultivation practices timely and effectively necessary, improve product quality, and enable a valuable return for producers. Due to the recent global focus on sustainable development (global and European targets at 2030 and 2050 e.g. SDGs, Paris Agreement targets, European Green Deal, 'Farm to Fork' strategy progressive and complete decarbonization of the system to make food systems more 'sustainable', reduction of at least 50% of the use and risk of plant protection products), the challenges facing agriculture today include the management and use of resources during the production process while ensuring a sustainable natural environment [1]. The use of technological innovation in agriculture, through the sustainable and economical use of mechanization, is an important resource to address these needs, contributing to the advancement of agricultural supply chains by moving them towards farming regimes where inputs related to the use of synthetic chemicals are minimized and/or rationalized [2].

According to the latest UN estimates, the world population will increase from the current 6.8 billion to around 10 billion in 2050, with a consequent impact on the agricultural sector, which will have to make the most of the arable land that is currently being cultivated and is currently being used, also in terms of its management, to meet the growing demand for products of agricultural origin (www.fao.org) [3]. The use of robotics in agriculture would allow a better management of the production process, filling the demand for labour, which is often lacking today, and safeguarding human health especially for crop operations related to the application of plant protection products [4, 5].

Recent studies report the successful application of robotics in agriculture. For example, [6] report the development of VINBOT, a robot capable of autonomous navigation between vineyard rows through a deliberative/reactive hybrid control system. Such systems may allow the management of robots capable of performing certain cultivation operations autonomously, such as seeding under optimal conditions of distance and depth [7] Other studies, on the other hand, report the development of more advanced robots, architected in modules that can perform monitoring of soil and vegetation characteristics and be used for the application of plant protection products [8]. This application is the most common and effective strategy to control diseases. However, the excessive use of pesticides, can have a negative impact on the environment and human health. Furthermore, the frequent use of pesticides can lead to the development of resistance [9]. There is therefore a growing interest in innovative strategies based on the use of abiotic or biotic factors to stimulate plant defence mechanisms [10]. One non-chemical tool that farmers could use to manage diseases is ultraviolet (UV) radiation, which will be discussed in this review.

2 Materials and Methods

To conduct this review, a search of the scientific literature was carried out through various information sources, including online databases, peer-reviewed journals, conference proceedings and web resources, such as Web of Science, Scopus and google scholar. The following keywords and their synonyms were used to guide the search: 'agricultural robots', 'UVC rays', 'agricultural treatment', 'agricultural disinfection', 'precision agriculture'. The search was limited to articles published within the last ten years to ensure the relevance of the information gathered.

Articles identified through the search underwent a selection process to ensure that only the most relevant and high-quality studies were included. Articles were initially screened according to their titles and abstracts to determine their suitability. Subsequently, the selected articles were examined in more detail to assess their consistency with the review objectives and methodological quality.

The focus of the research is on agricultural robots equipped with UVC ray delivery systems, positioning sensors and safety devices for the protection of human operators. UVC rays are generally generated by UV lamps specifically designed for agricultural applications, with optimal wavelengths and powers for treating crops without causing significant damage.

3 Using UVC Rays

The UVC (ultraviolet C-type) effect is a powerful weapon in the fight against fungal pathogens, playing a crucial role in disinfecting contaminated environments and surfaces. UVC radiation, with wavelengths between 200 and 280 nm, can damage the DNA and RNA of microorganisms, including pathogenic fungi, preventing their ability to replicate and causing cell death. The physical approach of UV-C irradiation is an alternative technique to conventional plant protection products. Fungal development can be halted by irradiating crops with UV-C light. Several studies have already successfully described the control of powdery mildew in strawberries [11, 12]. However, powdery mildew can overcome UV-C-induced DNA damage through a DNA repair mechanism by photolysis [12]. Since photolyase (an enzyme that specifically binds DNA strands damaged by exposure to ultraviolet radiation) uses UV-A and blue wavelengths as an energy source [13], dark-applied UVC is more damaging due to the inactivity of this repair mechanism. This technology is based on the principle that exposure to UVC radiation interferes with the genetic material of pathogens, compromising their structural and functional integrity. This damage prevents the ability of the fungus to proliferate and cause infection. The application ofthe UVC effect is particularly effective in settings such as hospitals, laboratories, food production facilities and other environments where the presence of fungal pathogens may pose a threat to human health or product quality.

The use of UVC devices for disinfection has gained popularity for its ability to offer a non-chemical and efficient method to combat the growth of pathogenic fungi. This technology can be employed in various forms, including UVC lamps, portable devices or room-integrated disinfection systems. However, it is important to carefully consider dosage and exposure time, as proper implementation of the UVC effect is essential to ensure effectiveness and safety. In summary, the UVC effect presents itself as a promising disinfection strategy to combat fungal pathogens, offering an innovative and versatile approach in protecting public health and ensuring environmental safety.

For several years now, studies have been initiated within the Fruit and Vegetable Physiology team (UMR Qualisud, University of Avignon and the Vaucluse countries) in collaboration with the Plant Pathology Unit (INRA PACA) to assess the possibility of using ultraviolet C (UV-C) rays as a tool to protect plants against pathogens such as Botrytis cinerea or Sclerotinia sp. [14, 15]. UV-C is high-energy electromagnetic radiation whose wavelength spectrum extends from 200 to 280 nm. Radiation with a wavelength of 254 nm is used for its germicidal properties, particularly bactericidal [16]. They act by denaturing the DNA of microorganisms.

3.1 Effect on Strawberries

Previous research on ultraviolet (UV) light for the control of fungal diseases has been hampered by the high level of plant damage. Strawberries (Fragaria ananassa Duch.) are grown under different environmental conditions using different production systems to make the fruit available all year round. This creates challenges for the control of pathogens and pests. Grey mould (Botrytis cinerea), anthracnose (Colletotrichum spp.) and powdery mildew (Podosphaera aphanis) and arthropods (e.g. two-spotted spider mite and spotted wing drosophila) can cause severe crop losses if not managed during production and after harvest. Fungicide applications begin in the field before flowering to protect flowers because infection at this stage accounts for up to 80% of fruit decay at and after harvest [17] and insecticides are applied as needed during the growing season. Frequent pesticide applications have led to an accumulation of resistance in targeted and non-targeted strawberry pathogens and arthropods. The development of resistance makes chemical control of diseases and pests unsustainable. Ultraviolet light (UV-C) (100 to 285 nm) kills microorganisms, plants and animals.

In Belgium, [18] demonstrated the killing power of UV-C on fungi and other fungal pathogens in the greenhouse strawberry system, but leaf burns also occurred. Therefore, UV-C light has not been widely used in crop production due to its harmful effects on plants at the doses required to kill microorganisms and arthropod pests with daytime UV-C application. In Japan, [19] reported the promotion of resistance induction in strawberry plants against fungal pathogens with 30-min UV-B irradiation treatments at night. [12, 20] reported that fungi causing grey mould, anthracite and powdery mildew, and two-spotted spider mite and greenhouse whitefly, respectively, were controlled with low doses of UV-C (72 J m^{-2}) if darkness followed immediately after irradiation. Darkness prevented the activation of the light-induced DNA photolyase repair mechanism of DNA damaged by UV-C irradiation [21] and significantly increased the lethality of the UV-C irradiation treatment. The inclusion of a specific dark period after UV-C irradiation (254 nm) allowed a substantial reduction in exposure time (s) and doses (J m^{-2}) [12] compared to the effective doses reported by [18]. At these low doses, UV-C had no negative effects on strawberry plant phytotoxicity, pollen germination, growth tube, leaf chlorophyll content, photosynthetic activity, fruit set and fruit quality [12]. Recently, it has been reported that shorter UV-C at 222 nm kills viruses and bacteria [22].

3.2 Autonomous Vehicle for the Field Application of UV-C Technology

In 2018, design work was initiated to develop an autonomous vehicle capable of treating field-grown strawberries with UV-C light (Fig. 1) and the following steps were taken:

1) develop a robotic chassis and transmission with geometry capable of navigating strawberry plantations with speed control;

2) to develop a UV-C light beam that can be used on an autonomous platform with height detection, a modular mechanism for mounting the light field and integrated with an autonomous control/power supply system.

Fig. 1. Autonomous vehicle capable of treating field-grown strawberries with UV-C light

The series comprised several 55 W 254 nm UV-C lamps and reflectors mounted under a shield. A unique flexible reflective surface was designed to provide a UV-C reflectivity of over 80%. The modularity of the robot allowed the configuration of the UV-C light field to be adjusted for treatment in 0.9m increments and to treat farms from 0.2 to 1.4 hectares during a 5-h run [25, 29]. To ensure that the robot platform could provide adequate treatment under a variety of field conditions using appropriate mobility and navigation systems, three robots were constructed and evaluated at three sites. Two robots were operated at night. The third robot was operated during the day due to the work restriction imposed by the COVID-19 pandemic. At all three sites, the robots were scheduled to operate three times a week from the pre-bloom phase for 10 weeks. Records were maintained on robot performance and plots were checked weekly for disease, parasitic arthropods and plant development.

4 UVC Robots

4.1 Icarus

ICARO X4, the revolutionary Italian agricultural robot, transforms the wine sector by using UV-C radiation to devitalise pathogenic fungi such as Uncinula necator (Schwein.) Burrill and Plasmopara viticola (Berk. & M.A. Curtis) Berl. & De Toni. After years of research, in 2021, it was developed by Maschio Gaspardo and Free Green Nature, promising a more sustainable agriculture, drastically reducing pesticide use and environmental impact by up to 90%. Its function is to protect vines from damage caused by downy and powdery mildew, using two foldable and adaptable side panels composed of UV-C radiation emitters (Fig. 2). Such radiation has germicidal effects and is capable of devitalising the micro-organisms present on the surface of the leaves or clusters. It also stimulates the plant's natural defences, causing the vine to produce fungicidal and fungistatic substances that can block the development of pathogenic fungi (https://www.freegreennature.it/).

Fig. 2. Icaro X4 is equipped with two foldable and adaptable side panels consisting of UV-C emitters.

This technology not only reduces the use of pesticides but also contributes to lower CO2 emissions, marking an important step towards greener and more sustainable agriculture. Equipped with foldable UV-C panels, ICARO X4 is able to navigate autonomously between the rows of vines, ensuring effective and safe treatment of the plants. Capable of covering up to 15 hectares of vineyards, ICARO X4 reduces the need for chemical fungicides by 70%, highlighting its commitment to reducing the use of harmful substances [26, 27]. Furthermore, thanks to its electric motors, ICARO X4 can operate for up to 72 consecutive hours without recharging. For companies requiring greater autonomy, there is the option of equipping the robot with a diesel engine, although this choice could compromise the robot's sustainability profile [28].

4.2 Thorvald

At the heart of advanced agricultural technology, the Thorvald robot, developed by Saga Robotics, is emerging as a key pillar for sustainable and precision agriculture. This autonomous robotic system is designed to perform a range of vital tasks, notably UV-C treatment aimed at controlling powdery mildew in delicate crops such as strawberries and grapes.

Thorvald uses UV-C energy to eradicate powdery mildew, a harmful fungus that affects many agricultural crops. The efficacy of this treatment is remarkably high, and UV-C has proven effective not only in eliminating powdery mildew, but also in boosting plant resistance against Plasmopara viticola, another problematic pathogen. However, it is important to note that this method is not effective against all pathogens.

In vineyards, for example, the application of UV-C up to 200 J/m^2 twice a week had no negative effects on plant development, physiology, yield or crop quality [23]. This non-chemical approach to disease control is particularly beneficial as it reduces dependence on fungicides and other chemical treatments, contributing to a healthier and more sustainable environment (Fig. 3).

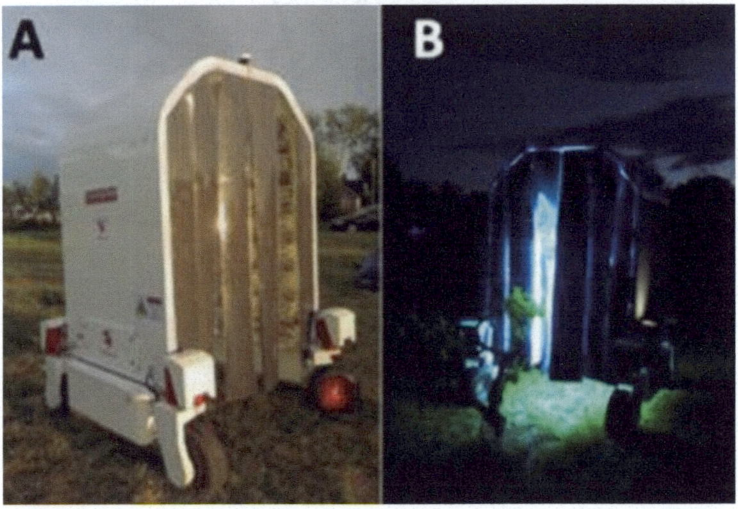

Fig. 3. UV lamp arrays (A) applied on Thorvald robots for night-time (B) applications in vineyards. Source: Use of Germicidal UV Light to Suppress Grapevine Diseases and Arthropod Pests

A revolutionary aspect of Thorvald's use is its autonomy. The robot carries out treatments at night, when UV-C radiation is most effective and least disturbed by daylight conditions. This feature not only optimises treatment efficacy but also allows significant labour savings compared to traditional fungicide application methods.

The use of Thorvald in crop treatment brings significant environmental benefits. Switching from chemical to light-based treatments reduces the ecological impact of farming, decreasing pollution and improving the health of the soil and surrounding waters. Moreover, the efficiency of this technology promises a substantial reduction in costs, making UV-C treatment not only an environmentally friendly but also a cost-effective option.

A striking example of Thorvald's effectiveness can be seen in the Pantano vineyard in Italy, where weekly UV-C treatment of the vines has proven highly effective in controlling powdery mildew. Here, Thorvald operates independently during the nights, the optimal period for applying the treatment, which proves to be just as effective as standard fungicide programmed but with less environmental impact.

5 Conclusions

Studies such as those conducted on strawberries and vines demonstrate the importance of continuing research and adaptation of UV-C technologies.

This solution is therefore not only an advanced technological solution for modern agriculture but also a bulwark for the promotion of more sustainable and environmentally friendly cultivation methods. With each vineyard and field that makes use of its technology, a step is taken towards a future in which agriculture is both productive and ecological.

References

1. Shahmohamadloo, R.S., Febria, C.M., Fraser, E.D., Sibley, P.K.: The sustainable agriculture imperative: a perspective on the need for an agrosystem approach to meet the United Nations Sustainable Development Goals by 2030. Integr. Environ. Assess. Manage. (2021)
2. Srinivasa, A.: Handbook of Precision Agriculture. Hayworth Press, Haworth (2006)
3. Saiz-Rubio, V., Rovira-Más, F., Millot, C.: Performance improvement of a vineyard robot through its mechanical design. Paper Presented at the 2017 ASABE Annual International Meeting (2017). https://doi.org/10.13031/aim.201701120
4. Ammoniaci, M., Kartsiotis, S.P., Perria, R., Storchi, P.: State of the art of monitoring technologies and data processing for precision viticulture. Agriculture **11**(3), 201 (2021)
5. Schimmelpfennig, D.E., Ebel, R.M.: Sequential adoption and cost savings from precision agriculture. J. Agric. Resour. Econ. **41**, 97–115 (2016)
6. Guzmán, R., et al.: Autonomous hybrid GPS/reactive navigation of an unmanned ground vehicle for precision viticulture-VINBOT. Intervitis Interfructa Hortitechnica-Technology for wine, juice and special crops (2016)
7. Naik, N.S., Shete, V.V., Danve, S.R.: Precision agriculture robot for seeding function. In: 2016 International Conference on Inventive Computation Technologies (ICICT), vol. 2, pp. 1–3. IEEE (2016)
8. Sarri, D., Martelloni, L., Rimediotti, M., Lisci, R., Lombardo, S., Vieri, M.: Testing a multi-rotor unmanned aerial vehicle for spray application in high slope terraced vineyard. J. Agric. Eng. **50**, 38–47 (2019)
9. Komárek, M., Čadková, E., Chrastný, V., Bordas, F., Bollinger, J.C.: Contamination of vineyard soils with fungicides: a review of environmental and toxicological aspects. Environ. Int. **36**(1), 138–151 (2010)
10. Conrath, U.: Priming of induced plant defense responses. Adv. Bot. Res. **51**, 361–395 (2009)
11. Van Delm, T., Melis, P., Stoffels, K., Baets, W.: The effect of long-day treatment on runners and inflorescences on everbearing strawberry cultivar'Capri. In: XXIX International Horticultural Congress on Horticulture: Sustaining Lives, Livelihoods and Landscapes (IHC2014), vol. II 1117, pp. 285–290 (2014)
12. Janisiewicz, W.J., Takeda, F., Nichols, B., Glenn, D.M., Jurick, W.M., II., Camp, M.J.: Use of low-dose UV-C irradiation to control powdery mildew caused by Podosphaera aphanis on strawberry plants. Can. J. Plant Path. **38**(4), 430–439 (2016)
13. Suthaparan, A., Pathak, R., Solhaug, K.A., Gislerød, H.R.: Wavelength dependent recovery of UV-mediated damage: tying up the loose ends of optical based powdery mildew management. J. Photochem. Photobiol. B. **178**, 631–640 (2018)
14. Ouhibi, C., Attia, H., Nicot, P., Urban, L., Lachaâl, M., Aarrouf, J.: Effect of UV-C radiation on resistance of romaine lettuce (Lactuca sativa L.) against botrytis cinerea and sclerotinia minor. J. Phytopathol. **163**(7–8), 578–582 (2015)

15. Forges, M., et al.: Impact of UV-C radiation on the sensitivity of three strawberry plant cultivars (Fragaria x ananassa) against Botrytis cinerea. Sci. Hortic. **240**, 603–613 (2018)
16. Bank, B.J., Slavings, R.L., Biddle, B.J.: Effects of peer, faculty, and parental influences on students' persistence. Sociol. Educ. 208–225 (1990)
17. Bulgler, M.A., Ellis, M.A., Madden, L.V.: Influence of temperature and wetness duration on infection of strawberry flowers by Botrytis cinerea and disease incidence of fruit originating from infected flowers. Phytopathology **77**, 1225–1230 (1987)
18. Van Delm, T., Melis, P., Stoffels, K., Beats, W.: Control of powdery mildew by UV-C treatment in commercial strawberry production (2014)
19. Zhang, Y., Maas, J. (eds.): Proceedings of the 7th International Strawberry Symposium; Beijing, China. February 18–22, 2012, vol. 1049, pp. 679–684. ISHS Acta Hort, Leuven (2012)
20. Kobayashi, M., et al.: Supplemental UV radiation controls rose powdery mildew disease under the greenhouse conditions. Environ. Control. Biol. **51**(4), 157–163 (2014)
21. Leskey, T.C., Short, B.D., Ludwick, D.: Comparison and refinement of integrated pest management tactics for Halyomorpha halys (Hemiptera: Pentatomidae) management in apple orchards. J. Econ. Entomol. **113**(4), 1725–1734 (2020)
22. Pathak, J., et al.: Oxidative stress and antioxidant defense in plants exposed to ultraviolet radiation. Reactive oxygen, nitrogen and sulfur species in plants: production, metabolism, signaling and defense mechanisms, pp. 371–420 (2019)
23. Kim, D.K., Kang, D.H.: Effect of surface characteristics on the bactericidal efficacy of UVC LEDs. Food Control **108**, 106869 (2020)
24. Gadoury, D.M., et al.: Use of germicidal UV light to suppress grapevine diseases and arthropod pests. In: BIO Web of Conferences, vol. 50, p. 01002. EDP Sciences (2022)
25. Bulgakov, V., et al.: Investigation of the parameters of a conveyor for transportation of bulk agricultural materials with a bladed working body. In: Farm Machinery and Processes Management in Sustainable Agriculture XII International Scientific Symposium (2024)
26. Paciolla, F., Pace, N., Barile, G., Patimisco, P., Valenza, F., Pascuzzi, S.: Autonomous navigation simulation of an agricultural robot during soil fertilization in open fields. In: 2023 IEEE International Workshop on Metrology for Agriculture and Forestry (MetroAgriFor), pp. 640–645 (2023)
27. Paciolla, F., Patimisco. P., Farella, A., Quartarella, T., Pascuzzi, S.: Performance evaluation of hybrid-electric architectures for agricultural tractors. In: Farm Machinery and Processes Management in Sustainable Agriculture XII International Scientific Symposium (2024)
28. Pascuzzi, S., Łyp-Wrońska, K., Gdowska, K., Paciolla, F.: Sustainability evaluation of hybrid agriculture-tractor powertrains. Sustainability **16**(3), 1184 (2024)
29. Farella, A., Paciolla, F., Quartarella, T., Pascuzzi, S.: Agricultural unmanned ground vehicle (UGV): a brief overview. In: Farm Machinery and Processes Management in Sustainable Agriculture XII International Scientific Symposium (2024)

Impact of the Integration of Mechanical Weeding on Sugar Beet Crop and Weed Control in a 4-Year Rotation

Fabienne Rabier[1](✉), François Henriet[1], Stanislaw Parafiniuk[2], Elena Pitchugina[1], and Quentin Limbourg[1]

[1] Walloon Agricultural Research Centre, Gembloux, Belgium
f.rabier@cra.wallonie.be
[2] University of Life Sciences, Lublin, Poland

Abstract. The use of herbicides has been called into question due to their impact on health and the environment and, in particular, water resources. European legislation establishes a framework for the use of pesticides (Directive 2009/128/CE). This restriction in potential uses is also exacerbated by a fall in the number of herbicides with new modes of action available on the market in conjunction with an increase of weed resistance to herbicides.

Non-chemical alternatives exist, they are based on crop management or direct techniques for controlling weeds. A diverse range of physical solutions are being developed and are expected to evolve rapidly through the integration of new technologies. Their use and effectiveness depend on the crop and pedoclimatic conditions and must be tested over several years.

In this context, this paper presents partial results of a 4-year trial, which aims to assess the integration of mechanical weeding in field crop. The trial consisted in a crop succession of sugar beet, maize and winter wheat. Three different treatments for weed control were compared during four years: chemical weeding, mixed weeding and mechanical weeding only.

The results highlight the relevance of assessing weeding practices over several years. The first year of the trial showed that the 3 types of treatment produced similar results in terms of weed numbers/m^2 and sugar yield, although there were differences in weed control efficiency. This situation changed from the second year to the disadvantage of mechanical weed control: lower efficiency and sugar yield and an increase in the number of weeds/m^2 at the final count.

Keywords: weeding · mechanical weeding · sugar beet · assessment · effectiveness

1 Introduction

The cultivation of sugar beets is an important crop in Belgium, with 56,306 hectares planted in 2023 [1]. It ranks as the fourth most significant crop in the country. Sugar beets are well-suited to the Belgian pedoclimatic context, allowing them to achieve high

E. Lorencowicz et al. (Eds.): FMPMSA 2024, LNCE 609, pp. 387–397, 2024.
https://doi.org/10.1007/978-3-031-70955-5_42

yields and sugar content while benefiting from the expertise of a local industry (farmers and agroindustry). This crop exhibits slow juvenile growth, making it particularly sensitive to weed competition, which affects its development and ultimately its yield. Consequently, sugar beets require significant herbicide use, averaging 9.7 kg of active substances per hectare [2]. They rank second among the most herbicide-intensive industrial crops, following potatoes. Additionally, herbicides face resistance issues, especially from common fat hen (*Chenopodium album*) [3] and spreading orach (*Atriplex patula*) [4], along with a reduction in available active ingredients.

In addition to their toxicity, protection plant products (PPPs) have a proven impact on the environment: air, soil, surface water and groundwater [5]. Herbicides applied pre-emergence and/or early in the season are particularly prone to leaching into water resources. According to the key indicators for the Walloon environment [6], pesticides (especially herbicides and their degradation products) and nitrates have led to the downgrading of 35% of groundwater bodies. Non-chemical alternatives exist; they are based on crop management practices or direct weed control techniques, among them, mechanical weeding is clearly identified. A diverse range of physical solutions are being developed [7, 8] and are evolving rapidly through the integration of new technologies [9, 10]. Annual trials have demonstrated the feasibility of controlling weeds in integrated systems combining mechanical weeding with chemical treatments for sugar beets [11–13]. However, mechanical weeding's effectiveness remains limited within rows [14] and is highly dependent on pedoclimatic conditions. Therefore, evaluating the integration of mechanical weeding over several years within crop rotations is essential to consider the impact of weed control methods on weeds population and crops performance over the years. The overall objective of this study was to assess three weed management methods on key crops in Belgium. The results presented in this article focus on sugar beets within a 4-year rotation.

2 Material and Method

2.1 Experimental Design

The trial is located on a 5.15 ha agricultural plot in Ernage, Belgium (loamy soil). A 4-year crop rotation of sugar beets, maize and winter wheat was established (Fig. 1). These crops were chosen because they are significant in the Walloon region and allow for mechanical weeding integration.

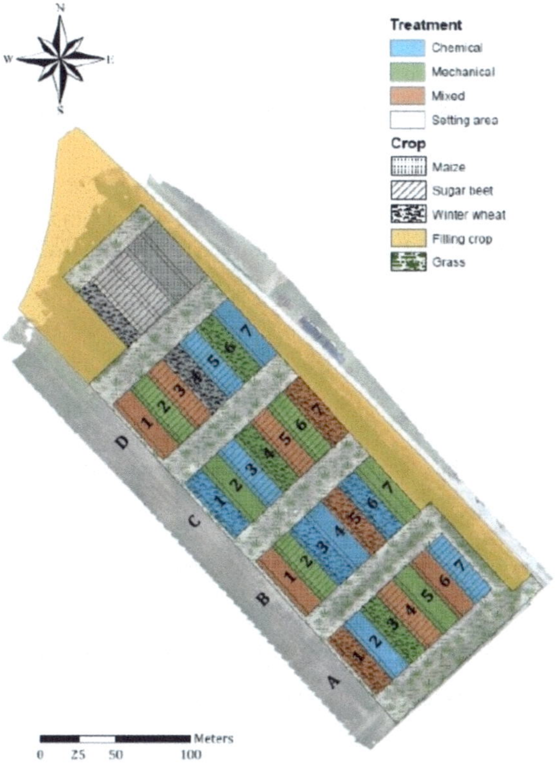

Fig. 1. Experimental plot in Ernage (2023)

Three weed control methods (Table 1) were represented each year for the three crops, with three replications within the experimental design.

Table 1. Description of weeding treatments

Weeding treatment	Description
Chemical	Conventional weed management using herbicides
Mixed	Mixed weed management that reduces the use of herbicides by partially integrating mechanical weeding or spot spraying (2023) when possible
Mechanical	Weed management using mechanical weeding tools (hoe and flex-tine harrow)

For mechanical weeding, three different tools were used:

- a 3-m harrow (Einböck), which works across the entire width independently of the crop row;

- a front-mounted hoe was used for the initial intervention on sugar beets with 2–4 leaves;
- a mounted hoe equipped with a camera (3 m, Carré Econet model). When the beet stage allowed (6–8 leaves), the hoe could be equipped with fingers to work within the row.

Field spraying was carried out with a classical sprayer (24 m, Kuhn Metris) and spot spraying (mixed weeding treatment only in 2023) with a 6 m mounted sprayer with nozzles spaced 4 cm apart and weed detection system (ARA from Ecorobotix). The Table 2 shows the treatment's description.

Table 2. Weeding treatments from 2020 to 2023

Year	Chemical	Mixed	Mechanical
2020	5 × herbicides spraying	4 × herbicides spraying and 3 × hoeing	1 × false seed bed before sowing 5 × hoeing
2021	5 × herbicides spraying	4 × herbicides spraying and 2 × hoeing	1 × false seed bed before sowing 1 × harrowing and 3 × hoeing
2022	6 × herbicides spraying	5 × herbicides spraying and 2 × hoeing	1 × false seed bed before sowing 1 × harrowing and 3 × hoeing
2023	7 × herbicides spraying	1 × false seed bed before sowing 3 × herbicides spraying, 3 × herbicides spot spraying and 1 × hoeing	1 × false seed bed before sowing 3 × hoeing

The chemical weeding was based on the recommendation of the Belgium Royal Institute for sugar beet (fractional, successive and closely spaced applications of combinations of active ingredients). The percentage reductions in herbicide use achieved by weed control treatments compared with chemical weed control are shown in Table 3.

The reduction in the use of herbicides in the mixed treatment was limited by the presence of potato regrowth and thistles, which required specific extra spraying. In addition, because the beet row was too dirty, the integration of the mechanical weeding has not been pushed forward. In 2023, the integration of spot spraying technology enabled a more significant reduction in the use of herbicides, with an average reduction of 87.2% for the three-spot spraying compared with broadcast spraying.

The crops were treated identically in all other respects (Table 4).

Table 3. Reduction in herbicides use (%) in comparison with chemical weeding

Year	Mixed	Mechanical
2020	20%	100%
2021	16%	100%
2022	22%	100%
2023	41%	100%

Table 4. Cropping information (sugar beet for 4 years of trials)

Year	Sowing[a]	Fertilization	Protection	Harvest
2020	1/04/2020	Total nitrogen: 70 U 1 × bore spraying	2 × insecticides spraying 1 × fungicide spraying	14/10/2020
2021	02/04/2021	Total nitrogen: 95 U 1 × bore spraying	1 × insecticide spraying 1 × fungicide spraying	12/10/2021
2022	30/03/2022	Total nitrogen: 115 U 1 × bore spraying	2 × insecticides spraying 1 × fungicide spraying	26/09/2022
2023	04/05/2023	Total nitrogen: 80 U 1 × bore spraying	1 × insecticide spraying 1 × fungicide spraying	07/11/2023

[a] identical soil preparation (cultivator and moving harrow) and same variety (Lisanna KWS 110 000 seeds/ha, 45 cm).

2.2 Number of Weeds

Weed counting (total number per species) occurs 3 to 4 times during the weeding season. The first count takes place after emergence, and the last count occurs just before beet row closure. Counting plots, covering an area of 0.81 m^2, are established before weeding operations begin. These plots are geolocated to ensure consistency across years and are positioned to avoid interference with weeding and wheel impact. Each plot includes 9 counting subplots: 3 in the untreated control area and 6 in the treatment modality, resulting in 18 repetitions per weeding modality each year.

2.3 Weeding Effectiveness

Effectiveness (%) represents the evaluation of the treatment's action on the weeds. It is calculated by comparing the number of weeds counted for the weeding treatment with the untreated control. This was done in each sample square, for the total number of weeds, using the following formula:

$$Ef_{P1} = 100 - (100 \times \frac{AdvP1}{AdvT}) \tag{1}$$

where Ef $_{P1}$= effectiveness of plot 1 in %, AdvP1 = number of weeds/m^2 in plot 1, and AdvT = number of weeds/m^2 in the nearest untreated control.

2.4 Yield Measurement

For each plot, measurements of sugar beet gross yield (weight and harvested area for 12 rows of beets × 50 m) and sugar content (IRBAB laboratory) were conducted.

2.5 Statistical Analysis

The statistical analysis was performed using SAS software package (SAS 9.4 TS level 1M1). In order to carry out the statistical processing, a transformation arcsin was applied to the effectiveness variable. The effects of treatments were subjected to analysis of variance (ANOVA) using proc glm SAS procedure where the significance level was set to 0.05. When the homogeneity of variance assumption is not met, the Welch's test was applied for performing an ANOVA analysis. Post hoc multiple comparisons using the Student–Newman–Keuls (SNK) method were used to determine which pairs of the three group means differed.

2.6 Climatic Information

The information used comes from summaries of the Belgian Crop Growth Monitoring System [15].

2.7 Geolocation and Mapping

Measurements are geolocated using a GNSS receiver mounted on a portable telescopic support. The receiver, a SOKKIA GRX3 model, equipped with RTK technology, provides a horizontal precision of 0.5 cm and vertical precision of 1.0 cm. Data processing and mapping are performed using ArcGIS software (ArcMap module).

3 Results and Discussion

3.1 Climatic Conditions

The four years of trials allowed testing the three weeding modalities under contrasting climatic conditions. The success of mechanical weeding is closely tied to meteorological conditions from sowing to beet row closure, typically occurring between April and June. Adequate weather conditions are crucial during and after treatment. The window for favorable outcomes is limited by soil conditions at the time of passage and drying conditions afterward, ensuring effective weed desiccation [16]. As shown in Table 5, the year 2021 was not adapted to mechanical weeding integration due to difficulties in finding suitable windows for tool passage and unfavorable low temperatures for destroying uprooted weeds. In contrast, 2020 and 2022 presented ideal conditions with dry and warm/sunny April and May.

Table 5. Climatic conditions during weeding seasons (2020–2023)

Year	Weather conditions during the weeding season (from April to June) [15]	Average Yield in Belgium [17]
2020	April: warm temperatures (+1 to +4 °C) and low rainfall (20 to 60% of normal amounts). May: slightly cooler than normal (−0.5 °C) and significant rainfall deficit (5 to 55% of normal observed amounts). June normal	84,7 t/ha - 17,39%
2021	April very cold (−2 to 3.5 °C below normal) and dry (55 to 85% of normal). May and June cold (−1.5 to 3.5 °C) with abundant, regular rainfall (till 150% normal)	83,2 t/ha - 17,03%
2022	April: cool temperatures, insufficient rainfall (80% of normal amounts). May: total rainfall < normal, concentrated in a few days, high temperatures (+1 to 2 °C above normal depending on the region). June could be considered normal	90,9 t/ha - 17,33%
2023	April cold (−1.4 °C compared with normal) and wet (delayed sowing). May normal temperature (average 14 °C), normal rainfall but concentrated in the first 15 days. Exceptionally hot June	Final figures not yet available

3.2 Number of Weeds

The main weed encountered in the trial were (EPPO codes): CHEAL, STEME, POLAV, POLPE, POLCO, SOLNI, GALAP, MATCH and FUMOF. The Table 6 presents the weed density at the final counting (beet row closure). For each year (2020–2023), a comparison between the three treatments on this parameter was made (different letters mean a statistical difference).

Table 6. Average value of final counting of weeds/m^2 for the 3 weeding treatments, n = 18 (2020 to 2023)

Year	Chemical	Mixed	Mechanical
2020	3.50 (A)	0.62 (A)	7.82 (A)
2021***	5.08 (A)	3.29 (A)	35.60 (B)
2022***	9.05 (A)	19.00 (A)	50.27 (B)
2023***	3.50 (A)	5.28 (A)	40.26 (B)

$P < 0.05$ *, $P < 0.01$ **, $P < 0.001$***

These results highlight that the mechanical modality exhibits increasing weed infestation over time, averaging from 7.82 to 40.26 weeds/m^2 in 2023, with high variability. Chemical and mixed weeding maintains more stable weed presence over time.

The statistical analysis shows an effect of the weeding treatment each year with a greater presence of weeds at row closure for the mechanical method. This difference increases with the years.

3.3 Weeding Effectiveness

The Table 7 presents the average values of effectiveness depending on year and treatment. For each year, the comparison between the three treatments was made (horizontal reading, different capital letters mean statistical difference).

Table 7. Effectiveness (%) average value for the weeding treatments (2020 to 2023)

Year	Chemical***	Mixed***	Mechanical***
2020***	88.43 (A, **a**)	96.88 (B, **a**)	74.79 (C, **ab**)
2021***	68.89 (A, **b**)	59.52 (A, **b**)	24.77 (B, **c**)
2022***	96.41 (A, **a**)	99.13 (A, **a**)	84.68 (B, **a**)
2023***	96.68 (A, **a**)	92.87 (A, **a**)	59.70 (B, **b**)

P < 0.05 *, P < 0.01 **, P < 0.001***

In 2020, a difference is observed between the three treatments and from 2021, there is a difference between mechanical weeding and the other two.

Mechanical weeding consistently shows lower efficacy. In 2022 and 2023, mixed and chemical treatments exhibit similar and high efficiencies (>90%). In 2021, lower average efficiencies for each treatment (especially mechanical and mixed) result from implementation challenges (inadequate weather conditions) for mechanical tools and, to a lesser extent, difficulties in correctly positioning chemical application (rain and wind).

Additionally, the same analysis was conducted to assess the influence of the year for each type of weed control (vertical reading of the Table 7, statistical difference when different lower-case blot letters). The results show that mechanical weeding varies the most from one year to another, while for mixed and chemical treatments, only 2021 stands out (very significant difference).

3.4 Yield

As Gross yield and sugar yield follow similar trends, only sugar yields are presented. The impact of weeding on yield can be estimated by calculating the difference between the average sugar yield of the modality and the control yields (without weeding). As shown in Fig. 2, all weeding modalities significantly enhance yield in average: +9.75 t/ha, + 6.18 t/ha and +5.72 t/ha for chemical, mixed and mechanical weeding respectively.

One-way analysis of variance was made to evaluate, each year, whether a significant difference is observed between the 3 weeding methods (Table 8, horizontal reading,

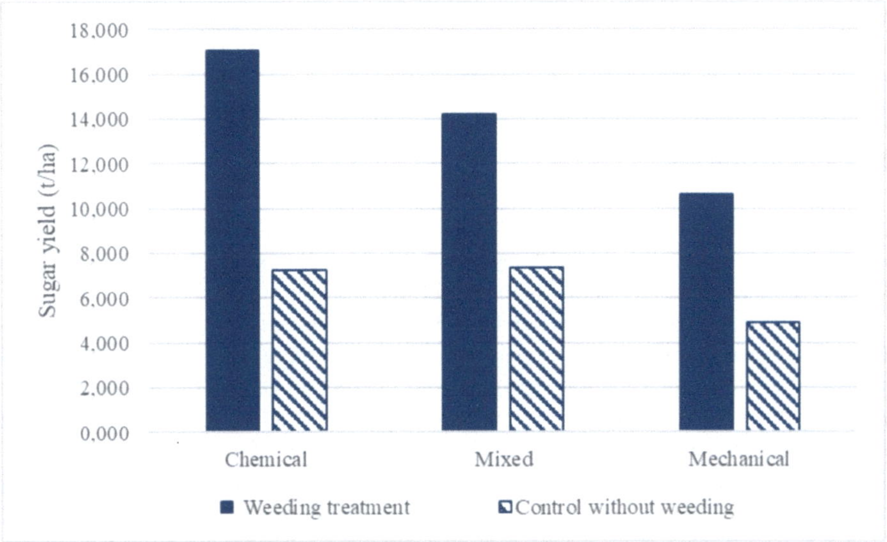

Fig. 2. Sugar yield difference (average value 2020–2023) between the 3 weeding treatments and controls without weeding

statistical difference when different capital letters), as well as the impact of the year on sugar yields of each weeding treatment (vertical reading of the Table 8, statistical difference when different lower-case blot letters).

Table 8. Sugar yield (t/ha), mean value and standard deviation, depending on the weeding treatment from 2020 to 2023

Year	Chemical*	Mixed**	Mechanical***
2020	18.73 (A, **a**)	19.08 (A, **a**)	17.25 (A, **a**)
2021***	17.63 (A, **a**)	17.65 (A, **ab**)	11.82 (B, **b**)
2022*	14.39 (A, **b**)	14.86 (A, **b**)	8.38 (B, **bc**)
2023***	17.37 (A, **a**)	14.72 (A, **b**)	4.97 (B, **c**)

$P < 0.05$ *, $P < 0.01$ **, $P < 0.001$***

After one year of trial (harvest 2020), the sugar yields obtained for the three weed control methods are similar. From 2021, a significant difference is observed between mechanical weeding on the one hand and other treatments in the other. This underscores the importance of multi-year evaluations, revealing contrasting situations on sugar yield and, thus, profitability of the crop.

Concerning the behavior of a weeding method over the years, the most notable finding is the substantial decline in mechanical modality yield, dropping from an average of 17.25 tons of sugar per hectare to 4.97 between 2020 and 2023. The mixed weeding shows also a decrease in sugar yield over time with a significant difference from 2022 onwards. The difference observed in 2022 for chemical weed control, can be explained by the greater presence of beet yellow virus in the plots concerned (the variety Lisanna is sensitive to the yellow virus).

4 Conclusion

The results highlight the relevance of assessing weeding practices over several years. This makes it possible to consider changes in the number of weeds over time as a function of the type of weeding associated with annual weather conditions.

The first year of the trial showed that the 3 types of treatment produced similar results in terms of weed numbers/m^2 and sugar yield, although there were differences in weed control efficiency. This situation changed from the second year to the disadvantage of mechanical weed control: lower efficiency and sugar yield and an increase in the number of weeds/m^2 at the final count.

Measurements taken over 4 years show that chemical and mixed weed control give stable results and control weed growth over time. Mechanical weeding, on the other hand, produced contrasting results from year to year, and the situation deteriorated with increasing weed pressure, which had an impact on weed control efficiency and, ultimately, crop yield.

The results for the mixed treatment were not significantly different from those for the chemical treatment. This can be explained by the relatively limited reduction in the quantity of herbicide used in this treatment (from 2020 to 2022). In the future, the mixed weeding should systematically incorporate precision spraying (band or spot) in order to address the problem of weed development in the crop row while reducing the quantities of herbicides used.

This article only presents the agronomic results observed for the sugar beet crop, which should be supplemented by an evaluation of the other 2 crops in the rotation (maize and winter wheat) and an analysis of the profitability of the complete rotation.

Acknowledgments. The authors would like to thank the technical teams of CRAW and the PVBC and Walloon Region for their financial support.

References

1. Statbel Homepage. https://statbel.fgov.be. Agricultural figures for 2023. Accessed 15 Apr 2024
2. CORDER-CRP: Estimation quantitative des utilisations de produits phytopharmaceutiques par les différents secteurs d'activités. Rapport d'étude (2020)
3. Mechant, E., De Marez, T., Hermann, O., Olsson, R., Buckle, R.: Target site resistance to metamitron in *Chenopodium album* L. J. Plant Dis. Protect. Spec. Issue XXI, 37–40 (2008). ISSN 1861-405
4. De Cauwer, B., Cardinael, A., Claerhout, S., Manderyck, B., Reheul, D.: Differential sensitivity of *Atriplex patula* and *Chenopodium album* to sugar beet herbicides: a possible cause for the upsurge of *Atriplex patula* in sugar beet fields. Eur. Weed Res. Soc. **58**, 99–111 (2018)

5. Calvet, R., Barriuso, E., Bedos, L., Benoit, P., Charnay, M.P., Coquet, Y.: Les pesticides dans le sol – conséquences agronomiques et environnementales. France Agricole Référence scientifique (2005)
6. DEE- DEMNA: Diagnostic environnemental de la Wallonie (2024)
7. Cloutier, D.C., Van Der Weide, R.Y., Peruzzi, A., Leblanc, M.L.: Chapter 8 Mechanical weed management. In: Non-Chemical Weed Management. CAB (2007)
8. Pannacci, E., Lattanzi, B., Tei, F.: Non-chemical weed management strategies in minor crops: a review. Crop Prot. **96**, 44–58 (2017)
9. Fennimore, S.A., Slaughter, D.C., Siemens, M.C., Leon, R.G., Saber, M.N.: Technology for automation of weed control in specialty crops. Weed Technol. **30**(4), 823–837 (2016)
10. Esposito, M., Crimaldi, M., Cirillo, V., Sarghini, F., Maggio, A.: Drone and sensor technology for sustainable weed management: a review. Chem. Biol. Technol. Agric. **18**(8), 1–11 (2021)
11. IRBAB-PVBC: Techniques culturales betteravières, FAR-Memo 2024. Le betteravier, mensuel CBB 5, 14–23 mars 2024 (2024)
12. Cioca Parasca, S., Spaeth, M., Rusu, T., Bogdan, I.: Mecanical weed control: sensor-based Inter-row hoeing in sugar beet (*Beta vulvaris* L.) in the transylvanian depression. Agronomy **14**, 176 (2024)
13. Kunz, C., Schröllkamp, C., Koch, H.J., Eßer, C., Schulze, L.P., Gerhards, R.: Potentials of post-emergent mechanical weed control in sugar beet to reduce herbicide inputs. Landtechnik **70**(3), 67–81 (2015)
14. Rabier, F., Stas, M., Manderyck, B., Huyghebaert, B., Limbourg Q.: Assessment of the integration of mechanical weeding for weed control in sugar beet-growing. In: Farm Machinery and Processes Management in Sustainable Agriculture, presented in IX International Scientific Symposium, pp. 330–335. University of Life Sciences in Lublin, Polland (2017)
15. BCGMS, Homepage. https://bcgms.be. Agrometeorological bulletins from 2020 to 2023. Accessed 10 May 2024
16. Abras, M., Rabier, F., Limbourg, Q.: Comparaison de techniques de désherbage mécanique en culture de froment biologique. In: 1967–2017 50 ans de partenariat entre agriculteurs et chercheurs. Presented at the Livre Blanc Céréales Gembloux, Belgique (2017)
17. CBB, Homepage. https://cbb.be. Average production of sugar beet in Belgium harvests 2020 to 2022. Accessed 10 May 2024

Possibility of Combined Control of Leaf Miners and Cereal Leaf Beetles in Winter Wheat Crops

Kamila Roik(✉), Anna Tratwal, and Marcin Baran

Institute of Plant Protection – National Research Institute, Władysława Węgorka 20, 60 – 318 Poznań, Poland

K.Roik@iorpib.poznzn.pl

Abstract. Recent years have shown that in addition to the well-known and commonly occurring pests of cereals, species that previously did not pose serious problems are appearing in increasing numbers. This phenomenon is primarily due to climate change as well as cultivation technologies (e.g., crop simplification) and a large share of cereals in crop rotation.

One of common cereal pests in Poland are flies of the *Agromyzidae* family, whose harmful stage are the larvae. Leaf miner larvae, feeding on the leaves' parenchyma, cause characteristic damages (mines). Their widespread occurrence contributes to the reduction in the leaves' assimilation surface. The flight of leaf miner imagines coincides with the development of leaf beetles (*Oulema* spp), another dangerous cereal crops pest.

The aim of the research was to monitor the seasonal flight dynamics of the leaf miner flies using yellow sticky traps (*Agromyzidae*) and the developmental stages of the cereal leaf beetle (CLB, *Oulema* spp.), taking into account the developmental stages of the host plant (winter wheat).

Keywords: leaf miners · Agromyzidae · winter wheat · leaf beetles

1 Introduction

Systematic monitoring of agrophages is a very important element of integrated plant protection. It is a basic activity aimed at identifying threats to crops from pest organisms. Based on the data from crop vetting, we can decide whether a chemical treatment is necessary and determine the optimal date for its implementation (Walczak 2010; Tratwal 2018).

For about 20 years, the main pests of cereals in Poland have been horsetails (*Oulema* spp.), aphids (*Aphididae*), and locally the cereal pest *Haplodiplosis marginata* Roser, syn. *Haplodiplosis equestris* Wagner)".

In recent years, damage caused by other, lesser-known pests can be increasingly observed. Over the years, they have not posed a significant threat, but the importance of some of them has increased, due to simplifications in agrotechnology, climate change and increasing acreage under cereal crops, among other factors. Among such pests are flies (mini-beetles) belonging to the family *Agromyzidae*. In Poland, there are more than

E. Lorencowicz et al. (Eds.): FMPMSA 2024, LNCE 609, pp. 398–403, 2024.
https://doi.org/10.1007/978-3-031-70955-5_43

a dozen species of leafminers on cereals, which can locally appear in large numbers. The harmful stage is the larvae, and the damage caused by their feeding reduces the quality and height of the crop (Walczak 1995).

2 Materials and Methods

Field experiments were carried out in the area of the Agricultural Experimental Station of the Institute of Soil Science and Plant Cultivation - National Research Institute in Baborówko (RZD IUNG - PIB Baborówko) and the Research Station for the Variety Testing in Słupia Wielka (SDOO Słupia Wielka) during eight growing seasons - 2008/2009 - 2015/2016.

Adult fly trapping was carried out in winter wheat plantations of the Bogatka variety using yellow sticky boards to determine the flight dynamics of leaf miners in each locality. At the same time, once a week, in the plots where the boards were located, the occurrence and developmental stage (eggs, larvae along with the determination of their size) of the CLB were assessed.

During the 2008/2009 - 2014/2015 growing seasons, chemical treatments against leaf miner were carried out in both localities and their effectiveness was evaluated.

The effectiveness of insecticide treatments on the designated dates, according to the established criteria (Z1, Z2, Z3), was evaluated on the basis of the number of mines in the experimental plots, the abundance of fiddlers and the yield obtained within each combination.

3 Results and Discussion

The flight dynamics of leaf miners in winter wheat plantations were largely similar in each year, but with some differences. One reason for the observed variability may probably be differences in the course of weather conditions in the years studied. In all years of the study, slightly higher average monthly temperatures and slightly higher monthly precipitation totals were recorded in the RZD IUNG - PIB Baborówko, hence the possible difference in the abundance of trapped leaf miner flies on yellow traps between localities (Tables 1 and 2).

Based on the results obtained from the trapping of leaf miners using the yellow sticky traps technique and observations on the development of CLB in the winter wheat plantation, the timing of treatments against leaf miners was determined according to the established criteria (Tables 3 and 4).

The effectiveness of the treatments carried out at the designated dates, according to the established criteria, was evaluated on the basis of the number of stalks with leaves damaged by leaf miners, the number of cereal leaf beetle and the yield obtained within each combination. The accuracy of the selection of chemical treatment dates was evaluated by estimating the decrease in the number of leaf miners and cereals leaf beetles in the analyzed combinations (Tables 5 and 6).

In experiments carried out in 2009–2015 on plots in SDOO Słupia Wielka and RZD IUNG - PIB Baborówko, grain was harvested from the entire area of individual plots using a plot harvester. The obtained yield (kg) was converted into decimals per hectare assuming a standard grain moisture content of 14% (Table 7).

4 Summary

Conducting monitoring with yellow sticky traps made it possible to capture the fly infestation on the plantation and determine the optimal date of control while observing the CLB larvae in the plantations under study, where their current developmental stage was recorded. Studies by other scientists have confirmed that performing a treatment at the optimal time for a given agrophage is more effective, while a chemical treatment performed at an inappropriate time is uneconomical and unnecessarily burdens the environment (Walczak 1998, 2003, Wrzodak et al. 2015). Based on the results, 3 treatment terms were determined according to the established criteria. After the experiment, we can conclude that after performing chemical treatments on the 3 dates, a decrease in the number of mines was noticed for the second date and the third date. The results of this research will enrich the knowledge on the reduction of the abundance and harmfulness of leaf miners and cereal leaf beetles in cereal plantations in the context of integrated cereal protection. An important achievement of this research is the determination of the optimum timing for the control of leaf miners, since, as we know in practice, chemical control of leaf miners is carried out too late and consequently leads to an increase in the costs associated with the protection of cereals, which does not always pay off in the form of a saved yield.

Table 1. Number of leaf miners and weather factors in the years of the study for IUNG - PIB Baborówko RZD (for the period April-May-June).

Year	Number of leaf miners		Temperature		Precipitation	
	Averge	Standard deviation	Average	Standard deviation	Average	Standard deviation
2009	26,26	32,30	14,22	2,30	20,60	20,39
2010	18,17	17,53	13,31	4,21	15,78	17,27
2011	15,53	14,53	16,07	3,94	7,31	7,26
2012	16,14	16,08	14,31	4,62	21,94	19,89
2013	12,46	13,87	14,28	5,45	20,34	28,35
2014	15,72	16,62	14,24	3,59	16,08	7,11
2015	12,23	10,94	13,83	4,88	9,28	8,55

Table 2. Number of leaf miners and weather factors in the years of the study for SDOO Słupia Wielka (for the period April-May-June).

Year	Number of leaf miners		Temperature		Precipitation	
	Averge	Standard deviation	Average	Standard deviation	Average	Standard deviation
2009	5,96	6,43	13,84	2,46	16,77	18,58
2010	10,52	8,46	12,70	4,06	16,03	16,05
2011	8,79	8,76	14,93	4,13	12,13	13,64
2012	7,30	7,21	13,69	5,08	16,98	20,15
2013	8,98	9,95	13,52	5,44	22,13	28,79
2014	13,24	12,73	13,66	3,36	13,87	16,37
2015	13,44	13,41	12,61	4,03	10,56	6,83

Table 3. Dates of chemical control of leaf miners in Baborówko during the study years

Dates of chemical control	Growing seasons						
	2008/2009	2009/2010	2010/2011	2011/2012	20012/2013	20013/2014	2004/2015
Z 1	05.V.2009	21.V.2010	19.V.2011	15.V.2012	20.V.2013	07.V.2014	15.V.2015
Z 2	20.V.2009	04.VI.2010	30.V.2011	25.V.2012	27.V.2013	23.V.2014	01.VI.2015
Z 3	27.V.2009	14.VI.2010	06.VI.2011	30.V.2012	06.VI.2013	05.VI.2014	05.VI.2015

Table 4. Dates of chemical control of leaf miners in Słupia Wielka during the study years

Dates of chemical control	Growing seasons						
	2008/2009	2009/2010	2010/2011	2011/2012	20012/2013	20013/2014	2004/2015
Z 1	05.V. 2009	21.V.2010	18.V.2011	14.V.2012	20.V.2013	30.IV.2014	13.V.2015
Z 2	20.V.2009	02.VI.2010	26.V.2011	21.V.2012	24.V.2013	21.V.2014	28.V.2015
Z 3	27.V.2009	08.VI.2010	01.VI.2011	28.V.2012	03.VI.2013	02.VI.2014	03.VI.2015

Explanations:
Z 1 - treatment in the period when a clear increase (several times) in the number of caught flies with yellow sticky traps is observed and at the same time the egg-laying stage of fiddlers is found,
Z 2 - treatment in the period of further numerous flights of leafminers, when at the same time most of the eggs have been laid and the oldest larvae of leafminers reach the size of about 2 mm,
Z 3 - treatment in the period of a large number of mink flight, when at the same time a mass hatching of larvae is observed, and the oldest fiddler larvae reach a size of about 4 mm, which is the basis for signaling control of fiddlers.

Table 5. Descriptive statistics for decreases in the number of leaf mines chemical treatments.

Experimental object	Average	N-number of observations	**Standard deviation**	Min	Max
K	3,45	56	5,93	−10	22
Z1	1,77	56	2,43	−4	7
Z2	9,71	56	6,32	1	28
Z3	8,55	56	6,93	−1	26
Total	5,87	224	6,56	−10	28

Table 6. Descriptive statistics for decreases in the number of cereal leaf beetles chemical treatments.

Experimental object	Average	N-number of observations	Standard deviation	Min	Max
K	−3,45	56	8,27	−14	13
Z1	5,84	56	4,21	−1	18
Z2	4,18	56	4,73	−3	16
Z3	3,34	56	5,47	−13	14
Total	2,48	224	6,83	−14	18

Table 7. Yields from different combinations of protection in the localities and years of the study expressed in dt/ha.

Place	Combination	Yield (dt/ha) -average						
		Growing seasons						
		2008/2009	2009/2010	2010/2011	2011/2012	2012/2013	2013/2014	2014/2015
Baborówko	Control	69,30	85,00	57,50	43,2	61,35	73,47	65,27
	Z 1	70,10	85,00	59.50	43,7	60,95	74,17	64,57
	Z 2	70,30	90,50	59,50	43,9	65,93	75,10	66,58
	Z 3	70,90	88,00	57,00	43,7	61,80	73,97	64,72
Słupia Wlk.	Control	77,30	90,50	101,00	52,00	72,50	86,36	97,42
	Z 1	80,33	86,00	103,50	52,16	71,50	89,09	96,66
	Z 2	83,67	101,30	104,50	52,30	79,83	89,84	98,18
	Z 3	85,83	97,30	100,00	50,80	73,16	86,60	95,60

References

Tratwal, A., Baran, M.: The role of guidelines on pest monitoring and warning systems in integrated pest management. J. Plant Protect. Res. **58**(3), 211–214 (2018). https://doi.org/10.24425/122941

Walczak, F.: Miniarki (Agromyzidae) – Szkodniki Upraw Zbożowych. Materiały 35 Sesji Nauk. Inst. Ochr. Roślin cz. **2**, 15–18 (1995)

Walczak, F.: System for warning and control of leaf mining flies and leaf beetles on cereal crops. J. Plant Prot. Res. **38**(1), 65–69 (1998)

Walczak, F.: Monitoring agrofagów dla potrzeb integrowanej ochrony roślin uprawnych. [Agrophages monitoring for plant integrated control needs]. Fragmenta Agron. **27**(4), 147–154 (2010)

Walczak, F., Roik, K.: Monitoring of leaf miners (*Agromyzidae*) and leaf beetles (Oulema Spp.) on winter wheat for the short-term forecasting. J. Plant Protection Res. **50**(4), 484–488 (2010)

Wrzodak, R., Woszczyk, K., Lewandowski, A., Rybczyński, D.: Ustalenie optymalnego terminu wykonania zabiegu zwalczania śmietki kapuścianej (Delia radicum L.) na podstawie monitoringu. Prog. Plant Prot./Post. Ochr. Roślin **55**(3), 369–374 (2015)

The Impact of the Use of Nanoparticles and Pulsed Electric Field on Oilseeds

Karolina Sokal(✉) and Magdalena Kachel

Department of Machine Operation and Production Processes Management, Faculty of Production Engineering, University of Life Sciences in Lublin, 28 Głęboka St., 20-612 Lublin, Poland
karolina.sokal@up.lublin.pl

Abstract. Sustainable agriculture involves practices ensuring a balance between production efficiency and environmental protection.

Coating rapeseed seeds with silver or copper nanoparticles can have a positive effect, increasing their germination rate as well as stem and root length. Reducing the use of chemical fertilizers in agriculture can benefit the environment by reducing greenhouse gas emissions and water pollution. Additionally, the application of Pulsed Electric Field can contribute to increase germination in a shorter time than normal. This is caused by the opening of cavities in the plasma membrane and cell wall and increasing the flow of polar particles in and out.

The aim of the study is to investigate the germination capacity of rapeseed seeds and sunflower seeds after PEF and nanoparticles application compared to a control sample and their effect on stem and root length. Nanoparticles were used at concentrations of 25 and 50 ml/l. Then, the seeds were exposed to electric current at 24 kV, intensity 1 kv/cm, 24 cm distance between electrodes, frequency 20 Hz, and pulse width 7μs. These values were constant for all seeds. The variable was the amount of energy supplied (kJ/kg). For rapeseed seeds and sunflower seeds 1; 5,5 kJ/kg. Control samples were carried out in water and another in a solution with nanoparticles. The germination for rapeseed seeds for AgNP and CuNP at a concentration of 25 and 50 ml/l at 1 and 5,5 kJ/kg was 95,0; 94,3; 88,3; 94,3; 93,3; 94,3; 96,0; 95,7; 93,3; 94,3% respectively. However, for sunflower seeds 91,7; 94,3; 96,3; 52,7; 93,3; 90,7; 96,3; 71,7; 89,7; 91,0% respectively.

Keywords: Sustainable Agriculture · Nanoparticles · Pulsed Electric Field · Gemination · Oilseeds

1 Introduction

Sustainable agriculture involves practices that balance production efficiency with environmental protection. In the current situation, there is a need to meet the nutritional needs of a continuously growing world population. Along with the rapid development of urban agglomerations, new innovative technologies for improving seedling growth should be applied. Coating rapeseed seeds with silver nanoparticles (AgNP) or copper nanoparticles (CuNP) may have a positive impact by increasing the speed of germination as well as the length of the stem and root. Limiting the use of chemical fertilizers

E. Lorencowicz et al. (Eds.): FMPMSA 2024, LNCE 609, pp. 404–410, 2024.
https://doi.org/10.1007/978-3-031-70955-5_44

in agriculture can benefit the environment by reducing greenhouse gas emissions and water pollution. Another technique is the application of Pulsed Electric Field (PEF), which can contribute to increasing germination capacity in a shorter time than usual. This occurs due to the opening of empty spaces in the cell membrane and cell wall, as well as increased flow of polar molecules inside and outside. The positive impact of PEF intensity largely depends on the type of seed coat and plant species. This technique can also affect seedling germination [1–3].

2 Methods

Rapeseed and sunflower seeds were soaked in tap water (control sample) and tap water with silver and copper nanoparticles for 15 min to achieve optimal conductivity. The seeds were weighed on an analytical balance, 200 g for rapeseed and then 800 ml of water were added, while for sunflower, 250 g of seeds and 750 ml of water were used. Nanoparticles were applied at concentrations of 25 and 50 ml/l. The prepared samples were transferred to a measuring tray. Then, the seeds were subjected to electric current at a voltage of 24 kV, intensity of 1 kV/cm, electrode distance of 24 cm, frequency of 20 Hz, and pulse width of 7 μs. These values were constant for all seeds. The variable was the amount of energy delivered (kJ/kg). For rapeseed and sunflower seeds, 1 and 5.5 kJ/kg were applied. Control samples were prepared in water, and subsequent ones in nanoparticle solutions.

The seeds were sown in three repetitions in Petri dishes, with 100 seeds each. After a week, the germination capacity was determined, i.e., the number of seeds that did not germinate was counted, and the length of stems and roots of seedlings was measured.

3 Research Results

3.1 Assessment of the Germination Capacity of Rapeseed and Sunflower Seeds

The results of the germination capacity assessment of rapeseed seeds are presented in Table 1. Germination was conducted under laboratory conditions. The germination capacity of seeds in the control group for 1 kJ/kg was 95%, while for 5.5 kJ/kg it was 94.3%. The germination capacity results after applying silver nanocolloid at a concentration of 25 ml/l and energy of 1 kJ/kg were 88.3%, while for copper under the same parameters, it was 96%. Compared to the control group with the same delivered energy, the germination energy is higher when using copper nanoparticles by 1%, while when using silver nanoparticles, it is lower by 6.7%. In the case of applying silver nanocolloid at a concentration of 25 ml/l and energy of 5.5 kJ, the germination capacity was at the same level as the control group, at 94.3%, while when using copper nanocolloid, an increase in germination capacity of 95.7% was observed, which is 1.4% higher compared to the control group. Results after soaking the seeds in silver and copper nanocolloids at concentrations of 50 ml/l with 1 kJ/kg for copper and silver showed a germination capacity of 93.3%. Compared to the control (1 kJ/kg), the results are 1.7% lower. It can be concluded that with the applied energy of 1 kJ/kg at a concentration of 50 ml/l, nanocolloids have a negative effect on rapeseed seedling growth. In the case of using copper and silver nanocolloids at a concentration of 50 ml/l with 5.5 kJ/kg, no changes in stimulation to germination were observed compared to the control group at 94.3%.

Table 1. The influence of the concentration of metal nanocolloids and pulsed electric field on the germination of rapeseed seeds

	Germination capacity									
Sample	Control 1kJ/kg	Control 5,5kJ/kg	Silver 25ml/l 1kJ/kg	Silver 25ml/l 5,5kJ/kg	Silver 50ml/l 1kJ/kg	Silver 50ml/l 5,5kJ/kg	Copper 25ml/l 1kJ/kg	Copper 25ml/l 5,5kJ/kg	Copper 50ml/l 1kJ/kg	Copper 50ml/l 5,5kJ/kg
1	95	95	88	94	93	95	95	95	93	94
2	96	95	90	95	93	93	96	97	95	96
3	94	93	87	94	94	95	97	95	92	93
Average	95,0	94,3	88,3	94,3	93,3	94,3	96,0	95,7	93,3	94,3
Standard Deviation	1,00	1,15	1,53	0,58	0,58	1,15	1,00	1,15	1,53	1,53

The results of the germination capacity assessment of sunflower seeds are presented in Table 2. Germination was conducted under laboratory conditions. The germination capacity of seeds in the control group for 1 kJ/kg was 91.7%, while for 5.5 kJ/kg it was 94.3%. The results after applying silver and copper nanocolloids at concentrations of 25 ml/l and with delivered energy of 1 kJ/kg were 96.3% for silver and 96.3% for copper. Compared to the control group with the same delivered energy (1 kJ), the germination capacity is higher. In the case of using silver nanocolloid at a concentration of 25 ml/l at 5.5 kJ/kg and copper nanocolloid at the same concentration at 5.5 kJ/kg, a significant decrease in germination capacity was observed, with 52.7% for silver and 71.7% for copper compared to the control. In the final result, considering the applied concentration of nanocolloids (25 ml/l) and the delivered energy, it can be concluded that 5.5 kJ/kg energy negatively affects germination stimulation. When soaking the seeds in silver and copper nanocolloids at a concentration of 25 ml/l with 1 kJ/kg, a positive effect of the applied energy and nanocolloids compared to the control was observed. The results after applying silver and copper nanocolloids at a concentration of 50 ml/l and with 1 kJ/kg energy were 93.3% and 89.7%, respectively. For silver, this value is higher than the control by 1.7%, while for copper, it is lower by 2%. It can be stated that with the applied energy of 1 kJ/kg, silver nanocolloids positively affect the germination of sunflower seeds compared to the control and copper nanocolloid. In the case of using silver and copper nanocolloids at a concentration of 50 ml/l with 5.5 kJ/kg, a decrease in

Table 2. The impact of metal nanocolloid concentration and pulsed electric field on sunflower seed germination.

	Germination capacity									
Sample	Control 1kJ/kg	Control 5,5kJ/kg	Silver 25ml/l 1kJ/kg	Silver 25ml/l 5,5kJ/kg	Silver 50ml/l 1kJ/kg	Silver 50ml/l 5,5kJ/kg	Copper 25ml/l 1kJ/kg	Copper 25ml/l 5,5kJ/kg	Copper 50ml/l 1kJ/kg	Copper 50ml/l 5,5kJ/kg
1	92	93	95	50	93	92	96	70	89	90
2	92	95	98	53	94	90	96	72	90	91
3	91	95	96	55	93	90	97	73	90	92
Average	91,7	94,3	96,3	52,7	93,3	90,7	96,3	71,7	89,7	91,0
Standard Deviation	0,58	1,15	1,53	2,52	0,58	1,15	0,58	1,53	0,58	1,00

stimulation to germination of 90.7% and 91% was observed, respectively. These values are 3.7% and 3.3% lower compared to the control group.

4 Assessment of the Length of Stems and Roots of Rapeseed and Sunflower Seedlings

4.1 Rapeseed

Table 3 shows the lengths of rapeseed seedling stems on the seventh day of the experiment, concerning the determination of seed germination capacity. In the treatments where seed soaking in silver nanoparticle solutions at concentrations of 25 and 50 ml/l with energies of 1 kJ/kg and 5.5 kJ/kg was applied, an increase in stem length of 36.9%, 46.7%, 18.9%, 43.8% respectively, was observed compared to the seedlings in the control group. In the treatments where seed soaking in copper nanoparticle solutions at a concentration of 25 ml/l with energies of 1 kJ/kg and 5.5 kJ/kg was applied, an increase in stem length compared to the control was observed at 48.4% and 36.2%, respectively. After applying copper at a concentration of 50 ml/l with 1 kJ/kg, an increase in stem length of 34.4% compared to the control treated with the same energy was observed, while with 5.5 kJ/kg, a decrease in stem length of 23.8% was observed. For silver, the largest increase in length was achieved by seedlings soaked in concentrations of 25 ml/l at 5.5 kJ/kg, with an increase of 46.7% compared to the control. For copper, the largest increase in length was achieved by seedlings soaked in concentrations of 25 ml/l at 1 kJ/kg, with an increase of 48.4% compared to the control.

Table 3. The impact of metal nanocolloid concentration and pulsed electric field on the length of rapeseed seedling stems.

	Stem length									
Sample	Control 1kJ/kg	Control 5,5kJ/kg	Silver 25ml/l 1kJ/kg	Silver 25ml/l 5,5kJ/kg	Silver 50ml/l 1kJ/kg	Silver 50ml/l 5,5kJ/kg	Copper 25ml/l 1kJ/kg	Copper 25ml/l 5,5kJ/kg	Copper 50ml/l 1kJ/kg	Copper 50ml/l 5,5kJ/kg
1	40	35	55	51	47	50	59	48	55	28
2	40	36	56	52	49	50	60	48	54	25
3	42	34	56	51	49	51	62	47	55	27
Average	40,7	35,0	55,7	51,3	48,3	50,3	60,3	47,7	54,7	26,7
Standard Deviation	1,15	1,00	0,58	0,58	1,15	0,58	1,53	0,58	0,58	1,53

In Table 4, the lengths of rapeseed seedling roots on the seventh day of the experiment, regarding the determination of seed germination capability, are presented. Significant increases in root length were observed in the treatments where seeds were soaked in silver nanocolloid solutions at concentrations of 25 and 50 ml/l with energies of 1 kJ/kg and 5.5 kJ/kg compared to the control groups by: 26.1%, 71.2%, 60.2%, 103.0% respectively. After soaking the seeds in copper nanocolloid solutions at a concentration of 25 ml/l, a decrease in root length of 41.0% was observed at 1 kJ/kg, while a significant increase of 72% was observed at 5.5 kJ/kg compared to the control groups. After soaking the seeds in a copper nanocolloid solution at a concentration of 50 ml/l, a slight shortening of the root by 4.3% was observed at 1 kJ/kg, while a slight elongation of 3.0% was observed at 5.5 kJ/kg compared to the control groups.

Table 4. The influence of metal nanocolloid concentration and pulsed electric field on the length of rapeseed seedling roots.

	Root length									
Sample	Control 1kJ/kg	Control 5,5kJ/kg	Silver 25ml/l 1kJ/kg	Silver 25ml/l 5,5kJ/kg	Silver 50ml/l 1kJ/kg	Silver 50ml/l 5,5kJ/kg	Copper 25ml/l 1kJ/kg	Copper 25ml/l 5,5kJ/kg	Copper 50ml/l 1kJ/kg	Copper 50ml/l 5,5kJ/kg
1	53	43	67	74	85	89	32	74	50	45
2	55	45	68	77	86	89	33	77	53	45
3	53	44	68	75	87	90	30	76	51	46
Average	53,67	44,00	67,67	75,33	86,00	89,33	31,67	75,67	51,33	45,33
Standard Deviation	1,15	1,00	0,58	1,53	1,00	0,58	1,53	1,53	1,53	0,58

4.2 Sunflower

In Table 5, the lengths of sunflower seedling stems on the seventh day of the experiment, regarding the determination of seed germination capability, are presented. In the cases where seed soaking was done in a silver nanocolloid solution with a concentration of 25 ml/l at 1 kJ/kg, an increase in stem length compared to the control was observed, by 22.6%. However, at the same concentration but at 5.5 kJ/kg, a decrease in stem length by 38.2% was observed compared to the control. After applying silver nanocolloids at a concentration of 50 ml/l at 1 kJ/kg and 5.5 kJ/kg, an increase in stem length was observed compared to the control; for 1 kJ/kg, there was an increase of 11.3%, while for 5.5 kJ/kg, a significant increase of 46.8% was observed. Upon applying copper nanocolloid at a concentration of 25 ml/l at 1 kJ/kg, a decrease in stem length of 25.9% compared to the control was observed, whereas at an energy level of 5.5 kJ/kg, an increase of 6.9% compared to the control was noted. After applying copper nanocolloid at a concentration of 50 ml/l, a decrease in stem length of 35.4% at 1 kJ/kg was observed, while an increase of 4.6% at 5.5 kJ/kg was observed.

Table 5. The influence of the concentration of metal nanocolloids and pulsed electric field on the steam length of sunflower seedlings.

	Stem length									
Sample	Control 1kJ/kg	Control 5,5kJ/kg	Silver 25ml/l 1kJ/kg	Silver 25ml/l 5,5kJ/kg	Silver 50ml/l 1kJ/kg	Silver 50ml/l 5,5kJ/kg	Copper 25ml/l 1kJ/kg	Copper 25ml/l 5,5kJ/kg	Copper 50ml/l 1kJ/kg	Copper 50ml/l 5,5kJ/kg
1	70	56	85	34	77	85	50	60	45	60
2	72	58	88	37	80	83	54	62	45	60
3	70	59	87	36	79	86	53	63	47	61
Average	70,7	57,7	86,7	35,7	78,7	84,7	52,3	61,7	45,7	60,3
Standard Deviation	1,15	1,53	1,53	1,53	1,53	1,53	2,08	1,53	1,15	0,58

In Table 6, the lengths of rapeseed seedling roots on the seventh day of the experiment, concerning the determination of seed germination ability, are presented. In the case of seeds soaked in silver nanocolloid solutions at a concentration of 25 ml/l and 1 kJ/kg, an increase in root length compared to the control was observed, by 33.7%, while at 5.5 kJ/kg, there was a decrease in length by 7.4%. Upon soaking the seeds in silver nanocolloid solutions at a concentration of 25 ml/l and energies of 1 kJ/kg and 5.5 kJ/kg, an increase in root length of 32% and 4.4%, respectively, compared to the control was observed. For seeds soaked in copper nanocolloid solutions at a concentration of 25 ml/l and 1 kJ/kg, a slight increase in root length of 1.1% compared to the control was observed. However, at 5.5 kJ/kg, a significant decrease in root length of 41% compared to the control was observed. For seeds soaked in copper nanocolloid solutions at a concentration of 50 ml/l and 1 kJ/kg, a decrease in root length of 2.2% compared to the control was observed, while at 5.5 kJ/kg, a slight increase of 0.4% was observed.

Table 6. The influence of the concentration of metal nanocolloids and pulsed electric field on the root length of sunflower seedlings.

	Root length									
Sample	Control 1kJ/kg	Control 5,5kJ/kg	Silver 25ml/l 1kJ/kg	Silver 25ml/l 5,5kJ/kg	Silver 50ml/l 1kJ/kg	Silver 50ml/l 5,5kJ/kg	Copper 25ml/l 1kJ/kg	Copper 25ml/l 5,5kJ/kg	Copper 50ml/l 1kJ/kg	Copper 50ml/l 5,5kJ/kg
1	60	75	80	70	80	80	63	45	60	77
2	62	78	83	73	78	78	60	47	59	75
3	59	76	79	69	81	81	60	43	58	78
Average	60,3	76,3	80,7	70,7	79,7	79,7	61,0	45,0	59,0	76,7
Standard Deviation	1,53	1,53	2,08	2,08	1,53	1,53	1,73	2,00	1,00	1,53

5 Conclusion

5.1 Rapeseed

Germination

Silver nanocolloids decrease germination ability at 1 kJ/kg energy or maintain it at the same level at 5.5 kJ/kg compared to the control. Copper nanocolloids at a concentration of 25 ml/l positively affect germination ability at both energies compared to the control. For 1 kJ/kg, it is higher by 1%, whereas for 5.5 kJ/kg, it is higher by 1.4%. Silver and copper nanocolloids, when applied at a concentration of 50 ml/l at 5.5 kJ/kg, do not affect the stimulation of germination ability.

Steam and Root Lengths

Nanoparticles positively affect the elongation of rapeseed steams. However, at a concentration of 50 ml/l at 5.5 kJ/kg, a 23.8% decrease in steam length compared to the control was observed. The optimal concentration of silver and copper nanocolloids, where an increase in rapeseed steam length compared to the control was observed, is 25 ml/l at 1 kJ/kg energy. Silver nanocolloids positively affect root elongation in rapeseed seedlings.

The greatest increase in root length was observed for silver nanocolloids at a concentration of 50 ml/l at 5.5 kJ/kg, compared to the control by 103.0%. Copper nanocolloids negatively affect root elongation at 1 kJ/kg energy.

5.2 Sunflower

Germination
Silver and copper nanocolloids, when applied at a concentration of 25 ml/l at 1 kJ/kg, positively affect germination ability. Applying these parameters contributed to a 4.7% increase in germination for both nanocolloids. Silver nanocolloids at a concentration of 25 ml/l and 1 kJ/kg have a positive effect on germination compared to silver at higher concentration but the same delivered energy. Silver and copper nanocolloids at 5.5 kJ/kg and a concentration of 25 ml/l negatively affect sunflower seed germination ability, 41.7% less for silver and 22.7% less for copper compared to the control. Silver nanocolloids at a concentration of 50 ml/l at 1 kJ/kg positively affect germination by 1.7% compared to the control. Applying copper nanocolloid (50 ml/l, 1 kJ/kg) and silver and copper nanocolloids at a concentration of 50 ml/l at 5.5 kJ/kg negatively affect sunflower seed germination compared to controls with the same delivered energy, respectively by 2; 3.7; 3.3%. Lower nanocolloid concentrations (25 ml/l) at 1 kJ/kg stimulate germination more compared to using nanocolloids at a concentration of 50 ml/l at 1 kJ/kg, 3% higher for silver and 6.7% higher for copper compared to the control.

Steam and Root Lengths.
The lowest value for silver was obtained for a concentration of 25 ml/l and 5.5 kJ/kg, resulting in a 38.2% decrease in steam length compared to the control. For copper, the lowest values were obtained for a concentration of 25 ml/l, 1 kJ/kg, and 50 ml/l, 1 kJ/kg. Copper adversely affects sunflower steam elongation at both concentrations. The greatest increase in steam length compared to the control was obtained for silver nanocolloids at a concentration of 50 ml/l, 5.5 kJ/kg, with stems being longer than the control by 46.8%. Silver nanocolloids at a concentration of 25 ml/l at 5.5 kJ/kg negatively affect root elongation. For silver nanocolloids, the lowest root length value was obtained for a concentration of 25 ml/l at 5.5 kJ/kg (7.4% lower than the control), while for copper, the lowest value was also obtained for a concentration of 25 ml/l at 5.5 kJ/kg (41% lower than the control). The highest value compared to the control was obtained for silver nanocolloids at a concentration of 25 ml/l at 1 kJ/kg energy, with roots being longer than the control by 33.7%.

References

1. Shorstkii, I., Mirshekarloo, M.S., Koshevoi, E.: Application of pulsed electric field for oil extraction from sunflower seeds: electrical parameter effects on oil yield. J. Food Process Eng. **40** (2017). https://doi.org/10.1111/jfpe.12281
2. Guderjan, M., Elez-Martínez, P., Knorr, D.: Application of pulsed electric fields at oil yield and content of functional food ingredients at the production of rapeseed oil. Innov. Food Sci. Emerg. Technol. **8**, 55–62 (2007). https://doi.org/10.1016/j.ifset.2006.07.001
3. Rifna, E.J., Ratish Ramanan, K., Mahendran, R.: Emerging technology applications for improving seed germination. Trends Food Sci. Technol. **86**, 95–108 (2019). https://doi.org/10.1016/j.tifs.2019.02.029

Development of Poultry Meat Production and Consumption Levels in the Light of Environmental Sustainability Goals

Anna Szeląg-Sikora[1,2](✉), Aneta Oleksy-Gębczyk[3], Katarzyna Kowalska-Jarnot[4], Jakub Sikora[1,2], and Joanna Stuglik[2]

[1] University of Agriculture in Krakow, al. Adama Mickiewicza 21, 31-120 Krakow, Poland
anna.szelag-sikora@urk.edu.pl

[2] Cavalry Captain Witold Pilecki State University of Małopolska in Oświęcim, Oświęcim, Poland

[3] University of Applied Sciences in Nowy Sącz, Nowy Sącz, Poland

[4] Department of Management and Accounting, School of Economics and Computer Science in Cracow, Krakow, Poland

Abstract. The study, by conducting a survey on a sample of 1180 randomly selected Polish consumers, showed whether there is market interest in poultry meat. The first research area was to verify the level of meat consumption in general, as well as poultry meat. In the study population, only 1% of people declared that they do not eat meat at all. Around half (49%) eat meat every day or almost every day, the other large group (42%) every two or three days. Another research objective was to verify what factors influence the purchase of poultry meat. Among the decision variables, respondents mentioned: ease of preparation, health reasons, price, easy access, habit, nutritional value, taste, among others. The analysis of the results included grouping variables, i.e. age, gender, education of the respondents. The scope of the study also includes a literature analysis considering the location of poultry meat production and consumption in the light of sustainable agricultural development.

Keywords: Poultry meat production · the consumer · consumer decision-making · sustainability

1 Introduction

Every human economic activity should be carried out in accordance with the principles of sustainable development. This is particularly true of livestock production, which, due to its biological nature, has a strong impact on the natural environment. It should not be forgotten, however, that animal production is supported by plant production, which is also important in the pursuit of sustainable development [1, 2]. Animal husbandry poses many threats, such as excessive ammonia and methane emissions to the atmosphere, impoverishment of biodiversity in ecosystems, eutrophication of water bodies or problems with rational management of animal excreta. Meat has become a controversial

E. Lorencowicz et al. (Eds.): FMPMSA 2024, LNCE 609, pp. 411–418, 2024.
https://doi.org/10.1007/978-3-031-70955-5_45

topic in public debates because it covers many dimensions of sustainability, including economic, social, environmental, health and animal welfare issues. Meat has a much larger environmental and climate footprint than plant-based foods and can be associated with negative health impacts. Technological options can help increase sustainable meat production, but changes in consumption are also needed. Particularly in high-income countries, where people consume an average amount of meat, significant reductions will be important. There is therefore a constant need to seek a way of managing animal husbandry that ensures the sustainability of the wider environment [3]. However, meat is an important source of protein in the diet of modern humans [4]. Average meat consumption worldwide - although highly geographically diverse - is increasing and currently stands at 33.7 kg per capita. According to FAO projections, by 2029, average meat consumption will increase to 34.94 kg per capita, an increase of 3.7% [5]. The drivers of this increase in consumption are the growth of the world population, the development of the global economy, progressive urbanisation processes and increased social welfare. The most consumed meat species worldwide is poultry, which overtook pork five years ago. It is gaining popularity due to consumer trends: health benefits, lower calories, lower price and ease of preparation [6, 7].

Despite the positive outlook, the coming years may be challenging for the industry, due to [8]:

- the likely achievement of maximum domestic consumption,
- popularity of meat substitutes competing for the same consumer attention [9].
- increasing imports of agricultural products from Ukraine, especially in the poultry sector, which are characterised by competitive prices

The fact that excessive meat consumption has been challenged on health [10] and sustainability [11] grounds may also be a barrier to development. For this reason, alternatives to conventional meat consumption, such as more sustainably produced meat or meat alternatives, are being considered. According to the PWC report, almost 50% of consumers, both in Poland and across Europe, plan to reduce their meat consumption. This trend is promoted by discounters offering meat substitutes (they produce almost three quarters of the sales value of this category). In 2022, more than 60% of Poles had access to meat substitutes, with almost half of them having tried these products, mainly out of curiosity. More than 40% consume meat substitutes regularly [8]. Poland is one of the European leaders in terms of meat production, ranking first in poultry production, fourth in pork production and sixth in beef [12]. It also stands out against the EU in terms of meat consumption structure. Surveys show that 89% of Poles consume meat more than once a week, and are almost twice as likely to consume pork as poultry, while beef is consumed only occasionally [13]. The aim of this article is to verify what type of purchasers consume poultry and at what level, taking into account criteria such as age, gender, residence, income level and education.

2 Research Methodology and Analysis of Results

The study is based on primary data. Using the author's survey questionnaire, 1180 consumers from different regions of Poland were interviewed using the CAWI (Computer Assisted Web Interviewing) method. The socio-demographic composition of the respondents is shown in Fig. 1.

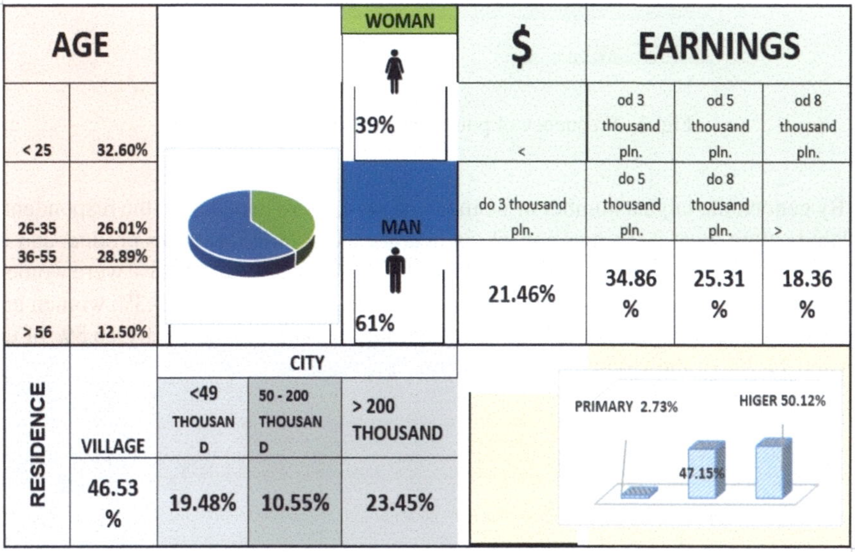

Fig. 1. Demographic composition of respondents

The first research area was to verify the level of consumption of poultry meat (Fig. 2) In the study population, only 2% of respondents declared that they do not eat it at all. 15% eat poultry meat daily or almost daily, 53% two or three times a week (making a total of 68% eating poultry meat frequently and very frequently). A further 23% eat poultry meat once a week (making a total of 91% of those eating poultry meat regularly, although with varying frequency).

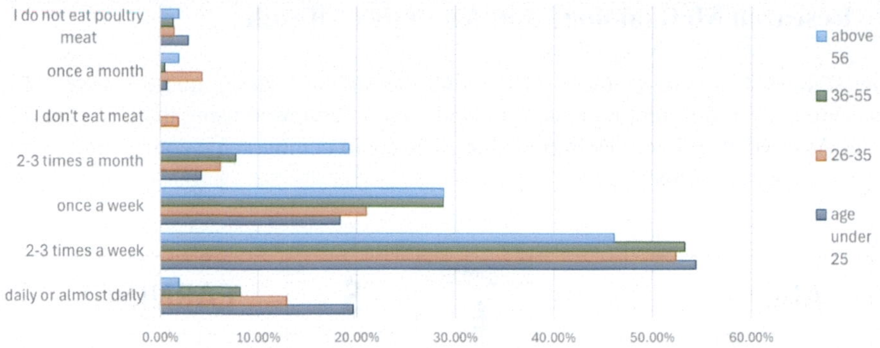

Fig. 2. Frequency of poultry meat consumption by age

By gender, the largest number of women and men, more than half of the respondents, declared eating meat 2–3 times a week, indicating the popularity of this product and its important place in the weekly menu. Of these, 54.9% were men and 51.8% were women. The second outstanding group consumes poultry meat once a week: 24.3% women and 20.3% men. Every day or almost every day, poultry meat is consumed by 16.8% of the men and 11.2% of the women surveyed (Fig. 3).

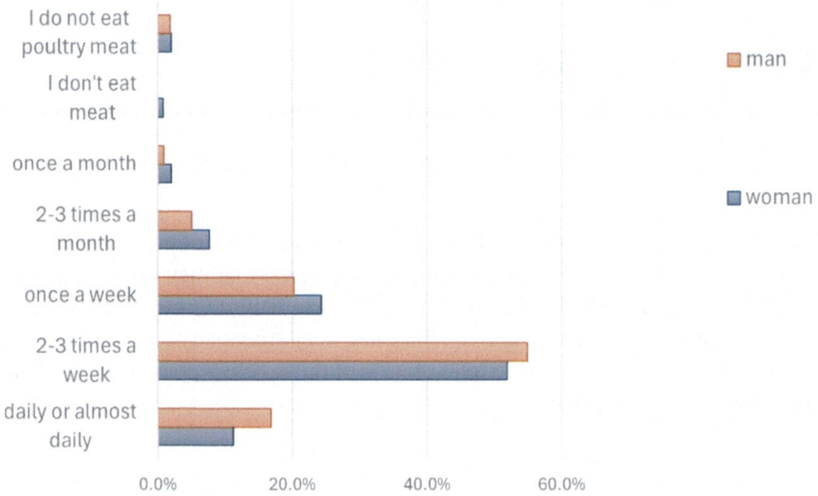

Fig. 3. Frequency of poultry meat consumption by gender

Respondents with primary, secondary and tertiary education most often consume poultry meat 2–3 times a week. Respondents with primary education 54.5%, secondary education 55.9% and tertiary education 50.2% respectively. In second place, all three groups indicated that they consume meat once a week, including those with primary education 27.3%, secondary education 20.1% and tertiary education 25.0%. Every day or almost every day, the group with secondary education eats meat 16.4%, higher education

10.9% and lower education 9.1%. Among those who do not eat meat at all, the largest group are respondents with a university education, but this is only 0.7% (Fig. 4).

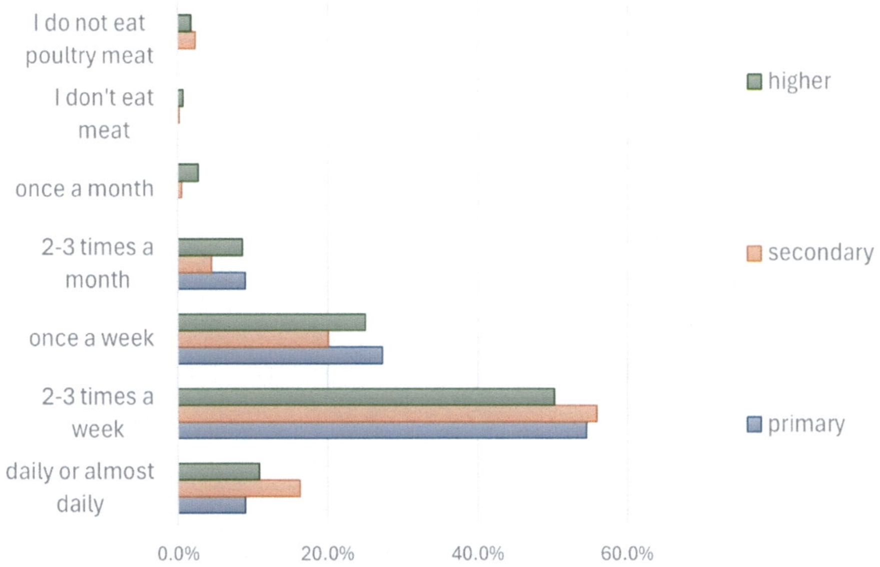

Fig. 4. Frequency of poultry meat consumption by level of education

Taking into account the criterion of place of residence, the largest group of respondents indicated that they eat poultry meat 2–3 racy per week. These are respondents living in rural areas and in cities with 50,000 to 200,000 inhabitants. This is almost 60% of respondents (Fig. 5).

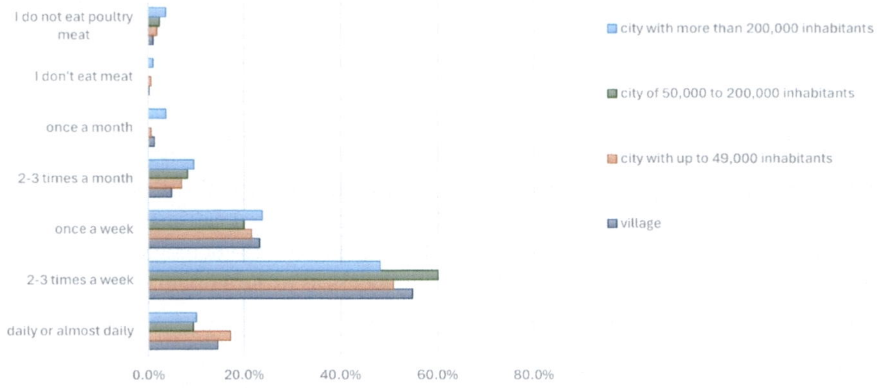

Fig. 5. Frequency of poultry meat consumption by place of residence

Taking into account the criterion of income, respondents were divided into groups earning up to PLN 3 000, between PLN 3 000 and 5 000, between PLN 5 000 and 8 000 and above PLN 8 000. It is noteworthy that in each income group, the highest indications related to consumption of poultry meat at the level of 2–3 times a week. Among those earning up to PLN 3 000, this was 50.9% of respondents; among those earning between PLN 3 000 and 5 000, it was 55.0% of respondents; among those earning between PLN 5 000 and 8 000, 54.4%; and among those earning more than PLN 8 000, 50.0% of respondents (Fig. 6).

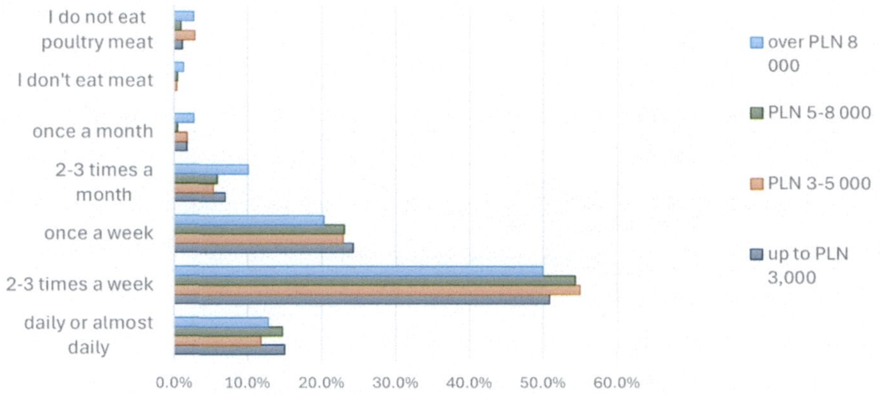

Fig. 6. Frequency of poultry meat consumption by net monthly income

3 Summary

Intensive economic development has resulted in many adverse ecological consequences. This also applies to agriculture, including livestock production, which is a significant emitter of greenhouse gases into the atmosphere. The development of new sustainable technologies is key to meeting climate and environmental challenges in the 21st century [14, 15]. The drive to meet societal needs and demands for high quality products on offer should be aimed at shaping food safety throughout the food chain, from producer to consumer, taking into account European Union food law and Common Agricultural Policy guidelines [16, 17]. However, it should not be forgotten that it is consumer preferences that determine the formation of the food market. The analysis of the research results unequivocally indicates the important role of poultry meat in the nutrition of the surveyed respondents. The grouping variables adopted (socio-demographic) showed variation in the decision/purchase/consumption process as to the amount of poultry meat purchased/consumed. From the analysis it can be concluded:

- Global economic development is leading to an increase in demand for food, including demand for poultry meat and changes in consumption patterns for meat and meat products. Meeting the demand for meat is driving the intensification of livestock production, with a consequent increase in animal populations and environmental pressures.

- Growing trends of healthy lifestyles, increased environmental awareness and concern for animal welfare mean that consumers care more and more about the quality and origin of meat.
- The Polish meat industry needs to focus on maintaining the highest quality standards, as well as adapting to evolving standards and regulations, if it wants to meet the expectations of theincreasingly demanding consumers

References

1. Niemiec, M., Chowaniak, M., Sikora, J., Szeląg-Sikora, A., Gródek-Szostak, Z., Komorowska, M.: Selected properties of soils for long-term use in organic farming. Sustainability **12**, 2509 (2020). https://doi.org/10.3390/su12062509
2. Komorowska, M., et al.: Closed-loop agricultural production and its environmental efficiency: a case study of sheep wool production in Northwestern Kyrgyzstan. Energies **15**, 6358 (2022). https://doi.org/10.3390/en15176358
3. Mroczek, K., Rudy, M., Stanisławczyk, R., Mroczek, J.R.: Food production and consumption in development consumption (2018). https://repozytorium.ur.edu.pl/items/e4a85816-a658-4c54-9f3f-ea045c572c1e
4. Boateng, E.F., Nasiru, M.M., Agyemang, M.: Meat: valuable animal-derived nutritional food (2020). https://www.researchgate.net/publication/340573356_Meat_Valuable_Animal-Derived_Nutritional_Food_A_Review
5. https://foodfakty.pl/konsumpcja-miesa-na-swiecie-wzrasta-chociaz-jest-silnie-zrozni cowana
6. Marangoni, F., Cricelli, C., Ferrara, N., Corsello, G.: Role of poultry meat in a balanced diet aimed at maintaining health and wellbeing: an Italian consensus document (2015). https://www.researchgate.net/publication/279226850_Role_of_poultry_meat_in_a_balanced_diet_aimed_at_maintaining_health_and_wellbeing_An_Italian_consensus_document
7. Oleksy-Gebczyk, A.: Preferences and consumer choices: a case of Polish markets for goods and services. Asian J. Appl. Econ./Appl. Econ. J. **30**(2) (2023)
8. PVC Report: State of the Meat Market in Poland in 2023. https://www.strategyand.pwc.com/pl/pl/publikacje/2023/stan-rynku-miesa-w-polsce-2023.html
9. Andreani, G, Marti, A., Sogari, G., Froldi, F.: Plant-based meat alternatives: technological, nutritional, environmental, market, and social challenges and opportunities (2023). https://www.researchgate.net/publication/367174459_Plant-Based_Meat_Alternatives_Technolog ical_Nutritional_Environmental_Market_and_Social_Challenges_and_Opportunities
10. Connolly, G., Clark, C.M., Campbell, R., Byers, A., Reed, J, Campbell, W.: Poultry consumption and human health: how much is really known? A systematically searched scoping review and research perspective (2022). https://www.ncbi.nlm.nih.gov/pmc/articles/PMC9776623/. Accessed 09 May 2024
11. Parlasca, M.: Quaim, M.: Meat Consumption and Sustainability (2022). https://www.resear chgate.net/publication/360181099_Meat_Consumption_and_Sustainability
12. Report on the State of the Meat Market in Poland in 2023. https://www.strategyand.pwc.com/pl/pl/publikacje/2023/stan-rynku-miesa-w-polsce-2023.html
13. CSO Supply to the domestic market and consumption of certain consumer goods per capita in 2022, OECD-FAO Agricultural Outlook 2023–2032, Kantar Polska and BLIX Group study "Meat market and vegetable alternatives according to the Polish smart shopper"
14. Szporysz, M., Sztuka, M., Tabor, S., Molenda, K., Kuboń, M.: Cluster analysis in assessment of organic farms sustainability. Part II results of research. Agric. Eng. **24** (2020). https://doi.org/10.1515/agriceng-2020-0008

15. Gródek-Szostak, Z., et al.: Promotion of RES in a technology transfer network. A case study of the enterprise europe network. Energies **13**, 3445 (2020). https://doi.org/10.3390/en13133445
16. Krupiński, J., Horbańczuk, K.O., Kołacz, R., Litwińczuk, Z., Niemiec, K., Zięcik, J.: Strategic directions of development of animal disease production compliance, protection of the environment and animal welfare are required (2011). https://www.google.com/search?client=firefox-b-d&q=translator
17. Gródek-Szostak, Z., Suder, M., Kusa, R., Sikora, J., Niemiec, M.: Effectiveness of instruments supporting inter-organizational cooperation in the RES market in Europe. A case study of enterprise Europe network. Energies **13**, 6443 (2020). https://doi.org/10.3390/en13236443

Effect of Fertilization with Digestate on Crop Yields: A Review

Małgorzata Szwed(✉), Milan Koszel, and Artur Przywara

Department of Machine Operation and Production Processes Management, University of Life Sciences in Lublin, Akademicka 13, 20-950 Lublin, Poland
malgorzata.szwed@up.lublin.pl

Abstract. The article presents the results of a study of the use of digestate as an organic fertilizer on crop yields. Nowadays, the use of digestate as a fertilizer is a promising concept in agriculture to increase crop yields. This publication focuses on analyzing recent research on the effects of digestate fertilization on crop quality and yield. It discusses how digestate is produced, methods of applying it as a fertilizer and its effects on plants. The increase in environmental pollution in recent years has had a tremendous impact on raising the issue related to sustainable agricultural practices and thus increasing the use of natural fertilizers. Digestate is a by-product obtained from the methane fermentation process, during the production of biogas. It contains in its composition irreplaceable nutrients necessary for plant growth and development. The article discusses various aspects related to fertilization techniques with digestate, such as its chemical composition, impacts on soil and plants. In addition, the results of scientific research related to the use of digestate as an alternative fertilizer source for field crops are presented. The conclusions that will be drawn from the review aim to provide a comprehensive view of the issue of fertilization with digestate and to identify the potential benefits and challenges of its use in agriculture.

Keywords: Digestate · Digestate Application Techniques · Sustainable Agriculture · Environmental Protection

1 Introduction

Anaerobic digestion results in a by-product called anaerobic digestate, which can be used in agriculture as an organic additive, but mainly as an alternative fertiliser source due to its high nitrogen concentration, mainly in mineral form [2]. It is worth noting that the chemical composition of the digest can vary depending on the type of feedstock and the process carried out in the biogas plant. The use of the digestate as a fertiliser contributes to the reuse of organic matter and minerals, which can improve the profitability of crop production by reducing the costs associated with the purchase of fertilisers [16]. In order to optimise the extraction of methane potential and the flow of material and nutrients of the digestate, physical separation of the material into solid and liquid fractions, characterised by different properties, can be used.

E. Lorencowicz et al. (Eds.): FMPMSA 2024, LNCE 609, pp. 419–425, 2024.
https://doi.org/10.1007/978-3-031-70955-5_46

The liquid fraction, which can be spilled in the fields, meets the requirements of the Ministry of the Environment's Regulation on the Recovery of R-10, or reprocessed as a process liquid, which allows it to be used for various industrial purposes. The aim of this article is to present the possibility of using the digestate as a fertigant and its application techniques.

2 Materials and Methods

Digestate is a by-product produced in an agricultural biogas plant during bio-gas production [9]. The composition of the biogas and thus the poferment will largely depend on the substrates used to create them [11].

The percentage composition of substrates can vary depending on the type of substrate, the proportion of mixed feedstock, and the fermentation conditions. In general, the organic substrate mix consists of: Water - Usually makes up the majority of the volume of the mix, usually about 80–95%, Organic matter - Organic components make up a significant portion of the substrates, usually 5–20%. Minerals - Mineral content is usually small, but important for the fermentation process, accounting for about 1–5%.

The main components include: food scraps, residues from food production (starch, cellulose), sludge from food processing (dairies, breweries), vegetable residues, residues from oilseed crops [5].

The digestate consists primarily of: undigested residue of organic matter, minerals (in amounts comparable to their content in the substrates) and biomass of methane bacteria. The main characteristics of digestate include the amount of dry matter, pH, content of organic matter, nitrogen, phosphorus, calcium and magnesium, and the presence of heavy metals [4]. The average chemical composition of the digestate is shown in Table 1.

Table 1. Average composition of solid and liquid fractions from Polish biogas plants [10].

Test characteristic	Solid fraction	Liquid fraction
Dry matter content s.m.o. (%)	22–27	2.7–4.3
Organic dry matter content s.m.o. (%)	89–94.5	58–62
Total N (%)	0.4–0.8	0.29–0.75
N-NH4 (%)	0.08–0.52	0.28–0.38
P (%)	0.1–0.28	0.03–0.05
K (%)	0.12–0.69	0.05–0.62
Ca (%)	0.22–0.43	0.05–0.07
Mg (%)	0.06–0.17	0.01–0.02

2.1 Results

In a study carried out on the soil nutrient content before the experiment, [15] found adequate phosphorus (P) levels for maize, while potassium (K) levels were slightly below expectations, especially in 2016. The same was true for magnesium (Mg). These deficiencies were partially compensated by the application of BDS (solid biogas digestate fraction) fertiliser. The content of heavy metals (Pb, Cd) was below the standards for organic fertilisers, but the copper content exceeded these standards.

The total mineral nitrogen content of the soil was sufficient for maize. The nitrogen source structure was similar in 2014 and 2015, but in 2016 the main nitrogen source was fertiliser. The total nitrogen in BDS was comparable to manure. The C:N ratio in BDS indicated a high mineralization potential.

The yield response to BDS application followed a quadratic regression model, reaching an optimum yield of 2.19 t ha^{-1}, which was significantly higher than in the control group. These results confirm the high fertilising value of BDS as a substitute for mineral fertilisers.

During an experiment conducted [6] to evaluate the effect of digestate from the biogas plant in Gorajec on maize cultivation, a significant effect of this digestate on plant yield was demonstrated. The plant response to this sludge was assessed by analysing the average weight of the maize plant, the average length of the stalk, the average number of large cobs per plant and the average weight of the large cobs. In addition, small cobs were also examined.

The yield of plants harvested from the plots where the post-ferment was applied was 11% higher than that of the control plants, while the yield of the plots with the 8% post-ferment exceeded that of the control plots by 12%. The indications 'with digestate 6%' and 'with digestate 8%' in the context of maize cultivation refer to the concentration of digestate applied as fertiliser.

In the study conducted by the author, different maize fields were fertilised with different concentrations of digestate: some fields were fertilised with 6% digestate and others with 8%. This means that the concentration of the digest in relation to the other components (e.g. water or other dilution medium) was 6% and 8%, respectively. This dosage is intended to investigate the effect of different concentrations of digestate on the growth, development and yield of maize. The results of the experiment were able to show how different concentrations of digestate affect these parameters, which is important for optimising the use of this type of fertiliser in agricultural practice.

In the research conducted during the experiment [1], the potential of the digestate as a fertiliser in horticultural crops was evaluated over a period of two years. The effect of the addition of digestate on soil fertility and plant yields was analysed, comparing its fertilising capacity with that of mineral fertiliser and a traditional organic fertiliser such as cattle manure. The soil physico-chemical changes induced by the digest tended to decrease gradually over time, leading to a small residual effect. The digest provided a significant amount of ammonium nitrogen, which was rapidly nitrified, thus becoming directly available to crops in the short term. In addition, the addition of the digestate resulted in an increase in the amount of available phosphorus in the soil, suggesting that its use in agricultural practice should be based not only on the nitrogen supplied, but also on the phosphorus it provides.

2.2 Methods of Digestate Application

Equipment used for the application of raw slurry and separated liquid can also be used for the spreading of digestate. Machines used for manure spreading can also be used to spread the solid form of the digestate. To ensure optimal use of the digestate as a fertiliser, it should be applied during the growing season of the crop, using equipment that ensures even distribution over the field and precise application rates. This approach also minimises ammonia volatilisation [12] (Fig. 1).

Fig. 1. Shallow injection into grassland [12]

In their article, the authors [17] compared manure with digestate, pointing out that digestate can be used directly as fertilizer, just like manure. They emphasized the value of digestate as an alternative to conventional fertilizers because of its ability to comprehensively promote plant growth.

The researchers presented in their study [12] a comparative analysis of four methods of applying manure to crop fields. These methods are: spreading by hose (drag hose), spreading by drag shoes, injection method (injection) and spreading by splash plate. Analysis of slurry spreading has shown that the hose and drag methods ensure even distribution of the substance, which is conducive to increasing the efficiency of nutrient utilization by plants. In contrast, the injection method, while equally effective, is associated with an increased risk of mechanical damage to crops. The disc method proves to be the least homogeneous, which can affect the non-uniformity of nutrient availability to crops. The aspect of ammonia volatilization is particularly important due to its impact on air quality and greenhouse gas emissions. In this context, the injection method stands out as having the lowest risk, in contrast to the disk method, which has a high risk. The risk of crop contamination indicates the potential impact of application methods on the quality and safety of agricultural products. Spray plate, by its nature, poses the highest risk, which requires careful consideration in the context of consumer agriculture.

3 Discussion

In order to better utilize the methane potential and nutrients of the digestate, research has been reviewed on different fractions of this material, such as solid and liquid, which have different properties. The liquid fraction of the digestate can be bottled in fields, meeting the requirements of relevant recovery regulations. Alternatively, it can be reprocessed as a process liquid, allowing it to be used for various industrial purposes. The application of waste organic fertilizer had a significant effect on increasing the yields of the plants studied, and differences in dry matter yields were evident especially in the first and second years of use. The effect of the level of biodigester fertilization on the increase in iron ion content was found only in the biomass of perennial ryegrass in the second year of cultivation, while no significant differences were observed in the chemical composition of the other species studied [13]. The application of digest improved soil quality by reducing bulk density, increasing saturated hydraulic conductivity, soil moisture retention capacity and aggregate stability. Studies have shown that the digestate has a beneficial effect on these soil properties compared to a control area that did not undergo such a process [14].

Based on studies conducted [7], it can be observed that the use of poferment as a fertilizer in the cultivation of winter oilseed rape has significant benefits for both yield quality and soil. The authors found that the application of digestate at a rate of 50,000 L/ha significantly increases the fat and protein content of rapeseed, resulting in better yield quality [8]. In addition, the study showed improved soil parameters and increased plant resistance to biotic and abiotic stresses, which is particularly important in the context of sustainable agricultural development. These results are consistent with the observations of other researchers, who also note the positive effects of using digestate as part of recycling practices in agriculture. The economic and environmental benefits of this approach can significantly impact sustainable agricultural production.

On the other hand, the analysis of the article [3] shows that digestate, in terms of heavy metal content, mostly exhibits trace amounts of these elements, making it safe for use as fertilizer in agriculture. The study conducted by the authors showed that even when the substrates used in a biogas plant contain certain amounts of heavy metals, the process of anaerobic digestion and final processing into digestate does not lead to the accumulation of these metals in alarming amounts. Specifically, lead concentrations in the digestate were well below 5 mg/kg, which is below the limit considered safe for soil and plants. This suggests that the digestate can be effectively used as an organic fertilizer, providing nutrients without the risk of heavy metal contamination of the soil.

4 Conclusion

The analysis undertaken in the article on the use of digestate as an organic fertilizer shows the significant potential of this substance in the context of sustainable agriculture and environmental protection. The studies analyzed show the benefits of using digestate, including improved soil structure, increased soil fertility due to organic matter and nutrients, as well as the continued development of soil microorganisms. The use of digestate can help reduce the costs associated with the purchase of mineral fertilizers, which is particularly important in the face of rising fertilizer prices and restrictions on access to fertilizers.

At the same time, experiments on different fractions of digestate, such as solid and liquid, highlight the variety of applications for each. It has been shown that solid fractions can be used for energy production and can be a valuable feed additive, while liquid fractions are used in field irrigation, providing plants with essential nutrients. Emphasizing the value of digestate as a fertilizer is in line with the results presented, where the solid fraction of digestate, after proper certification, can be used as an organic fertilizer. This highlights the importance of further research into the composition and properties of the digestate, in order to optimize its use and minimize the potential negative effects of its use. The cited studies prove that digestate can be effectively used as an alternative to mineral fertilizers, as demonstrated in studies on its effects on yields of corn and perennial grasses. In addition, the potential of digestate as a valuable product in the operation of biogas plants has been confirmed, highlighting its importance as a plant-supporting fertilizer.

The use of digestate in agriculture can be key to achieving sustainability goals, positively influencing crop productivity and environmental protection. This is in line with efforts to close the cycle of matter in nature and promote circular economics. However, it is necessary to continue research to fully understand the impact of digestate on various aspects of agriculture, including soil, plants and the environment, in order to maximize benefits and minimize potential risks of negative impacts on the ecosystem.

References

1. Alburquerque, J.A., et al.: Agricultural use of digestate for horticultural crop production and improvement of soil properties. Eur. J. Agron. **43**, 119–128 (2012)
2. Askri, A., Laville, P., Trémier, A., Houot, S.: Influence of origin and post-treatment on greenhouse gas emissions after anaerobic digestate application to soil. Waste Biomass Valor **7**(3), 293–306 (2016)
3. Derehajło, S., Tymińska, M., Skibko, Z., Borusiewicz, A., Romaniuk, W.: Heavy metal content in substrates in agricultural biogas plants. Agric. Eng. **27**(1), 315–328 (2023)
4. Dubský, M., Chaloupková, Š., Kaplan, L., Vondráčková, S., Tluštoš, P.: Use of solid phase of digestate for production of growing horticultural substrates. Horticult. Sci. (Prague) **46** (2019)
5. Hanafiah M., et al.: biogas production from different substrates under anaerobic conditions. In: 3rd International Conference on Chemical, Agricultural and Medical Sciences (CAMS-2015), 10–11 December 2015, Singapore (2015)
6. Kornak, N.: Uprawa kukurydzy z zastosowaniem Pofermentu z biogazowni rolniczej (2020)
7. Koszel, M., Lorencowicz, E.: Agricultural use of biogas digestate as a replacement for fertilizers. Agric. Agricult. Sci. Procedia **7**, 119–124 (2015)
8. Koszel, M., et al.: Impact of digestate application as a fertilizer on the yield and quality of winter rape seed. Agronomy **10**(878), 1–18 (2020)
9. Kowalczyk-Juśko, A., Szymańska, M.: Poferment: Nawozem dla rolnictwa. FnRRPR, Warszawa (2015)
10. Kowalczyk-Juśko, A.: Nawozowe wykorzystanie masy pofermentacyjnej (2018)
11. Kowalczyk-Juśko, A., et al.: Energy value estimation of silages for substrate in biogas plants using artificial neural network. Energies **15**(4) (2020)

12. Lukehurst, C.T., Frost, P., Al Seadi, T., Biosantech. A.S.: Utilisation of digestate from biogas plants as biofertiliser. IEA Bioenergy Task 37 (2010)
13. Martyniak, D., Wiewióra, B., Żurek, G.: Wpływ bioodpadu z pofermentu biogazowni na plonowanie oraz skład chemiczny traw wieloletnich. Biuletyn Instytutu Hodowli i Aklimatyzacji Roślin **297/298**, 31–37 (2022)
14. Möller, K.: Effects of anaerobic digestion on soil carbon and nitrogen turnover, N emissions, and soil biological activity: a review. Agron. Sustain. Dev. **35**, 1021–1041 (2015)
15. Przygocka-Cyna, K., Grzebisza, W.: Biogas digestate – benefits and risks for soil fertility and crop quality – an evaluation of grain maize response. Open Chem. **16**, 258–271 (2018)
16. Sienkiewicz, S., Wierzbowska, J., Kovacik, P., Krzebietke, S., Zarczynski, P.: Digestate as a substitute of fertilizers in the cultivation of Viginia fanpetals. Fresenius Environ. Bull. **27**(6), 3970–3976 (2018)
17. Sobczak, A., Chomać-Pierzecka, E., Kokiel, A., Różycka, M., Stasiak, J., Soboń, D.: Economic conditions of using biodegradable waste for biogas production, using the example of Poland and Germany. Energies **15**(14) (2022). The Lubelskie Province/Wykorzystanie zasobów energii odnawialnej w planach gospodarki niskoemisyjnej wybranych gmin województwa lubelskiego. Economic and Regional Studies **12**(1), p. 33–44

The Role of Monitoring and Signaling in Integrated Crop Protection

Anna Tratwal(✉), Marcin Baran, and Kamila Roik

Institute of Plant Protection – National Research Institute, Władysława Węgorka 20, 60–318 Poznań, Poland
A.Tratwal@iorpib.poznzn.pl

Abstract. Integrated plant protection is a way of protecting plants from harmful organisms by using all available methods of plant protection, especially non-chemical methods, in a way that minimizes the risk to human health, animal health and the environment. Integrated pest management makes full use of knowledge of organisms that are harmful to plants (in particular, their biology and harmfulness) in order to determine the optimal timing for taking action against these organisms. It also takes advantage of the natural occurrence of beneficial organisms, including predators and parasites of organisms harmful to plants, and uses their introduction. Thus, integrated pest management reduces the use of chemical plant protection products to the minimum necessary and thus reduces pressure on the environment and protects the biodiversity of the agricultural environment.

Keywords: integrated protection · pest monitoring · signaling · On-line Pest Warning System

1 Introduction

There are many factors of increasing pressure from agrophages in agricultural crops. The main reasons for this situation are the increase in the area under cultivation (for example, the acreage under oilseed rape in Poland in 2022 exceeded one million hectares), the intensification of production (and the associated simplification of crops) and, above all, climate changes. Of course, agrophages do not occur simultaneously and their severity varies from season to season depending on agroclimatic conditions. Sometimes there are few or almost none; or, conversely, some of them appear unexpectedly and massively in a given year. Such situations occur more and more often, unfortunately. Therefore, growers should be prepared for them and have appropriate solutions to limit their pressure-both preventive (including the selection of resistant and tolerant varieties) and intervention (including safe, effective plant protection products, including biological).

E. Lorencowicz et al. (Eds.): FMPMSA 2024, LNCE 609, pp. 426–431, 2024.
https://doi.org/10.1007/978-3-031-70955-5_47

2 Tools in Monitoring and Signaling

Important tools used in integrated pest management:

1. Methodologies of integrated plant protection of individual crops.
2. Pathogen harmfulness thresholds - thresholds determine when the use of chemical plant protection products becomes economically viable, i.e. at what abundance of an organism harmful to plants, the losses it can cause, exceed the cost of its chemical control.
3. Decision support systems (DSS) in plant protection – DSS, based on knowledge of the biology of pest organisms, indicates the optimal timing of chemical plant protection treatments.
4. Monitoring and signaling of agrophages - systematic field observations that are the basis for deciding on the need for chemical treatment [1].

3 Discussion

The task of plant protection is to protect plants from pest organisms, so as to reduce yield losses. This means that in the absence of the occurrence of an agrophage or in its low intensity, plant protection treatment is unnecessary. On the other hand, the lack of protection in the case of the massive appearance of a harmful species can even lead to a total loss of the crop and thus the expenses incurred by the farmer. Such a situation makes it possible to understand the importance of conducting systematic observations over the occurrence and severity of crop agrophages. The data obtained from such observations are the basis for making decisions on the need to carry out protective treatments, both chemical and various non-chemical. Conducting constant observations over the appearance and intensity of agrophages is particularly important now, when, in addition to avoiding crop losses, economic and environmental factors must be taken into account, as well as the obligation to carry out plant protection based on the principles of integrated pest management.

Although non-chemical methods (of a preventive character) make it possible to significantly reduce the level of subsequent chemization, at present, in most crops (especially major species), chemical protection still remains the main method of their protection against agrophages. Decisions on the implementation of plant protection treatments should be based on monitoring the occurrence of harmful organisms, taking into account the thresholds of economic harmfulness. Conducting regular monitoring and signaling makes it possible to predict and determine the occurrence of pest organisms on the plantation. With properly conducted signaling of the appearance of agrophages, it is possible to determine the appropriate date and need to carry out the necessary chemical treatment. This allows for increased efficiency and effectiveness of the protection carried out, minimizing the negative impact of plant protection products on the environment. Integrating different methods of reducing risks from agrophages requires a great deal of knowledge about crop plants, their biology, as well as about agrophages and their development cycles. It is essential to know the threat thresholds for individual agrophages in crops, conditioning decisions on chemical protection.

Agrophage monitoring, which is a basic activity for knowing the threats to crops from agrophages, is based on systematic observations carried out on field crops, orchards, vegetable gardens and other places of crop production and storage throughout the country.

Monitoring enables:

- detecting the appearance of a pest, or disease outbreaks,
- prediction of the occurrence of such an aggravation of the disease or such a stage of the pest to be controlled.

Based on experience, it was found that the optimal timing of treatment often has the greatest impact on the effectiveness of pest reduction of agrophages.

Determining the optimal timing of treatment requires knowledge of the agrophages that occur on plants, knowledge of their biology and ecology, regions of occurrence or climatic conditions that affect the rate of their development. Thus, scientific research is of great importance, as a result of which methods of observation of crop pests are developed, modernized and improved, which are published in the form of manuals, methodologies and guides [2].

On specific plantations, the appearance of diseases or pests or their developmental stages is observed at different times not only on a national scale, but also on a provincial or district scale. Hence monitoring of agrophages for short-term forecasting, referring to small areas or even a specific plantation, often underestimated by producers or advisors, is of such great importance in integrated pest management.

A good example to confirm the need for monitoring on a specific plantation and individually for an agrophage are two species of cereal pests. These are cereal leaf beetle (Oulema spp.), for which the determination of the optimal date of treatment depends on observing a mass hatching of larvae. Another pest is the saddle gall midge (*Haplodiplosis equestris*) - effective control depends on observing a mass raid of adult forms or presence egg deposits on leaves.

In the case of cereal leaf beetle (Oulema spp.), systematic field observations should be done in the spring, at the time when the beetles are regenerative feeding time. The finding of 10–15 beetles per 1 m^2 in winter wheat crops is the basis for predicting a higher abundance of larvae, the main harmful stage of cereal leaf beetle, in the observed plantation.

Systematic vetting should also be carried out at the time of their hatching. They should give an answer as to the location of their mass occurrence and estimated abundance, so that you can get an idea of whether the density of the pest exceeds the value of the threshold of economic harmfulness, which is:

- 1–2 larvae per 1 stalks of winter wheat, triticale and rye,
- 1 larvae per 2–3 stalks of winter and spring barley, spring wheat, spring triticale and oats [3].

The second example is the sadle gall migde (*H. equestris*). On endangered plantations, i.e. those where overwintering larvae have been found, set up yellow Moericke dishes or hang yellow boards covered with glue is necessary. Then carry out systematic field observations (every 2 days is recommended), each time counting the caught flies in Moericke's dishes. Using yellow glue-covered traps, count the glued flies and note their number is recommended.

10 caught insects in a Moericke's dish or on a yellow trap (the value of the economic pest threshold) is a signal to carry out a chemical treatment. Chemical treatment should be done no later than 6–8 days after the detection of increased flight (peak flight) of the pest, when the plants are at the stage of shooting (development stage on the scale BBCH 3/30-39).

Regular plant pest monitoring is a vital part of integrated pest management. This basic activity is aimed at identifying any risks to crops posed by pests with a view to determining the phytosanitary status of crops. Plant pest monitoring helps determine the current phytosanitary status of crops for the purpose of selecting the best date to apply a protective treatment and issuing treatment alerts. By making adequate use of observational data on the emergence and rise in pest populations, it is possible to develop pest control strategies, minimize potential damage, and prevent the excessive and often unnecessary use of chemicals, as required under the Integrated Pest Management EU Directive [4].

Plant pest monitoring and reporting methods include the following sections:

1. Taxonomy
2. Pest description and biology
3. Images of damage
4. Influence of environmental factors on pest development
5. Protective treatment alerts:
    a. timing of control measures and extent of economic damage,
    b. methods for best timing of chemical control treatments.
6. Methods for assessing pest damage

An essential part of integrated pest management other than the choice of preventive measures are decisions about the application of chemical control. Integrated pest management, crop protection and production guidelines, methodologies and programs are no substitute for regular field inspections and due consideration of weather conditions and economic harm levels by farmers facing the decision to use chemical treatments.

In recent years, advanced research has been carried out to provide scientific background knowledge for short-term pest emergence forecasting. Much of this research centered on determining correlations between disease and pest development and weather conditions. Its findings have been used to establish support systems to, for example, forecast pathogens breakouts and the developmental stages of pests as a way to determine the best time for treatment. The instructions produced on that basis are to help farmers and plant protection consultants identify the need for plant protection treatment based on ecological principles, with due consideration of cost-benefit analysis and climate conditions [1].

Another extremely important element of integrated protection, closely related to monitoring, signaling is Pest Risk Assessment. Activities in this area mainly concern quarantine organisms but are nevertheless very important and must be kept in mind. Pest Risk Assessment is an internationally agreed procedure for assessing threats posed by pests that may invade and settle in new areas by either natural migration or with human involvement. In view of emerging phytosanitary risks caused by the growing volume of international trade in plant material and climate change, it is essential to update the risk assessments that cover existing pests and goods subject to phytosanitary regulations and add further assessments for new pests and goods that may require phytosanitary measures to ensure the crop and ecosystem safety across the European Union. A particular threat to phytosanitary safety is posed by emerging risks associated with new pests, new importation routes and new types of goods that are imported. To protect Poland against such threats, an early warning system has been established. Within its framework, a mechanism for early detection and response to emerging pest-related threats was created, modeled after solutions adopted in other European Union countries. Under this mechanism, IPP - National Research Institute staff carry out phytosanitary monitoring for emerging threats to crops in Poland based on data sourced from literature, the press, and phytosanitary service reports. Where data is found to be insufficient, additional laboratory tests are performed to facilitate assessments [5].

4 Summary

As for the demand of advisors, agricultural producers and access to information on current threats to crop plantations, in addition to inspections of our own plantations, we can be assisted by other available data, such as from different agricultural portals. One such service is the On-line Pest Warning System (www.agrofagi.com.pl) (Fig. 1). The information service is provided by the Institute of Plant Protection - NRI, in cooperation with other scientific and industry units, as well as agricultural advisory centers. The On-line Pest Warning System information service aims to disseminate information on crop monitoring results, principles and guidelines of integrated pest management. The

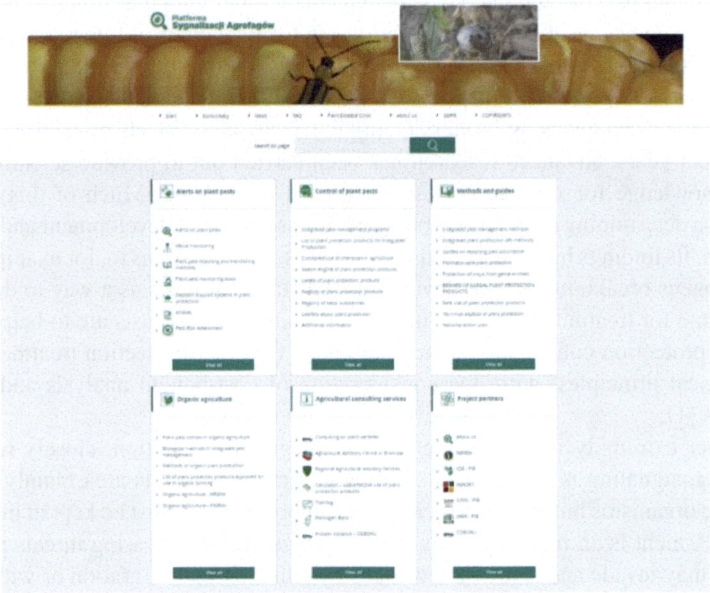

Fig. 1. On-line Pest Warning System – web page

monitoring covers 8 crops and almost 30 agrophages at more than 600 points throughout the country. The information is entered on an ongoing basis, is free of charge, and is available to any interested party [6].

Widespread access to disease and pest monitoring and signaling databases makes it easier for farmers to decide on the need for chemical treatment. This has an extremely high impact on protecting the environment from unnecessary chemical use, as well as the economic effect of reducing the number of chemical treatments.

References

1. Jakubowska, M., Tratwal, A., Magdalena, K.M.: The weather as an indicator for decision-making support systems regarding the control of cutworms in beets and cereal leaf beetles in cereals and their adoption in farming practice. Agron. Section Precis. Dig. Agric. **13**(3), 786 (2023)
2. Tratwal, A., Baran, M.: The role of guidelines on pest monitoring and warning systems in integrated pest management. J. Plant Protect. Res. **58**(3), 211–214 (2018)
3. http://www.agrofagi.com.pl
4. Directive 2009/128/EC of the European Parliament and of the Council of 21 October 2009 establishing a framework for Community action to achieve the sustainable use of pesticides. OJ L 309, 24.11.2009 Available online: https://eur-lex.europa.eu/eli/dir/2009/128/oj (Accessed on Apr 02, 2024)
5. https://www.agrofagi.com.pl/350,pest-risk-assessment
6. Tratwal, A., Baran, M.: Platforma sygnalizacji agrofagów – nauka praktyce rolniczej. Progr. Plant Protect. **59**(2), 119–125 (2019)

Pollen Concentration Data of *Xanthium* spp. Between 2014 and 2019 in Hungary

Viktor József Vojnich[1](✉) and Donát Magyar[2]

[1] Faculty of Agriculture, University of Szeged, Hódmezővásárhely 6800, Hungary
vojnich.viktor.jozsef@szte.hu
[2] Laboratory Department, National Centre for Public Health and Pharmacy, Budapest 1097, Hungary

Abstract. The species of *Xanthium italicum* and *X. spinosum* belong to the same subgenus group as ragweed (*Ambrosiinae*), they are both invasive, allergenic weeds. Pollen concentrations of *Xanthium* spp. Were analyzed in air samples collected with a 7-day Hirst-type (Burkard) pollen trap placed in the Northern Great Plain region (Jász-Nagykun-Szolnok county, Hajdú-Bihar county, Szabolcs-Szatmár-Bereg county) during the 6-year period 2014–2019 at the three county capitals (Szolnok, Debrecen, Nyíregyháza), in Hungary. The pollen season was defined as the day on which the sum of the daily average pollen concentration reaches 1% of the total amount, and the end of the season was defined when it reaches 99%. The highest annual total pollen count was detected in 2018 in Jász-Nagykun-Szolnok county (116 pollen), and the lowest in 2018 in Szabolcs-Szatmár-Bereg county (12). The highest amount of pollen grains was counted on August 24th (16 pollen/m^3) in Debrecen, in 2015. The lowest daily value was recorded on August 25th (4 pollen/m^3) in Nyíregyháza, in 2019. The results show that *Xanthium* spp. Species have relatively low pollen concentrations, which is striking compared to the closely related common ragweed: the highest daily pollen count of *A. artemisiifolia* in the same region (658) is more than 40 times that of *Xanthium* spp.; while the 6-year average of its total pollen count (12,863) is almost 263 times higher. Among the possible explanations, pollen production and the number of individuals are important. Pollen monitoring provides information on the size of *Xanthium* spp. Stands.

Keywords: Invasive and allergenic weeds · *Xanthium* spp. · pollen concentration · Northern Great Plain region · Burkard pollen trap · pollen monitoring data

1 Introduction

The genus *Xanthium* belongs to the family *Asteraceae*. The literature is divided on the number of species of the genus *Xanthium*, ranging from 2 to 25–30. The most accepted scientific sources distinguish only two species within the genus, thc spiny serpentine *X. spinosum* and *X. strumarium* (sensu lato) [1–4].

E. Lorencowicz et al. (Eds.): FMPMSA 2024, LNCE 609, pp. 432–439, 2024.
https://doi.org/10.1007/978-3-031-70955-5_48

Xanthium italicum (Italian cocklebur) is common in the floodplains of rivers. Their largest populations are found in areas subject to periodic flooding, usually in abandoned flooded fields with poor grassland fertility. They also occur sporadically further away from rivers, essentially along the entire Danube-Tisza (Hungary), mainly around canals.

Italian cocklebur is native to North America. Today it is a common weed in southern and central Europe. In Australia, it is found along the Hunter River. Occurs along major rivers in Ukraine, scattered in Western Europe (Alsace, France), and rare (Slovakia) or absent further north (Czech Republic, Poland). In Europe, a Mediterranean-continental element, mainly in the southern parts of the range of the *X. strumarium* (sensu lato) group. Borbás (1893) described, he mentions the occurrence of *X. italicum* as 'on the coasts: it has also made its home in the weedy places of Fiume and Buccari'. The wording suggests that it is not an old settler in these places. This is supported by the following idea, which, with reference to Ascherson, assumes that Italian cocklebur is of American origin. The most probable date of its establishment in Europe is the early 19th century, but there is no clear evidence of its occurrence before the 1822 description (Moretti). It appeared et al.-Duna in 1908 [5–10].

Xanthium taxa can hybridize with each other, so a hybrid of *X. strumarium* and *X. italicum* (*X.* × *widderi*) was found already in 1924, and its presence may cause further problems in the separation of taxa, which is not easy anyway.

István Győrffy found Italian cocklebur in 1922 on the island of Boszorkány in Szeged and at the mouth of the Mures. This year is considered the first record for the territory of present-day Hungary. Towards the end of the 1950s, and especially in the 1960s, a very intensive further spread of *X. italicum* was observed, not only along the Tisza and all its tributaries, but also along the Drava and the Danube. The species has spread from wave plains to floodplains and then to fields, mainly catch crops and other agricultural areas. By the end of the 1980s, it had infested more than 120,000 hectares of arable land, the most infested counties being Hajdú-Bihar, Békés, Jász-Nagykun-Szolnok and Csongrád, with these four counties accounting for about 80% of the total area recorded as infested. In the Transdanubian region, Tolna County was the most affected. Although *Xanthium italicum* cover continued to increase in most late summer agricultural crops according to the data of the V. Hungary Weed Surveys (2007–2008), its importance decreased compared to other weed species [7, 11, 12].

Italian cocklebur occurs on fresh, nutrient- and base-rich, humus loess and loess soils, ranging from sand-based soils with low mechanical demands to heavy clays. It prefers nutrient-rich sites, such as floodplains flooded with new alluvium from year to year or fertilized fields. Invasive *Xanthium spp.* Prefers somewhat fresher habitats than the native *Xanthium strumarium*, so cannot displace it in drier fields, with the difference in drought sensitivity being more significant in the early life stage (germination, seedling). *X. italicum* is relatively heat demanding and is associated with a suitable climate of thermophilic forests and wooded steppes.

Xanthium italicum (Fig. 1) is an increasingly important field weed, replacing mainly in the more salubrious areas the *Xanthium strumarium* (Fig. 2), which is itself an agriculturally dangerous weed. They cause most damage to maize, but are also a threat to sunflowers, sugar beet, potatoes and other catch crops grown on more compacted soils. The fruits and germplasm of the *Xanthium spp.* Are highly poisonous. Symptoms of poisoning appear within a few hours, and in severe cases, the animals die within a short time. Susceptible animals include pigs, poultry, sheep, goats, but especially horses and cattle [13, 14].

Xanthium spp. are invasive and allergenic weeds.

Fig. 1. Xanthium italicum

Fig. 2. Xanthium strumarium

2 Materials and Methods

The pollen concentration of *Xanthium spp.* Was measured at the three county capitals (Szolnok, Debrecen, Nyíregyháza) of the Northern Great Plain region (Jász-Nagykun-Szolnok county, Hajdú-Bihar county, Szabolcs-Szatmár-Bereg county) in Hungary (Fig. 3) with the 7-day Hirst-type (Burkard) pollen trap (Figs. 4 and 5), for the 6-year period 2014–2019.

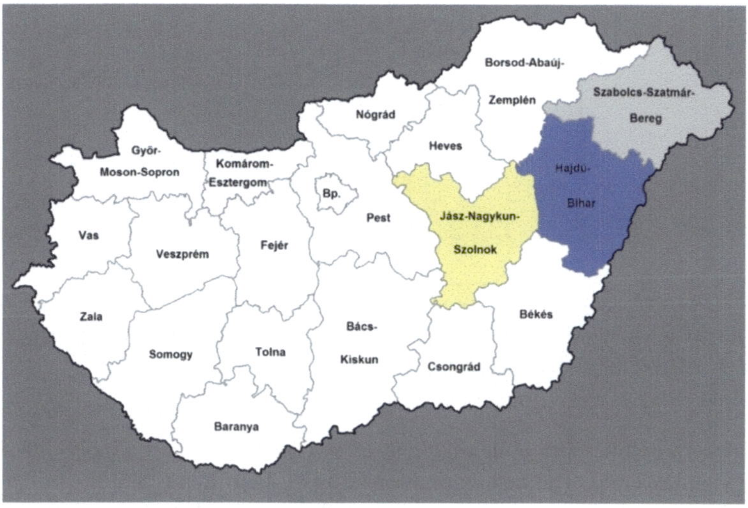

Fig. 3. Northern Great Plain region

Fig. 4. Hirst-type (Burkard) pollen trap (in Debrecen)

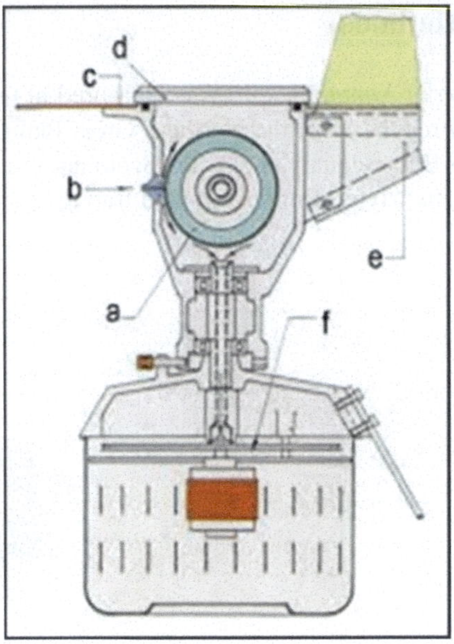

Fig. 5. Hirst-type (Burkard) pollen trap. A: trapping surface; B: orifice; C: rain shield; D: lid assembly; E: wind vane; F: vacuum pump

The device collides the pollen content of the intake air onto a tape with a sticky surface (Melinex tape). The tape is divided into two-hour bands and the pollen grains stained with basic fuchsine are counted in two 0.5 mm bands 6–6 mm from the edges at 400 × magnification. The results are given as a 24 h average, expressed in pollen grains/m^3 of air.

The pollen season was defined as the day on which the sum of the daily average pollen concentration reaches 1% of the final amount and the end of the season when it reaches 99% [15].

3 Results and Discussions

3.1 Pollen Data for Jász-Nagykun-Szolnok County

The highest pollen concentration during the monitoring period was measured in 2018. In Szolnok, the pollen trap caught a total of 116 *Xanthium spp.* Pollen grains. The highest amount of pollen grains was counted on the 25th August (11 pollen/m^3). The lowest pollen data was measured in 2016. Two years earlier, the pollen count of *Xanthium spp.* Was 61, when the highest daily amount (9 pollens/m^3) was detected on August 26th.

3.2 Pollen Data for Hajdú-Bihar County

In Debrecen, the highest total number of *Xanthium spp.* Pollen was 89 in 2015. The highest daily concentration was detected on August 28th (16 pollen/m^3). The lowest pollen grains were counted in 2018. The pollen data of *Xanthium spp.* Was 49, while the highest daily amount (6 pollen grains /m^3) was measured on August 25th.

3.3 Pollen Data for Szabolcs-Szatmár-Bereg County

The highest pollen concentration during the monitoring period was measured in 2014. In Nyíregyháza, the pollen trap caught a total of 70 *Xanthium spp.* Pollen grains. The highest amount of pollen grains were counted on the 30th August (10 pollen/m^3). The lowest pollen data was measured in 2018. Four years later, the pollen count of *Xanthium spp.* Was 12, when the highest daily amount (4 pollens/m^3) was detected on August 25th.

The data on the daily pollen number (highest and lowest) of *Xanthium spp.* Are shown Fig. 6. And Fig. 7.

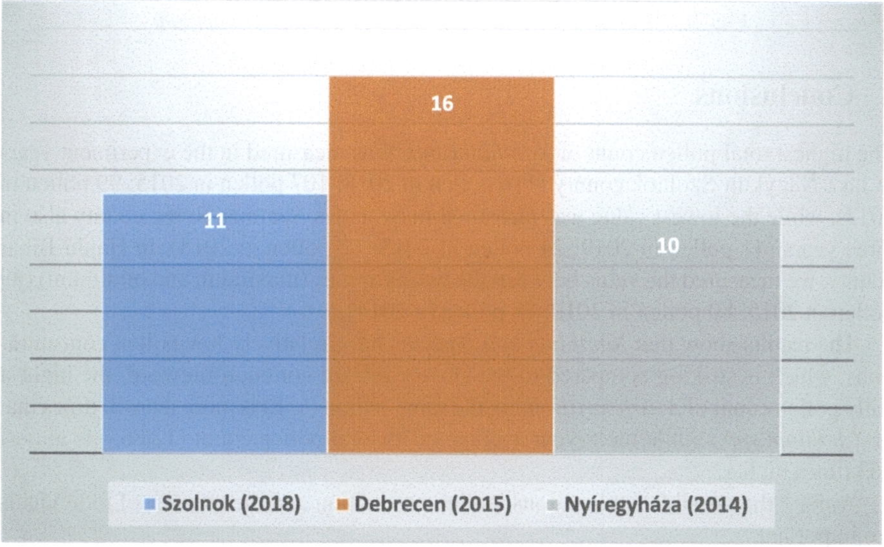

Fig. 6. The highest daily pollen concentrations of *Xanthium spp.*

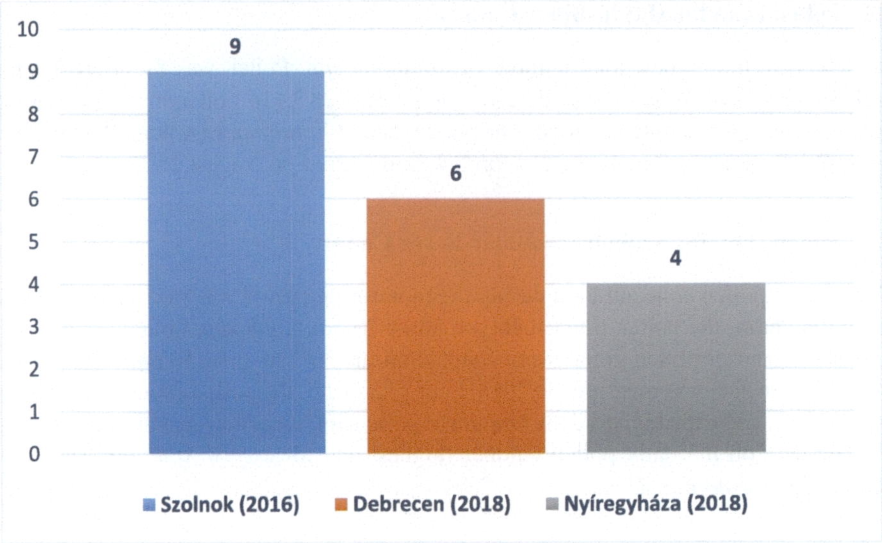

Fig. 7. The lowest daily pollen concentrations of *Xanthium spp.*

4 Conclusions

The highest total pollen count of *Xanthium spp.* Was measured in the experiment years in Jász-Nagykun-Szolnok county (116 pollen in 2018; 107 pollen in 2015; 99 pollen in 2014), while the lowest value was measured in Szabolcs-Szatmár-Bereg county also in three years (41 pollen in 2019; 24 pollen in 2015; 12 pollen in 2018). In Hajdú-Bihar county, we measured the value between the two extremes (maximum and minimum) (89 pollen in 2015; 80 pollen in 2019; 78 pollen in 2014).

The results show that *Xanthium spp.* Species have relatively low pollen concentrations, which is striking compared to the closely related common ragweed: the highest daily pollen count of *A. artemisiifolia* in the same region (658) is more than 40 times that of *Xanthium spp.*; while the 6-year average of its total pollen count (12,863) is almost 263 times higher.

Among the possible explanations, pollen production and the number of individuals are important.

Although both genera are wind-pollinated (i.e. capable of producing significant amounts of pollen), it cannot be excluded that the *Xanthium spp.* Vetch has a lower pollen production capacity, but further research is needed to confirm this hypothesis.

Acknowledgment. The authors would like to thank *Vivien Pál* for her work in producing pollen data from the National Centre for Public Health and Pharmacy.

References

1. Blais, P.A., Lechowicz, M.J.: Variation among populations of *Xanthium strumarium* (*Compositae*) from natural and ruderal habitats. Am. J. Bot. **76**, 901–908 (1989)

2. Borbás, V.: A szerbtövis hazája és vándorlása. MTA Matem. Természettudományi Közlemények **25**, 487–580 (1893)
3. Böszörményi, A., Bagi, I.: *Xanthium italicum* Mor. dominálta vegetációfolt fejlődésdinamikájának vizsgálata a Tisza hullámterén. Kitaibelia **6**, 45–50 (2001)
4. Böszörményi, A., Bagi, I.: Olasz szerbtövis. In: Botta-Dukát Z., Mihály, B. (szerk.): Biológiai inváziók Magyarországon. Özönnövények II. A KvVM Természetvédelmi Hivatalának Tanulmánykötetei, 10: 193–245 (2006)
5. Böszörményi, A.: Az olasz szerbtövis (*Xanthium italicum* Moretti) invázióját elősegítő magbiológiai tulajdonságok. Botanikai Közlemények **94**, 19–25 (2007)
6. Dinelli, G., Bonetti, A., Viggiani, P.: Genetic structure and mating system of Italian *Xanthium strumarium* complex. Weed Sci. **51**, 69–77 (2002)
7. Győrffy, I.: (1922) Xanthium echinatum Murr (*X. italicum* Moretti) prope Szeged. Magyar Botanikai Lapok, 20, 64
8. Horváth, K., Radványi, B., Szabó, L., Varga, L.: Szerbtövis fajok (*Xanthium spp.*). In: Benécs-Bárdi, G., Hartman, F., Radvány, B., Szentey, L. (szerk.): Veszélyes 48 – Veszélyes, nehezen irtható gyomnövények és az ellenük való védekezés. Mezőföldi Agrofórum Kft, Szekszárd, 121–126 (2005)
9. Karamán, J., Novák, R.: Magyarország szántóföldjeinek gyomnövényzete. Ötödik országos szántóföldi gyomfelvételezés (2007–2008), FVM, Budapest, 50–56 (2009)
10. Löve, D., Dansereau, P.: Biosystematic studies on *Xanthium*: taxonomic appraisal and ecological status. Can. J. Bot. **37**, 173–208 (1959)
11. Maksymowych, R.: Analysis of Growth and Development of *Xanthium*, p. 220. Cambridge University Press, Cambridge New York (1990)
12. McMillan, C.: Photoperiodic adaptation of *Xanthium strumarium* in Europe, Asia minor, and northern Africa. Can. J. Bot. **52**, 1779–1791 (1974)
13. Priszter, S.Z.: Megjegyzések adventív növényeinkhez. 11. A Xanthium saccharatum Wallr. Em. Widder Magyarországon. Botanikai Közlemények **58**, 171–174 (1971)
14. Weaver, S.E., Lechowicz, M.J.: The biology of Canadian weeds 56 Xanthium strumarium L. Canadian. J. Plant Sci. **63**, 211–225 (1983)
15. Ziska, L.H., et al.: Temperature-related changes in airborne allergenic pollen abundance and seasonality across the northern hemisphere: a retrospective data analysis. Lancet Planet. Health **3**(3), e124–e131 (2019). https://doi.org/10.1016/S2542-5196(19)30015-4

Assessment of the Effectiveness of Selected Disinfectants Against Biological Agents Identified on the Hands of Cattle Breeders

Maciej Wilk, Mateusz Gancarz, Sebastian Jaguszewski, Mateusz Ossowski, Łukasz Wlazło, Katarzyna Karpińska(✉), and Bożena Nowakowicz-Dębek

University of Life Sciences in Lublin, Lublin, Poland
katarzyna.karpinska@up.lublin.pl

Abstract. The purpose of this study was to determine the contamination of the skin of the palms of workers employed on a dairy cattle farm and to evaluate the disinfectant efficacy of selected preparations against isolated microorganisms. Samples were collected by direct swabbing method from employees while performing their duties on the cattle farm. Subsequently, identification to the species level of the obtained isolates was carried out using culture and biochemical methods. For the labeled bacteria, MIC (minimal inhibitory concertation) and MBC (minimal bactericidal concentrations) determinations were performed against four selected disinfectants using the broth microdilution method. The presence of opportunistic microorganisms belonging to both Gram-negative and Gram-positive bacteria, including spore-forming flora, was determined in the biological material. Bactericidal activity (MBC/MIC $\leq$ 4) was demonstrated for all disinfectants against bacteria used in the study, except for P2 disinfectant against *Serratia rubidae.* Gram-negative bacteria were less sensitive to the tested preparations compared to Gram-positive bacteria, with MIC values ranging from 1 to 4 mg/mL and 0.25 to 4 mg/mL, respectively. The P2 disinfectant showed the best antimicrobial properties against both Gram-positive and Gram-negative bacteria. In view of the increasing resistance of microorganisms to biocides, it is necessary to monitor the resistance of pathogens and environmental microorganisms to the cleaning and disinfecting preparations used, as a prevention of the transmission of pathogenic microorganisms to the environment and occupational diseases among farmers.

Keywords: biological agents · resistance · cattle farm · disinfection

1 Introduction

EFSA (European Food Safety Authority) controls food production in all stages of the food chain in EU countries. Its task is to provide supervision over production defined as "from farm to fork". The quality of the food obtained is shaped not only by environmental conditions, but also by the hygiene of the employees involved in the process. There is a relationship between product safety and the presence of microorganisms in manufactured food, therefore critical point identified and eliminated should be monitored at each

E. Lorencowicz et al. (Eds.): FMPMSA 2024, LNCE 609, pp. 440–447, 2024.
https://doi.org/10.1007/978-3-031-70955-5_49

stage (MacLeod et al., 2023; 2. Brzostek-Kasprzak, 2011; Todd et al., 2010; Regulation EC, 2002). One such point is the hygiene of the skin of the hands of those working in production maintained through systematic washing of workers' hands. This is a basic activity at work, but also an important element reflecting the hygiene conditions of the entire agricultural community. This activity is referred to as the primary barrier preventing the transfer of microorganisms and preventing cross-contamination of food (Foddai et al., 2016; De Aceituno et al. 2015). According to the FDA (Food and Drug Administration), hand hygiene in animal production is a particularly important aspect because farmers' hands can be a vector for transmitting pathogens in their environment. The sanitary condition of the hands of those working in various industries is closely dependent on the type of employment and the tasks performed. It is possible that the degree of environmental pollution may correspond to microbiological agents found on workers' hands (Prince-Guerra et al. 2020; Foddai et al., 2016; Todd et al., 2010; Aiello et al., 2008). This hypothesis is confirmed by the research of Prince-Guerra et al., (2020), who observed different microorganisms in farmers harvesting melons than in those harvesting peppers or tomatoes. Mihalache et al., (2021) tested the effectiveness of removing bacteria using soap and water. The authors indicate that if water and soap are not available, antibacterial wipes can be used. They are characterized by high killing effectiveness against microorganisms found on hands. Similarly, Foddai et al., (2016), indicates that water and soap seem to be effective in removing organic contaminants and microorganisms from hands. The use of alcohol-based products promotes the inactivation of microorganisms, but their effectiveness is lower against non-enveloped viruses. Additionally, the presence of organic matter may reduce the inactivation of microorganisms on the hands. Consistent application of hand hygiene at critical points in the microbial transmission chain plays an important role in reducing the incidence of gastrointestinal diseases and the microbiological safety of the food chain (Aiello et al., 2008). Inherent in the industrial farm environment is the presence of numerous microbiological hazards in addition to contaminants of a chemical nature. The widespread occurrence of many microorganisms, including pathogenic strains that often create biofilms that are difficult to remove, is a cause for public health concerns for workers associated with the agricultural industry. Hand hygiene is not always carried out effectively, and hands are the main vector of pathogen transmission. The degree of their colonization by microorganisms depends, among others, on hygienic conditions at the farm level, microbiological purity of feed, and the health of the animals themselves. Microbiological contamination on products may appear as secondary contamination from various stages of production, and their contamination with pathogens may pose a zoonotic risk and pose a threat to consumers (Goldsztejn et al., 2022; Young et al., 2020; Aiello et al., 2008).

Therefore, the aim of the research was to identify biological factors present on farmers' hands and to determine their sensitivity to selected disinfectants.

2 Material and Methods

Hand swabs were taken from breeders and workers employed at the cattle farm. Swabs were collected with sterile swabs with Amies transport medium in a transport tube (Medlab-Products, Poland). The collected material was placed in a thermal bag and immediately transported to the laboratory of the Department of Animal Hygiene and Environmental Threats of the University of Life Sciences in Lublin, Poland. Swabs were sown on a previously prepared microbiological agar medium containing 5% sheep blood. The samples were incubated for 24 h, and then the grown colonies of microorganisms were morphologically assessed, and Gram staining was performed, which allowed the isolate to be classified into the appropriate family of microorganisms. The initial macro and microscopic assessment allowed the selection of isolates, which were then isolated in pure culture on TSA (tryptone-soy-agar; BTL, Poland) and intended for further stages of species identification using API biochemical tests (BioMerieux, Poland). For the identified microorganisms, the effectiveness of selected disinfecting and cleaning preparations was analyzed using the microdilution method in MHB broth (Mueller-Hinton Broth) using 96-well plates. Minimum inhibitory concentrations (MIC) and minimal bactericidal concentrations (MBC) were assessed. Four disinfectants were used:

P1 - ethyl alcohol 89%, myristyl alcohol 1–2.5%, butanone 0.5–1%, glycerin 0.25 0.5%

P2 - propan-2-ol 60%, chlorhexidine digluconate 0.5%

P3 - ethanol 63.7%, propan-2-ol 6.3%

P4 - ethanol 83.7%, myristyl alcohol 1–2.5%

Disinfectant preparations were dissolved in DMSO (dimethyl sulfoxide) at a concentration of 200 mg/mL, and then a stock solution of the tested substances was prepared at a concentration of 32 mg/mL in MHB medium. Bacterial inoculum was prepared in sterile saline with an optical density of 0.5 McFarland. Positive control (containing only bacteria) and negative control (without inoculum) were performed in the last two wells. MIC was defined as the lowest concentration at which no bacterial growth was observed after 24 h incubation at 36–37 °C. Reading was performed at a wavelength of 570 nm using an ELx808 microplate reader (Bio-Tek Instruments, Inc., USA). The MBC was defined as the concentration at which no bacterial growth was observed after seeding and incubation on MHA medium for 18–24 h at 36–37 °C. An MBC/MIC ratio ≤ 4 was taken as an indicator of bactericidal activity, and an MBC/MIC ratio > 4 as an indicator of bacteriostatic activity.

The consent of the Local Ethics Committee was obtained to conduct the research (Resolution No. UKE/28/2024).

3 Results and Discussion

Both Gram-positive and Gram-negative microorganisms were identified from swabs collected from farmers' hands (Table). The Gram-positive bacteria included: *Staphylococcus lentus*, *Cellulosimicrobium cellulans*, *Bacillus sphaericus*, *Staphylococcus* sp.. These bacteria are considered to be common in the environment, but they have allergenic and opportunistic properties, and in some cases they may lead to the development of infections of the respiratory tract, ears or eyes. Especially in the case of constant exposure or weakening of the immune system. However, in the group of Gram-negative bacteria, the occurrence of *Pseudomonas fluorescens*, *Neisseria bacilliformis*, *Acinetobacter calcoaceticus*, *Serratia rubidaea* and *Pseudomonas alcaliqens* was determined. The effectiveness of the inhibitory and biocidal effects of four commercially available preparations intended for washing and hygienic disinfection of hand skin was assessed for the obtained microorganisms. The results are presented in Table 1.

Table 1. MIC (mg/mL), MBC (mg/mL) values of the analyzed disinfectant preparations against the detected bacterial strains

Microorganisms	P1			P2			P3			P4		
	MIC	MBC	MBC/MIC	MIC	MBC	MBC/MIC	MIC	MBC	MBC/MIC	MIC	MBC	**MBC/MIC**
Gram-positive bacteria												
Staphylococcus lentus	32	32	1	0,5	0,5	1	32	> 32	> 1	32	32	1
Cellulosimicrobium cellulans	1	2	2	0,25	0,25	1	32	> 32	> 1	4	4	1
Bacillus sphaericus	8	32	4	0,25	0,5	2	32	32	1	32	32	1
Bacillus sphaericus	> 32	> 32	> 1	0,25	0,25	1	32	32	1	32	32	1
Staphylococcus sp.	16	16	1	4	4	1	32	32	1	32	32	1
Staphylococcus sp.	16	16	1	2	2	1	16	16	1	16	16	1
Gram-negative bacteria												
Pseudomonas fluorescens	32	32	1	1	4	4	32	32	1	32	32	1
Neisseria bacilliformis	16	16	1	1	2	2	16	16	1	16	16	1
Acinetobacter calcoaceticus	32	32	1	4	4	1	32	32	1	16	32	2
Serratia rubidaea	32	32	1	2	16	8	16	32	2	16	16	1
Pseudomonas alcaliqens	32	> 32	> 1	4	4	1	32	32	1	16	32	2

MIC - minimum inhibitory concentration, the lowest concentration of the tested substance that inhibits the growth of microorganisms.
MBC – minimum bactericidal concentration, the lowest concentration of the tested substance at which 99.9% of bacteria are killed.
MBC/MIC: MBC/MIC $\leq$ 4 – bactericidal effect of the compound; MBC/MIC > 4 – bacteriostatic effect of the compound.

The experiment showed that the disinfectants used had bactericidal properties against the bacterial isolates used in the study (MBC/MIC $\leq$ 4). Only in the case of agent P2 against *Serratia rubidaea* the MBC/MIC was 8 and showed bacteriostatic activity (MBC/MIC > 4). The values shown in Table 1 indicate that the most effective antimicrobial activity was characterized by agent P2 against both Gram-positive and Gram-negative bacteria. Gram-negative bacteria were less sensitive to the tested preparations compared to Gram-positive bacteria. The MIC values of the P2 disinfectant ranged from 0.25 to 4 mg/mL against Gram-positive bacteria, and against Gram-negative bacteria from 1 to 4 mg/mL. Analyzing the results, it can be seen that preparations P3 and P3 had the least effective antimicrobial activity - MIC values for both types of bacteria ranged from 16 to 32 mg/mL. For some bacteria, it was not possible to determine MBC values for the tested preparations, these were: *Staphylococcus lentus*, *Cellulosimicrobium cellulans*, *Bacillus sphaericus* and *Pseudomonas alcaliqens*.

The use of veterinary disinfectant preparations, especially in livestock, and especially those for skin disinfection, is one of the most effective ways to reduce biological agents of animal origin. This is of the greatest importance especially regarding multidrug-resistant microorganisms (Wales et al., 2021). Numerous scientific works prove that contact with subinhibitory doses of various biocidal agents used systematically can induce bacterial adaptation to the substances in question, as well as promote the emergence of antibiotic resistance and an increased ability to form biofilms (Capita et al., 2019). This phenomenon is dangerous both for the public health of workers occupationally exposed to biological hazards, as well as a challenge to food safety. Therefore, an important part of the monitoring of biological hazards is research to determine the effective composition of disinfectant-washing preparations and the principles of adherence to the hand hygiene protocol of workers in food processing environments. In the agricultural environment, there is the possibility of spreading pathogens during food production. Hand washing is a basic practice used in preventing diseases transmitted this way. With limited access to water and soap in agriculture, alcohol-based disinfectant products are useful, as are disinfectant wipes. Such products can reduce the number of microorganisms on your hands. However, their effect is significantly limited when the hands are greasy or dirty (De Aceituno et al., 2015). However, disinfectants have no residual effect, compared to antimicrobial soap with antimicrobial properties. However, they are not a universal agent for general use (Todd et al., 2010). Prince-Guerra et al. (2020), found the concentration of microorganisms on the hands of farmers harvesting vegetables ranging from 0.88 to 5.1 log10 CFU/hand. The number of microorganisms differed significantly between the collected products. However, Geresu et al. (2021), indicate that hands may be a vector of significant microbiological contamination. In their research among employees of the agri-food workers, they showed that Salmonella isolates, including those resistant to antibiotics, were identified on the hands of 8.57% of slaughterhouse workers. The authors report that such a high rate of hand contamination was due to lack of hygiene among employees. According to Mihalache et al. (2023), it is possible to reduce public health exposure by broadly educating workers on proper hand hygiene. In the study by Weber et al. (2003), various antiseptics and soapy water were tested against *Bacillus athrophaeus* (a substitute for *B. anthracis*), previously contaminating the participants' hands with an inoculum of *B. atrophaeus*. The highest reduction effectiveness was achieved with soapy water, preparation with 2% chlorhexidine gluconate and antibacterial towels soaked in hypochlorite. The use of an anhydrous agent containing 61% ethyl alcohol proved ineffective against the strain used in the study. Determining the effectiveness of disinfection with preparations of similar alcohol content (preparation P2 and P3) conducted in our own research did not confirm the results presented by Mihalache et al. (2023). In our own tests, the P2 preparation showed the highest effectiveness of those tested. This indicates that it is not the alcohol content itself that is the key determinant of biocidal effectiveness, but the remaining bioactive ingredients. The authors also showed that prolonged use of soap, chlorhexidine and alcohol did not increase the reduction of *B. athrophaeus*. Only in the case of towels with hypochlorite, longer use was associated with a greater reduction of this bacterium on hands. Shamsul et al. (2016), report that an important indicator of hand hygiene practices among workers is the presence of the so-called indicator microorganisms - coliform bacteria, which are considered the etiological factor of various occupational diseases. The authors analyzed the microbiological contamination of the skin of slaughterhouse workers' hands in terms as *E. coli* bacteria and bacteria from the *Enterobacteriaceae* family. It was shown that the frequency of *Escherichia coli* O157:H7 bacteria during work was 9.7%, while the prevalence of bacteria other than O157:H7 was 35.5 and 13%, respectively, in the case of all analyzed hand skin swabs. However, our own research did not detect the presence of this group of bacteria, which indicates a good sanitary and hygienic condition of the hands of the workers of the tested farm. Numerous studies indicate that animals may also be a source of highly pathogenic strains of microorganisms, including those with increased resistance to drugs and disinfectants. According to Zang et al. (2016), ESBL (extended-spectrum b-lactamase) strains - *Escherichia coli*, can be transmitted to pigsty workers directly from believe. The authors showed that ESBL-carrying *E. coli* isolates had the same genotype and antibiotic resistance profile, which were isolated from swabs from both pigs and workers. This confirms the transmission of these pathogens between animals and workers (Zhang et al., 2016). A significant threat that appears more and more often in scientific reports is the transmission of *Staphylococcus aureus* originating from the animal breeding environment to workers and people involved in animal production. This microorganism is subject to epidemiological monitoring due to its multidrug resistance, including resistance to methicillin - methicillin-resistant *S. aureus* (MRSA). Strains originating from animals are referred to as LA-MRSA, i.e. livestock-associated MRSA. Our study identified strains belonging to this genus but other than *S aureus.* Khairullah et al. (2024), examining microbiological contamination of the hands of cattle farm workers, from 39 samples of swabs taken from farmers' hands for the isolation of *S. aureus*, a positive result was recorded for 28 (71.8%) of which 2 isolates were multidrug resistant, and 14 of them were MRSA strains. The authors emphasize that special attention should be paid to the use of antibiotics in animals to limit the constantly growing resistance of microorganisms. A similar experiment was conducted by El-Mahlaw et al. (2017). The prevalence of *S. aureus* on the hands of dairy workers was (15.2%, 7/46 swabs). The authors suggest that workers may be the source of these bacteria in milk or cows' udders. Biological factors occurring in the environment of breeding farms pose a threat to the life and health of workers occupationally exposed to their influence. Continuous exposure combined with simultaneous hygienic errors can predispose to many unfavorable health effects, ranging from respiratory diseases, through inflammatory skin lesions, allergies, and infectious zoonotic diseases (Kizerwetter-Świda and Pławińska-Czarnak, 2017).

4 Conclusion

In order to improve the implementation of hygiene practices on farms, hand sanitation monitoring and employee training in hygiene principles are needed in addition to procedural changes.. It is reasonable to develop a system of hygiene practices and internal protocols in farm facilities that will facilitate regular hand hygiene. At the same time, due to the increasing resistance of microorganisms to biocidal agents, it is necessary to identify target sensitive areas and test the resistance of microorganisms to the most commonly used cleaning and disinfecting preparations, which will avoid the transmission of pathogenic microorganisms to the municipal environment and prevent the occurrence of zoonotic disease entities. Thus, based on the results, it is possible to develop a safe work instructions for the breeder as a necessary element of reducing exposure to biological factors and maintaining the optimal sanitary condition of the farm in the light of current legislation.

Funding. The research was funded by the Ministry of Science and Higher Education as part of the implementation of the program 'Student Scientific Circles Create Innovations' (No. SKN/SP/569532/2023).

References

MacLeod, C., et al.: Recommendations for hand hygiene in community settings: a scoping review of current international guidelines. BMJ Open **13**(6), e068887 (2023). https://doi.org/10.1136/bmjopen-2022-068887

Brzostek-Kasprzak, B., Kwasek, M., Michna, W., Obiedziński, M.W. Z badań nad rolnictwem społecznie zrównoważonym: Jakość i bezpieczeństwo żywności a zdrowie konsumenta (2011)

Todd, E.C., Michaels, B.S., Holah, J., Smith, D., Greig, J.D., Bartleson, C.A.: Outbreaks where food workers have been implicated in the spread of foodborne disease. Part 10. Alcohol-based antiseptics for hand disinfection and a comparison of their effectiveness with soaps. J. Food Protect. **73**(11), 2128–2140 (2010)

Regulation EC No. 178/2002 of the European Parliament and of the Council of 28 January 2002 laying down the general principles and requirements of food law, establishing the European Food Safety Authority and laying down procedures in matters of food safety

Foddai, A.C., Grant, I.R., Dean, M.: Efficacy of instant hand sanitizers against foodborne pathogens compared with hand washing with soap and water in food preparation settings: a systematic review. J. Food Prot. **79**(6), 1040–1054 (2016)

De Aceituno, A.F., Bartz, F.E., Hodge, D.W., Shumaker, D.J., Grubb, J.E., Arbogast, J.W., Leon, J.S.: Ability of hand hygiene interventions using alcohol-based hand sanitizers and soap to reduce microbial load on farmworker hands soiled during harvest. J. Food Protect. **78**(11), 2024–2032 (2015)

Prince-Guerra, J.L., et al.: Both handwashing and an alcohol-based hand sanitizer intervention reduce soil and microbial contamination on farmworker hands during harvest, but produce type matters. Appl. Environ. Microbiol. **86**(18), e00780-20 (2020). https://doi.org/10.1128/AEM.00780-20

Aiello, A.E., Coulborn, R.M., Perez, V., Larson, E.L.: Effect of hand hygiene on infectious disease risk in the community setting: a meta-analysis. Am. J. Public Health **98**(8), 1372–1381 (2008)

Mihalache, O.A., Borda, D., Neagu, C., Teixeira, P., Langsrud, S., Nicolau, A.I.: Efficacy of removing bacteria and organic dirt from hands—a study based on bioluminescence measurements for evaluation of hand hygiene when cooking. Int. J. Environ. Res. Public Health **18**(16), 8828 (2021)

Goldsztejn, M., Grenda, T., Kozak, B., Kozieł, N., Kwiatek, K.: Microbiological contamination of feed-current hazards and new challenges. Medycyna Weterynaryjna **78**, 6616 (2022)

Young, I., Chung, A., McWhirter, J., Papadopoulos, A.: Observational assessment of food safety behaviours at farmers' markets in Ontario, Canada: a cross-sectional study. Food Control **108**, 106875 (2020)

Wales, A.D., Gosling, R.J., Bare, H.L., Davies, R.H.: Disinfectant testing for veterinary and agricultural applications: a review. Zoonoses Public Health **68**(5), 361–375 (2021)

Capita, R., Vicente-Velasco, M., Rodríguez-Melcón, C., García-Fernández, C., Carballo, J., Alonso-Calleja, C.: Effect of low doses of biocides on the antimicrobial resistance and the biofilms of Cronobacter sakazakii and Yersinia enterocolitica. Sci. Rep. **9**(1), 15905 (2019)

Geresu, M.A., Desta, W.Z.: Carriage, risk factors, and antimicrobial resistance patterns of Salmonella isolates from raw beef in Jimma. Southw. Ethiopia Infect. Drug Resist. **14**, 2349–2360 (2021)

Mihalache, O.A., Teixeira, P., Langsrud, S., Nicolau, A.I.: Hand hygiene practices during meal preparation—a ranking among ten European countries. BMC Public Health **23**(1), 1315 (2023)

Weber, D.J., Sickbert-Bennett, E., Gergen, M.F., Rutala, W.A.: Efficacy of selected hand hygiene agents used to remove Bacillus atrophaeus (a surrogate of Bacillus anthracis) from contaminated hands. JAMA **289**(10), 1274–1277 (2003)

Shamsul, B.M.T., Adamu, M.T., Desa, M.N.M., Khairani-Bejo, S.: Prevalence of Escherichia Coli O157: H7 and Enterobacteriaceae on hands of workers in halal cattle abattoirs in peninsular Malaysia. Malays. J. Med. Sci. MJMS **23**(5), 65 (2016)

Zhang, H., Zhai, Z., Li, Q., Liu, L., Guo, S., Li, Q., Zhai, J.: Characterization of extended-spectrum β-lactamase–producing Escherichia coli isolates from pigs and farm workers. J. food Protect. **79**(9), 1630–1634 (2016)

Khairullah, A.R., Kurniawan, S.C., Sudjarwo, S.A., Effendi, M.H., Widodo, A., Moses, I.B., Ramandinianto, S.C.: Kinship analysis of mecA gene of methicillin-resistant Staphylococcus aureus isolated from milk and risk factors from the farmers in Blitar. Indonesia. Veterinary World **17**(1), 216 (2024)

El-Mahallawy, H.S., Elhariri, M., Elhelw, R., Hamza, D.: Zoonotic importance of some pathogenic bacteria isolated from small ruminants' milk and hands of dairy workers. Zagazig Veter. J. **45**(4), 305–313 (2017)

Kizerwetter-Świda, M., Pławińska-Czarnak, J.: Odzwierzęce szczepy Staphylococcus aureus oporne na metycylinę (LA-MRSA)–obecny stan wiedzy. Medycyna Weterynaryjna **73**(2), 92 (2017)

The Necessity of Sustainable Agriculture from the Perspective of the Importance and Risks of Agriculture for a Country: A Comparative Research

Yue Wu[1](✉), Balázs Pál[2], and Katalin Takács-György[3]

[1] Bánki Donát Faculty of Mechanical and Safety Engineering, Obuda University, Budapest, Hungary
wu.yue@bgk.uni-obuda.hu
[2] Wizz Air Hungary Limited, Budapest, Hungary
x-balazspal@wizzair.com
[3] Department of Business Development and Infocommunications, Keleti Faculty of Business and Management, Obuda University, Budapest, Hungary

Abstract. Food is necessary for everyone, which does not have a substitute. Agriculture is the source of food, both crops and meat. Agriculture directly decides food security and indirectly contributes to economic growth, social stability, and environmental conservation. However, modern agriculture is facing challenges in the face of the question of sustainable development. The aim of our research is to provide a comprehensive understanding of why agriculture is important for a country and what are the risks in agriculture for a country. Our study focused on China and South Korea because these two eastern Asian countries are important contributors to world food security and economics. We conducted this research based on a literature review and content analysis as the research methodology. The value of our research is to prove the necessity of sustainable agriculture from the perspective of the importance and risks of agriculture in a country such as China and South Korea. Agriculture is and will remain an important sector for a country's national security and beyond, which is missing or less discussed in previous research. In the face of diverse global security risks and threats trends, and agricultural vulnerabilities, sustainable agriculture is the only way to feed future generations. Beyond these research values, our comparative research provided the unique success, challenges, and global implications of agriculture from two different countries, which can also be used to compare other important agriculture countries, such as Poland in the EU.

Keywords: China · South Korea · Agriculture · Security challenges · Arable land · Aging population · Water · Digitalization

E. Lorencowicz et al. (Eds.): FMPMSA 2024, LNCE 609, pp. 448–461, 2024.
https://doi.org/10.1007/978-3-031-70955-5_50

1 Introduction

Nevertheless, agriculture's most important role is to provide food to people and raw materials (such as fiber) to industries. The role of agriculture is far beyond its basic functions. From a systematic perspective, the role of agriculture is usually regarded as economic, social, and environmental functions. In the face of the mission to feed the increasing population, agriculture encounters a series of risks.

Agriculture is the primary industry in terms of industry classification by economics, which is the dominant industry of economics, but it can be decreased as the secondary and tertiary industries increase [1]. The secondary sector (manufacturing process) relies on agriculture heavily, and the third sector (service: trade, transport, and tourism) depends on agriculture less heavily [2]. In developing countries, the productivity of agriculture is responsible for both rural and urban poverty [3]. Agriculture contributes 4% to the global gross domestic product (GDP) and can account for up to 25% in certain developing countries [4], which plays a crucial role. In some developing countries, agriculture contributes the most job opportunities than other sectors, such as industry, construction, transportation, etc. [5].

The role of agriculture for locals beyond economics where got less attention relatively [6, 7]. In terms of social function or social stability, the most important function of agriculture is maintaining food security [8]. In some countries, agriculture provides employment support to the population who do not have jobs or are unemployed in urban areas and who are old or unable to do farming. Agriculture is especially important in small-scale family farming [9]. For locals, agriculture also guarantees the local high-quality food, contributes to the community and quality of life, and maintains important traditions, heritage, and work ethic [6]. Nowadays, as more and more attention is paid to renewable energy, agriculture is an important source of renewable energy, which is called bioenergy, ensuring some countries' energy security [10].

The third basic foundation of agriculture that is usually discussed is environmental function. For example, agriculture preserves wildlife [7], and the ecosystem provides bucolic views [6] and reduces GHG (CO2) and risks of floods [11]. Instead of the pollution from waste, waste can be collected, composted, and reused in agriculture [12].

No matter the critical role of agriculture in economics, society, and the environment, agriculture also threatens human health and planet development [4]. The current food system generates 30% GHG emissions (Greenhouse gas emissions). One-third of the produced food is lost and wasted annually worldwide. At the same time, agriculture is also facing long-lasting and new emerging risks and challenges [13]. In the coming years, agriculture will play a crucial role in feeding the ever-growing global population. By 2030, the population is projected to reach 8.6 billion, 9.8 billion by 2050, and 11.2 billion by 2100 [14]. As the middle class continues to grow, the demand for food will inevitably become more complex and nuanced [15, 16]. Extreme weather conditions such as floods, droughts, and erratic rainfall patterns have become more frequent, which negatively impacts on agriculture. Agriculture is highly dependent on natural resources, such as water, arable land, energy, etc., but there is a resource scarcity challenge [17–19]. Fluctuations in agricultural food prices are a common occurrence, but they have a significant impact on food security, farmers, and vulnerable populations [20, 21]. Digital agriculture can avoid some negative impacts from traditional agriculture [21], but we will face some security issues from the application, which we are not perfect at so far [22].

In the introduction, we discussed the three main roles of agriculture for a country and the general risks in agriculture. However, we will also highlight the importance of agriculture from the view of national security, which is a lack of discussion because food insecurity causes political instability and conflicts [23]. In this research, we aimed to prove the necessity of sustainable agriculture for a country from the perspective of the importance (focus on economics) and risks of agriculture (the threats brought by agriculture and the risks agriculture is facing) by the example of China and South Korea. We chose these two countries because China and South Korea are two important countries for global food security and economic development, and they share some common cultural features but are not the same. The comparative results based on two different countries will contribute to identifying the broader patterns and insights that might not be apparent within single-case studies, helping policymakers to adopt better practices and avoid potential pitfalls, providing cross-national and cultural insights, and so on. The value of our research goes beyond the current findings about the unique challenges and global impacts on agriculture. The method we used in this comparative research can also be used in other countries as well. For example, Poland is one of the largest agricultural countries in the EU, contributing significantly to the economy, food production, and employment from the point of view of agriculture [24]. At the same time, Poland's agriculture also gets significant funding from the EU project Common Agricultural Policy (CAP) for agricultural activities [25]. In such a country like Poland, agriculture plays an important role in domestic and regional, or even global, should get more attention on its structure and development.

2 Materials and Methodology

We conducted this research with a focus on the case of China and South Korea to better understand the necessity and importance of sustainable agriculture in the face of diverse global security risks, threats, trends, and agricultural vulnerabilities, using the following research methodologies: secondary research review and content analysis. Our study relies on credible and renowned publications, official international organizations (FAO, the UN, OECD economic surveys, World Bank etc.), and so on.

In the first chapter, we will discuss the importance of agriculture to the economy, the features of each other's agriculture in terms of arable land, contribution to GDP, employment rate, and future trends and objectives. In the second chapter, we will focus on the problems and dangers that agriculture in China and South Korea is now facing. These challenges and risks underscore the significance of agriculture in terms of its role in maintaining national security.

3 Results

China, also known as the People's Republic of China, is situated in the eastern region of the Asian continent and borders the western part of the Pacific Ocean. With an area of 9.6 million square kilometers, it is a vast country [26]. China has the biggest population of 1.45 billion [27] and has 5 million km2 of agricultural lands [28], 56% of total land area [29]. In 2021, the GDP in China will be 17 trillion dollars [28], and the GDP per capita will be 12 thousand dollars [30].

South Korea is the world's tenth biggest economy, with an estimated 105 trillion dollars, and the fourth largest in Asia, and is on track to become one of the richest countries in the future. Its GDP per capita is over $33,000, and it is predicted to reach ninth place by 2025. The Republic of Korea has a population of 52 million people. South Korea lacks the natural resources essential for agriculture. Mountains and hills cover two-thirds of the nation. Out of the country's total land area (100 210 km2) only 22% is arable land. Given Korea's export-oriented economy, it is critical for Korea to remain internationally competitive.

In the following part, we will discuss the the importance and challenges of agriculture in both two countries in detail.

3.1 The Contribution of Agriculture to the Economy in China and South Korea

A Case Study in China The GDP in agriculture, together with forestry and fishing, value added (% of GDP) is steadily decreasing from 41% in 1968 to 7% in 2020 [31]. The GDP in agriculture was 28%, and the employment share was 71% in 1978 in China [32]. The employment in agriculture of total employment in China has decreased steadily from 60% in 1991 to 24% in 2021 [33] (Fig. 1). The share of agriculture was much higher than that of other OECD countries in 2015 [32]. Although China's agricultural contribution to its GDP (Fig. 2) is declining, it is still a crucial sector as it feeds one-fifth of the world's population.

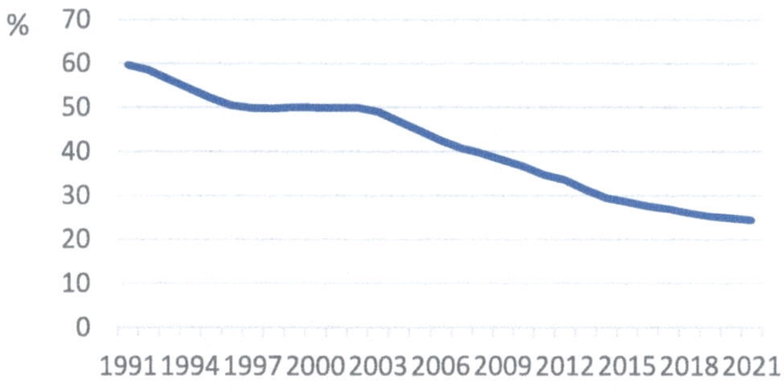

Fig. 1. *Employment in Agriculture (% of Total Employment)-China.* Source: The World Bank

It is impressive that China has managed to produce 25% of the world's grain and provide food for 20% of the global population (1.45 billion as of 2022 [34]), despite having less than 10% of the world's arable land [26]. China's arable land (119 million hectares [35]) is ranked after India, U.S., and Russia [36], taking 12% of its total land as of 2020 [37]. This achievement is significant for ensuring food and nutrition security, not just in China but also around the world. China holds the top position in global production of cereals, cotton, fruits, vegetables, meat, poultry, eggs, and fishery products [34, 38]. In

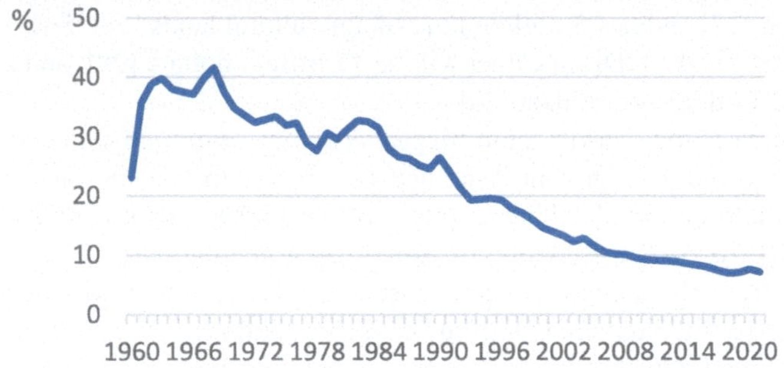

Fig. 2. Agriculture, forestry, and fishing, value added (% of GDP) in China. Source: The World Bank

the long-term view, the goal of China's rural area is to build "thriving businesses, pleasant living environment, social equity and civility, effective governance, and prosperity" [26].

A Case Study in South Korea In 1963, when the economy started to get better, most South Koreans were farmers. 63% of the people lived in the countryside. In the next 25 years, South Korea went from being a mostly rural farming country to a newly developed, urban country, and the number of people working in agriculture went down.

Agriculture in South Korea was plagued by several issues. South Korea is a hilly nation with just 22% arable land and less rainfall than most of its neighbors. Substantial land reform in the late 1940s and early 1950s gave rural peasants ownership of land. Individual holdings, on the other hand, were either too tiny (averaging one hectare, making agriculture inefficient and discouraging automation) or too spread out to give families a good chance of producing a large amount of food. The massive rise of metropolitan areas resulted in a quick loss in accessible cropland, while population increases and higher wages meant that food demand rapidly outstripped supply. As a consequence of these changes, by the late 1980s, about half of South Korea's requirements, namely wheat and animal feed corn, were imported. Agriculture remained the most sluggish sector of the economy when compared to the industrial and service sectors.

Korea's agricultural industry saw a lot of structural changes in a short period of time, coinciding with fast industrialization and economic expansion. From the 1950s through the 1970s, the government prioritized agricultural production and establishing self-sufficiency in basic staples, notably rice. Throughout the late 1980s and early 1990s, policy aims were to reform the industry and increase its competitiveness in response to agricultural market liberalization. Employment in the agricultural sector also decreased from 16% in 1991 to 5% in 2021.

With the growing liberalization of agriculture and food markets, agricultural policy in Korea embraced a broader range of goals, ranging from increased productivity to improved long-term agricultural sustainability. Rapid expansion and industrialization widened the income gap between rural and urban families. Since the first post-2000 decade, the focus has changed to a larger range of goals, such as revitalizing the rural economy, growing the export market, improving agricultural environmental performance, and boosting the food sector. Furthermore, international and bilateral trade agreements necessitated gradual structural changes in the agriculture sector.

Contribution of Agriculture to GDP: The shifting importance of agriculture reflects the fast expansion of the korean economy. Agriculture accounted for over a quarter of Korea's GDP and half of total employment until the 1970s. In 1970, agricultural output provided 27.4% of GDP, While the agricultural workforce accounted for 50.4% of total employment in the nation. Agricultural product exports accounted for 30.8% of total Korean exports in 1970. However, as industrialization proceeded, agriculture's proportion in the national economy shrank dramatically. In 2017, agricultural output contributed 2.0% of GDP and 4.8% of total employment, while agricultural exports represented 1.2% of total national exports. The rapid decline in the agricultural sector's proportion in the national economy is the outcome of an unparalleled degree of modernization. From 1970 to 2017, National GDP rose at an average annual rate of 11.8%, whilst agricultural GDP increased at a comparatively low annual rate of 5.7%. Agriculture's latest GDP proportion as of 2021 was 1,79%. Nonetheless, the agricultural sector continues to play a major part in the Korean national economy, accounting for a relatively substantial proportion of GDP and employing a huge rural population. Korean agriculture, in particular, is seen to contribute greatly to land usage and employment, social and economic stability, and rural life.

Agricultural Trade: Korean agriculture trade (exports plus imports) expanded gradually from $ 0.8 billion in 1970 to $ 39.1 billion in 2017. However, agricultural commerce's proportion in national trade fell from 30.7% in 1970 to 3.7% in 2017. This is due to the fact that non-agricultural commerce grew at a quicker pace than agricultural trade. It is partly due to the Korean government's growth policy of industrialization and non-agri-cultural exports since the 1970s. Korea is a particularly significant market for agricul-tural exporters of cereals, including feed grains and animal goods, fruits, and processed foods, with a population of around 51 million people and a fast-rising level of living. Agricultural imports account for around 6.7% of total national imports, while agricul-tural exports account for approximately 1.2% of total national exports. Agricultural product imports increased 65 times from $0.5 billion in 1970 to $32.3 billion in 2017. Agricultural exports climbed from US$ 0.3 billion to US$ 6.8 billion within the same time period. Because imports outnumber exports, the Korean agriculture trade imbal-ance grew dramatically throughout the same time, from US$ 0.3 billion to US$ 25.5 billion. It is mostly due to the development of agricultural trade liberalization, which has resulted in Korea being a significant net agro-food importer.

Imports rather than exports are projected to fuel major growth in Korean agricultural commerce. Furthermore, economic development and changes in social demography are predicted to enhance demand for a diverse variety of high-quality agricultural goods. As a consequence, agricultural imports are likely to rise as well. The growth in feed grains and meat imports, in particular, is predicted to continue as demand for meat products grows, while imports of tropical fruits and vegetables are also expected to climb. Furthermore, trade partners and agricultural trade commodities are likely to grow. As import and export trade partners have diversified, such developments are projected to increase further in accordance with the worldwide growth of market liberalization.

3.2 The Challenges in Agriculture

We have discussed the importance of agriculture in the economy of the mentioned countries. In this part, we will discuss the present agricultural challenges and their significance in China and South Korea. Agriculture is vital to providing food security and sustainability for people across the world. As digitalization pervades every element of our lives, different forms of digitalization have started to be used in agriculture, offering both benefits and risks that make it part of a nation's national security.

In the end, we will conclude that the necessity of adapting sustainable agriculture in light of the challenges in agriculture from the perspective of these two countries.

A Case Study in China *Increasing demand:* As the economic growth continues, more and more middle-class people are growing in China. The demand for value-added products such as meat, dairy, fish, and so on will increase [32]. China's population is projected to decline in the long term, but the peak is predicted to be 1.44 billion in 2029 [39]. This means that China's agriculture is responsible for the increasing population and demand for food quality.

Declining of working age: even though China has the largest population, the demographics have changed. The working age has been declining since 2015 [32] and will decline in agriculture. The declining rate of China's aging population annually is 1.73% [39]. The aging of the population speed is the highest in China and faster in rural areas. So far, more than half of the rural residents are elderly, and the number of aged farmers will increase (Fig. 3).

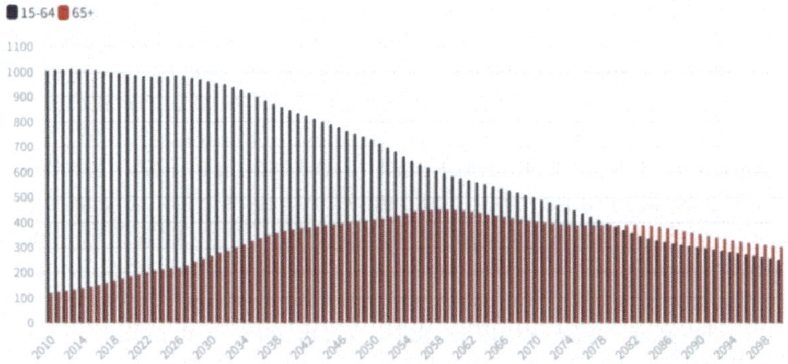

Fig. 3. China working-age population, 65 + population (million). Source: Shanghai Academy of Social Sciences

Resource-scarcity: China has a large amount of resources in total, but the per capita is scarce due to the huge population. It's worth noting that the arable land per person is only 8% per hectare, which is less than one-third of the average, and freshwater per capita is 2 thousand cubic meters, which is less than 10% of the average in OECD countries [32]. This information is crucial in understanding the current situation and the possible challenges China may face in the future.

Loss of farming land: China is feeding 1/5 of the world's population with less than 10% of the world's arable lands. With the extension of urbanization and industrialization, the farming land is losing. The arable land has decreased since 1991 (125 hectares) to 119 million hectares in 2020 [35]. Decades ago, the president of China, Xi, announced the "red line" for preserving 120 million hectares of arable lands [40]. Besides, soil fertility also decreases yearly.

Decreasing self-sufficiency: even though China has sufficient self-sufficiency as of 2000. With the rapid development of urbanization and the change in farmland to industrial land, China relies more and more on food imports [34]. In 2019, China has become the leading food import country. In 2022, China imported more than 23% of its food. The soybean consumption is imported more than 80% [40]. China's food self-sufficiency rate wate was 101% in 2000, and it fell to 76% in 2020, which is predicted to fall to 65% by 2035.

Food loss and waste: due to the weak infrastructure, such as inadequate storage and transportation techniques, as well as excessive processing, China wastes a minimum of 35 million metric tonnes of grain [41]. This wasted amount can feed 200 million people in a year. In recent years, China passed an anti-food waste law against avoiding food loss and waste [40].

Climate change: China is ranked second in the economic risk of flood after India [42]. Droughts also cause severe shortages of water for humans and animals [32]. The temperature is projected to increase by 2–6°C in Asia, which will increase the risks of pests and diseases.

Emerging challenges from digitalization: nevertheless, the application of digitalization will contribute to a more resilient agriculture. In China, digital technologies are accepted for use in agriculture [43]. The lack of knowledge in digitalization in agriculture or the lack of experts is hindering the popularity of the digital agriculture age [44]. As a niche of digital technologies, digital security issues in agriculture have come to the risk stage, which is still under research [22, 45, 46].

A Case Study in South Korea *Safe diversification:* Diversification of business partners and increased trade flows are predicted to enhance the likelihood of dangerous pests and foods entering Korea. As a result, the most significant duty in agricultural commerce has arisen as a more effective system establishment for safeguarding both national health and the natural environment from dangerous pests, illnesses, and foods. Another key policy agenda item is the preparation of measures to relieve the deteriorating trend in food self-sufficiency caused by the growth of agricultural market liberalization and the consequent rise in agricultural imports. As agriculture diminishes owing to the opening of agricultural markets, societal requirements for the multifunctionality of agriculture (such as environmental preservation, balanced national development, succession and development of traditional culture, food security, and so on) are rising. As a result, a significant difficulty for Korean agricultural trade policy is determining how to strike a balance between changes in the agricultural trade environment and the societal need for agriculture.

Water management: Water in river basins is completely or almost fully allocated due to high population density, terrain issues, regional water imbalances, and escalating demands. Droughts and other severe weather events are predicted to become more frequent, intense, and long-lasting as weather patterns alter due to global warming. To achieve long-term, sustainable water usage, South Korea must attempt to regulate demand via water efficiency and conservation measures. One approach will be to promote water-saving equipment in homes, companies, and industries. Another industry where water consumption should be carefully regulated is agriculture. Droughts and floods aren't the only natural disasters that result in crop or livestock losses and massive financial costs. Soil moisture sensors and climate-resilient agricultural methods, for example, may be utilized to assist in minimizing water usage while preserving crop yields.

Price volatility: One of the primary dangers in agricultural markets is price volatility. Volatile commodity prices may create policy uncertainty, expose smallholder farmers to greater risks, modify land usage, reduce forecast accuracy of future supply and demand for agricultural commodities, and boost agricultural output speculative activity. Price fluctuation in agricultural commodities is a big problem in South Korea, and it has been particularly noticeable in the vegetable market. Vegetable crops have been marked by very significant price volatility due to their weather-sensitive and cyclical nature. In response, the South Korean government has designed and executed many treasury assistance programs, including a price stability system, a production contract, and a reserve program, since the 1970s to regulate the vegetable market situation. Despite legislative attempts, price swings in South Korea's vegetable market remain. This has a significant influence on five important vegetable crops in South Korea: cabbage, radish, dried red pepper, garlic, and onion. Furthermore, the amplitude of price series variation has increased during the recent decade. The price series of cabbage and radish has increased somewhat, and their seasonal pattern has gotten more dramatic during the previous decade. Furthermore, prices for dried red pepper and garlic have risen significantly. In contrast to the previous pattern, Onion has witnessed a significant variance in price changes throughout the last decade. As a result, it has been suggested that the current policy be assessed and remedial strategies devised. As a result, the variables driving price volatility in the vegetable market have been revisited.

Aging population: The agricultural industry is facing a shrinking and aging farm population, as well as pressures to enhance production and fulfill social needs such as natural resources and environmental protection. Despite improvements, certain agricultural practices continue to contradict these goals. The high level of producer support, which is 2.7 times the OECD average, is dominated by market price support, which confuses producer decision-making, has the potential to damage the environment and natural resources, and impedes agricultural innovation and the sector's ability to adapt to climate change.

Shortage of skilled workforce: The scarcity of young and qualified people in agriculture, as well as the rising economic inequality, continue to be major challenges. To solve labor shortages and low-income challenges, more efforts are required to enhance rural infrastructure, generate job possibilities, ease social security payments for older farmers who choose to leave the field and diversify revenue sources. Given the labor scarcity in rural Korea, one of the primary policy priorities is building the ability of adolescents and women. Youthful farmers get financial assistance (targeted loan or fund) and farmland leasing to recruit youthful labor. Simultaneously, the government increases its expenditure on education and training services. A five-year action plan to support women farmers was launched in December 2020. Its goal is to increase women's empowerment in rural communities by encouraging business engagement and leadership.

Digitalization: Korea's "Smart Agriculture project" continues to push digital technologies. Despite increased public expenditure in agricultural research and development (R&D), the government-led R&D program could still meet the diverse requirements of stakeholders. To develop a more competitive, demand-driven R&D system, private-sector investment and involvement must be increased. Through the "Smart Agriculture Project," the government boosted its investment in information and communication technology (ICT). The initiative emphasizes the application of digital technologies at the farm level, such as big data, artificial intelligence technology, and real-time crop growth monitoring. The government anticipates that digital technology will increase predictability and reduce volatility, hence improving agricultural output and lowering production costs. Every coin has two faces. First, digital gadgets like drones, onboard equipment, or animal chips all use pricey technology, raising the risk of debt for farmers who are already struggling financially. Second, because of the expensive cost of digital technology, it is often reserved for large enterprises, excluding smaller farmers. Third, they run the danger of driving farmers out of their fields, resulting in a loss of know-how. If farmers record and display their production processes digitally and publicly in the future, this does not mean that public authorities or other parties will have unlimited access to this data. Legal frameworks must be updated to safeguard both corporate and operational data. Clear criteria, including the exclusion of data monopolies, must be developed to avoid the hazards and risks.

China's influence: Rising Chinese agricultural imports may pose a danger to Korean agriculture. Without even discussing geographical closeness or food culture similarities, it is apparent that Chinese agricultural imports would increase once the FTA (Free Trade Agreement) goes into force. Agricultural trade by item and area is more important to Korea than aggregate commerce. Garden produce such as chili peppers, garlic, onions, carrots, radish, and Korean cabbage is the economic lifeblood of Korean farm families. These items are often grown in China's Shandong Province, which has a climate, soil, and overall agricultural environment comparable to Korea. Because agricultural products from China are neither less expensive nor of higher quality than food cultivated in Korea, it is believed that the bilateral FTA would have a significant impact on the Korean agriculture industry. While the FTA will not have an immediate impact on fruits and animal products, they will eventually come under its direct effect in the medium-to-long term.

Import dependency: South Korea has one of the OECD's lowest rates of food self-sufficiency. During the epidemic, supply chain restrictions highlighted the country's dependence on food imports. Now, the Ukraine situation, along with India's restrictions on wheat exports, threatens to drive up domestic food costs even more. South Korea will and must strive to greatly increase its self-sufficiency rate in grain, rice, and other food commodities in order to mitigate the effects of growing food import costs and improve food security. Even after a sharp increase in global rice and wheat prices in 2008, successive Korean administrations were sluggish in recognizing local production as a significant vulnerability.

In 2011, the Lee Myung-bak government (2008–2013) committed to improving the country's meat and produce production and suggested creating an agency to guarantee adequate food procurement. This new body, however, never saw daylight. The COVID-19 epidemic and accompanying supply chain difficulties seem to have forced a more immediate reaction. In 2021, the South Korean Ministry of Agriculture, Food, and Rural Affairs (MFRA) incorporated food security into its yearly business plan. In addition to an early warning system to identify and react to international food price volatility, the government is now aiming to increase soybean and wheat consumption to become more self-sufficient. Despite the fact that President Yoon Suk-yeol's campaign agenda did not include food security, government bodies are likely to continue efforts to strengthen South Korea's food security, including the establishment of alternative import options and grain stockpiles to stabilize prices.

4 Discussion

4.1 Implications

Obviously, the importance of agriculture can be seen from three main aspects: economic, social, and environmental. However, the role of agriculture in national security should also be highlighted, particularly in the face of climate change challenges and resource scarcity. National political stability can be hurt, and military priority can be shifted if there is international competition for key agricultural elements, such as water and arable lands [23].

Besides the critical role of agriculture in four aspects, which is emphasized in our research, economic, society, environmental, and national security, the risks brought by agriculture (GHG emissions, food loss, and waste, etc.) [4] and the threats agriculture is facing also prove the necessity of sustainable agriculture.

In terms of agriculture, China and South Korea have comparable issues. The first problem is climate change [32, 42], which has an impact on various other sectors, such as arable land [35, 40], water management [32], and so on. And the issue related to national security from agriculture, such as self-sufficiency [34, 40]. This finding can also be found in Poland's agriculture, which is one of the important agricultural countries in the EU [47].

First, the rising scarcity of water is lowering the quantity of land to water, resulting in less productivity, less self-sufficiency, and more reliance on exports. Furthermore, both nations' urbanization adds to the loss of arable land, making the conservation of arable lands a major national security priority. The aging population [32, 39] is a shared concern in both nations; however, it is more pressing in South Korea due to the country's already declining population. Furthermore, a scarcity of qualified workers is forcing governments to spend more on digitization in order to compensate for new, more effective agricultural technologies. However, digitization poses hazards to data protection, which is becoming an increasingly vital component of a nation's national security [22, 44–46]. Some common challenges are also serious in Poland, such as the environmental and climate change challenge, water resource management [47], and so on, which require sustainable agriculture solutions [48].

4.2 Limitations and Suggestion

In this research, in order to investigate the necessity of sustainable agriculture, we discussed the importance of agriculture for China and South Korea, mainly from one side, economics, due to the page limitation. As we highlighted in the introduction, future researchers can also extend the relevant topic from other perspectives in a particular country, attracting more attention to sustainable agriculture and food security for the next generations. For example, discuss the necessity of sustainable agriculture from the perspective of the role of agriculture in the environment and society for the world's important food supply countries, such as Ukraine and Russia, which are at war. Targeting other agricultural countries that are not in conflict can also support the idea of sustainable agriculture. Last but not least, we also suggest that future researchers pay attention to the role of agriculture in national security, which we discussed but not thoroughly.

5 Conclusion

A country's agriculture feeds its population and the world population for now and in the future. Agriculture plays a vital role that directly impacts the livelihood of a nation's population and indirectly contributes to economic growth, social stability, and environmental conservation. However, in our research, we also highlight the role of agriculture in national security. Due to the irresistible risks and challenges in agriculture, we believe the future is sustainable agriculture from the viewpoint of two big economic countries, China and South Korea.

References

1. Industry | Definition, Sectors, & Facts | Britannica, https://www.britannica.com/technology/industry, last accessed 2022/11/29
2. Loizou, E., Karelakis, C., Galanopoulos, K., Mattas, K.: The role of agriculture as a development tool for a regional economy. Agric. Syst. **173**, 482–490 (2019). https://doi.org/10.1016/j.agsy.2019.04.002
3. Eswaran, M., Kotwal, A.: The Role of Agriculture in Development. Oxford University Press, USA (2006)
4. Overview of Agriculture and Food, https://www.worldbank.org/en/topic/agriculture/overview, last accessed 2023/05/18
5. Saydullaeva, S., Nurulllaeva, S., Muninova, M.A.: The role and importance of agriculture in the development of the national economy of the republic of uzbekistan. Int. J. Soc. Sci. Interdis. Res. **11**, 178–188 (2022)
6. Hilchey, D.L., Jr, G.W.G., Kay, D.L., Smith, R.D.: The Impact of Agriculture: It's More Than Economic (Part I). Rural New York Minute (2008)
7. Lang, S.S.: Agriculture's impact far more than economic, study says, https://news.cornell.edu/stories/2008/06/agricultures-local-impact-more-economic-study-finds, last accessed 2023/05/18
8. Pawlak, K., Kołodziejczak, M.: The role of agriculture in ensuring food security in developing countries: considerations in the context of the problem of sustainable food production. Sustainability. **12**, 5488 (2020). https://doi.org/10.3390/su12135488
9. Wang, X., Weaver, N., You, J.: The social security function of agriculture in China: the social security function of agriculture in China. J. Int. Dev. **25**, 1 (2013). https://doi.org/10.1002/jid.2827
10. Bielski, S., Marks-Bielska, R., Zielińska-Chmielewska, A., Romaneckas, K., Šarauskis, E.: Importance of agriculture in creating energy security—a case study of Poland. Energies **14**, 2465 (2021). https://doi.org/10.3390/en14092465

11. Agriculture and the environment - OECD, https://www.oecd.org/agriculture/topics/agriculture-and-the-environment/, last accessed 2023/05/20
12. Stentiford, E.I., Issaias, I., Pereira-Neto, J.T., Monge, G.: The importance of agriculture in global waste management (2004)
13. Deák Z.: Vulnerability of the Global Food Supply Chain. 21st International Conference on Management, Enterprise, Benchmarking. 207–215 (2023)
14. World population projected to reach 9.8 billion in 2050, and 11.2 billion in 2100, https://www.un.org/en/desa/world-population-projected-reach-98-billion-2050-and-112-billion-2100, last accessed 2022/10/26
15. Marr, B.: The Biggest Future Trends In Agriculture And Food Production, https://www.forbes.com/sites/bernardmarr/2022/01/28/the-biggest-future-trends-in-agriculture-and-food-production/, last accessed 2023/03/13
16. Challenges for modern agriculture, https://www.syngenta.com/en/innovation-agriculture/challenges-modern-agriculture, last accessed 2022/10/26
17. Ramankutty, N., et al.: Trends in global agricultural land use: implications for environmental health and food security. Annu. Rev. Plant Biol. **69**, 789–815 (2018). https://doi.org/10.1146/annurev-arplant-042817-040256
18. Maximillian, J., Brusseau, M.L., Glenn, E.P., Matthias, A.D.: Pollution and environmental perturbations in the global system. In: Environmental and Pollution Science, pp. 457–476. Elsevier, NJ (2019). https://doi.org/10.1016/B978-0-12-814719-1.00025-2
19. Rodell, M., et al.: Emerging trends in global freshwater availability. Nature **557**, 651–659 (2018). https://doi.org/10.1038/s41586-018-0123-1
20. FAO publications catalogue 2022. FAO (2022). https://doi.org/10.4060/cc2323en
21. Chaudhry, I., Suleman, R., Bhatti, A., Ullah, I.: Review: food price fluctuations and its influence on global food market. ASSAP. **2**, 21–33 (2021). https://doi.org/10.52700/assap.v2i1.33
22. Abbasi, R., Martinez, P., Ahmad, R.: The digitization of agricultural industry–a systematic literature review on agriculture 4.0. Smart Agric. Technol. **2**, 100042 (2022). https://doi.org/10.1016/j.atech.2022.100042
23. Lloyd, L.: Importance of agriculture to national security scrutinized | Food Business News | November 04, 2015 16:41, https://www.foodbusinessnews.net/articles/5324-importance-of-agriculture-to-national-security-scrutinized, last accessed 2023/05/17
24. Kowalczyk, S., Kwasek, M.: Agricultural sector in economy of Poland in 1950–2020. West Balk J Agric. Econom. Rural Develop. **2**, 77–98 (2020). https://doi.org/10.5937/WBJAE2002077K
25. Wojcieszak-Zbierska, M.M., Sadowski, A.: Use of union funds in developing agricultural entrepreneurship between 2014 and 2020 in Poland. Agriculture **14**, 83 (2023). https://doi.org/10.3390/agriculture14010083
26. China at a glance | FAO in China |, https://www.fao.org/china/fao-in-china/china-at-a-glance/en/, last accessed 2023/05/21
27. World Population Clock: 2023- Worldometer, https://www.worldometers.info/world-population/, last accessed 2023/05/22
28. Agricultural land (sq.km)-China, https://data.worldbank.org, last accessed 2023/05/22
29. Agricultural land (% of land area)-China, https://data.worldbank.org, last accessed 2023/05/22
30. GDP per capita (current US$) China, https://data.worldbank.org, last accessed 2023/05/22
31. Agriculture, forestry, and fishing, value added (% of GDP)-China, https://data.worldbank.org, last accessed 2023/05/22
32. OECD: Overview of the food and agriculture situation in China. In: Innovation, Agricultural Productivity and Sustainability in China. pp. 39–74. OECD (2018). https://doi.org/10.1787/9789264085299-5-en

33. Employment in agriculture (% of total employment)-China, https://data.worldbank.org, last accessed 2023/05/22
34. Ross, S.: 4 Countries that produce the most food, https://www.investopedia.com/articles/investing/100615/4-countries-produce-most-food.asp, last accessed 2023/05/21
35. Arable land (hectares)-China, https://data.worldbank.org, last accessed 2023/05/22
36. Ranking Of Countries With The Most Arable Land, //beef2live.com/story-ranking-countries-arable-land-0–108929, last accessed 2023/05/21
37. Arable land (% of land area)-China, https://data.worldbank.org, last accessed 2023/05/22
38. TractorGuru.IN: Top 10 Agriculture Production Countries - Statistics of 2022, https://tractorguru.in/blog/top-10-agriculture-production-countries-statistics-of-2022/, last accessed 2023/05/22
39. Peng, X.: China's population is about to shrink for the first time since the great famine struck 60 years ago. Here's what it means for the world, http://theconversation.com/chinas-population-is-about-to-shrink-for-the-first-time-since-the-great-famine-struck-60-years-ago-heres-what-it-means-for-the-world-176377, last accessed 2023/05/22
40. Wang, O.: 5 major concerns for China's food security, https://www.scmp.com/economy/china-economy/article/3169278/china-food-security-5-major-concerns-loss-fertile-land, last accessed 2023/05/22
41. Wang, Q.: Millions of tons of grain wasted yearly, https://africa.chinadaily.com.cn/china/2014-10/17/content_18757852.htm, last accessed 2023/05/22
42. Sadoff, C.W., et al.: Securing Water, Sustaining Growth. University of Oxford, Oxford, UK (2015)
43. Shao, L., Gong, J., Fan, W., Zhang, Z., Zhang, M.: Cost comparison between digital management and traditional management of cotton fields—evidence from cotton fields in Xinjiang China. Agriculture **12**, 1105 (2022). https://doi.org/10.3390/agriculture12081105
44. Tang, W., Lin, C., Cui, X.: Research on the current situation and development countermeasures of Wenzhou digital agriculture. IOP Conf. Ser. Earth Environ. Sci. **692**, 042039 (2021). https://doi.org/10.1088/1755-1315/692/4/042039
45. Chang, J.: 10 Cybersecurity Trends for 2022/2023: latest predictions you should know, https://financesonline.com/cybersecurity-trends/, last accessed 2023/03/14
46. Ferrag, M.A., Shu, L., Friha, O., Yang, X.: Cyber security intrusion detection for agriculture 4.0: machine learning-based solutions, datasets, and future directions. IEEE/CAA J. Autom. Sinica. **9**, 407–436 (2022). https://doi.org/10.1109/JAS.2021.1004344
47. Prandecki, K., Wrzaszcz, W., Zieliński, M.: Environmental and climate challenges to agriculture in poland in the context of objectives adopted in the european green deal strategy. Sustainability **13**, 10318 (2021). https://doi.org/10.3390/su131810318
48. Bassetière, P.M. de la: Agriculture in Poland: key policies and EU action plans, https://www.blue-europe.eu/analysis-en/short-analysis/agriculture-in-poland-key-policies-and-eu-action-plans/, last accessed 2024/04/29

Noise Effect of Tractors Purchased "Second Hand" upon the Agricultural Workers

Ivan Zahariev[1](✉) and Atanas Atanasov[2]

[1] Department of "Mechanization of Agriculture", Agricultural University - Plovdiv, 12 Mendeleev Blvd, Plovdiv, Bulgaria
i_zahariev@au-plovdiv.bg
[2] University of Ruse Angel Kanchev, Ruse, Bulgaria

Abstract. The article presents the results of the research and comparison of the noise produced by three tractors purchased "second-hand" with different engine powers and years of manufacture. The results of the measurements made with the same scheme and location of the measurement points are compared. Areas exceeding the 80 dB threshold and the need for workers to wear personal protective equipment are defined. The obtained results of the measurements were used to make recommendations for monitoring the change in noise pollution, to protect the health and working capacity of agricultural workers.

Keywords: Acoustic comparative · Measurement · Testing · Personal protective equipment

1 Introduction

Agricultural tractor has led to development of agriculture industry, however, same as other new technologies it produces noise pollution on fields and makes some problems for drivers and farmers as well [1].

Tractors are one of the most important sources of noise exposure in agriculture [2].

In agriculture, tractor noise and vibration are the main problems for workers who work with tractor-driven farm machinery [3].

The main sources of noise in agricultural tractors are the internal combustion engine and its systems, the transmission elements and pneumatic and hydro systems, the running system (wheel, chain or mixed type).

One of the main sources of noise in vehicles equipped with a diesel engine is the power unit itself [4].

From an acoustic point of view, diesel engines lag significantly behind positive ignition engines. Noise and vibrations have a negative impact on the psyche and physicality of both the driver and passengers in the car, as well as pedestrians and residents in areas with intense traffic [5].

Despite their shortcomings, reciprocating internal combustion engines continue to be the main engine in vehicles [6].

E. Lorencowicz et al. (Eds.): FMPMSA 2024, LNCE 609, pp. 462–476, 2024.
https://doi.org/10.1007/978-3-031-70955-5_51

Agricultural workers involved in work processes near running tractors or other energy sources are affected by the pollution produced by these sources [7].

Agriculture is among the professions with the highest risk of hearing loss [8].

The noise propagated in the neighboring environment is transmitted both through the air (so-called "airborne noise") and through solid materials and environments (so-called "body noise") [9].

Noise is widespread and particularly aggressive to human health [10].

Noise negatively affects:

- central nervous system – fatigue, disturbances in the psyche and memory, headache, disturbed sleep, irritability, disturbances in color perception, etc.;
- autonomic nervous system – increased vascular tone;
- cardiovascular system – increased blood pressure, abnormal heart rhythm;
- respiratory system – rapid and shallow breathing;
- digestive system – damage to the stomach, delayed passage of food through the digestive organs;
- endocrine system – changes in the amount of blood sugar, increase in basic metabolism, water retention in the body;
- hearing impairment – at more than 80 dB, irreversible damage to the auditory analyzer may occur, and at more than 120 dB, total deafness may occur suddenly [11].

The impact of noise in the work environment can put the worker in risky situations with direct consequences on safety, due to not hearing warning signals, a cry for help, and indirect - due to a decrease in the ability to concentrate attention, a decrease in dexterity and skills, an accelerated occurrence of fatigue, reduced quality of work, increased risk of accidents, etc. [12].

According to data published on the website of the Ministry of Health of the Republic of Bulgaria [13] the degree of risk of damage to human health under the influence of the "noise" factor in the environment is difficult to determine. Usually, this factor does not act in isolation, but participates in an extremely complex combination with other health risk factors, which can be chemical, physical, biological, psychological and those related to lifestyle, attacking the human organism during its entire life. Noise damage is becoming one of the features of modern life.

From the studies done, it was found that there is no organ in the human body that is spared from the harmful effects of noise. Some data from the World Health Organization shows the following:

- from 8 to 10 million people in the USA need hearing aids, as deafness is most common among the unemployed;
- according to research done in England, every third woman and every fourth man suffer from neurosis, which is the result of noise impact;
- studies in France show that every fifth patient of psychiatric hospitals has lost his mind due to the harmful effect of the ever-increasing noise in the living environment [14].

Therefore, it is important to know the different sources and parameters of noise, the factors that influence its spread and impact on human health [15, 16].

In the [17] defined five ways of attaching the machines to the tractor - rear, front, front, side and combined. The same conclusion was reached by [18] and [19].

The operations where there are agricultural workers (besides the tractor driver) are planting, sowing and others. The machines with which these operations are carried out are seedling machines and others. The most common case of aggregating the machine to the tractor in this type of operation is rear (Fig. 1a). The work process is technological. Requires slow movement across the field on the tractor with the attached machine.

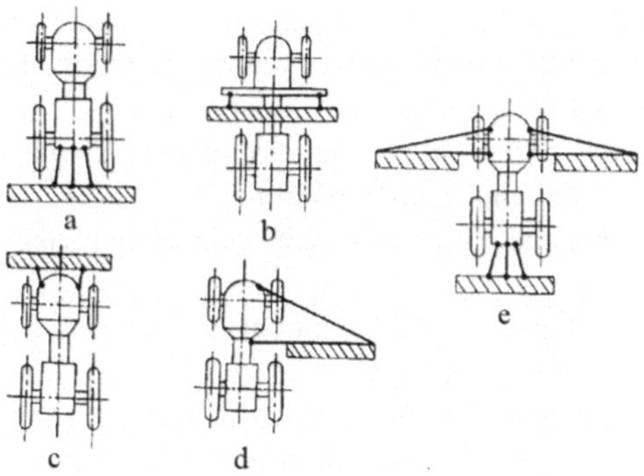

Fig. 1. Ways to aggregate mounted machines. a – rear; b – average; c – frontal (frontal); d – lateral; e – combined [19]

Determining the noise level in the area where agricultural workers are available is important for making management decisions related to their working capacity and the protection of physical and mental health.

For newly manufactured tractors, the noise levels are given in the operating instructions and are in accordance with the current legislative regulations.

With an increase in tractor's age and lack of maintenance higher sound levels are expected [1].

The purpose of the present study is to specify the level of noise created by tractors purchased "second-hand" (not new) and used in the farms of students studying in the specialty "Agricultural Engineering" at the "Mechanization of Agriculture" department of the Agricultural University – Plovdiv.

2 Material and Methods

A measurement is taken to determine the height above ground location of the measuring device. A minimum of 20 people sit on a seedling machine in a working position in a row on a random basis and the height of their hearing organ - the ear - is measured. The obtained results are averaged.

The measured tractor is inspected for loose bolts, nuts and other parts that will create additional noises. The amount of fuel and the level of oil and coolant for the engine are checked. It is positioned in an open area with a hard surface. The transmission control levers are placed in a neutral position. The parking brake is applied. It is secured against self-starting by specially designed elements. In the absence of those with materials at hand.

The tractor engine is started before starting the measurement and idling until normal operating coolant temperature is reached.

According [3] the measurement of the noise level is carried out according to the following scheme: Along the axis of the tractor at a distance of 1 m from the rear wall of the driver's cabin, the point that perceives the lake as the center of the scheme is marked (Fig. 2). Semicircles with diameters of 1 m, 2 m and 3 m are described around this centre. A linear beam is drawn through the center of the diagram, which is perpendicular to the axis of the tractor. From the left side of this beam, another 4 beams are drawn down at 45° intervals. At the intersections of these 5 rays with the semi-circles, the noise level is measured.

The measuring device is held in a horizontal position at the specified height from the ground surface.

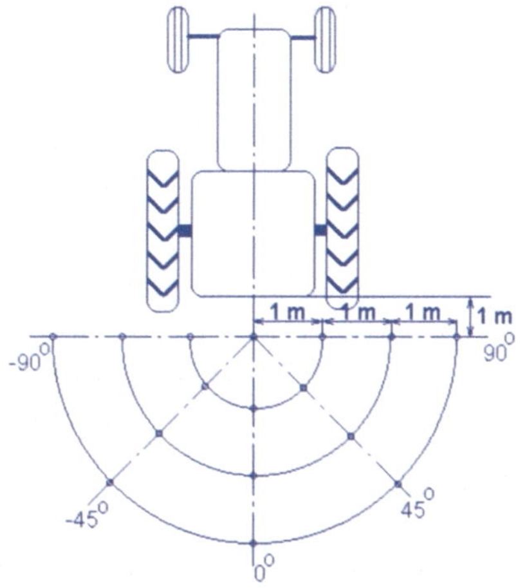

Fig. 2. Ways to aggregate mounted machines [3]

When the engine warms up, the "throttle control" lever is placed successively in 3 positions corresponding to the following engine speeds: 1000 RPM; 1500 RPM; 2000 RPM. The measurement starts from the lowest revolutions to the highest. For each of the 3 positions, record the noise data in triplicate for each point in the diagram in Fig. 2.

The noise measurements are performed with a noise meter owned by the Agrarian University-Plovdiv, brand All-sun Sound level meter EM2242 (Fig. 3), purchased on 06.11.2013, set to:

dBA - mode A - standard measurements (Mode A is used for standard noise level measurements - corresponds to sound pressure felt by the human ear);

SLOW – slow mode that measures in 1s (one second).

A windshield is placed to reduce the error caused by air movement (wind);

A new 9 V battery is inserted before each measurement.

The ambient air temperature is measured, which should be 15 ± 1 °C.

Fig. 3. Sound level meter All-sun Sound level meter EM2242 [20]

According to the methodology for determining the total sound power emitted into the environment by an industrial enterprise and determining the noise level in the place of impact of the Ministry of Environment and Water of the Republic of Bulgaria, measurements are not carried out at:

- Precipitation (rain, snow);
- Air humidity below 20% and above 90%;
- Air temperature below 5 °C;
- Wind speed over 3 m/s [21].

The data from the experiments are processed statistically using the method of regression analysis. The obtained results are presented with mathematical dependencies and visualized with graphs using the Statistika v7 software product.

3 Results and Discussions

The study to determine the height of the sound level meter above the earth's surface was conducted at the Agricultural University-Plovdiv, Department of Mechanization. Students are randomly invited to sit on the seats of a RSMB-4(8) seedling planter and a KSM-4 potato planter. The range of placement of the auricle opening is from 0.90 to 1.30 m. The installation height of the measuring device is assumed to be 1.10 m.

Data collection attempts took place near the owners' bases, namely in the villages of Momino Selo and Rogosh. The two villages are in the Plovdiv region.

Tractor N1 (CLAAS ARES 836 RZ) is owned by the student Kolyo Delev from the village of Momino selo, municipality. Rakovski, region Plovdiv. It was produced in 2006, with an engine power of 206 hp and at the time of measurement it has 9847 moto hours.

Tractors N2 (RENAULT ARES 735 RZ) and N3 (CLAAS AXION 850) are owned by the student Georgi Georgiev from the village of Rogosh, municipality. Maritsa, region Plovdiv.

Tractor N2 was manufactured in 2003, with an engine power of 205 hp and at the time of measurement it has 8940 moto hours.

Tractor N3 was manufactured in 2008, with an engine power of 260 hp and at the time of measurement it has 7073 moto hours.

The experimental data are tabulated, and noise surfaces are plotted at 1000 RPM, 1500 RPM and 2000 RPM for each of the three tractors tested. The areas exceeding the threshold of 80 dB and the need to wear personal protective equipment by the workers of the agricultural machine, which is in accordance with the ergonomic requirements for noise, are defined.

Tractor N1 (Table 1)

Table 1. Average values of measured noise, dB

Engine speed, RPM	Distance to the center of the scheme, m	Angle of rotation relative to the abscissa of the plan				
		−90°	−45°	0°	45°	90°
1000	1	73.9	78.1	75.2	77.5	76.7
1000	2	74.3	77.2	72.8	74.5	75.8
1000	3	74.7	74.6	71.7	73.5	76.8
1500	1	77.3	79.6	81.3	81.8	81.4
1500	2	78.6	80.4	77.6	79.8	78.9
1500	3	81.8	78.8	76.2	78.2	79.3
2000	1	77.6	85.1	82.8	82.8	82.2
2000	2	79.2	84.1	78.6	79.6	80.5
2000	3	82.6	81.1	77.2	78.0	78.3

Tractor N2 (Table 2)

Table 2. Average values of measured noise, dB

Engine speed, RPM	Distance to the center of the scheme, m	Angle of rotation relative to the abscissa of the plan,				
		−90°	−45°	0°	45°	90°
1000	1	74.8	74.9	75.2	73.6	74.8
1000	2	73.9	73.0	73.4	72.1	73.3
1000	3	72.1	69.0	70.0	69.2	71.4
1500	1	77.1	76.5	78.1	77.5	76.4
1500	2	76.6	76.4	76.7	76.9	76.8
1500	3	75.0	73.5	73.9	73.8	75.2
2000	1	81.5	82.2	82.1	78.5	77.3
2000	2	78.0	79.7	80.8	76.7	77.6
2000	3	75.5	75.8	76.3	73.7	75.4

Tractor N3 (Table 3)

Table 3. Average values of measured noise, dB

Engine speed, RPM	Distance to the center of the scheme, m	Angle of rotation relative to the abscissa of the plan,				
		−90°	−45°	0°	45°	90°
1000	1	77.6	77.6	77.6	75.9	77.1
1000	2	75.9	75.1	75.4	74.3	75.7
1000	3	76.5	73.0	74.2	73.8	76
1500	1	79.9	79.1	80.5	79.8	78.7
1500	2	78.6	78.5	78.7	79.2	79.2
1500	3	79.4	77.5	78.1	78.4	79.8
2000	1	84.3	84.9	84.5	80.8	79.7
2000	2	80.1	81.8	82.8	78.9	80.0
2000	3	79.9	79.8	80.5	78.3	80.0

After statistical processing, the following results presented in tabular form were obtained. The mathematical model for each of the three tractors was determined and graphically visualized at 1000 RPM, 1500 RPM and 2000 RPM in the following graphs. Tractor N1 (Table 4)

Table 4. Regression analysis results for tractor N1

Regression Summary for Dependent Variable: Y1 (N1) $R = 0.99911858$ $R^2 = 0.99823795$ Adjusted $R^2 = 0.99801204$ $F(5.39) = 4418.9$ $p < 0.000001$						
	Beta	Std.Err.	B	Std.Err.	t(39)	p-level
X1	1.909436	0.079540	0.095717	0.003987	24.0059	0.000000
X2	0.160125	0.060989	5.763363	2.195170	2.6255	0.012296
X3	−0.002535	0.006736	−0.003160	0.008400	−3.3763	0.008761
$X1^2$	−0.931549	0.078773	−0.000027	0.000002	−11.8257	0.000000
X1*X2	−0.180773	0.061214	−0.004180	0.001415	−2.9531	0.005306

Note: X1 – engine speed; X2 – distance from the axis of the plan; X3 – angle of rotation relative to the abscissa of the plan

The mathematical model has the form:

$$Y_{N1} = 0.096 * X1 + 5.76 * X2 - 0.03 * X3 - 0.00003 * X1^2 - 0.004 * X1 * X2 \quad (1)$$

Tractor N2 (Table 5)

Table 5. Regression analysis results for tractor N2

Regression Summary for Dependent Variable:
Y1 (N2) $R = 0.99923000$ $R^2 = 0.99846059$
Adjusted $R^2 = 0.99821752$ $F(6.38) = 4107.8$ $p < 0.00001$

	Beta	Std.Err.	B	Std.Err.	t(38)	p-level
X1	1.670656	0.111255	0.08081	0.005381	15.01644	0.000000
X2	0.473956	0.115921	16.45986	4.025770	4.08862	0.000217
X3	−0.007407	0.006379	−0.00891	0.007674	−2.16113	0.002831
$X1^2$	−0.786211	0.085619	−0.00002	0.000002	−9.18268	0.000000
$X2^2$	−0.217042	0.072880	−2.84311	0.954677	−2.97809	0.005029
X1*X2	−0.194472	0.058684	−0.00434	0.001309	−3.31386	0.002029

The mathematical model has the form:

$$Y_{N2} = 0.081 * X1 + 16.46 * X2 - 0.009 * X3 - 0.00002 * X1^2 - 2.843 * X2^2 - 0.004 * X1 * X2 \quad (2)$$

Tractor N3 (Table 6)

Table 6. Regression analysis results for tractor N

Regression Summary for Dependent Variable:
Y1 (N3) $R = 0.99909877$ $R^2 = 0.99819835$
Adjusted $R^2 = 0.99796737$ $F(5.39) = 4321.6$ $p < 0.000001$

	Beta	Std.Err.	B	Std.Err.	t(39)	p-level
X1	1.867588	0.080429	0.093911	0.004044	23.2203	0.000000
X2	0.198085	0.061670	7.151908	2.226622	3.2120	0.002642
X3	−0.006900	0.006812	−0.008630	0.008520	−1.0129	0.017339
$X1^2$	−0.889132	0.079653	−0.000026	0.000002	−11.1625	0.000000
X1*X2	−0.217391	0.061898	−0.005042	0.001436	−3.5121	0.001141

The mathematical model has the form:

$$Y_{N3} = 0.094 * X1 + 7.152 * X2 - 0.009 * X3 - 0.00003 * X1^2 - 0.005 * X1 * X2 \quad (3)$$

Comparison of visualization at 1000 RPM of the engine

Tractor N1 (Fig. 4)

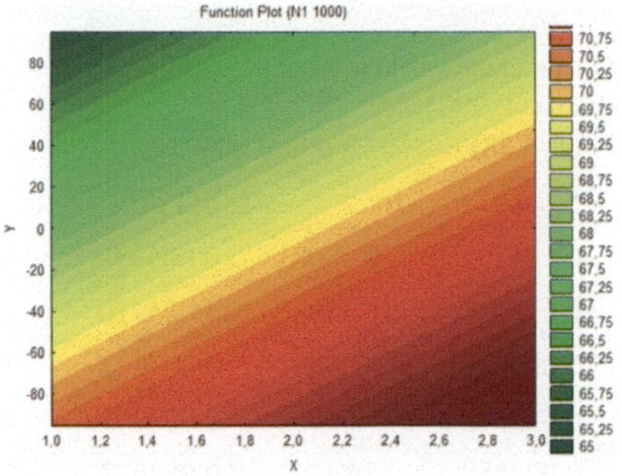

Fig. 4. Surface plot of the noise level in dB as a function of the distance to the center of the plot X and the angle of rotation relative to the abscissa of the plan Y at 1000 RPM. Note: X - distance from the center of the plan, m; Y – angle of rotation relative to the abscissa of the plan

Tractor N2 (Fig. 5)

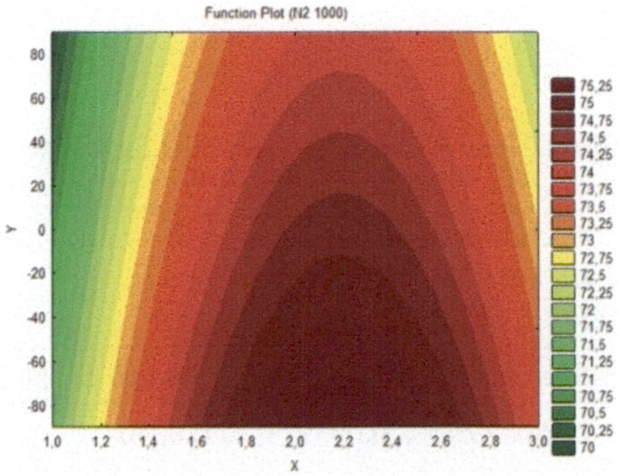

Fig. 5. Surface plot of the noise level in dB as a function of the distance to the center of the plot X and the angle of rotation relative to the abscissa of the plan Y at 1000 RPM.

Tractor N3 (Fig. 6)

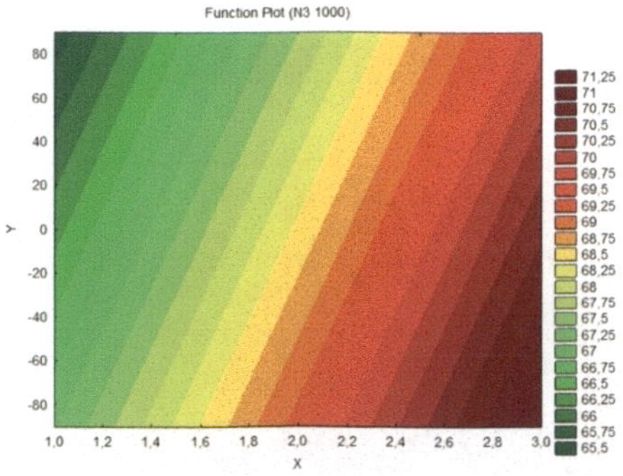

Fig. 6. Surface plot of the noise level in dB as a function of the distance to the center of the plot X and the angle of rotation relative to the abscissa of the plan Y at 1000 RPM.

The graphs show that for all three tractors the noise level is below 80 dB at 1000 RPM engine speed. Symmetry of the noise distribution along the tractor axis is not found.

The noise distribution of N1 and N3 tractors is similar in shape. In tractor N2, there is a distortion due to construction features made by previous owners.

At tractor N2, the highest noise level was observed at minus 90° at a distance of 2.2 m. This is completely logical in view of the fact that the wide-profile tractor tire with which the machine is used serves as a barrier (noise insulation). When moving away from the tire, we have a gradual increase in the noise level up to a distance of 2.2 m. The noise level then naturally decreases with distance from the noise source.

Comparison of visualization at 1500 RPM of the engine

Tractor N1 (Fig. 7)

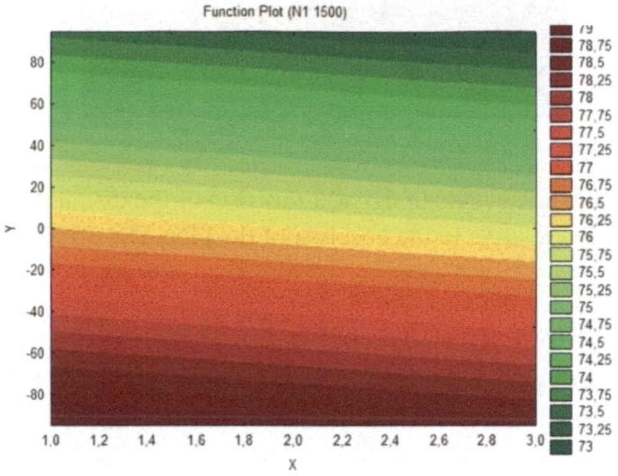

Fig. 7. Surface plot of the noise level in dB as a function of the distance to the center of the plot X and the angle of rotation relative to the abscissa of the plan Y at 1500 RPM.

Tractor N2 (Fig. 8)

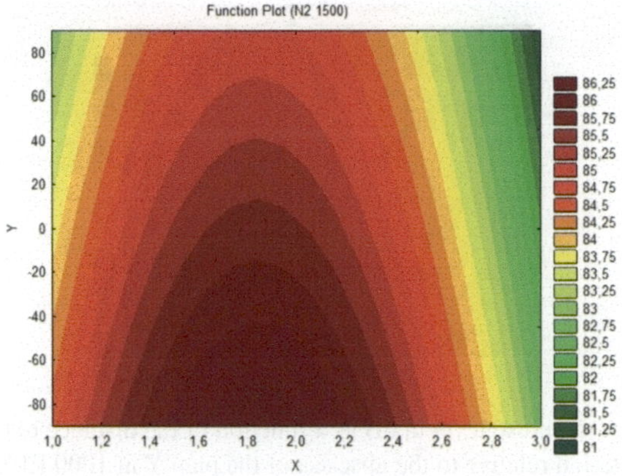

Fig. 8. Surface plot of the noise level in dB as a function of the distance to the center of the plot X and the angle of rotation relative to the abscissa of the plan Y at 1500 RPM.

Tractor N3 (Fig. 9)

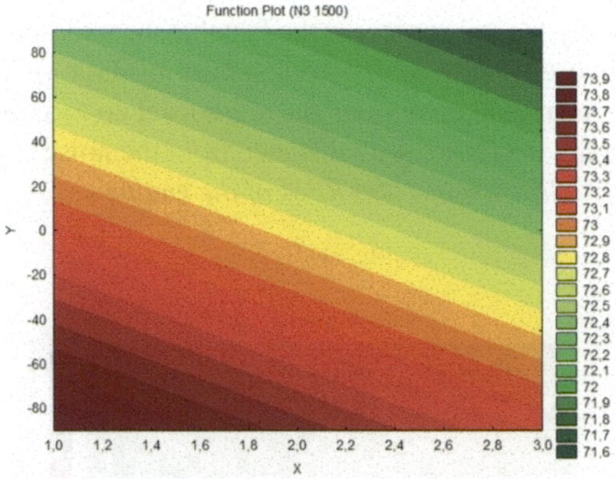

Fig. 9. Surface plot of the noise level in dB as a function of the distance to the center of the plot X and the angle of rotation relative to the abscissa of the plan Y at 1500 RPM.

At engine speeds of 1500 RPM, data below 80 dB is observed for N2 and N3 tractors. At Tractor N1, at certain points located 1 m at 0°, 45° and 90° on the right side and at 3 m at −90° and 2 m −45° on the left side, the noise exceeds 80 dB.

The excess of the norm in the points on the right side is due to their proximity to the engine exhaust manifold.

On the left side, we have an excess of the norm as far as the 2nd m −45° and the 3rd m at −90°, which is mainly due to the noise-absorbing properties of the tractor tire.

Comparison of visualization at 2000 RPM of the engine

Tractor N1 (Fig. 10)

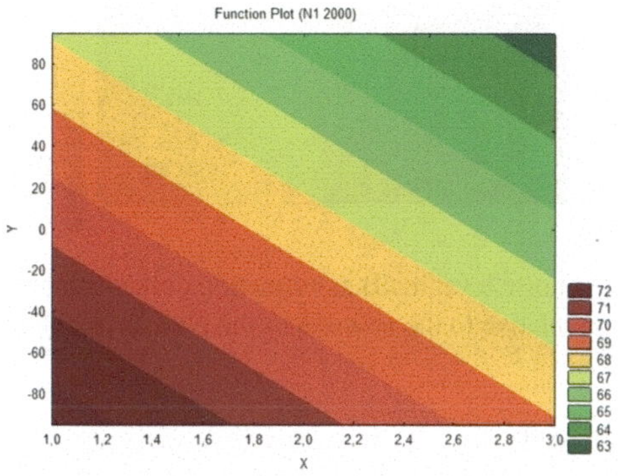

Fig. 10. Surface plot of the noise level in dB as a function of the distance to the center of the plot X and the angle of rotation relative to the abscissa of the plan Y at 2000 RPM.

Tractor N2 (Fig. 11)

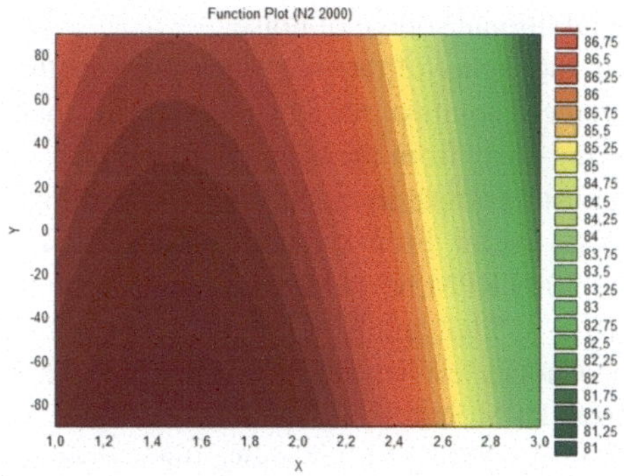

Fig. 11. Surface plot of the noise level in dB as a function of the distance to the center of the plot X and the angle of rotation relative to the abscissa of the plan Y at 2000 RPM.

Tractor N3 (Fig. 12)

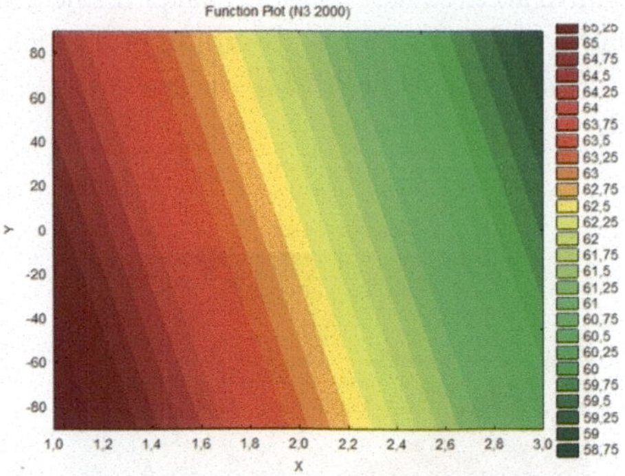

Fig. 12. Surface plot of the noise level in dB as a function of the distance to the center of the plot X and the angle of rotation relative to the abscissa of the plan Y at 2000 RPM.

At 2000 RPM engine speed.

The points without exceeding 80 dB on the Tractor N3 are the fewest (5 in number) as can be seen from the chart and accompanying table.

For Tractor N1, these points are 7, and the quietest tractor is Tractor N2 with 11 points below 80 dB.

These differences are due to design features and machine hours.

Another fact that the graphs show is that on the left side of the tractors the noise level is higher compared to the right side. The reason is the location of the spark arrester-noise suppressor, which is an additional source of noise.

As the distance from the center of the diagram increases, the noise level decreases except for some measurement points of the Tractor N1. This makes perfect sense given the fact that we are moving away from the source of the noise.

Similar studies of acoustic comparative analysis were made by [7] for two Bulgarian-made tractors and a CASE tractor. From the results obtained, it was found that engine speed of 1500 RPM and 2000 RPM noise levels were lower in tractors CLAAS ARES 836 RZ, RENAULT ARES-735 RZ and N3 CLAAS AXION 850.

Authors [1] obtained noise level results directly behind a Massey Ferguson 285 tractor of 76.5 dB at 1000 RPM and 88.37 dB at 2000 RPM. At 1000 RPM the noise level is comparable, while at 2000 RPM it is above the measurement made on all three tractors.

If we keep everyone working on the agricultural machine, the tractor is not subjected to a noise level above 40 dB, the workplaces of the machine must be equipped with personal protective equipment to preserve health and good working condition throughout the day. Cabins can be installed if possible.

According to Ghotbi et al. (2013) & Sümer et al. (2006) [1, 22] the sound pressure level was 2 to 5 dB lower at the driver's ear for combines with attached cabs after production and 9 dB lower for those without a booth.

According to Han et al. (2022) [23] interior cabin noise is reduced by 4–6 dB(A) over the entire operating range when caulking the cabin and lining the area between the engine room and the dashboard with PU foam sound-absorbing material.

It is necessary to make rest stops. The number of breaks and their duration depends on the noise level of the specific work, the age of the workers and their physical condition.

4 Conclusions

On the basis of the above, can be formulating the following conclusions:

1. The distance between the back of the tractor cab and the working place of the agricultural machine affects the perception of noise. When this distance increases, the noise level decreases, except for the points located in the -90° zone of the Tractor N1 (CLAAS ARES 836 RZ).
2. The noise level is not the same on the left and right sides of the tractor at equal distances. At this level, some design features of the tractor have an impact.
3. In order to realize the possibility of all workers being subjected to the same noise load, it is necessary that the workplaces of the agricultural machine are located on a curve that corresponds to the same noise level line, meeting the ergonomic requirements for noise.
4. In order not to expose workers to a noise level above 40 dB, the workplaces of the machine must be equipped with personal protective equipment. If possible, cabins can be installed.
5. With the growing lack of trained personnel for maintenance and repair of agricultural machinery, the technical condition of the machines is expected to deteriorate. As a consequence, noise pollution will also rise. It is necessary to monitor the level in dB in order to protect the health of workers.

Acknowledgments. The authors of the article express their gratitude to the Agricultural University - Plovdiv for the financial support provided under project 17–12 "Support of the publication activity of the teachers from AU-Plovdiv".

References

1. Ghotbi, M., Monazzam, M., Khanjani, N., Nadri Farshad, N., Fard, S.: Driver exposure and environmental noise emission of Massey Ferguson 285 tractor during operations with different engine speeds and gears. Afr. J. Agric. Res. **8**(8), 652–659 (2013). ISSN: 1991-637X 2013 Academic Journals (2013). https://doi.org/10.5897/AJAR12.435. http://www.academicjournals.org/AJAR
2. Monazzam, M., et al.: Tractor drivers and bystanders noise exposure in different engine speeds and gears. Iran. J. Mil. Med. **14**(2), 149–154 (2012)
3. Kehayov, D., Trifonov, A.: Noise effect of tractor stayer 942 upon the subsidiary agricultural workers. J. Environ. Protect. Ecol. **4**(4), 831–835 (2003)
4. Mihailov, V.: Vibroacoustic field of a diesel engine with direct injection. In: XX Scientific-Technical Conference with International Participation "eco Varna 2014", pp. 423–430 (2014). ISBN: 954-20-00030

5. Mihailov, V.: Study of the emission of noise from automobile diesel engines, Abstract. Technical University - Varna, Department of "Transport Engineering and Technologies", p. 1 (2016)
6. Kostov, R.: Roads for improving combined turbocharger engines. J. Tech. Univ. Plovdiv Fundam. Sci. Appl. **13**(8), 115–121 (2006). ISSN: 1310-8271
7. Sfîru, R., Atanasov, A., Muraru, V., Muraru, S.: Acoustic comparative analysis for tractors. INMATEH – Agric. Eng. **58**(2), 167–176 (2019). print ISSN: 2068-4215, on-line ISSN: 2068-2239
8. https://www.bio-fit.eu/bg/q7/lo7-application-of-biofertilizers?start=6. Accessed 22 Feb 2023
9. Lepoev, M.: Modern solutions of the upper construction of the railway in tunnels and metropolitan. In: XVII Scientific-Technical Conference with International Participation "ECO VARNA 2011, pp. 124–129 (2011). ISBN: 954-20-00030
10. Dimitrova, D.: Typological characteristics of reading comprehension tasks. Bul. Lang. Lit. **58**(3), 223–233 (2016)
11. Marinova, T., Mardirossian, G.: Some problems of the ecological assessment of real estate. Ann. Nat. Sci. Dept. **7**, 41–54 (2022). https://doi.org/10.33919/ansd.22.7.5
12. Mihailova, L.: Production noise and its impact on the human organism (2010). https://ns.neochim.bg/Upload/Proizvodstven_shum_Neochim.pdf. Accessed 20 Mar 2023
13. https://www.mh.government.bg/bg/informaciya-za-grazhdani/zdravosloven-nachin-na-zhivot/okolna-sreda/vliyanie-na-shuma/vredni-efekti/. Accessed 20 Feb 2023
14. Bankova, A.: Influence of transport noise on humans and ways to reduce its harmful impact. In: XIX Scientific-Technical Conference with International Participation "eco Varna 2013", pp. 62–67 (2013). ISBN: 954-20-00030
15. Hassall, J.R., Zaveri, K.: Acoustic noise measurements. Bruel & Kaer, Sweden (1988)
16. Kovachev, N.: Noise assessment of server operating in educational environment. In: Proceedings of University of Ruse-2022, vol. 61, Book 1.1, pp. 113–118 (2022). https://conf.uni-ruse.bg/bg/docs/cp22/1.1/1.1-18.pdf
17. Demirev, Z., Bratoev, K.: Mechanization of crop production I, manual for exercises. Ruse University "Angel Kanchev" Ruse, p. 4 (2011)
18. Dimitrov, Y., Velev, N.: Design, construction and calculation of the tractor, DI "Technika" Sofia, fig. 4.1., p. 78 (1981)
19. Dimitrov, J.: Tractors. Design and construction of the tractor. Technical University - Sofia, Sofia, p. 76 (1997). ISBN 954-438-184-8, fig. 4.1
20. https://www.all-sun.com/en/d.aspx?pht=1169. Accessed 21 Feb 2023
21. Methodology for determining the total sound power emitted into the environment by an industrial enterprise and determining the noise level in the place of impact of the Ministry of Environment and Water of the Republic of Bulgaria. https://www.moew.government.bg/static/media/ups/tiny/file/Noise/Metodiki/Metodika_shum_2012.pdf. Accessed 09 Apr 2023
22. Sümer, S., Say, S., Ege, F., Sabanci, A.: Noise exposed of the operators of combine harvesters with and without a cab. Appl. Ergon. **37**(6), 749–756 (2006)
23. Han, H.W., Im, W.H., Choi, H.J., Cho, S.J., Lee, S.D., Park, Y.J.: Effect of sound insulation on noise reduction in an agricultural tractor cab. Sci. Rep. **12**, 22038 (2022). https://doi.org/10.1038/s41598-022-26408-3

Reasons for not Using the Energy Potential of Residual Biomass Generated During the Contour Pruning of Oil-Bearing Roses as Heating Fuel

Ivan Zahariev[1(✉)] and Atanas Atanasov[2]

[1] Department of "Mechanization of Agriculture", Agricultural University - Plovdiv, 12 Mendeleev Blvd, Plovdiv, Bulgaria
i_zahariev@au-plovdiv.bg

[2] University of Ruse Angel Kanchev, Ruse, Bulgaria

Abstract. The article presents the results of a study of cut biomass generated during the annual contour pruning of the massifs planted with oil rose in the Republic of Bulgaria and the reasons for its non-use as fuel for heating. The monitoring period is 2015–2024. Pruning methods and ways of post-processing this biomass are described. Its processing and utilization, as well as the damages resulting from these actions of the owners and users of the arrays. Proposals have been made highlighting the benefits of a wider use of this biomass, not just the oleaginous inflorescences.

Keywords: Oil-bearing rose · Damask rose · energy · biological potential

1 Introduction

The oil-bearing rose is one of the symbols of the Republic of Bulgaria [1]. Rose production is a traditional branch of our economy, and the products produced from the rose color are widely used in medicine, cosmetics, the food industry and other sectors [2, 3]. For thousands of years, the butter rose has been grown in a place not known by the people as the "Rose Valley" or "Rose Valley". In the map of the touristic zoning of the Republic of Bulgaria, a large area is designated as "Rose Valley Region" (Fig. 1) [9].

According to the data of Zahariev & Kehayov [1] and references from the "Regional Directorates of Agriculture in Plovdiv, Pazardzhik and Stara Zagora" for 2015 [10] the occupied areas with oil rose in the three regions of Bulgaria were 27,488 da.

Massifs of oil-bearing roses have been planted in other municipalities. In addition, there is a discrepancy between the locations of the address registration of the owner (legal entity or individual) and of the oil-bearing rose plantation itself. Due to this fact, the references provided by the Regional Directorate of Agriculture include areas of owners registered in the given region/municipality. It follows from this that there are plantations that are not included in the provided references. In 2015, the same authors published an article in which it was recorded that about 40,000 da rose plantations are grown in

E. Lorencowicz et al. (Eds.): FMPMSA 2024, LNCE 609, pp. 477–487, 2024.
https://doi.org/10.1007/978-3-031-70955-5_52

Fig. 1. Tourist map of the Republic of Bulgaria [4–9]

Bulgaria [11]. According to research by Kehayov et al. as of 2016, rose plantations in Bulgaria occupy 40,000 da [12]. In 2019, the authors Zahariev & Kehayov in the article "EXPLOITATION-ECONOMIC ASPECTS OF SHREDDING OF PLANT-DERIVED BIOMASS FROM OIL-BEARING ROSE" [13] published an exact value of 39,260 da of rose plantations. In 2023, Zahariev published data on the amount of rose plantations for 2022, which are already 58,444 da [9, 14]. According to the same research, we have an increase in the occupied areas with oil-bearing rose from 2015 to 2022 by more than 112% despite the difficulties faced by producers, processors and traders of oil-bearing rose products.

During the monitoring period (2015–2024), the state of the oil-bearing rose areas and their development revealed an interesting trend - simultaneous planting and uprooting of young plantations with different locations and owners (land users). This is due to the lack of manpower, difficulties in realizing the harvested oil color and its purchase price, realization of the finished products and unfair competition from synthetically created substitutes, but with a lower price. On the other hand, the existing competition of natural rose oil, coming from countries with a lower standard of living and requirements for producers, and hence a lower price. Despite the difficulties associated with the cultivation and sale of the products of this technical crop - the oil-bearing rose, the plantations have increased.

Every year the oil-bearing rose needs contour pruning to stimulate the growth of new shoots and to limit its overall size to facilitate the rose garden. This pruning is carried out manually with hand shears (vine and orchard), with motorized brush cutters or mechanized (tractor with attached cutting device).

Kehayov et al. [12] collected data on young and old stands of oil rose. When processing the data from the plantations, they noticed that the row spacing in the arrays with young plantations (up to 3 years) was less than that of the old plantations (older than 3 years). In all plantations, technological roads have been laid for 200 m across the rows, which will be used as intermediate depots for storing the remaining biomass. The volume of rose sticks that remains after cutting the bushes in a section of 200 m is about 0.450–0.500 m^3. The data are placed in Table 1 and graphically interpreted in Fig. 2.

Table 1. Dimensions of rose plantings [12]

N	Type of plantation	Average row spacing - c, mm	Diameter of the uncut rose bush - d, mm	Light clearance between uncut plants - b, mm
1	Young	2,575	535	2,040
2	Old	2,850	1,141	1,709

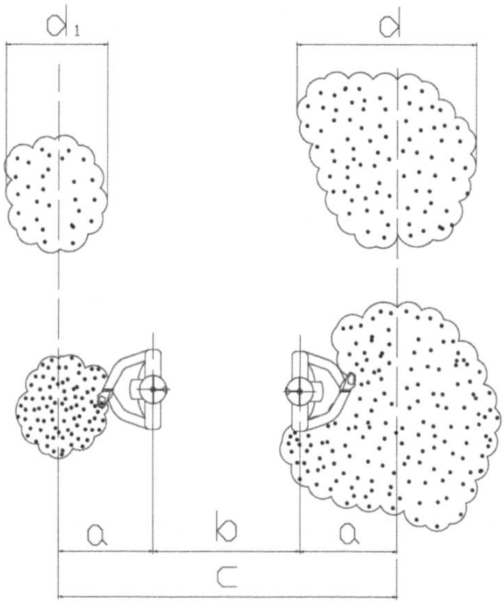

Fig. 2. Scheme of the row spacing when pruning the rose bushes: a- distance from the center of the plantation to the rose sticks (residual biomass), d_1 - trimmed rose bush [12]

According to Popov et al. [15] in 1 day rose plantings are grown on average from 450 to 500 bushes. With normal mineral fertilization, 24.5 m of cane growth is obtained per year from a bush with an average of about 0.01 m when cut.

Zahariev [9] found that in the areas with the greatest concentration of rose plantations (Plovdiv, Pazardizhik and Stara Zagora) a cut biomass of over 16,100 t will be generated at a humidity of ~10.32% (more plantations - more cut sticks of roses) and will result in the burned gross energy of 69.4 GWh in the native state (with a specified moisture content of 10.32%).

Zahariev [18] came to the conclusion that through natural ventilation and drying (in the absence of rainfall) within just 3 days, the leaf-stem biomass of oil-bearing rose lost more than 50% of its mass (moisture) in just 3 days. On the 5th day after cutting and removal, the moisture content of this biomass (18.17%) reaches the limits of air-dry condition - winter value (18–22%), and after the 14th, the moisture content is 14.38%, which is at the limit with air-dry condition - summer values (10–14%). During subsequent storage for three months in a dry, ventilated and closed storage room, the moisture content reaches 10% - the lower limit of air-dry condition - summer values.

The residual biomass of oil-bearing roses consists of young shoots, perennial stems, leaves, inflorescences and spiny spikes. In order to be collected and utilized, it should not be crushed and scattered on the field [16]. In addition, it needs to be in a form suitable for use for heating (pellets, briquettes or energy shavings/wood chips) and with an appropriate moisture content.

Biomass quality is determined by heat of combustion, moisture and ash content [17, 18].

The European Committee for Standardization CEN has issued a number of standards relating to the quality of solid biofuels. Since 2010 The European standards for wood chips (EN14961-4) (Table 2) and for wood pellets for non-industrial needs (EN14961-2) (Table 3) provide for the application of uniform rules for the production of wood biofuels in Europe [18–22].

Table 2. Sample of the European standard for the quality of wood chips EN14961-4 [18–22]

Class	A1	A2	B
Humidity, %	M10 $<$ 10; M25 $\leq$ 25	M35 $\leq$ 35	–
Ash, %	A1.0 $\leq$ 1.0	A1.5 $\leq$ 1.5	A3.0 $\leq$ 3.0
Calorie content, MJ/kg	Q13.0 $\geq$ 13.0	Q11.0 $\geq$ 11.0	–

Table 3. Excerpt from the European Wood Pellet Quality Standard EN14961-2 [18–22]

Class	EN plus –A1	EN plus –A1	EN - B
Humidity, %	$\leq$10	$\leq$10	$\leq$10
Ash, %	$\leq$0.70	$\leq$1.50	$\leq$3.50
Calorie content, MJ/kg	$\geq$16.5 $\div$ 19.0	$\geq$16.3 $\div$ 19.0	$\geq$16.0 $\div$ 19.0
Mechanical strength, %	$\geq$97.5	$\geq$97.5	$\geq$96.5

The authors Zahariev & Kehayov [11] in the article "Results of research on the burning of rose sticks" found that after the complete burning of the cut biomass generated during the contour pruning of the massifs of roses, the average ash value was 3.28% and fell into the class EN – B for pellets, approaching class B for wood chips. For shredded wood, this value is from 0.3 to 6% [23]. The rest of the indicators from their study are listed in Table 4.

Table 4. Indicators in the combustion of rose oil biomass [11]

Silicates, %	Ash content, %	Calorie content, J/g	CO_2 and CO content, mg/m^3	Energy, kWh/kg (kcal/kg)
3.04	3.28	17415.4	3.24/0.23	4.84 (4162)

The required energy to break down 1g of leaf-stem mass of rose oil is 16.555 J/g (45,973 10^{-7} kWh/g) [24].

Kehayov et al. [16] have identified three technologies for harvesting this biomass: Direct harvesting of residual plant biomass from oil rose in the state of energy chips; Direct harvesting of residual plant biomass from oil rose in the form of bales; Collection of residual plant biomass from oil rose in the form of pellets/briquettes directly from the field.

In 2021, Zahariev & Kehayov [25] have developed three more technologies: Transporting the remains of rose oil in the form of energy chips with motor vehicles for removable bodies and containers; Transportation of oil rose residues in the form of energy chips with specialized vehicles for transporting containers with gantry cranes; Transportation of rose oil residues in the form of energy chips with dump trucks and dump trailers.

From the above it is clear that there are researches and developments that provide prerequisites for active use of this annually generated locally plant biomass, it is suitable for use for heating. The reasons for not using the residual leaf-stem biomass of oil-bearing rose for heating purposes remain unspecified.

2 Material and Methods

The aim of the present study is to determine the reasons for the mass non-use of the annually generated plant biomass for heating purposes during the contour pruning, after the rose harvest, of the massifs with oil-bearing rose.

For this purpose, the salted areas are visited and conversations are held with owners, tenants and users of oil rose massifs in the areas with the greatest concentration, namely - Plovdiv region, Pazardzhik region and Stara Zagora region.

Each interviewee is asked the same questions:

Question 1: How do you prune?

Question 2: How do you remove the cut leaf-stem biomass from the inter-rows of the stand?

Question 3: How do you handle exported biomass?

Question 4: What difficulties do you encounter when handling this paint mass?

Question 5: State at least 6 reasons for the mass non-use of this paint mass for heating purposes?

The obtained results are processed statistically, presented in tabular and graphical form.

3 Results and Discussions

The following settlements were visited for data collection in the period 2015–2024:

Plovdiv Region: town of Brezovo, Zelenikovo village, Varben village, Hristo Danovo village, Vasil Levski village, Iganovo village, Dabene village, Pevtsite village, Moskovets village, town of Kalofer, town of Sopot, Stoletovo village, Kliment village, Karnare village, Karavelovo village, Begovo village, Panicheri village, Voynyagovo village, town of Banya, Anevo village, Staro Zhelezare village, Ivan Vazovo village, Sukhozem village, Pesnopoi village, Krasnovo village, Panicheri village, Starosel village and Malo Krushevo village.

Stara Zagora Region: Bratya Daskalovi village, town of Gurkovo, town of Kran, town of Maglizh, Panicherevo village, Dubovo village, Tulovo village, Konare village, Gabarevo village, Tuzha village, Gorno Sakhrane village, Dolno Sahrane village, Osetenovo village, Manolovo village, Turnicheni village, Turia village, Hrischeni village, Kolena village, Gorno Izvorovo village, Enina village, Cherganovo village, Dolno Izvorovo village, town of Shipka, town of Kazanlak, Sheynovo village and Koprinka village.

Pazardzhik Region: town of Bratsigovo, Byaga village, Isperihovo village, Rozovo village, Chernogorovo village, town of Vetren, Pishtigovo village, Byaga village, town of Panagyurishte, town of Peshtera, town of Strelcha, Krali Marko village and Varvara village.

238 people with cultivated area from 0.5 to 150 da of oil-bearing rose plantations participated in the survey. 31 people give up this crop mainly due to lack of labor to collect the oil-bearing flower and the low purchase price. The remaining 207 people answered as follows:

The following responses were received to the first question:

- Mechanized (tractor with attached cutting machine) - 77 people (37.20%);
- Manually with the help of electric or gasoline machines of the type of brush cutters, vine or fruit shears - 126 people (60.87%);
- By hand with hand-held vine or orchard shears – 4 people (1.93%).

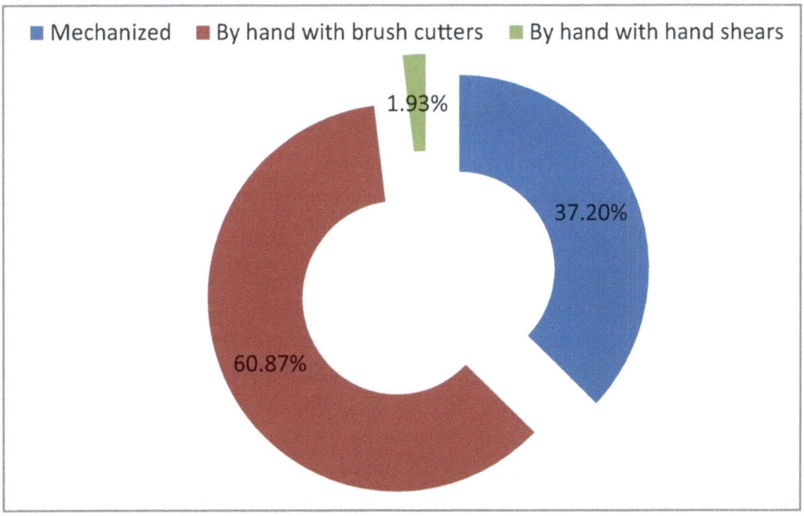

Fig. 3. Distribution according to the method of carrying out the contour pruning of the oil-bearing rose bushes

From Fig. 3. it can be seen that in the cultivation of oil-bearing roses in the Republic of Bulgaria manual labor is mainly used - 62.8% for performing the pruning operation.

The following responses were received to the second question:

- Mechanized - 84 people (40.58%);
- Manually – 13 people (6.28%);
- We do not export them. We grind them in between the rows and plow them (for fertilizer for green fertilization) - 110 people (53.14%).

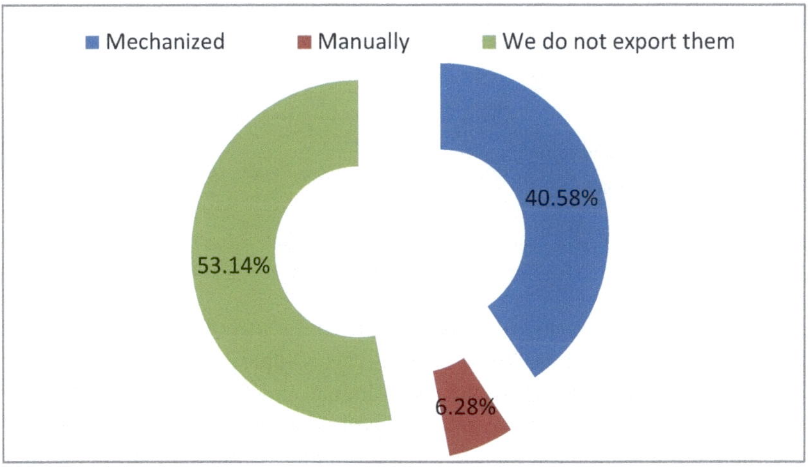

Fig. 4. Distribution according to the method of export of cut leaf-stem mass from the inter-rows

From Fig. 4 it is evident that more than half of the respondents do not export the cut leaf-stem mass from the inter-rows. This answer prevails in the Stara Zagora region.

97 people answered the third question. The remaining 110 did not respond due to the fact that they do not export this biomass.

- We transport it to a place for disposal - 15 people.
- We store the biomass in piles at the end of the array and wait for it to dry. – 82 people.

The second group give a different continuation of their answer according to the actions they take after the drying of this biomass. The continuation of the response can be grouped into two large groups.

34 fall into the first group. They leave the biomass in the chosen place for 2–4 years. During this time, decay processes begin and the deposited biomass slowly decomposes. The pile grows every year: 48 of the 82 burn them whenever possible.

On the fourth question, everyone is unanimous that, in the first place, the prickly spikes create the greatest difficulties and inconveniences.

The following responses were given to question 5, categorized into 11 categories and ranked by importance in Table 5.

Table 5. Answers to question 5, ranked by importance

N	Reason
1	Presence of prickly spines
2	Small thickness of stems
3	Lack of interest in this biomass (only oil colors are important))
4	Lack of shredding and hauling equipment
5	Dispersion of arrays
6	Impossibility of use at home
7	Absence of a processing plant for pellets and briquettes or unexpressed interest in such type of plant for this biomass
8	Lack of time and funds for further processing of biomass
9	Presence of leaves and inflorescences
10	A small amount of array
11	Bad roads to the massifs

From Table 5 notes that the biggest obstacle to the use of oil rose biomass is the prickly thorns and the small thickness of the stems compared to the widely used plant mass of woody species. The third answer is logical due to the fact that the oil rose is grown for its oil flowers. In addition, different energy sources are used for heating purposes and residual biomass is seen as problematic waste. The lack of equipment and organization for shredding, hauling, processing and utilization also has a significant impact. The dispersion of the massifs and the poor road infrastructure further influence the non-use of this annually generated product.

4 Conclusions

Judging by the volume of oil rose areas, the annually generated amount of residual biomass and characteristics allowing it to be used as heating fuel, the developed harvesting and utilization technologies presented above, the following conclusions can be formulated:

1. In order for residual biomass from contour pruning to be used as heating fuel, it must first be removed from the inter-rows and not shredded for green manure.
2. The biggest obstacle to the use of oil rose biomass is its natural prickly spikes, the size of the stems, the weak mechanization in places and the still low financial culture of part of the population, which sees a problem and costs, not a solution and additional income.
3. By proper organization of collection and transportation to existing briquetting or pelletizing plants, this biomass can replace to some extent wood, which in turn would be used for other purposes.
4. Through the processing of the biomass, funds would be saved for the municipalities and the General Directorate "Fire Safety and Protection of the Population" for extinguishing dropped fires with the existing practice of burning at the end of the massif.
5. The regions with massifs of oil rose are visited by tourists from all over the world during the rose harvest, and not only. The existing practice of leaving the removed biomass to rot on the sides of the massifs, apart from threatening the view with the roses in bloom, is a breeding ground for enemies that are transferred to the plantation. This, in turn, leads to plant disease, lower yields and additional costs for preparations and plant protection measures.
6. By using the residual biomass from oil-bearing rose harvested annually at the regional level, our country will be more independent from imported energy sources.

References

1. Zahariev, I., Kehayov, D.: Distribution of the areas occupied by oil rose plantations in Bulgaria, collection/reports/of the 6th national scientific conference 2015 for students. Doctoral students and young scientists, pp. 111–115 (2015). ISSN 1314-9547
2. Zapryanova, G., Ivanova, A.: The Bulgarian olive rose - the beautiful face of Bulgaria before the world, Shumen university "Bishop Konstantin Preslavski." SCI. PAP. Coll. - Dobrich **X**, 29–36 (2018)
3. Stanchev, S.: Mechanized cultivation of oil rose - status, problems, perspectives, Sp. Agricultural Mechanization (2015)
4. Marinova, E.: Nine regions are included in the new tourist map of Bulgaria (2014). https://www.investor.bg/a/332-ikonomika-i-politika/176474-devet-rayona-vlizat-v-novata-turisticheska-karta-na-balgariya. Accessed 16 Mar 2023

5. Ministry of Tourism of the Republic of Bulgaria, Concept for Tourist Zoning of Bulgaria, p. 50, fig.9 (2015). http://www.tourism.government.bg/sites/tourism.government.bg/files/uploads/raionirane/koncepcia.pdf. Accessed 07 Dec 2017
6. BG GLOBE New tourist map of Bulgaria with 9 main regions (2015). http://bgglobe.net/%D0%BD%D0%BE%D0%B2%D0%B8%D0%BD%D0%B8/%D0%BD%D0%BE%D0%B2%D0%B0-%D1%82%D1%83%D1%80%D0%B8%D1%81%D1%82%D0%B8%D1%87%D0%B5%D1%81%D0%BA%D0%B0-%D0%BA%D0%B0%D1%80%D1%82%D0%B0-%D0%BD%D0%B0-%D0%B1%D1%8A%D0%BB%D0%B3%D0%B0%D1%80%D0%B8%D1%8F-%D1%81-9-%D0%BE%D1%81%D0%BD%D0%BE%D0%B2%D0%BD%D0%B8-%D1%80%D0%B0%D0%B9%D0%BE%D0%BD%D0%B0. Accessed 07 Dec 2017
7. Zahariev, I.: Technique and technology for utilization of plant residues from the production of oil rose in Bulgaria. Dissertation, Agrarian University Plovdiv, Plovdiv (2018)
8. Ivanova, K.: 9 regions boost tourism, article in standartnews.com (2021). https://www.standartnews.com/biznes/9-rayona-davat-tlask-na-turizma-467344.html. Accessed 16 Mar 2023
9. Zahariev, I.: Energy potential of the residues obtained during contour pruning of oil-bearing rose plantations in the republic of Bulgaria in year 2022. In: X International Scientific and Technical Online Conference "Problems of Modern Energy and Automation in the Nature Management System" (Theory, Practice, History, Education), pp. 21–27 (2023).http://econference.nubip.edu.ua/index.php/pmeas/pmeas
10. Reports on the occupied areas with oil rose plantations as of 2015 from the Regional Directorates of Agriculture in Pazardzhik, Plovdiv and Stara Zagora (2015)
11. Zahariev, I., Kehayov, D.: Results of research on the burning of rose sticks. RUSE UNIVERSITY SCIENTIFIC PAPERS, vol. 54, no. 1.1, pp. 123–125 (2015). http://conf.uni-ruse.bg/bg/docs/cp15/1.1/1.1-18.pdf. ISSN 1311-3321
12. Kehajov, D., Zahariev, I., Komitov, G.: Development of machine for removing rose rods from area between rows. Eur. J. Tech. Nat. Sci. **4**, 27–29 (2016). https://doi.org/10.20534/EJTNS-16-4-27-29, http://ppublishing.org/upload/iblock/0c5/EJT-4_2016.pdf. ISSN 2414-2352
13. Zahariev, I., Kehajov, D.: Exploitation-economic aspects of shredding of plant-derived biomass from oil-bearing rose. In: International May Conference on Strategic Management – IMCSM19, 24–26 May 2019, Bor, Serbia, vol. XV, no. 1, pp. 372–376 (2019). https://drive.google.com/file/d/18f5D0yQe-VJx64Y7yjzSr6M2pSczdItk/view. ISSN 2620-0597
14. Register of rose growers, according to Art. 3 of the Oil-bearing Rose Law, current as of 15.05.2023. https://www.mzh.government.bg/bg/sektori/rastenievadstvo/. Accessed 28 Sept 2023
15. Popov, A., Pavlov, K., Popov, P.: Plant breeding. Zemizdat, Sofia (1968)
16. Kehayov, D., Zahariev, I., Mitkov, I.: Technologies for the use of residual biomass from the production of oil rose in Bulgaria. In: FRI-8.121-1-AMT&ASVM-11, Proceedings of University of Ruse, vol. 56, book 1.1, pp. 67–72 (2017)
17. Sabeva, G., Tsvetkov, N., Boycheva, S., Gadzhanov, P., Stankov, N.: Studies on determining the fuel-technical characteristics of plant biomass products. In: EMF Scientific Conference 2012, pp. 63–70 (2012)
18. Zahariev, I.: Determination of moisture and drying time of residual biomass from oil-yearing rose during natural drying. In: 31st International Symposium Management of Energy, Industrial and Environmental Systems, 10–11 November 2023, pp. 57–60 (2023). ISSN 1313-2237
19. Binev, I., Georgieva, N.: Energy utilization of wood biomass. In: International Conference on Technics, Technologies and Education ICTTE 2013, 30–31 October 2013, Yambol, Bulgaria (2013)
20. Asikainen, A., Pulkkinen, P.: Comminution of logging residues with evolution 910R chipper, MOHA chipper truck, and Morbark 1200 tub grinder. J. For. Eng. **9**, 47–53 (1998)

21. European Standard (EN14961) for wood chips and hog fuel, Eija Alakangas, Forest bioenergy (2010)
22. Spinelli, R., Ivorra, L., Magagnotti, N., Picchi, G.: Performance of a mobile mechanical screen to improve the commercial quality of wood chips for energy. Biores. Technol. **102**, 7366–7370 (2011)
23. Dinev, D.: Renewable energy sources in agriculture. St. Zagora (2006)
24. Zahariev, I., Kehayov, D.: Results of hardness research and energy required for destruction of the residues from oil-bearing rose production in republic of Bulgaria. In: Scientific Papers Series E Land Reclamation, Earth Observation & Surveying, Environmental Engineering, vol. IX, pp. 259–265, (2020). ISSN 2285-6064
25. Zahariev, I., Kehayov, D.: Transport-manipulation technologies for collection of vegetable residues from Rosa damascene mill production in republic of Bulgaria for the purpose of following use for energy needs. Scientific Papers Series E Land Reclamation, Earth Observation & Surveying, Environmental Engineering, vol. X, pp. 338–342 (2021). ISSN 2285-6064

Monitoring Quality Parameters of Wood Pellets Available on the Polish Market

Grzegorz Zając, Andrzej Kuranc(✉), and Joanna Szyszlak-Bargłowicz

University of Life Science in Lublin, Lublin, Poland
andrzej.kuranc@up.lublin.pl

Abstract. Wood pellets are a type of biofuel that is traded on the global market. In Poland, the production and use of pellets as a biofuel is rising among individual consumers and the industrial energy sector. Unlike other forms of biomass, pellets are known for their high energy density, consistent quality, and ease of storage and transportation over long distances. The parameters that determine the energy suitability of solid biofuels include calorific value, ash content, and sulfur content. For pellets used in individual heating, they should have a high calorific value (≥16.5 MJ·kg $^{-1}$), low ash content (≤0.7% for A1 or ≤1.2% for A2 or ≤2% for B), and low sulfur content (≤0.04% for A1 and A2 or ≤0.05% B). It's important to monitor the quality of pellets by considering the critical values of the technical and elemental analysis results. This study aimed to determine the selected quality parameters of the wood pellets available on the market. The authors tested 40 brands of pellets obtained from various manufacturers. The calorific value of the tested pellets ranged from 15.52 MJ·kg^{-1} to 18.28 MJ·kg^{-1}, where only 3 samples did not meet the requirements. The ash content varied from 0.25% to 3.08% - 2 samples did not meet the requirements. Also, two samples did not meet the requirements regarding moisture content, and the content in the tested pellets was in the range of 3.76% to 12.19%. The sulfur content ranged from 0.011% to 0.533% - 17 samples did not meet the requirements. Monitoring of selected wood pellet parameters showed that the quality of pellets available on the market varies, and some products do not meet the standards despite manufacturers' declarations.

Keywords: Biomass · Wood pellets · Quality of solid biofuels

1 Introduction

Changes in the fossil fuel market caused by the COVID-19 pandemic, the outbreak of the war in Ukraine, and the European Union's policy have resulted in restrictions on access to hard coal. The energy system, small enterprises, and households use this energy raw material. Hence, there is a need to replace this fuel with biofuels.

Biofuels derived from biomass offer a promising alternative to traditional fuels. They are locally available, have a positive environmental impact, and lower global warming potential than fossil fuels [1]. Solid biofuels are a cost-effective substitute, and their widespread availability makes them an attractive option [2]. However, using biomass for energy production can be challenging due to its heterogeneity, low energy density,

E. Lorencowicz et al. (Eds.): FMPMSA 2024, LNCE 609, pp. 488–496, 2024.
https://doi.org/10.1007/978-3-031-70955-5_53

and high moisture content, which can complicate handling, transport, and storage [3–5]. A solution to these issues is to increase density and standardize the material, making it homogeneous and easy to use with a higher energy density [6]. Wood pellets are a common form of solid biofuel used in single-family households. In Poland, the ban on hard coal and firewood in some regions has increased the use of wood pellets [7]. Production of wood pellets in Poland has varied between 0.7 and 1.4 million tons from 2017 to 2019. According to international statistics from Eurostat 2021, Poland does not import wood pellets. Statistics (ENplus 2021) show that as of 2021, Poland had:

- 52 producers with the international ENplus certificate,
- 42 traders with that certificate,
- 30 entities with the DINplus certificate,
- 5 entities were waiting for a certificate number to be granted.

It is crucial to assess the quality of solid biofuels to ensure their correct and sustainable use, especially in the case of pellet production, where there is a wide range of possible raw materials for granulation [8]. Over the years, many articles have been published on assessing and improving the quality of pellets from various biomass raw materials, providing the latest knowledge on this extensive and interesting topic.

1.1 Pellet Quality Standards

The International Organization for Standardization (ISO) has developed pellet quality standards. In particular, the requirements for pellets as a biofuel are included in the EN ISO 17225-1 standard for general quality requirements, EN ISO 17225-2 for wood pellets for industrial and domestic uses and EN ISO 17225-6 for non-wood pellets. In the past several European countries have developed their own regulations and standards for granulate quality certification. The mentioned standards are the Austrian standard NORM M 7135, the Swedish standard SS 187120, the German standards DIN 51731 and DIN EN 15270, the Italian standard CTI-R04/05 and the French ITEBE recommendation [9].

The classified standard for wood pellets (EN ISO 17225-2) is related to using pellets for industrial and non-industrial applications. According to this standard, the best quality class is A1, representing virgin woods and chemically untreated wood residues with low ash and nitrogen content. Pellets with a slightly higher ash and nitrogen content are described as A2. In addition, there is also a quality class B. This class covers chemically processed industrial by-products and residues [10]. The classification of granulates for industrial applications is also reported under ISO 17225-2. This classification provides three different quality classes (I1, I2 and I3), which are slightly more stringent than classes A1, A2, and B for pellets for home use. The introducing of this standard at the European level was an essential step for the sector, providing greater transparency of pellet quality and achieving greater consistency in global markets.

High-quality pellets are necessary to obtain an optimal combustion process and avoid problems related to the operation of heating devices. High-quality fuels facilitate the regulation of the combustion process and thus reduce investment and operating costs. This is especially important in the case of small-scale boilers intended for home use, as their sensitivity to pellet quality is much greater than in the case of high-power boilers.

Larger, industrial-sized pellet boilers may use lower-quality pellets because they have advanced systems for controlling combustion and purification [11].

This work presents a general view of the quality of wood pellets available on the Polish market. The research aimed to assess the quality of wood pellets intended for use in individual households. It was tested to determine whether the pellets were of good quality and met the requirements of the applicable ENplus certificates. Forty pellets from different producers were tested. The scope of the research included determining the calorific value and determining the moisture content, ash and sulfur.

2 Materials and Methods

2.1 Research Material

The research assumed that the actual quality of pellets that reach the user was to be determined; hence, no samples were taken directly from pellet manufacturers from the production lines. For the purposes of the research, 40 different brands of pellets were purchased from Polish producers in 2023 at fuel depots, retail chains, or online.

According to the manufacturer's declarations, all pellet brands had the EN-plus® A1 certificate. It was a pellet with a diameter of 6 mm in commercial packaging – 15 kg bags. The study did not distinguish the type of biomass of which the pellets were produced because it could not always be determined based on the information provided in the product descriptions.

2.2 Methods

Test samples were prepared according to EN 14780: 2017 Solid Biofuel Standard—Sample Preparation to maintain representativeness.

The material for testing was ground using a laboratory grinder and then sieved to obtain appropriate dimensions.

Pellet samples were successively determined for moisture, ash, sulfur and calorific value, and the results were given as an average of three repetitions.

Moisture content was determined by the recommendations of the EN ISO 18134-1 standard, Solid bio-fuels - Determination of moisture content - Part 1: Reference method. The LECO TGA701 thermogravimetric analyzer was used in the research.

Ash content measurements were carried out using the LECO TGA701 thermogravimetric analyzer, following the recommendations of EN ISO 18122 EN 14775:2009 Solid biofuels - Determination of ash content standard.

The sulfur content was determined using the LECO 628S analyzer, the EN ISO 16994 Solid biofuels - Determination of total sulfur and chlorine standard content.

The net calorific value was determined by the EN ISO 18125 Solid Biofuels - Determination of calorific value standard. Combustion heat measurements were made using a semi-automatic isoperibol calorimeter LECO AC600.

3 Results and Discussion

Table 1 and histograms (Figs. 1, 2, 3 and 4) present the test results for the tested pellets' calorific value, sulfur, ash and moisture content.

Table 1. Test results of pellet samples.

Code	Net calorific value [MJ·kg^{-1}]	Ash [%]	Sulfur [%]	Moisiture [%]
P#01	17.28	0.34	0.057	6.90
P#02	17.14	0.76	0.043	7.97
P#03	16.57	0.56	0.056	6.40
P#04	16.84	0.25	0.076	5.59
P#05	17.26	0.45	0.049	7.98
P#06	18.27	0.57	0.248	3.85
P#07	18.28	0.53	0.159	3.76
P#08	17.90	0.8	0.099	5.09
P#09	17.38	0.53	0.084	7.79
P#10	17.24	0.55	0.034	8.20
P#11	17.73	0.38	0.035	4.64
P#12	18.00	0.57	0.028	5.92
P#13	18.14	0.41	0.044	4.17
P#14	17.19	0.57	0.066	8.11
P#15	17.35	0.47	0.015	7.32
P#16	17.64	0.47	0.304	7.32
P#17	16.68	1.01	0.052	7.52
P#18	17.48	0.86	0.146	6.78
P#19	17.83	0.44	0.043	5.69
P#20	17.97	0.49	0.025	6.15
P#21	17.19	0.57	0.044	4.11
P#22	17.63	0.59	0.165	4.11
P#23	17.45	0.53	0.015	6.86
P#24	16.34	2.09	0.079	10.14
P#25	17.31	0.43	0.133	6.83
P#26	16.95	1.53	0.533	7.64
P#27	17.25	0.61	0.012	7.91
P#28	17.80	1.39	0.066	7.23
P#29	17.51	0.7	0.045	7.16
P#30	17.12	3.08	0.091	8.57
P#31	15.65	0.27	0.018	6.25
P#32	15.52	1.58	0.023	12.19
P#33	17.45	0.39	0.013	6.73
P#34	17.77	0.56	0.043	5.76
P#35	17.43	0.59	0.042	7.12
P#36	17.73	0.43	0.036	5.29
P#37	17.84	0.43	0.021	5.29
P#38	17.08	0.49	0.012	6.34
P#39	17.47	0.68	0.014	6.81
P#40	17.30	0.79	0.011	6.53

Legend: A1 class (white), A2 class (orange), B class (pink), No class (grey)

According to the EN ISO 17225-2 standard, the net calorific value of pellets should be higher than 16.5 MJ·kg^{-1} for all the three quality classes (A1, A2 and B) of pellets for domestic use.

The calorific value of the tested pellets ranged from 15.52 MJ·kg^{-1} to 18.28 MJ·kg^{-1}. Among the tested pellets, 3 samples did not meet the standard's requirements in terms of calorific value. The value of this parameter was below the required 16.50 MJ·kg^{-1}. There were two samples of pellets with a calorific value lower than the required by approximately 1 MJ·kg^{-1} (15.65 MJ·kg^{-1}, 15.52 MJ·kg^{-1}), and one sample with a value slightly lower - 16.34 MJ·kg^{-1} (Table 1). The calorific value is the most critical factor determining the quality of fuels. The low calorific value of biofuels - below 16 MJ·kg $^{-1}$ - causes deterioration of the combustion process and reduces the product's commercial value.

Analyzing the distribution of the results obtained (Fig. 1), it can be seen that among the analyzed pellets, 18 pellets with a calorific value within the range of 17 MJ·kg^{-1}–17.50 MJ·kg^{-1} dominated, 12 samples had a calorific value in the range of 17.5 MJ·kg^{-1}–18 MJ·kg^{-1}. These values are consistent with the declarations provided by manufacturers on the product label. Three samples had a calorific value above 18 MJ·kg^{-1}, which proves their high quality. The variation in calorific value results mainly from the type of wood used to produce pellets (coniferous, deciduous) and the amount of impurities. The moisture content of the material also had a significant impact on the results obtained.

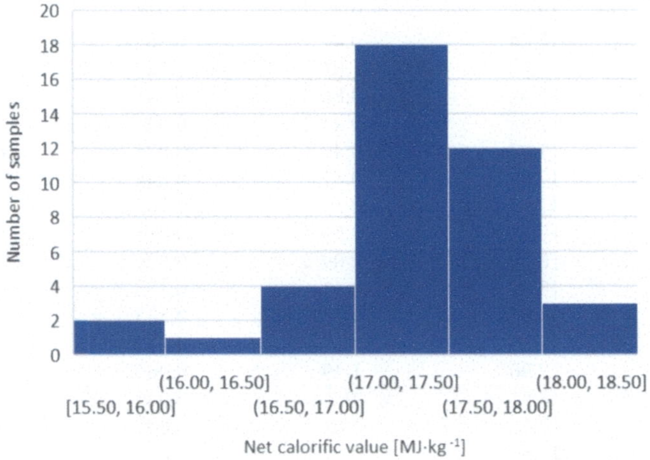

Fig. 1. Distribution of the net calorific value of the tested pellets.

Referring to the EN ISO 17225-2 standard requirements, the ash content (in dry matter) for quality class A1 should be less than 0.7%, for the second class - A2 less than 1.2% and less than 2% for class B. The ash content in the tested pellets varied from 0.25% to 3.08% (Table 1).

Of the tested samples, 30 met the expected requirements for class A1 pellets (Fig. 2). Another five samples met the requirements of class A2, and another three samples met the criteria of class B. However, these values cannot be considered as the highest quality pellets because the producers declared the quality to be class A1. In 2 samples, the ash content was too high to meet the quality requirements. An ash content above 1.5% indicates the presence of mineral impurities such as bark, sand additives and other contaminants. Excessive amount of ash results in more waste production and the operation of boiler equipment and increases the risk of slagging. Determining the type of impurities contributing to the increased ash value would require additional tests.

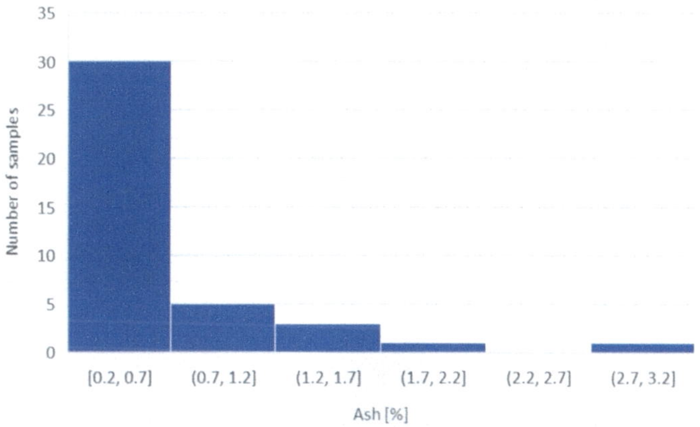

Fig. 2. Distribution of ash content of tested pellets.

Also, the sulfur content was, in many cases, too high for the requirements specified in the EN ISO 17225-2 standard. The content of this element in the tested samples ranged from 0.011% to 0.533%. Only 15 samples met the standard requirements for pellets A1 and A2 (less than 0.04%). The next 8 samples met the requirements for pellets class B (less than 0.05%). However, 17 samples did not meet the requirements of any class. This is particularly concerning because the combustion of fuels with high sulfur content results in emissions of sulfur oxides into the atmosphere [12]. Low-power heating devices are usually not equipped with exhaust gas conditioning systems - exhaust gas desulfurization. Dust removal devices are also rare. Therefore, a critical quality parameter of fuels is low sulfur and ash content.

Moisture content is one of the most important factors influencing the efficiency and stability of the combustion process. The moisture content in the tested pellets ranged from 3.76% to 12.19%. Regarding the EN ISO 17225-2 standard, only 2 samples failed to meet the requirements for all classes (less than 10%). Analyzing the distribution of this parameter in the tested pellets (Fig. 4), it can be seen that the moisture content of 11 samples was in the range of 4–6%, and 22 samples were in the range of 6–8%. This moisture content was indicated on the product's packaging and declared by most producers.

The preferred moisture content for wood pellets is up to 10%. According to [13], deviations above this value reduce the calorific value of the biofuel, and with a moisture content above 15%, biological decomposition of wood pellets may occur. Therefore, the extreme moisture content in the tested pellets of 12.19% may be very unfavorable.

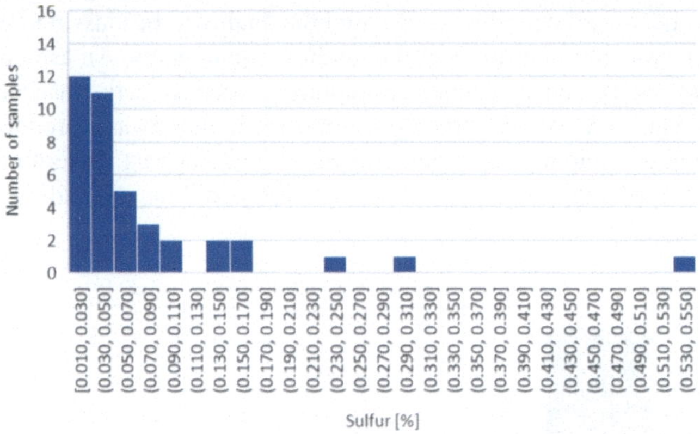

Fig. 3. Distribution of sulfur content of tested pellets.

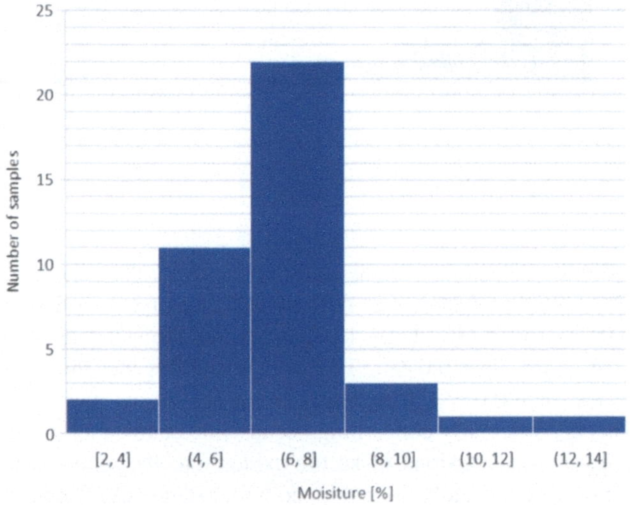

Fig. 4. Moisture content distribution of the tested pellets

Among the 40 tested pellets, only 12 samples met all the analyzed quality requirements appropriate for class A1 and declared by the manufacturer. Another 19 samples did not meet one of the requirements; the next seven did not meet two requirements. There was also one sample that did not meet three requirements and one that did not meet four quality criteria. This was mainly due to exceedances in sulfur and ash content. To a lesser extent, it concerned higher moisture content and lower calorific value. Analyzing the obtained results in terms of meeting the quality requirements for pellets for home use (taking into account class B criteria), all requirements were met by 21 samples, one exceedance of the limit values was recorded for 16 samples, two exceedances for two samples and four exceedances for 1 sample.

Kamperidou [14] obtained similar results from pellet quality monitoring. Out of 20 tested samples, 10 did not meet the manufacturer's declaration, and the most critical feature contributing to the lack of appropriate pellet classification was mechanical durability and, to a lesser extent also, bulk density and ash content. Lykidis and Mantanis [15] examined 10 samples of pellets from the Greek market and found that only 6 of them could be classified as class A1, two samples were in class A2, one in class B, and one sample was considered not to meet the requirements of any quality class. The ash content was the feature that influenced the quality of the tested pellets. Schön et al. [16] tested 42 wood pellet samples from Europe. They found that only 8 of 42 samples did not meet the ISO 17225-2 standard requirements, mainly regarding S and Cl content, fines and bulk density. Duca et al. [17] tested 130 pellet samples and found that only slightly more than half of the samples met the requirements of class A1.

4 Conclusions

Considering the dynamics of the production development and the use of solid biofuels, the standards for these energy carriers must be constantly refined. These standards should be rigorous especially for biofuels intended for household use. They should be strict about heterogeneous materials as biofuels, including a ban on biofuels made from demolition or construction waste because the quality of pellets is directly related to their physical, chemical and mechanical properties. The properties influence the emission of harmful compounds resulting from fuels combustion and the technical possibilities of using them in various heating installations.

The tested wood pellets were, in most cases (21 out of 40), of satisfactory quality and met the requirements in the quality classes for pellets for home and commercial use. However, manufacturers declared quality class A1 on the labels, and only 12 out of 40 samples met these requirements. The exceedances found were mainly concerned with sulfur and ash contents. These results may question the credibility of the information contained in pellet packaging. Pellet producers should pay more attention to the quality and origin of the raw material from which pellets are made as well as they should pay attention to the quality of pellets after production. Moreover, control authorities should implement more reliable and frequent inspections at sale points. Such actions would undoubtedly increase the users' conviction to biofuels and the use of pellets could grow.

References

1. Verma, V.K., Bram, S., De Ruyck, J.: Small scale biomass heating systems: standards, quality labelling and market driving factors–an EU outlook. Biomass Bioenergy **33**, 1393–1402 (2009)
2. Nabiałczyk, M., Musiał, D., Bala-Litwiniak, A.: Analiza właściwości fizykoche-micznych materiału z biomasy w formie pelletu z łupiny orzecha włoskiego. In: Dudek, A., Staniewska E. (eds.) Potencjał innowacyjny w inżynierii materiałowej i zarządzaniu produkcją, pp. 367–374. Wyd. Politechniki Częstochowskiej, Częstochowa (2023)
3. Rentizelas, A.A., Tolis, A.J., Tatsiopoulos, I.P.: Logistics issues of biomass: the storage problem and the multi-biomass supply chain. Renew. Sustain. Energy Rev. **13**, 887–894 (2009)

4. Kaliyan, N., Morey, R.V.: Factors affecting strength and durability of densified biomass products. Biomass Bioenergy **33**, 337–359 (2009)
5. Kuranc, A., Stoma, M., Rydzak, L., Pilipiuk, M.: Durability Assessment of wooden pellets in relation with vibrations occurring in a logistic process of the final product. Energies **13**, 5890 (2020)
6. Sánchez López, J., Curt Fernández de la Mora, M., Sanz Gallego, M., Fernandez Gonzalez, J.: A proposal for pellet production from residual woody biomass in the island of Majorca (Spain). AIMS Energy **3**, 480–504 (2015)
7. Stala-Szlugaj, K.: Rynek paliw stałych dla gospodarstw domowych w Polsce. Zesz. Nauk. Inst. Gospod. Surowcami Miner. Energią PAN **110**, 65–74 (2022)
8. Picchio, R., et al.: Pellet production from woody and non-woody feedstocks: a review on biomass quality evaluation. Energies **13**, 2937 (2020)
9. García-Maraver, A., Popov, V., Zamorano, M.: A review of European standards for pellet quality. Renew. Energy **36**, 3537–3540 (2011)
10. Alakangas, E.: Solid biofuels for energy: a lower greenhouse gas alternative. Green Energy Technol. **28**, 1–3 (2014)
11. García, R., Gil, M.V., Rubiera, F., Pevida, C.: Pelletization of wood and alternative residual biomass blends for producing industrial quality pellets. Fuel **251**, 739–753 (2019). https://doi.org/10.1016/j.fuel.2019.03.141
12. Szyszlak-Barglowicz, J., et al.: Chemical properties of selected agri-food industry waste products in the aspect of their use for energetics purposes. Przem. Chem. **97**, 779–783 (2018). https://doi.org/10.15199/62.2018.5.26
13. Jelonek, I., Jelonek, Z.: Ocena jakości paliw na przykładzie węgli kamiennych i pelletów drzewnych. Gór. Odkryw. **59**, 69–77 (2018)
14. Kamperidou, V.: Quality analysis of commercially available wood pellets and correlations between pellets characteristics. Energies. **15** (2022). https://doi.org/10.3390/en15082865
15. Lykidis, C., Mantanis, G.: Evaluation of the quality of wood pellets found on the market (2015)
16. Schön, C., Mack, R., Hartmann, H.: Quality assessment of wood pellets for residential heating systems and combustion behaviour in a pellet stove. Presented at the Proceedings of the 27th European Biomass Conference and Exhibition, Lisbon, Portugal (2019)
17. Duca, D., Riva, G., Pedretti, E.F., Toscano, G.: Wood pellet quality with respect to EN 14961-2 standard and certifications. Fuel **135**, 9–14 (2014)

Author Index

E. Lorencowicz et al. (Eds.): FMPMSA 2024, LNCE 609, pp. 497–499, 2024.
https://doi.org/10.1007/978-3-031-70955-5